Decoherence and the Appearance
of a Classical World in Quantum Theory

E. Joos   H. D. Zeh   C. Kiefer   D. Giulini
J. Kupsch   I.-O. Stamatescu

Decoherence and the Appearance of a Classical World in Quantum Theory

Second Edition

With 40 Figures and 4 Tables

 Springer

Erich Joos
Rosenweg 2, 22869 Schenefeld, Germany

H. Dieter Zeh
Universität Heidelberg, Institut für Theoretische Physik
Philosophenweg 19, 69120 Heidelberg, Germany

Claus Kiefer
Universität zu Köln, Institut für Theoretische Physik
Zülpicher Strasse 77, 50937 Köln, Germany

Domenico Giulini
Universität Freiburg, Physikalisches Institut
Hermann-Herder-Strasse 3, 79104 Freiburg, Germany

Joachim Kupsch
Technische Universität Kaiserslautern, Erwin-Schrödinger-Strasse 46
67663 Kaiserslautern, Germany

Ion-Olimpiu Stamatescu
Forschungsstätte der Ev. Studiengemeinschaft (FESt)
Schmeilweg 5, 69118 Heidelberg, Germany
and
Universität Heidelberg, Institut für Theoretische Physik
Philosophenweg 16, 69120 Heidelberg, Germany

Library of Congress Cataloging-in-Publication-Data

Decoherence and the appearance of a classical world in quantum theory / E. Joos ... [et al.]. – 2nd ed.
p. cm. Includes bibliographical references. ISBN 3-540-00390-8 (acid-free paper) 1. Quantum theory.
2. Quantum field theory. 3. Coherent states. I. Joos, E. (Erich), 1955- QC174.12.D42 2003   530.12-dc21

ISBN 3-540-00390-8   2nd Edition   Springer-Verlag Berlin Heidelberg New York
ISBN 3-540-61394-3   1st Edition Springer-Verlag Berlin Heidelberg New York

Springer-Verlag Berlin Heidelberg New York
a member of BertelsmannSpringer Science+Business Media GmbH

http://www.springer.de

© Springer-Verlag Berlin Heidelberg 1996, 2003
Printed in Germany

Data prepared by the authors using a Springer TeX macro package
Cover design: design &production GmbH, Heidelberg

Printed on acid-free paper   55/3141/XT   5 4 3 2 1

Preface to the Second Edition

When we were preparing the first edition of this book, the concept of decoherence was known only to a minority of physicists. In the meantime, a wealth of contributions has appeared in the literature – important ones as well as serious misunderstandings. The phenomenon itself is now experimentally clearly established and theoretically well understood in principle. New fields of application, discussed in the revised book, are chaos theory, information theory, quantum computers, neuroscience, primordial cosmology, some aspects of black holes and strings, and others.

While the first edition arose from regular discussions between the authors, thus leading to a clear "entanglement" of their otherwise quite different chapters, the latter have thereafter evolved more or less independently. While this may broaden the book's scope as far as applications and methods are concerned, it may also appear confusing to the reader wherever basic assumptions and intentions differ (as they do). For this reason we have rearranged the order of the authors: they now appear in the same order as the chapters, such that those most closely related to the "early" and most ambitious concept of decoherence are listed first. The first three authors (Joos, Zeh, Kiefer) agree with one another that decoherence (in contradistinction to the Copenhagen interpretation) allows one to eliminate primary classical concepts, thus neither relying on an axiomatic concept of observables nor on a probability interpretation of the wave function in terms of classical concepts. While the fourth and fifth authors (Giulini and Kupsch) still regard the probability interpretation (expressed by means of expectation values, for example) as basic for an interpretation of the formalism, the sixth author (Stamatescu) critically reviews collapse models, which put all quantum probabilistic aspects solely into their (novel) dynamics.

Evidently, collapse models are meant to describe real physical states in terms of wave functions. They attempt to explain definite measurement outcomes by means of speculative stochastic dynamical terms added to the Schrödinger equation. Environmental decoherence is instead based on entanglement as a crucial element of reality, as a consequence of a universal unitary dynamics. Subsystems then no longer possess their own states, but may be effectively described by "apparent" ensembles of wave functions. Since the latter are essentially the same as those expected to result from collapse models, they effectively describe the measurement process *dynamically* in quantum mechanical terms. However, if one is *not* willing to accept an Everett-type interpretation, an essential element of description, such as an objective collapse, is still missing – although it now seems clear that such a collapse is required neither at the microscopic level nor in a measurement device.

Although the assumptions and methods used by the last three authors are perfectly able to describe the phenomena of decoherence, the first three authors nonetheless feel that neither observables and expectation values nor a collapse of the wave function should and need be separately postulated in a fundamental way. In their opinion, the universality and completeness of the quantum formalism (without presuming any classical concepts) need not be called into question. In spite of these differences on issues of principle, all authors agree that the applications and methods studied in Chaps. 6 to 8 are valuable contributions to the theory of environmental decoherence.

The authors are again grateful to FESt for its hospitality and continuous support of the project. This second edition would not have been possible without the ongoing invaluable activities of this institution.

Heidelberg,
December 2002

Erich Joos
H. Dieter Zeh
Claus Kiefer
Domenico Giulini
Joachim Kupsch
Ion-Olimpiu Stamatescu

Preface to the First Edition

The central theme of this book is the problem of how to explain within quantum theory the classical appearance of our macroscopic world. This topic is of fundamental importance not only for the status of quantum theory but also for the philosophical discussion concerning our view of the physical world. The book originated from the regular Workshop on Foundations of Quantum Theory organized for the past several years at the *Forschungsstätte der Evangelischen Studiengemeinschaft* (FESt, or *Protestant Institute for Interdisciplinary Research*) by one of the authors (I.O.S.). These meetings assembled people with longstanding activities in the field of decoherence and people who within their fields of research became interested in the decoherence program as a quantitative approach to basic questions in quantum theory.

Although we have tried to write a "coherent" book, our reader will soon notice that our conceptions vary on some basic notions. Characteristic are our different opinions on the relevance of mathematical concepts for the interpretation of quantum mechanics and hence different inclinations to make active use of these concepts in physical arguments. In some cases this leads us to view or stress differently some fundamental physical concepts, such as 'state' and 'observable'. As a consequence, the reader will find expressed not only mutually complementing, but also partially competing views and proposals on certain issues. We made no attempt to hide any aspect of our debate, which is fuelled by our conviction that in any serious attempt to fully understand the physics and formalism of decoherence one has to address all (or more) of the issues raised in this book.

Since we agreed that all of us should be given the freedom to choose their own emphasis and presentation – even at the price of some redundancies – we decided to indicate the various authorships chapterwise and, where required, sectionwise. In this sense they should be regarded (and – whenever it appears appropriate – even quoted) as independent contributions. On the other hand, each single topic in the book was discussed extensively in content and presentation by all authors until a certain degree of approval was reached. This makes a mutual influence clearly apparent at many places. We like to think of the reader as a participant in our discussion and hope that he or she will benefit from our presentation of different aspects of a common central theme.

The research project that led to this book was dedicated to a subject that in the authors' opinion is unduly neglected at German universities and research institutions. Indeed, this project would not have been possible without the participation and support of FESt as part of its interdisciplinary research activity that promotes projects in various fields. These include research in, and the dialogue between, natural sciences, humanities, ethics, and theology. We hope that our efforts will stimulate similar workshops or research programs elsewhere.

We sincerely thank FESt for its unfailing support of, and confidence in, our work.

We are further indebted to Prof. Wolf Beiglböck, to Dr. Angela Lahee, and to Tilman Bohn for competent advice and assistence in the completion of the book and for help in preparing the manuscript.

E.J. is indebted to his wife for her consistent, unwavering support and encouragement during the dark times.

Heidelberg   Domenico Giulini
April 1996   Erich Joos
                                                                 Claus Kiefer
                                                              Joachim Kupsch
                                                         Ion-Olimpiu Stamatescu
                                                                H. Dieter Zeh

Contents

1 Introduction

E. Joos

What distinguishes classical from quantum objects? What is the precise structure of the transition from quantum to classical? Is this transition smooth and harmless, or does it rather involve a sudden, abrupt change of concepts?

The efforts to find an unambiguous and clear-cut answer to these questions have persisted since the formulation of the theory in the mid-twenties.[1] While quantum mechanics was originally conceived as a theory of atoms, it has shown an ever increasing range of applicability, making it more and more evident that the formalism describes some "true" and general properties of Nature. Today there seem to be no phenomena which contradict quantum theory – perhaps with the sole exception that there are definite ("classical") phenomena at all!

Despite this spectacular success of quantum theory, there is still no consensus about its interpretation. The main problems center around the notions of "observation" and "measurement". The founders of quantum mechanics insisted that measurement results had necessarily to be expressed in classical ("everyday") terms. How can this be reconciled with a description in terms of wave functions (quantum states)? Where is the borderline between the two kinematical concepts (if any)? The early Bohr–Einstein debate showed that – because of the uncertainty relations – the kinematical concepts of classical mechanics cannot be generally applied to quantum objects. Therefore they have no direct counterpart in quantum mechanics; in particular, there are no classical trajectories. On the other hand, the validity of the uncertainty relations was required also in the macroscopic domain in order to save the theory from Einstein's attacks. The orthodox (Copenhagen) interpretation thus uses two incompatible kinematical concepts for the description of physical systems – an unacceptable state of affairs in view of the fact that macro-objects are believed to be made out of atoms! From a conservative point of view such a theory may well be called inconsistent.

It is often claimed that the difference between quantum and classical physics is just a matter of "new statistical rules". The true problem, however,

---

[1] An anthology including many essays of considerable interest has been compiled by Wheeler and Zurek (1983). Other recommended reviews are Jammer (1974), d'Espagnat (1995) and Auletta (2001). Tegmark and Wheeler (2001) provide a lucid account of the present situation. For further information see also *www.decoherence.de*.

is this: what are the fundamental kinematical objects and why do statistical arguments apply to them at all? Since the work of John Bell on the Einstein-Podolsky-Rosen "paradox" it is is well-known that entangled quantum states have properties which defy any local statistical explanation.

Obviously, the problem of the "classical limit" is at the heart of the interpretation problem. Most textbooks suggest that classical mechanics is in some sense contained in quantum mechanics as a special case, similar to the limit of small velocities in relativity. Then, for example, the center-of-mass motion of a macroscopic body would be described by a narrow wave packet, well localized in both position and momentum. The spreading of the wave packet according to the Schrödinger equation is indeed negligible for large mass, so that the Ehrenfest theorems seem to allow a derivation of Newtonian mechanics as a limiting case. These standard arguments are insufficient for several reasons. It remains unexplained why macro-objects come only in narrow wave packets, even though the superposition principle allows far more "nonclassical" states (while micro-objects are usually found in energy eigenstates). Measurement-like processes would *necessarily* produce nonclassical macroscopic states as a consequence of the unitary Schrödinger dynamics. An example is the infamous Schrödinger cat, steered into a superposition of "alive" and "dead". Such states would of course not merely describe ensembles of macrostates, but entirely new states, in the same sense as the superposition of a K meson and its antiparticle, which leads to the "new" particles K_{long} and K_{short}.

It is now increasingly being realized that the conventional treatments of the classical limit are flawed for a simple reason: they do not represent any realistic situation. The assumption of a closed macroscopic system (and thereby the applicability of the Schrödinger equation) is by no means justified in the situations which we find in our present universe – not even in the sense of an approximation. Objects we usually call "macroscopic" are interacting with their natural environment in such a strong manner that they can never be considered as isolated, even under extreme conditions. Most molecules, for example, are already "macroscopic" in this dynamical sense. This observation opens up a new approach to the understanding of classical properties within the framework of quantum theory.

The description of open quantum systems is, however, entirely different from the corresponding situation in classical theories. The formalism of quantum theory uses the concept of a density matrix for characterizing parts of a larger system (which in general has to be described by an *entangled* quantum state). It is the properties of this density matrix that now appear to be able (in a sense to be discussed in detail) to explain the classical behavior of macroscopic objects. In particular, the annoying superpositions of macroscopically different properties can be shown to disappear from these density matrices on an extremely short timescale. This process is called "decoherence". For example, the position of a dust particle becomes "classical" through scattering of a vast number of air molecules and photons, acting together like a continually active position monitor (and thereby creating ever-increasing entanglement). Phase relations between different positions are continually destroyed (more precisely, delocalized into the environment).

The relevance of this irreversible coupling to the environment seems now to have become widely accepted (Zurek 1991, 1993b, 2003, Blanchard *et al.* 2000), although the physical meaning of decoherence is still discussed controversially – if it is contemplated at all. The mechanisms of decoherence are different from (though related to) those responsible for the approach of thermal equilibrium. In fact, decoherence precedes dissipation in being effective on a much faster timescale, while it requires initial conditions which are essentially the same as those responsible for the thermodynamic arrow of time.

Decoherence has found attention in many areas of physics. Despite of many – often technically sophisticated – applications one should be aware of the fact that the original motivation for studying the consequences of quantum entanglement was to understand quantum theory and, in particular, the quantum measurement process (Zeh 1970, Zurek 1981, Joos and Zeh 1985). This quest for understanding was also the main incentive for writing this book. Consequently, we tried to avoid diving all too deeply into complicated calculations which often hide conceptual simplicity. The reader may find remaining details in the references.

Quantum entanglement is often viewed as an exceptional situation. This is quite wrong. Granted, experiments showing quantum nonlocality require a careful preparation of a few entangled degrees of freedom. On the other hand, as we will vividly show in this book, entangled states are the most natural situation we have to expect if quantum theory is universally valid – in particular in the macroscopic domain, where entanglement leads to the appearance of a classical world.

This derivation of classical concepts in a *purely quantum framework* dissolves many interpretational quirks of the orthodox Copenhagen interpretation. As described mainly in Chaps. 2–4, it allows for a consistent, unique picture of reality in terms of quantum states. In this sense, there is indeed a smooth transition from the micro- to the macroworld. How this can be achieved is the central theme of this book.

The persistent interpretation problems of quantum theory are often sidestepped by promoting some kind of "operational" approach. At a certain stage in the chain of physical interactions the "rules of the game" are applied, for example, by calculating expectation values of certain "observables". Obviously, this pragmatic approach *does* work, although the reason for this fact remains obscure. In fact, decoherence seems to be able to give an explanation of why a "Heisenberg cut" can be applied as an effective description beyond a certain threshold without running into conceptual inconsistencies which are often circumscribed by mere words such as "complementarity".

The following Chap. 2 recalls and analyzes the fundamental concepts of quantum theory as they will be used in the main part of this book. Special attention is given to a careful description of the phenomena and problems which led to the program of decoherence. Also, some important conceptual differences to other approaches, some of which are used in later chapters, are pointed out.

In Chap. 3 we analyze the principal mechanisms of environmental decoherence. We will find that even objects (such as molecules) which are usually regarded as "microscopic" can acquire classical properties very rapidly due to the formation of quantum correlations with their environment. This arising entanglement has also important consequences for the dynamical behavior of such objects. We will in particular analyze the origins of the quantum Zeno effect as well as classical rate equations. Decoherence by the environment also plays a major role in the transition from quantum to classical chaos and a – mostly negative – part in attempts to build some kind of "quantum computer". This field has enormously stimulated the interest in the properties of (entangled) quantum states. Unfortunately the interpretational problems are often begged by formalisms and by mixing up quantum and classical concepts.

Chapter 4 extends this treatment to field theory and quantum gravity. It will be shown how decoherence leads to the appearance of classical fields and the emergence of a classical spacetime structure in a quantized theory of gravitation.

There have been attempts aiming at a reformulation of quantum theory in terms of "consistent histories". Since the concept of histories (as sequences of "events") has much in common with "classical trajectories", decoherence is expected to play a major role in this framework. The connection of consistent histories with decoherence is the topic of Chap. 5.

The absence of interference between subsets of the state space is often called a "superselection rule". These restrictions of the superposition principle are traditionally related to symmetries of the state space. Chapter 6 contains an analysis of symmetries of physical systems, redundancy of description (e. g., in gauge theories) and the relation to superselection rules and decoherence. The reader should be aware that this approach uses a different concept of "states" compared to the previous chapters, a fact that may lead to a somewhat different understanding of decoherence.

The phenomenological treatment of open systems in quantum theory is of course not new. Many sophisticated formalisms have been developed for describing the behavior of open systems, usually in studies of thermodynamic and dissipative processes. In Chap. 7 we review some of the essentials of these formalisms and their relation to decoherence.

In the main part of this book, decoherence is derived on the basis of quantum correlations between a system and its environment, that is, as a consequence of quantum nonlocality. However, the ensemble described by a subsystem density matrix is not "real", since it only corresponds to an "improper mixture", thus leaving the problem of measurement essentially unsolved (unless one is willing to accept some variant of the Everett interpretation). There have been several attempts to modify standard quantum theory, aiming at realistic local descriptions. These models are intended to give an explicit dynamical description of the collapse of the wave function, in most cases in the form of a stochastic dynamics replacing the deterministic Schrödinger equation. An overview of these models is given in Chap. 8.

In our last Chap. 9 we will discuss miscellaneous topics related to decoherence.

As already mentioned, the major intention of this book is to analyze and compare fundamental physical concepts. We have therefore tried to keep technicalities to a minimum. Some details which we found useful to complete the presentation are deferred to appendices. The comprehensive reference list should allow the reader to easily delve deeper into any field of interest.

2 Basic Concepts and Their Interpretation

H. D. Zeh

2.1 The Phenomenon of Decoherence

2.1.1 Superpositions

The superposition principle forms the most fundamental kinematical concept of quantum theory. Its universality seems to have first been postulated by Dirac as part of the definition of his "ket-vectors", which he proposed as a complete[1] and general concept to characterize quantum states, regardless of any basis of representation. They were later recognized by von Neumann as forming an abstract Hilbert space. The inner product (also needed to define a Hilbert space, and formally indicated by the distinction between "bra" and "ket" vectors) is not part of the kinematics proper, but required for the probability interpretation, which may be regarded as dynamics (as will be discussed). The third Hilbert space axiom (closure with respect to Cauchy series) is merely mathematically convenient, since one can never decide empirically whether the number of linearly independent physical states is infinite in reality, or just very large.

According to this kinematical superposition principle, *any* two physical states, $|1\rangle$ and $|2\rangle$, whatever their meaning, can be superposed in the form $c_1|1\rangle + c_2|2\rangle$, with complex numbers c_1 and c_2, to form a *new physical state* (to be distinguished from a *state of information*). By induction, the principle can be applied to more than two, and even an infinite number of states, and appropriately generalized to apply to a continuum of states. After postulating the linear Schrödinger equation in a general form, one may furthermore conclude that the superposition of two (or more) of its *solutions* forms again a solution. This is the *dynamical* version of the superposition principle.

Let me emphasize that this superposition principle is in drastic contrast to the concept of the "quantum" that gave the theory its name. Superpositions obeying the Schrödinger equation describe a deterministically evolving

---

[1] This *conceptual* completeness does not, of course, imply that all degrees of freedom of a considered system are always known and taken into account. It only means that, within quantum theory (which, in its way, is able to describe all known experiments), no more complete description of the system is required or indicated. Quantum mechanics lets us even understand *why* we may neglect certain degrees of freedom, since gaps in the energy spectrum often "freeze them out".

continuum rather than discrete quanta and stochastic quantum jumps. According to the theory of decoherence, these *effective* concepts "emerge" as a consequence of the superposition principle when universally and consistently applied (see, in particular, Chap. 3).

A dynamical superposition principle (though in general with respect to real coefficients only) is also known from classical waves which obey a linear wave equation. Its validity is then restricted to cases where these equations apply, while the quantum superposition principle is meant to be universal and exact (including speculative theories – such as superstrings or M-theory). However, while the physical meaning of classical superpositions is usually obvious, that of quantum mechanical superpositions has to be somehow determined. For example, the interpretation of a superposition $\int dq\, e^{ipq} |q\rangle$ as representing a state of momentum p can be derived from "quantization rules", valid for systems whose classical counterparts are known in their Hamiltonian form (see Sect. 2.2). In other cases, an interpretation may be derived from the dynamics or has to be based on experiments.

Dirac emphasized another (in his opinion even more important) difference: all non-vanishing components of (or projections from) a superposition are "in some sense contained" in it. This formulation seems to refer to an *ensemble* of physical states, which would imply that their description by formal "quantum states" is *not* complete. Another interpretation asserts that it is the (Schrödinger) *dynamics* rather than the concept of quantum states which is incomplete. States found in measurements would then have to *arise* from an initial state by means of an indeterministic "collapse of the wave function". Both interpretations meet serious difficulties when consistently applied (see Sect. 2.3).

In the third edition of his textbook, Dirac (1947) starts to explain the superposition principle by discussing one-particle states, which can be described by Schrödinger waves in three-dimensional space. This is an important application, although its similarity with classical waves may also be misleading. Wave functions derived from the quantization rules are defined on their classical configuration space, which happens to coincide with normal space only for a single mass point. Except for this limitation, the two-slit interference experiment, for example, (effectively a two-state superposition) is known to be very instructive. Dirac's second example, the superposition of two basic photon polarizations, no longer corresponds to a spatial wave. These two basic states "contain" all possible photon polarizations. The electron spin, another two-state system, exhausts the group SU(2) by a two-valued representation of spatial rotations, and it can be studied (with atoms or neutrons) by means of many variations of the Stern–Gerlach experiment. In his lecture notes (Feynman, Leighton, and Sands 1965), Feynman describes the maser mode of the ammonia molecule as another (very different) two-state system.

All these examples make essential use of superpositions of the kind $|\alpha\rangle = c_1 |1\rangle + c_2 |2\rangle$, where the states $|1\rangle$, $|2\rangle$, and (all) $|\alpha\rangle$ can be observed as *phys-*

ically different states, and distinguished from one another in an appropriate setting. In the two-slit experiment, the states $|1\rangle$ and $|2\rangle$ represent the partial Schrödinger waves that pass through one or the other slit. Schrödinger's wave function can itself be understood as a consequence of the superposition principle in being viewed as the amplitudes $\psi_\alpha(q)$ in the superposition of "classical" configurations q (now represented by corresponding quantum states $|q\rangle$ or their narrow wave packets). In this case of a system with a known classical counterpart, the superpositions $|\alpha\rangle = \int dq\, \psi_\alpha(q)|q\rangle$ are assumed to define *all* quantum states. They may represent new observable properties (such as energy or angular momentum), which are not simply functions of the configuration, $f(q)$, only as a nonlocal *whole*, but not as an integral over corresponding local densities (neither on space nor on configuration space).

Since Schrödinger's wave function is thus defined on (in general high-dimensional) configuration space, increasing its amplitude does not describe an increase of intensity or energy density, as it would for classical waves in three-dimensional space. Superpositions of the intuitive product states of composite quantum systems may not only describe particle exchange symmetries (for bosons and fermions); in the general case they lead to the fundamental concept of *quantum nonlocality*. The latter has to be distinguished from a mere *extension* in space (characterizing extended classical objects). For example, molecules in energy eigenstates are incompatible with their atoms being in definite quantum states by themselves. Although the importance of this "entanglement" for many observable quantities (such as the binding energy of the helium atom, or total angular momentum) had been well known, its consequence of violating Bell's inequalities (Bell 1964) seems to have surprised many physicists, since this result strictly excluded all local theories conceivably underlying quantum theory. However, quantum nonlocality appears paradoxical only when one attempts to interpret the wave function in terms of an ensemble of local properties, such as "particles". If reality were *defined* to be local ("in space and time"), then it would indeed conflict with the empirical actuality of a general superposition. Within the quantum formalism, entanglement also leads to decoherence, and in this way it *explains* the classical appearance of the observed world in quantum mechanical terms. The application of this program is the main subject of this book (see also Zurek 1991, Mensky 2000, Tegmark and Wheeler 2001, Zurek 2003, or www.decoherence.de).

The predictive power of the superposition principle became particularly evident when it was applied in an ingenious step to postulate the existence of superpositions of states with different particle numbers (Jordan and Klein 1927). Their meaning is illustrated, for example, by "coherent states" of different photon numbers, which may represent quasi-classical states of the electromagnetic field (cf. Glauber 1963). Such dynamically arising (and in many cases experimentally confirmed) superpositions are often misinterpreted as representing "virtual" states, or mere probability amplitudes for the occurrence of "real" states that are assumed to possess definite particle number. This would be as mistaken as replacing a hydrogen wave function by the probability distribution $p(\mathbf{r}) = |\psi(\mathbf{r})|^2$, or an entangled state by an *ensemble* of product states (or a two-point function). A superposition is in general observably different from an ensemble consisting of its components with *any* probabilities.

Another spectacular success of the superposition principle was the prediction of new particles formed as superpositions of K-mesons and their antiparticles (Gell-Mann and Pais 1955, Lee and Yang 1956). A similar model describes the recently confirmed "neutrino oscillations" (Wolfenstein 1978), which are superpositions of energy eigenstates.

The superposition principle can also be successfully applied to states that may be generated by means of symmetry transformations from asymmetric ones. In classical mechanics, a symmetric Hamiltonian means that each asymmetric *solution* (such as an elliptical Kepler orbit) implies other solutions, obtained by applying the symmetry transformations (e.g. rotations). Quantum theory requires *in addition* that all their superpositions also form solutions (cf. Wigner 1964, or Gross 1995; see also Sect. 9.4). A complete set of energy eigenstates can then be constructed by means of *irreducible linear representations* of the dynamical symmetry group. Among them are usually symmetric ones (such as s-waves for scalar particles) that need not have a counterpart in classical mechanics.

A great number of novel applications of the superposition principle have been studied experimentally or theoretically during recent years. For example, superpositions of different "classical" states of laser modes ("mesoscopic Schrödinger cats") have been prepared (Davidovich *et al.* 1996), the entanglement of photon pairs has been confirmed to persist over tens of kilometers (Tittel *et al.* 1998), and interference experiments with fullerene molecules were successfully performed (Arndt *et al.* 1999). Even superpositions of a macroscopic current running in opposite directions have been shown to exist, and confirmed to be different from a state with *two* (cancelling) currents – just as Schrödinger's cat superposition is different from a state with *two* cats (Mooij *et al.* 1999, Friedman *et al.* 2000). Quantum computers, now under intense investigation, would have to perform "parallel" (but not just *spatially* separated) calculations, while forming one superposition that may later have a coherent effect (Sect. 3.3.3.2). So-called quantum teleportation (Sect. 3.4.2) requires the *advanced preparation* of an entangled state of distant systems (cf. Busch *et al.* 2001 for a consistent description in quantum mechanical terms – see also Sect. 3.4.2). One of its components may then later be selected by a *local* measurement in order to determine the state of the other (distant) system.

Whenever an experiment was technically feasible, all components of a superposition have been shown to *act* coherently, thus proving that they *exist* simultaneously. It is surprising that many physicists still seem to regard superpositions as representing some state of ignorance (merely characterizing unpredictable "events"). After the fullerene experiments there remains but a minor step to discuss *conceivable* (though hardly realizable) interference experiments with a conscious observer. Would he have one or many "minds" (being aware of his path through the slits)?

The most general quantum states seem to be superpositions of different classical fields on three- or higher-dimensional space.[2] In a perturbation expansion in terms of free "particles" (wave modes) this leads to terms corresponding to Feynman diagrams, as shown long ago by Dyson (1949). The path integral describes a *superposition* of paths, that is, the propagation of wave functions according to a generalized Schrödinger equation, while the individual paths under the integral have no *physical* meaning by themselves. (A similar method could be used to describe the propagation of classical waves.) Wave functions will here always be understood in the generalized sense of *wave functionals* if required.

One has to keep in mind this universality of the superposition principle and its consequences for *individually* observable physical properties in order to appreciate the meaning of the program of decoherence. Since quantum coherence is far more than the appearance of spatial interference fringes observed statistically in series of "events", decoherence must not simply be understood in a classical sense as their washing out under *fluctuating* environmental conditions.

2.1.2 Superselection Rules

In spite of this success of the superposition principle it is evident that not *all* conceivable superpositions are found in Nature. This led some physicists to postulate "superselection rules", which restrict this principle by axiomatically excluding certain superpositions (Wick, Wightman, and Wigner 1970, Streater and Wightman 1964). There are also attempts to *derive* some of these superselection rules from other principles, which can be postulated in quantum field theory (see Chaps. 6 and 7). In general, these principles

---

[2] The empirically correct "pre-quantum" configurations for fermions are given by spinor fields on space, while the apparently observed particles are no more than the consequence of decoherence by means of *local* interactions with the environment (see Chap. 3). Field amplitudes (such as $\psi(\mathbf{r})$) seem to form the general arguments of the wave function(al) Ψ, while space points $\mathbf{r}$ appear as their "indices" – not as dynamical position *variables*. Neither a "second quantization" nor a wave-particle dualism are required on a fundamental level. N-particle wave functions may be obtained as a non-relativistic approximation by applying the superposition principle (as a "quantization procedure") to these apparent particles instead of the correct pre-quantum variables (fields), which are not directly observable for fermions. The concept of particle permutations then becomes a *redundancy* (see Sect. 9.4). Unified field theories are usually expected to provide a general (supersymmetric) pre-quantum field and its Hamiltonian.

merely exclude "unwanted" consequences of a general superposition principle by hand.

Most disturbing in this sense seem to be superpositions of states with integer and half-integer spin (bosons and fermions). They violate invariance under 2π-rotations (see Sect. 6.2.3), but such a non-invariance has been experimentally confirmed in a different way (Rauch *et al.* 1975). The theory of supersymmetry (Wess and Zumino 1971) *postulates* superpositions of bosons and fermions. Another supposedly "fundamental" superselection rule forbids superpositions of different charge. For example, superpositions of a proton and a neutron have never been directly observed, although they occur in the *isotopic spin* formalism. This (dynamically broken) symmetry was later successfully generalized to SU(3) and other groups in order to characterize further intrinsic degrees of freedom. However, superpositions of a proton and a neutron may "exist" within nuclei, where isospin-dependent self-consistent potentials may arise from an *intrinsic symmetry breaking*. Similarly, superpositions of different charge are used to form BCS states (Bardeen, Cooper, and Schrieffer 1957), which describe the intrinsic properties of superconductors. In these cases, definite charge values have to be projected out (see Sect. 9.4) in order to describe the observed physical objects, which do obey the charge superselection rule.

Other limitations of the superposition principle are less clearly defined. While elementary particles are described by means of wave functions (that is, superpositions of different positions or other properties), the moon seems always to be at a definite place, and a cat is either dead or alive. A general superposition principle would even allow superpositions of a cat and a dog (as suggested by Joos). They would have to define a "new animal" – analogous to a K_{long}, which is a superposition of a K-meson and its antiparticle. In the Copenhagen interpretation, this difference is attributed to a strict conceptual separation between the microscopic and the macroscopic world. However, where is the border line that distinguishes an n-particle state of quantum mechanics from an N-particle state that is classical? Where, precisely, does the superposition principle break down?

Chemists do indeed know that a border line seems to exist deep in the microscopic world (Primas 1981, Woolley 1986). For example, most molecules (save the smallest ones) are found with their nuclei in definite (usually rotating and/or vibrating) classical "configurations", but hardly ever in superpositions thereof, as it would be required for energy or angular momentum eigenstates. The latter are observed for hydrogen and other *small* molecules. Even chiral states of a sugar molecule appear "classical", in contrast to its parity and energy eigenstates, which correctly describe the otherwise analogous maser mode states of the ammonia molecule (see Sect. 3.2.4 for details). Does this difference mean that quantum mechanics breaks down already for very small particle number?

Certainly not in general, since there are well established superpositions of many-particle states: phonons in solids, superfluids, SQUIDs, white dwarf stars and many more! All properties of macroscopic bodies which can be calculated quantitatively are consistent with quantum mechanics, but not with any microscopic classical description. As will be demonstrated throughout the book, the theory of decoherence is able to *explain* the apparent differences between the quantum and the classical world under the assumption of a *universally valid* quantum theory.

standard procedure insufficient

The attempt to derive the absence of certain superpositions from (exact or approximate) conservation laws, which forbid or suppress transitions between their corresponding components, would be insufficient. This "traditional" explanation (which seems to be the origin of the name "superselection rule") was used, for example, by Hund (1927) in his arguments in favor of the chiral states of molecules. However, small or vanishing transition rates require *in addition* that superpositions were absent initially for all these molecules (or their constituents from which they formed). Similarly, charge conservation by itself does *not* explain the charge superselection rule! Negligible wave packet dispersion (valid for large mass) may prevent initially presumed wave packets from growing wider, but this initial condition is quantitatively insufficient to explain the quasi-classical appearance of mesoscopic objects, such as small dust grains or large molecules (see Sect. 3.2.1), or even that of celestial bodies in chaotic motion (Zurek and Paz 1994). Even the required initial conditions for conserved quantities would in general allow one only to exclude *global* superpositions, but not local ones (Giulini, Kiefer and Zeh 1995).

So how can superselection rules be explained within quantum theory?

2.1.3 Decoherence by "Measurements"   *SUPERSELECTION!?*   *measure*

interference ... as d... ears

Other experiments with quantum objects have taught us that interference, for example between partial waves, disappears when the property characterizing these partial waves is *measured*. Such partial waves may describe the passage through different slits of an interference device, or the two beams of a Stern–Gerlach device ("*Welcher Weg* experiments"). This loss of coherence is indeed required by mere logic once measurements are assumed to lead to definite results. In this case, the frequencies of events on the detection screen measured in coincidence with a *certain* (that is, measured) passage can be counted separately, and thus have to be added to define the total probabilities.[3] It is therefore a *plausible* experimental result that the interference disappears also when the passage is "measured" without registration

fair

---

[3] Mere logic does *not* require, however, that the frequencies of events on the screen which follow the observed passage through slit 1 of a two-slit experiment, say, are the same as those without measurement, but with slit 2 closed. This distinction would be relevant in Bohm's theory (Bohm 1952) if it allowed nondisturbing measurements of the (now *assumed*) passage through one definite slit (as it does *not* in order to remain indistinguishable from quantum theory). The

Heisenberg and Bohr

of a definite result. The latter may be *assumed* to have become a "classical fact" as soon the measurement has irreversibly "occurred". A quantum phenomenon may thus "become a phenomenon" without being *observed*. This is in contrast to Heisenberg's remark about a trajectory coming into being by its observation, or a wave function describing "human knowledge". Bohr later spoke of objective irreversible events occurring *in the counter*. However, what precisely is an irreversible quantum event? According to Bohr this event can *not* be dynamically analyzed.

Analysis within the quantum mechanical formalism demonstrates nonetheless that the essential condition for this "decoherence" is that complete information about the passage is carried away in some *objective physical* form (Zeh 1970, 1973, Mensky 1979, Zurek 1981, Caldeira and Leggett 1983, Joos and Zeh 1985). This means that the state of the environment is now *quantum correlated* (entangled) with the relevant property of the system (such as a passage through a specific slit). This need *not* happen in a controllable way (as in a measurement): the "information" may as well form uncontrollable "noise", or anything else that is part of reality. In contrast to statistical correlations, quantum correlations characterize *real* (though nonlocal) quantum states – not any lack of information. In particular, they may describe *individual* physical properties, such as the non-additive total angular momentum $\mathbf{J}^2$ of a composite system at any distance.

necessary for decoh.

meaning of decoh-erence

Therefore, one cannot *explain* entanglement in terms of a concept of information (cf. Brukner and Zeilinger 2000 and see Sect. 3.4.2). This terminology would mislead to the popular misunderstanding of the collapse as a "mere increase of information" (which requires an initial ensemble describing ignorance). It would indeed be a strange definition if "information" determined the binding energy of the He atom, or prevented a solid body from collapsing. Since environmental decoherence affects individual physical states, it can *neither* be the consequence of phase averaging in an ensemble, *nor* one of phases fluctuating uncontrollably in time (as claimed in some textbooks). Entanglement exists, for example, in the static ground state of relativistic quantum field theory, where it is often erroneously regarded as vacuum *fluctuations* in terms of "virtual" particles.

contrast with classical correlations

When is unambiguous "information" carried away? If a macroscopic object had the opportunity of passing through two slits, we would always be able to convince ourselves of its choice of a path by simply opening our eyes in order to "look". This means that in this case there is plenty of light that con-

ENTANGLE-MENT CAN NOT BE REDUCED TO CONCEPT OF INFORMATION

---

fact that these two quite different situations (closing slit 2 or measuring the passage through slit 1) lead to exactly the same subsequent frequencies, which differ entirely from those that are *defined* by this theory when not measured or selected, emphasizes its extremely artificial nature (see also Englert *et al.* 1992, or Zeh 1999). The predictions of quantum theory are here simply reproduced by leaving the Schrödinger equation unaffected and universally valid, identical with Everett's assumptions (Everett 1957). In both these theories the wave function is (for good reasons) regarded as a *real* physical object (cf. Bell 1981).

tains information about the path (even in a controllable manner that allows us to "look"). Interference between different paths never occurs, since the path is evidently "continuously measured" by light. The common textbook argument that the interference pattern of macroscopic objects be too fine to be observable is entirely irrelevant. However, would it then not be sufficient to dim the light in order to reproduce (in principle) a quantum mechanical interference pattern for macroscopic objects?

This could be investigated by means of more sophisticated experiments with mesoscopic objects (see Brune *et al.* 1996). However, in order to precisely determine the subtle limit where measurement by the environment becomes negligible, it is more economic first to apply the established theory which is known to describe such experiments. Thereby we have to take into account the quantum nature of the environment, as discussed long ago by Brillouin (1962) for an information medium in general. This can usually be done easily, since the quantum theory of interacting systems, such as the quantum theory of particle scattering, is well understood. Its application to decoherence requires that one averages over all unobserved degrees of freedom. In technical terms, one has to "trace out the environment" after it has interacted with the considered system. This procedure leads to a quantitative theory of decoherence (cf. Joos and Zeh 1985). Taking the trace is based on the probability interpretation applied to the environment (averaging over all possible outcomes of measurements), even though this environment is *not* measured. (The precise physical meaning of these formal concepts will be discussed in Sect. 2.4.)

Is it possible to explain *all* superselection rules in this way as an effect induced by the environment[4] – including the existence and position of the border line between microscopic and macroscopic behavior in the realm of molecules? This would mean that the universality of the superposition principle could be maintained – as is indeed the basic idea of the *program of decoherence* (Zeh 1970, Zurek 1982; see also Chap. 4 of Zeh 2001). If physical states are thus exclusively described by wave functions rather than points in configuration space – as originally intended by Schrödinger *in space* by means of narrow wave packets instead of particles – then no uncertainty relations apply to quantum *states* (apparently allowing one to explain probabilistic aspects): the Fourier theorem applies to *certain* wave functions.

As another example, consider two states of different charge. They interact very differently with the electromagnetic field even in the absence of radiation: their Coulomb fields carry complete "information" about the total charge *at any distance*. The quantum state of this field would thus decohere a superposition of different charges if considered as a quantum system in a *bounded* region of space (Giulini, Kiefer, and Zeh 1995). This instantaneous

---

[4] It would be sufficient, for this purpose, to use an *internal* "environment" (unobserved degrees of freedom), but the assumption of a closed system is in general unrealistic.

action of decoherence at an arbitrary distance by means of the Coulomb field gives it the appearance of a kinematic effect, although it is based on the dynamical law of charge conservation, compatible with a *retarded* field that would "measure" the charge (see Sect. 6.4.1).

There are many other cases where the unavoidable effect of decoherence can easily be imagined without any calculation. For example, superpositions of macroscopically different electromagnetic fields, $f(\mathbf{r})$, may be defined by an appropriate field functional $\Psi[f(\mathbf{r})]$. Any charged particle in a sufficiently narrow wave packet would then evolve into several separated packets, depending on the field f, and thus become entangled with the quasi-classical state of the quantum field (Kübler and Zeh 1973, Kiefer 1992, Zurek, Habib, and Paz 1993; see also Sect. 4.1.2). The particle can be said to "measure" the state of the field. Since charged particles are in general abundant in the environment, no superpositions of macroscopically different electromagnetic fields (or different "mean fields" in other cases) are observed under normal conditions. This result is related to the difficulty of preparing and maintaining "squeezed states" of light (Yuen 1976) – see Sect. 3.3.3.1. Therefore, the field appears to *be* in one of its classical states (Sect. 4.1.2).

In all these cases, this conclusion requires that the quasi-classical states (or "pointer states" in measurements) are robust (dynamically stable) under natural decoherence, as pointed out already in the first paper on decoherence (Zeh 1970; see also Diósi and Kiefer 2000).

A particularly important example of a quasi-classical field is the metric of general relativity (with classical *states* described by spatial geometries on space-like hypersurfaces – see Sect. 4.2.1). Decoherence caused by all kinds of matter can therefore explain the absence of superpositions of macroscopically distinct spatial curvatures (Joos 1986b, Zeh 1986, 1988, Kiefer 1987), while *microscopic* superpositions would describe those hardly ever observable gravitons.

Superselection rules thus arise as a straightforward consequence of quantum theory under realistic assumptions. They have nonetheless been discussed mainly in mathematical physics – apparently under the influence of von Neumann's and Wigner's "orthodox" interpretation of quantum mechanics (see Wightman 1995 for a review). Decoherence by "continuous measurement" seems to form the most fundamental irreversible process in Nature. It applies even where thermodynamical concepts do *not* (such as for individual molecules – see Sect. 3.2.4), or when any exchange of heat is entirely negligible. Its time arrow of "microscopic causality" requires a Sommerfeld radiation condition for microscopic scattering (similar to Boltzmann's chaos), *viz.*, the absence of any dynamically relevant *initial* correlations, which would define a "conspiracy" in common terminology (Joos and Zeh 1985, Zeh 2001).

2.2 Observables as a Derived Concept

Measurements are usually described by means of "observables", formally represented by hermitean operators, and introduced in addition to the concepts of quantum states and their dynamics as a fundamental and independent ingredient of quantum theory. However, even though often forming the starting point of a formal quantization procedure, this ingredient may not be separately required if all *physical* states are perfectly described by general quantum superpositions and their dynamics. This interpretation, to be further explained below, complies with John Bell's quest for the replacement of observables with "beables" (see Bell 1987). It was for this reason that his preference shifted from Bohm's theory to collapse models (where wave functions are assumed to completely describe *reality*) during his last years.

Let $|\alpha\rangle$ be an arbitrary quantum state (perhaps experimentally prepared by means of a "filter" – see below). The *phenomenological* probability for finding the system in another quantum state $|n\rangle$, say, after an appropriate measurement, is given by means of their inner product, $p_n = |\langle n \mid \alpha\rangle|^2$, where both states are assumed to be normalized. The state $|n\rangle$ represents a specific measurement. In a position measurement, for example, the number n has to be replaced with the continuous coordinates x, y, z, leading to the "improper" Hilbert states $|\mathbf{r}\rangle$. Measurements are called "of the first kind" if the system will again be found in the state $|n\rangle$ (except for a phase factor) whenever the measurement is immediately repeated. *Preparations* can be regarded as measurements which *select* a certain subset of outcomes for further measurements. n-preparations are therefore also called n-filters, since all "not-n" results are thereby excluded from the subsequent experiment proper. The above probabilities can be written in the form $p_n = \langle\alpha \mid P_n \mid \alpha\rangle$, with a special "observable" $P_n := |n\rangle\langle n|$, which is thus *derived* from the kinematical concept of quantum *states* and their phenomenological probabilities to "jump" into other states in certain situations.

Instead of these special "n or not-n measurements" (with fixed n), one can also perform more general "n_1 or n_2 or ... measurements", with all n_i's mutually exclusive ($\langle n_i|n_j\rangle = \delta_{ij}$). If the states forming such a set $\{|n\rangle\}$ are pure and exhaustive (that is, complete, $\sum P_n = \mathbb{1}$), they represent a basis of the corresponding Hilbert space. By introducing an arbitrary "measurement scale" a_n, one may construct *general* observables $A = \sum |n\rangle a_n \langle n|$, which permit the definition of "expectation values" $\langle\alpha \mid A \mid \alpha\rangle = \sum p_n a_n$.[5] In the special case of a yes-no measurement, one has $a_n = \delta_{nn_0}$, and expectation values become probabilities. Finding the state $|n\rangle$ during a measurement is then also expressed as "finding the value a_n of an observable".[6] A unique change

---

[5] The popular textbook argument that observables must be hermitean in order to have real expectation values is successful but wrong. The essential requirement for an observable is its diagonalizability, which allows even the choice of a complex scale a_n if convenient.

of scale, $b_n = f(a_n)$, describes the *same* physical measurement; for position measurements of a particle it would simply represent a coordinate transformation. Even a measurement of the particle's potential energy is equivalent to a position measurement (up to degeneracy) if the function $V(\mathbf{r})$ is *given*.

According to this definition, quantum expectation values must not be understood as mean values in an ensemble that represents ignorance of the precise state. Rather, they have to be interpreted as probabilities for *potentially arising* quantum states $|n\rangle$ – regardless of the latters' interpretation. If the set $\{|n\rangle\}$ of such potential states forms a basis, any state $|\alpha\rangle$ can be represented as a superposition $|\alpha\rangle = \sum c_n |n\rangle$. In general, it neither forms an n_0-state nor any not-n_0 state. Its dependence on the complex coefficients c_n requires that states which differ from one another by a numerical factor must be different "in reality". This is true even though they represent the same "ray" in Hilbert space and cannot, according to the measurement postulate, be distinguished operationally. The states $|n_1\rangle + |n_2\rangle$ and $|n_1\rangle - |n_2\rangle$ could not be physically different from another if $|n_2\rangle$ and $-|n_2\rangle$ were the *same* state. While operationally meaningless in the state $|n_2\rangle$ by itself, any numerical factor would become relevant in the case of *recoherence*. (Only a *global* factor would be "redundant".) For this reason, projection operators $|n\rangle\langle n|$ are insufficient to characterize quantum states (cf. also Mirman 1970).

The expansion coefficients c_n, relating physically meaningful states – for example those describing different spin directions or different versions of the K-meson – must in principle be determined (relative to one another) by appropriate experiments. However, they can often be derived from a previously known (or conjectured) classical theory by means of "quantization rules". In this case, the classical configurations q (such as particle positions or field variables) are *postulated* to parametrize a basis in Hilbert space, $\{|q\rangle\}$, while the canonical momenta p parametrize another one, $\{|p\rangle\}$. Their corresponding observables, $Q = \int dq\, |q\rangle q\langle q|$ and $P = \int dp\, |p\rangle p\langle p|$, are required to obey commutation relations in analogy to the classical Poisson brackets. In this way, they form an important *tool* for constructing and interpreting the specific Hilbert space of quantum states. These commutators essentially determine the unitary transformation $\langle p \mid q\rangle$ (e.g. as a Fourier transform e^{ipq}) – thus more than what could be defined by means of the projection operators $|q\rangle\langle q|$ and $|p\rangle\langle p|$. This algebraic procedure is mathematically very elegant and appealing, since the Poisson brackets and commutators may represent generalized symmetry transformations. However, the *concept* of observables

---

[6] Observables are axiomatically *postulated* in the Heisenberg picture and in the algebraic approach to quantum theory. They are also presumed (in order to define fundamental expectation values) in Chaps. 6 and 7. This may be pragmatically appropriate, but appears to be in conflict with attempts to describe measurements and quantum jumps dynamically – either by a collapse (Chap. 8) or by means of a universal Schrödinger equation (Chaps. 1–4).

(which form the algebra) can be derived from the more fundamental one of state vectors and their inner products, as described above.

Physical states are assumed to vary in time in accordance with a dynamical law – in quantum mechanics of the form $i\partial_t|\alpha\rangle = H|\alpha\rangle$. In contrast, a measurement device is usually defined regardless of time. This must then also hold for the observable representing it, or for its eigenbasis $\{|n\rangle\}$. The probabilities $p_n(t) = |\langle n \mid \alpha(t)\rangle|^2$ will therefore vary with time according to the time-dependence of the physical states $|\alpha\rangle$. It is well known that this (Schrödinger) time dependence is formally equivalent to the (inverse) time dependence of observables (or the reference states $|n\rangle$). Since observables "correspond" to classical *variables*, this time dependence appeared suggestive in the Heisenberg–Born–Jordan algebraic approach to quantum theory. However, the absence of *dynamical states* $|\alpha(t)\rangle$ from this Heisenberg picture, a consequence of insisting on *classical* kinematical concepts, leads to paradoxes and conceptual inconsistencies (complementarity, dualism, quantum logic, quantum information, and all that).

An environment-induced superselection rule means that certain superpositions are highly unstable with respect to decoherence. It is then impossible in practice to construct measurement devices for them. This *empirical* situation has led some physicists to *deny the existence* of these superpositions and their corresponding observables – either by postulate or by formal manipulations of dubious interpretation, often including infinities. In an attempt to circumvent the measurement problem (that will be discussed in the following section), they often simply *regard* such superpositions as "mixtures" once they have formed according to the Schrödinger equation (cf. Primas 1990b).

While *any* basis $\{|n\rangle\}$ in Hilbert space defines formal probabilities, $p_n = |\langle n|\alpha\rangle|^2$, only a basis consisting of states that are not immediately destroyed by decoherence defines "realizable observables". Since the latter usually form a genuine subset of *all* formal observables (diagonalizable operators), they must contain a nontrivial "center" in algebraic terms. It consists of those which commute with all the rest. Observables forming the center may be regarded as "classical", since they can be measured simultaneously with *all* realizable ones. In the algebraic approach to quantum theory, this center appears as part of its axiomatic structure (Jauch 1968). However, since the condition of decoherence has to be considered quantitatively (and may even vary to some extent with the specific nature of the environment), this algebraic classification remains an approximate and dynamically emerging scheme.

These "classical" observables thus characterize the subspaces into which superpositions decohere. Hence, even if the superposition of a right-handed and a left-handed chiral molecule, say, *could* be prepared by means of an appropriate (very fast) measurement of the first kind, it would be destroyed before the measurement may be repeated for a test. In contrast, the chiral states of all individual molecules in a bag of sugar are "robust" in a normal environment, and thus retain this property *individually* over time intervals which by far exceed thermal relaxation times. This stability may even be increased by the quantum Zeno effect (Sect. 3.3.1). Therefore, chirality appears not only classical, but also as an approximate constant of the motion that has to be taken into account in the definition of thermodynamical ensembles (see Sect. 2.3).

The above-used description of measurements of the first kind by means of probabilities for transitions $|\alpha\rangle \rightarrow |n\rangle$ (or, for that matter, by corresponding observables) is phenomenological. However, measurements should be described *dynamically* as interactions between the measured system and the measurement device. The observable (that is, the measurement basis) should thus be derived from the corresponding interaction Hamiltonian and the initial state of the device. As discussed by von Neumann (1932), this interaction must be diagonal with respect to the measurement basis (see also Zurek 1981). Its diagonal matrix elements are operators which act on the quantum state of the device in such a way that the "pointer" moves into a position appropriate for being read, $|n\rangle|\Phi_0\rangle \rightarrow |n\rangle|\Phi_n\rangle$. Here, the first ket refers to the system, the second one to the device. The states $|\Phi_n\rangle$, representing different pointer positions, must approximately be mutually orthogonal, and "classical" in the explained sense.

Because of the dynamical superposition principle, an initial superposition $\sum c_n|n\rangle$ does *not* lead to definite pointer positions (with their empirically observed frequencies). If decoherence is neglected, one obtains their *entangled superposition* $\sum c_n|n\rangle|\Phi_n\rangle$, that is, a state that is different from all potential measurement outcomes $|n\rangle|\Phi_n\rangle$. This dilemma represents the "quantum measurement problem" to be discussed in Sect. 2.3. Von Neumann's interaction is nonetheless regarded as the first step of a measurement (a "pre-measurement"). Yet, a collapse seems still to be required – now in the measurement device rather than in the microscopic system. Because of the entanglement between system and apparatus, it would then affect the total system.[7]

If, in a certain measurement, a whole subset of states $|n\rangle$ leads to the same pointer position $|\Phi_{n_0}\rangle$, these states can not be distinguished by this measurement. According to von Neumann's interaction, the pointer state $|\Phi_{n_0}\rangle$ will now be correlated with the *projection* of the initial state onto the subspace spanned by this subset. A corresponding *collapse* was therefore postulated by Lüders (1951) in his generalization of von Neumann's "first intervention" (Sect. 2.3).

---

[7] Some authors seem to have taken the phenomenological collapse in the *microscopic system* by itself too literally, and therefore disregarded the state of the measurement device in their measurement theory (see Machida and Namiki 1980, Srinivas 1984, and Sect. 9.1). Their approach is based on the assumption that quantum states must always exist for all systems. This would be in conflict with quantum nonlocality, even though it may be in accordance with early interpretations of the quantum formalism.

In this dynamical sense, the interaction with an appropriate measuring device *defines* an observable. The formal time dependence of observables according to the Heisenberg picture would now describe a time dependence of the states diagonalizing the interaction Hamiltonian with the device, paradoxically controlled by the intrinsic Hamiltonian of the *system*.

The question whether a certain formal observable (that is, a diagonalizable operator) can be physically realized can only be answered by taking into account the unavoidable environment. A macroscopic measurement device is *always* asssumed to decohere into its macroscopic pointer states. However, environment-induced decoherence by itself does not (yet) solve the measurement problem, since the "pointer states" $|\Phi_n\rangle$ may be assumed to *include* the total environment (the "rest of the world"). Identifying the thus arising global superposition with an *ensemble* of states, represented by a statistical operator ρ, that merely leads to the same *expectation values* $\langle A \rangle = \text{tr}(A\rho)$ for a *limited* set of observables $\{A\}$ would beg the question. This argument is nonetheless found wide-spread in the literature. For example, Haag (1992) used it to select the subset of all *local* observables.

In Sect. 2.4, statistical operators ρ will be *derived* from the concept of quantum states as a tool for calculating expectation values, whereby the latter are defined, as described above, in terms of probabilities for the appearance of new states in measurements. In the Heisenberg picture, ρ is often regarded as in some sense representing the *ensemble of potential "values"* for all observables that are postulated to formally replace all classical variables. This interpretation is suggestive because of the (incomplete) formal analogy of ρ to a classical phase space distribution. However, "prospective values" are physically meaningful only if they characterize prospective states. Note that Heisenberg's uncertainty relations refer to *potential* (mutually exclusive) measurements – not to variables characterizing the physical states.

2.3 The Measurement Problem   *meaning of ρ*

The superposition of different measurement outcomes, resulting according to a Schrödinger equation when applied to the total system (as discussed above), demonstrates that a "naive ensemble interpretation" of quantum mechanics in terms of incomplete knowledge is ruled out. It would require that a quantum state (such as $\sum c_n |n\rangle |\Phi_n\rangle$) represents an ensemble of some as yet unspecified fundamental states, of which a sub-ensemble (for example represented by the quantum state $|n\rangle |\Phi_n\rangle$) may be "picked out by a mere increase of information". If this were true, then the sub-ensemble resulting from this measurement could in principle be traced back in time by means of the Schrödinger equation in order to determine also the initial state more completely (to "postselect" it – see Aharonov and Vaidman 1991 for an inappropriate attempt to do so). In the above case this would lead to the initial quantum state $|n\rangle |\Phi_0\rangle$ that is *physically different* from – and thus inconsistent

epistemic reading of $\sum c_n |n\rangle |\Phi_n\rangle$ as an ensemble of possibilities' is wrong!

with – the superposition $(\sum c_n|n\rangle)|\Phi_0\rangle$ that had been prepared (whatever it *means*).

In spite of this simple argument, which demonstrates that an ensemble interpretation would require a complicated and miraculous nonlocal "background mechanism" in order to work consistently (cf. Footnote 3 regarding Bohm's theory), a merely statistical interpretation of the wave function seems to remain the most popular one because of its pragmatic (though limited) value. A general and more rigorous critical discussion of problems arising in various ensemble interpretations may be found in d'Espagnat's books (1976 and 1995), for example.

A way out of this dilemma within quantum mechanical concepts requires one of two possibilities: a modification of the Schrödinger equation that explicitly describes a collapse (also called "spontaneous localization" – see Chap. 8), or an Everett type interpretation, in which all measurement outcomes are assumed to exist in one formal superposition, but to be *perceived* separately as a consequence of their dynamical autonomy resulting from decoherence. While this latter suggestion has been called "extravagant" (as it requires myriads of co-existing quasi-classical "worlds"), it is similar in principle to the conventional (though nontrivial) assumption, made tacitly in all classical descriptions of observation, that consciousness is *localized* in certain semi-stable and sufficiently complex *subsystems* (such as human brains or parts thereof) of a much larger external world. Occam's razor, often applied to the "other worlds", is a dangerous instrument: philosophers of the past used it to deny the existence of the interior of stars or of the back side of the moon, for example. So it appears worth mentioning at this point that environmental decoherence, derived by tracing out unobserved variables from a universal wave function, readily describes precisely the *apparently* observed "quantum jumps" or "collapse events" (as will be discussed in great detail in various parts of this book).

decoherence explains jumps

The *effective* dynamical rules which are used to describe the observed time dependence of quantum states represent a "dynamical dualism". This was first clearly formulated by von Neumann (1932), who distinguished between the unitary evolution according to the Schrödinger equation (remarkably his "zweiter Eingriff" or "second intervention"),

$$i\hbar \frac{\partial}{\partial t} |\psi\rangle = H |\psi\rangle \quad , \tag{2.1}$$

valid for isolated (absolutely closed) systems, and the "reduction" or "collapse of the wave function",

$$|\psi\rangle = \sum c_n|n\rangle \to |n_0\rangle \tag{2.2}$$

(remarkably his "*first* intervention"). The latter was meant to describe the stochastic transitions into new state $|n_0\rangle$ during measurements (Sect. 2.2). This dynamical discontinuity had been anticipated by Bohr in the form of "quantum jumps", assumed to occur between his discrete atomic electron

"*dynamical dualism*"

Von Neumann two part description ① $|\psi\rangle = \sum c_n|n\rangle \to |n_0\rangle$ *measure* ② $\partial_t|\psi\rangle = H|\psi\rangle$ "*evolution*" *orbits*. Later, the *time-dependent* Schrödinger equation (2.1) for interacting systems was often regarded merely as a method of calculating probabilities for similar (individually unpredictable) discontinuous transitions between different energy eigenstates (static *quantum* states) of atomic systems (Born 1926).[8]

In scattering theory, one usually probes only *part* of quantum mechanics by restricting consideration to "free" asymptotic states and their phenomenological probabilities (disregarding their entangled superpositions). Quantum correlations between them then appear statistical ("classical"). Occasionally even the unitary scattering amplitudes $\langle m_{out}|n_{in}\rangle = \langle m|\,S\,|n\rangle$ are confused with the *probability* amplitudes $\langle \phi_m|\psi_n\rangle$ for finding a state $|\phi_m\rangle$ in an initial one, $|\psi_n\rangle$, in a measurement. In his general S-matrix theory, Heisenberg temporarily speculated about deriving the latter from the former. Since *macroscopic* systems never become asymptotic because of their unavoidable interaction with their environment, they cannot be described by an S-matrix at all.

The unacceptable Born-von Neumann dynamical dualism was evidently the major motivation for an ignorance interpretation of the wave function. It attempts to explain the collapse *not* as a dynamical process occurring in the system, but as an increase of *information* about it. This would be represented by the reduction of an ensemble of *possible* states. While the classical description of ensembles uses a similar dualism, a corresponding interpretation in quantum theory leads to the severe (and apparently fatal) difficulties indicated above. They are often circumvented by the invention of new formal "rules of logic and statistics", which are *not* based on any interpretation in terms of ensembles or incomplete knowledge.

If the state of a *classical* system is incompletely known, and the corresponding point p,q in phase space therefore replaced by an ensemble (a probability distribution) $\rho(p,q)$, this ensemble can be "reduced" by an additional observation. For this purpose, the system must interact in a controllable manner with an external "observer" who holds the information (cf. Szilard 1929). The latter's physical memory state must thereby change in dependence on the property-to-be-measured, without disturbing the system in the ideal case (negligible "recoil"). According to *deterministic* dynamical laws, the ensemble entropy of the combined system, which initially contains the entropy corresponding to the unknown microscopic quantity, would remain constant if it were defined to include the entropy characterizing the final ensemble of different outcomes. However, since the observer is assumed to "know" (to be

---

[8] Thus also Bohr (1928) in a subsection entitled "Quantum postulate and causality" about "the quantum theory": "...its essence may be expressed in the so-called quantum postulate, which attributes to any *atomic process* an essential discontinuity, or rather individuality, completely foreign to classical theories and symbolized by Planck's quantum of action" (my italics). The later revision of these early interpretations of quantum theory (required by the important role of entangled quantum states for much larger systems) seems to have gone unnoticed by many physicists.

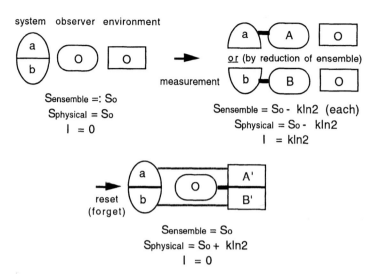

Fig. 2.1. Entropy relative to the state of information in an ideal *classical* measurement. Areas represent *sets* of microscopic states of the subsystems (while those of uncorrelated combined systems would be represented by their direct products). During the first step of the figure, the memory state of the observer changes deterministically from 0 to A or B, depending on the state a or b of the system to be measured. The second step depicts a subsequent reset, required if the measurement is to be repeated with the *same* device (Bennett 1973). A' and B' are effects which must thereby arise in the thermal environment in order to preserve the total ensemble entropy in accordance with presumed microscopic determinism. The "physical entropy" (*defined* to add for subsystems) measures the phase space of all microscopic degrees of freedom, including the property to be measured, while *depending on given* macroscopic variables. Because of its presumed additivity, this physical entropy neglects all remaining statistical correlations (dashed lines, which indicate *sums* of products of sets) for being "irrelevant" in the future – hence $S_{\text{physical}} \geq S_{\text{ensemble}}$. I is the amount of information held by the observer. The minimum initial entropy, S_0, is $k \ln 2$ in this simple case of two equally probable values a and b.

aware of) his own state, this ensemble is reduced correspondingly, and the ensemble entropy defined *with respect to his state of information* is lowered.

This situation is depicted by the first step of Fig. 2.1, where ensembles of states are represented by areas. In contrast to many descriptions of Maxwell's demon, the observer (regarded as a device) is here subsumed into the ensemble description. *Physical entropy*, unlike ensemble entropy, is usually understood as a *local* (additive) concept, which neglects long range correlations for being "irrelevant", and thus approximately defines an *entropy density*. Physical and ensemble entropies are equal if there are no correlations. The information I, given in the figure, measures the reduction of entropy corresponding to the increased knowledge of the observer. This description is consistent with classical concepts, where a real physical state is represented by a point in the diagram, while physical entropy may be characterized by means of "representative ensembles" (cf. Zeh 2001).

This picture does not necessarily require a *conscious* observer (although it may ultimately rely upon him). It applies to any macroscopic measurement device, since physical entropy is not only defined to be local, but also as a function of "given" macroscopic properties (which thus define representative ensembles). The dynamical part of the measurement transforms "physical" entropy (here the ensemble entropy of the microscopic variables) deterministically into entropy of lacking information about controllable macroscopic properties. Before the observation is taken into account (that is, before the "or" is applied), both parts of the ensemble after the first step add up to give the ensemble entropy. When it *is* taken into account (as done by the numbers given in the figure), the ensemble entropy is reduced according to the information gained by the observer.

Any registration of information by the observer must use up his memory capacity ("blank paper"), which represents non-maximal entropy. If the same measurement is to be repeated, for example in a cyclic process that could be used to transform heat into mechanical energy (Szilard 1929), this capacity would either be exhausted at some time, or an equivalent amount of entropy must be absorbed by the environment (for example in the form of heat) in order to *reset* the measurement or registration device (second step of Fig. 2.1). The reason is that two *different* states cannot deterministically evolve into the same final state (Bennett 1973).[9] This argument is based on an arrow of time of "causality", which requires that all correlations possess *local causes* in their past (no "conspiracy"). The irreversible formation of "irrelevant" correlations then explains the increase of *physical* (local) entropy, while the ensemble entropy is conserved.

The insurmountable problems encountered in a statistical interpretation of the *wave function* (or of any other superposition, such as $|a\rangle + |b\rangle$) are reflected by the fact that there is no ensemble entropy that would represent the unknown property-to-be-measured (see the first step of Fig. 2.2 or 2.3 – cf. also Zurek 1984). The "ensemble entropy" is now *defined* by the "corresponding" expression $S_{ensemble} = -k\mathrm{tr}\{\rho \ln \rho\}$ (but see Sect. 2.4 for the meaning of the density matrix ρ). If the entropy of observer plus environment were numerically the same as in the classical case of Fig. 2.1, the total initial ensemble entropy would be lower; for equal initial probabilities of a and b (as assumed in the figures) it is now given in terms of the previous values by $S_0 - k \ln 2$. It would even vanish for pure states ϕ and χ, respectively, of observer and environment: $(|a\rangle + |b\rangle)|\phi_0\rangle|\chi_0\rangle$. The global Schrödinger evolution

---

[9] Bennett did *not* define physical entropy to include that of the microscopic ensemble a,b, since he regarded this variable as "controllable" – in contrast to the thermal (ergodic or irrelevant) property A',B'.

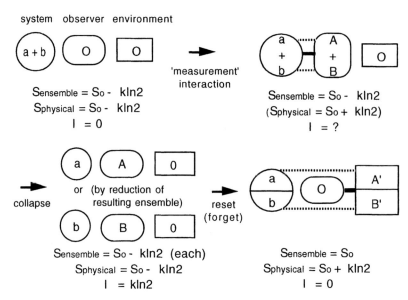

system observer environment

'measurement'
interaction

Sensemble = So - kln2
Sphysical = So - kln2
I = 0

Sensemble = So - kln2
(Sphysical = So + kln2)
I = ?

collapse

or (by reduction of
resulting ensemble)

reset
(forget)

Sensemble = So - kln2 (each)
Sphysical = So - kln2
I = kln2

Sensemble = So
Sphysical = So + kln2
I = 0

Fig. 2.2. Quantum measurement of a *superposition* $|a\rangle + |b\rangle$ by means of a *collapse* process, here assumed to be triggered by the macroscopic pointer position. The initial entropy is smaller by one bit than in Fig. 2.1 (and may in principle vanish), since there is no initial *ensemble* a, b for the property to be measured. Dashed lines *before* the collapse now represent quantum entanglement. (Compare the ensemble entropies with those of Fig. 2.1!) Increase of physical entropy in the first step is appropriate only if the arising entanglement is *regarded as irrelevant*. The collapse itself is often divided into *two* steps: first increasing the ensemble entropy by replacing the superposition with an ensemble, and then lowering it by reducing the ensemble (applying the "or" – for macroscopic pointers only). The increase of ensemble entropy, observed in the final state of the Figure, is a consequence of this first step of the collapse. It brings the entropy up to its classical initial value of Fig. 2.1

(depicted in Fig. 2.3) would then be described by three dynamical steps,

$$
\begin{aligned}
(|a\rangle + |b\rangle)\,|\phi_0\rangle\,|\chi_0\rangle &\to (|a\rangle\,|\phi_A\rangle + |b\rangle\,|\phi_B\rangle)\,|\chi_0\rangle \\
&\to |a\rangle\,|\phi_A\rangle\,|\chi_{A''}\rangle + |b\rangle\,|\phi_B\rangle\,|\chi_{B''}\rangle \\
&\to (|a\rangle\,|\chi_{A'A''}\rangle + |b\rangle\,|\chi_{B'B''}\rangle)\,|\phi_0\rangle \quad , \qquad (2.3)
\end{aligned}
$$

with an "irrelevant" (inaccessible) final quantum correlation between system and environment as a relic from the initial superposition. In this unitary evolution, the two "branches" recombine to form a *nonlocal* superposition. It "exists, but it is not there". Its local unobservability characterizes an "apparent collapse" (as will be discussed). For a genuine collapse (Fig. 2.2), the final correlation would be statistical, and the ensemble entropy would increase, too.

a genuine non-unitary collapse rule would resolve
measurement problem but no such mechanism
has ever been found

2 Basic Concepts and Their Interpretation   27

As mentioned in Sect. 2.2, the general interaction dynamics that is required to describe "ideal" measurements according to the Schrödinger equation (2.1) is derived from the special case where the measured system is prepared in an eigenstate $|n\rangle$ before measurement (von Neumann 1932),

$$|n\rangle|\Phi_0\rangle \rightarrow |n\rangle|\Phi_n\rangle \quad . \tag{2.4}$$

Here, $|n\rangle$ corresponds to the states $|a\rangle$ or $|b\rangle$ used in the figures, the pointer positions $|\Phi_n\rangle$ to the states $|\phi_A\rangle$ and $|\phi_B\rangle$. (During non-ideal measurements, the state $|n\rangle$ would change, too.) However, applied to an initial superposition, $\sum c_n|n\rangle$, the interaction according to (2.1) leads to an entangled superposition,

$$\left(\sum c_n|n\rangle\right)|\Phi_0\rangle \rightarrow \sum c_n|n\rangle|\Phi_n\rangle \quad . \tag{2.5}$$

As explained in Sect. 2.1.1, the resulting superposition represents an *individual physical state* that is different from all components appearing in this sum. While decoherence arguments teach us (see Chap. 3) that neglecting the environment of (2.5) would be absolutely unrealistic for macroscopic pointer states $|\Phi_n\rangle$, this superposition remains nonetheless valid if Φ is defined to include the "rest of the universe", such as $|\Phi_n\rangle = |\phi_n\rangle|\chi_n\rangle$, with an environmental state $|\chi\rangle$. This powerful consequence of the Schrödinger equation holds regardless of all complications, such as decoherence and other, in practice irreversible, processes (which need not even be known). Therefore, it does seem that the measurement problem can only be resolved if the Schrödinger dynamics (2.1) is supplemented by a nonunitary collapse (2.2).

Specific proposals for such a process will be discussed in Chap. 8. Remarkably, however, there is no empirical evidence yet on where the Schrödinger equation may have to be modified for this purpose (see Joos 1987a, Pearle and Squires 1994, or d'Espagnat 2001). On the contrary, the dynamical superposition principle has been confirmed with fantastic accuracy in spin systems (Weinberg 1989, Bollinger *et al.* 1989).

The Copenhagen interpretation of quantum theory insists that the measurement outcome has to be described in fundamental classical terms rather than as a quantum state. While according to Pauli (in a letter to Einstein: Born 1969), the appearance of an electron position is "a creation outside of the laws of Nature" (*eine ausserhalb der Naturgesetze stehende Schöpfung*), Ulfbeck and Bohr (2001) now claim (similar to Ludwig 1990 in his attempt to derive "the" Copenhagen interpretation from fundamental principles) that it is the *click in the counter* that appears "out of the blue", and "without an event that takes place in the source itself as a precursor to the click". Together with the occurrence of this, thus *not dynamically analyzable*, irreversible event in the counter, the wave function is then claimed to "lose its meaning" (precisely where it would otherwise describe decoherence!). The Copenhagen interpretation is often hailed as the greatest revolution in physics, since it rules out the general applicability of the concept of objective physical reality. I

Copenhagen nonsense

28   H. D. Zeh

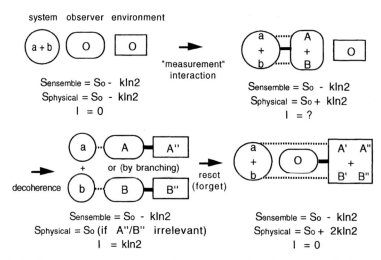

Fig. 2.3. Quantum measurement of a superposition by means of "branching" caused by decoherence (see text). The increase of physical entropy during the second step applies if the distinction between environmental degrees of freedom A'', B'', responsible for decoherence, is "irrelevant" (uncontrollable). After the last step, *all* entanglement has irreversibly become irrelevant in practice. Since the whole superposition is here assumed to "exist" forever (and may have future consequences *in principle*), the branching is meaningful only with respect to a *local* observer.

am instead inclined to regard it as a kind of "quantum voodoo": irrationalism in place of dynamics. The theory of decoherence describes events in the counter by means of a universal Schrödinger equation as a fast and for all practical purposes irreversible *dynamical* creation of entanglement with the environment (see also Shi 2000). In order to remain "politically correct", some authors have recently even *re-defined* complementarity in terms of entanglement (cf. Bertet *et al.* 2001), although the latter has never been a crucial element of the Copenhagen interpretation.

The "Heisenberg cut" between observer and observed has often been claimed to be quite arbitrary. This cut represents the borderline at which the probability interpretation for the occurrence of events is applied. However, shifting it too far into the microscopic realm would miss the readily observed quantum aspects of certain large systems (SQUIDs etc.), while placing it beyond the detector would require the latter's decoherence to be taken into account anyhow. As pointed out by John Bell (1981), the cut has to be placed "far enough" from the measured object in order to ensure that our limited capabilities of investigation (such as those of keeping the measured system isolated) prevent us from discovering any inconsistencies with the assumed classical properties or a collapse.

As noticed quite early in the historical debate, the cut may even be placed deep into the human observer, whose consciousness, which may be provisionally located in the cerebral cortex, represents the final link in the observational chain. This view can be found in early formulations by Heisenberg, it was favored by von Neumann, later discussed by London and Bauer (1939), and again supported by Wigner (1962), among others. It has even been interpreted as an *objective* influence of consciousness on physical reality (e.g. Wigner *l.c.*), although it may be consistent with the formalism only when used with respect to *one* final observer, that is, in a strictly subjective (though partly objectivizable) sense (Zeh 1971).

The "indivisible chain between observer and observed" is physically represented by a complex interacting medium, or a causal chain of intermediary systems $|\chi^{(i)}\rangle$, in quantum mechanical terms symbolically written as

$$\left|\psi_n^{\text{system}}\right\rangle \left|\chi_0^{(1)}\right\rangle \left|\chi_0^{(2)}\right\rangle \cdots \left|\chi_0^{(K)}\right\rangle \left|\chi_0^{\text{obs}}\right\rangle$$

$$\rightarrow \left|\psi_n^{\text{system}}\right\rangle \left|\chi_n^{(1)}\right\rangle \left|\chi_0^{(2)}\right\rangle \cdots \left|\chi_0^{(K)}\right\rangle \left|\chi_0^{\text{obs}}\right\rangle$$

$$\vdots$$

$$\rightarrow \left|\psi_n^{\text{system}}\right\rangle \left|\chi_n^{(1)}\right\rangle \left|\chi_n^{(2)}\right\rangle \cdots \left|\chi_n^{(K)}\right\rangle \left|\chi_n^{\text{obs}}\right\rangle \quad , \tag{2.6}$$

instead of the simplified form (2.4). While an initial superposition of the observed system now leads to a *superposition* of such product states (similar to (2.5)), we know empirically that a collapse must be "taken into account" by the conscious observer before (or at least when) the information arrives at him as the final link. If there are several chains connecting observer and observed (for example via other observers, known as "Wigner's friends"), the correctly applied Schrödinger equation warrants that each individual component (2.6) describes consistent ("objectivized") measurement results (cf. Zeh 1973). From the subjective point of view of the final observer, all intermediary systems ("Wigner's friends" or "Schrödinger's cats", including their environments) could well remain in a superposition of drastically different situations until he observes (or communicates with) them!

Environment-induced decoherence means that an avalanche of other causal chains unavoidably branch off from the intermediary links of the chain as soon as they become macroscopic (see Chap. 3). This *might* even trigger a genuine collapse process (to be described by hypothetical dynamical terms), since the many-particle correlations arising from decoherence would render the total system prone to such as yet unobserved, but nevertheless conceivable, non-linear many-particle forces (Pearle 1976, Diósi 1985, Ghirardi, Rimini, and Weber 1986, Tessieri, Vitali, and Grigolini 1995; see also Chap. 8). Decoherence by a *microscopic* environment has been experimentally confirmed to be *reversible* in a process now often called "quantum erasure of a measurement" (see Herzog et al. 1995). In analogy to the concept of particle creation, reversible decoherence may be regarded as "virtual decoherence".

reversible decoherence

decoherence in brain

"Real" decoherence, which gives rise to the familiar classical appearance of the macroscopic world, is instead characterized by its unavoidability and ir-reversibility *in practice*.

In an important contribution, Tegmark (2000) was able to demonstrate that neuronal and other processes in the brain also become quasi-classical because of environmental decoherence – see Sect. 3.2.5. (Successful neuronal models are indeed classical.) This seems to imply that at least *objective* aspects of human thinking and behavior can be described by conceptually classical (though not necessarily deterministic) models of the brain. Since no precise "localization of consciousness" within the brain has been confirmed yet, the neural network (just as the retina, say) may still be part of the "external world" with respect to the unknown ultimate observer system (Zeh 1979). Because of Tegmark's arguments, this problem may not affect an *objective* theory of observation any longer.

physical consciousness theory

However, even "real" decoherence in the sense of above must be distinguished from the concept of a genuine collapse, which is defined as the *disappearance* of all but one components from reality (thus representing an irreversible *law*).[10] As pointed out before, a collapse could well occur later in the observational chain, and possibly remain less fine-grained than decoherence. It should nonetheless be detectable in other situations if it follows dynamical rules. Environment-induced decoherence alone (the dynamically arising strong correlations with the rest of the world) leads to the possibly sufficient consequence that, in a world with no more than few-particle forces, certain "robust" states $|\chi_n^{\text{obs}}\rangle$ are not affected by what goes on in the other branches that would have formed according to the Schrödinger equation.

In order to represent a subjective observer, such a physical system must be in a definite state with respect to properties of which he/she/it is aware. The salvation of a psycho-physical parallelism of this kind was von Neumann's main argument for the introduction of his "first intervention" (the collapse). As a consequence of the mentioned dynamical independence of the different individual components of type (2.6) in their superposition, one may instead associate *all* arising factor wave functions $|\psi_n^{\text{obs}}\rangle$ (different ones in each component) with *separate* subjective observers, that is, with independent states of consciousness. This description, which avoids a collapse as a new dynamical law, is essentially Everett's "relative state interpretation" (so called, since the worlds *observed* by these different observer states are described by their respective relative factor states). While also called a "many worlds interpretation", it describes *one* quantum universe. Because of its (essential and

EVERETT

---

[10] Proposed decoherence mechanisms involving event horizons (Hawking 1987, Ellis, Mohanty and Nanopoulos 1989) would either have to *postulate* such a fundamental violation of unitarity, or merely represent a specific kind of environmental decoherence (entanglement beyond the horizon) – see Sect. 4.2.5. The most immediate consequence of quantum entanglement is that an exactly unitary evolution can only be consistently applied to the whole universe.

'Unitarity as a universal thing'

non-trivial) reference to conscious observers, it would more appropriately be called a "multi-consciousness" or "many minds interpretation" (Zeh 1970, 1971, 1979, 1981, 2000, Albert and Loewer 1988, Lockwood 1989, Squires 1990, Stapp 1993, Donald 1995, Page 1995).[11]

Because of their dynamical independence, none of these different observers (or, in another language, different arising "versions" of the same observer) can find out by experiments whether or not the other components resulting from the Schrödinger equation have survived. This dynamical consequence leads to the *impression* that all "other" components have ceased to exist as soon as decoherence became irreversible for all practical purposes. So it remains a pure matter of taste whether Occam's razor is applied to the wave function (by adding appropriate but not directly detectable collapse-producing nonlinear terms to its dynamical law), or to the dynamical law (by instead adding myriads of unobservable Everett components to our conception of "reality"). Traditionally (and mostly successfully), consistency of the law has been ranked higher than simplicity of the facts.

Fortunately, the dynamics of decoherence can be discussed without having to make this choice. A collapse (real or apparent) just has to be *taken into account* in order to describe the dynamics of that (partial) wave function which represents our *observed quasi-classical world* (the time-dependent component which contains "our" observer states $|\chi_n^{\mathrm{obs}}\rangle$). Only *specific* dynamical collapse models could be confirmed or ruled out by experiments, while Everett's relative states, on the other hand, may *in principle* depend on the definition of the observer system.[12] (No *other* "systems" have to be specified in principle. Their density matrices, which describe decoherence and quasi-classical concepts, are merely convenient.)

---

[11] As Bell (1981) once pointed out, Bohm's theory would instead require consciousness to be psycho-physically coupled to certain *classical* variables (which this theory *postulates* to exist). These variables are probabilistically related to the wave function by means of a conserved statistical initial condition. Thus one may argue that the "many minds interpretation" merely eliminates Bohm's unobservable and therefore meaningless intermediary classical variables and their trajectories from this psycho-physical connection. This is possible because of the dynamical autonomy of a wave function that evolves in time according to a universal Schrödinger equation, and independently of Bohm's classical variables. The latter cannot, by themselves, carry memories of their "surrealistic" histories. Memories are solely in the quasi-classical wave packets that effectively guide them, while the other myriads of "empty" Everett world components (criticized for being "extravagant" by Bell) *exist* as well in Bohm's theory. Barbour (1999), in his theory of timelessness, proposed in effect a *static* Bohm theory. It eliminates the latter's formal classical trajectories, while preserving a concept of memories without a history ("time capsules" – see also Chap. 6 of Zeh 2001).

[12] Another aspect of this observer-relatedness of the observed world is the concept of a *presence*, which is not part of *physical* time. It reflects the *empirical fact* that the subjective observer is local in space *and* time.

probabilities in Everett

Born Rule

In contrast to Bohm's theory and stochastic collapse models, nothing has been said yet (or postulated) about *probabilities* of measurement outcomes. For this purpose, the Everett branches have to be given statistical weights in a way that appears *ad hoc* again. However, these probabilities are meaningful to an observer only as frequencies in *series* of equivalent measurements, which lead to multiple branchings. Graham (1970) was able to show that the norm of the superposition of all those Everett branches which contain frequencies of outcomes that substantially *differ* from Born's probabilities vanishes in the limit of infinite series. Although the definition of the norm, which is used in this argument, is *precisely equivalent* to Born's probabilities, it is preferred in comparison with other choices of a norm by its unique property of being conserved under the Schrödinger dynamics.

To give an example: an isolated decaying quantum system may be described by a superposition of its metastable initial state with those of all its decay channels. On a large but finite region of space and time the total wave function may then *approximately* decrease exponentially and coherently in accordance with the Schrödinger equation (with very small late-time deviations from exponential behavior caused by the dispersion of the outgoing waves). For a system that decays by emitting electromagnetic radiation into a reflecting cavity, a *superposition of different decay times* has in fact been confirmed in the form of coherent "state vector revival" (Rempe, Walther and Klein 1987). An even more complex experiment exhibiting coherent state vector revival was performed by means of spin waves (Rhim, Pines and Waugh 1971). In general, however, the decay fragments would soon be "monitored" by surrounding matter. The resulting state of the environment must then *depend on* the decay time, such that the superposition will decohere into dynamically independent components corresponding to *different* (approximately defined) decay times for all nuclei. From the point of view of a local observer, each individual nucleus may be assumed to decay discontinuously (even when the decay is not actually observed). This situation does *not* allow coherent state vector revival any more. Instead, it leads to an exactly exponential statistical decay law, valid shortly after the decaying state was produced (see Joos 1984 and Sect. 3.3.1).

A similar decoherence mechanism explains the quasi-discontinuous collisions observed for a Brownian particle under a light microscope (which leads to almost immediate decoherence). In turn, scattered molecules contribute to the decoherence.

As long as the information about the decay has not yet reached the observer,

$$\left(\sum_n c_n \left| \psi_n^{\text{system}} \right\rangle \left| \chi_n^{(1)} \right\rangle \left| \chi_n^{(2)} \right\rangle \cdots \left| \chi_n^{(K)} \right\rangle \right) \left| \chi_0^{\text{obs}} \right\rangle, \qquad (2.7)$$

he may as well assume (from his subjective point of view) that the nonlocal superposition still exists. According to the formalism, Schrödinger's cat (represented by $\left| \chi^{(2)} \right\rangle$, say) would then "become" dead *or* alive only when the observer becomes aware of it. On the other hand, the property described by the state $|\psi_n^{\text{system}}\rangle$ (just as the cat's status of being dead or alive, $\left|\chi_n^{(2)}\right\rangle$) can be *assumed* to have become "real" as soon as decoherence has become irreversible in practice. (In the case of strong decoherence the time of the cat's death can be approximately determined later by forensic investigation of the state of the cat's body, which thus contains even *controllable* information in physical form.) In the same way, decoherence must corrupt any controllable entanglement that might lead to a violation of Bell's inequalities (as it does – see Venugopalan, Kumar and Ghosh 1995b). All predictions which a local observer can check are consistent with the assumption that the system was in *one* of the states $|\psi_n^{\text{system}}\rangle$ (with probability $|c_n|^2$) before the second measurement (see also Sect. 2.4). This justifies the interpretation that the cat is *determined* to die or not yet to die as soon as irreversible decoherence has occurred somewhere in the causal chain (which will in general be the case even *before* the poison is applied).

2.4 Density Matrix, Coarse Graining, and "Events"

The theory of decoherence uses some (more or less technical) auxiliary concepts. Their physical meaning will be recalled and discussed in this section, since it is essential for a correct understanding of what will be achieved.

In *classical* statistical mechanics, incomplete information about the real physical state of a system is described by means of "ensembles", that is, by probability distributions, such as $\rho(p,q)$, on the space of states (in Fig. 2.1 indicated by areas of uniform probability). Such ensembles are often called "thermodynamic" or "macroscopic states". Corresponding mean values of *functions of state*, $a(p,q)$, with respect to these ensembles are then given by the expression $\int dp\,dq\,\rho(p,q)a(p,q)$. An ensemble $\rho(p,q)$ may be recovered from the mean values of a complete set of state functions (such as all δ-functions), while a smaller set (that can be realized in practice) determines only a "coarse-grained" probability distribution.

If all members of an ensemble are assumed to obey the same Hamiltonian equations, their probability distribution ρ evolves according to the Liouville equation,

$$\frac{\partial \rho}{\partial t} = \{H, \rho\} \quad , \tag{2.8}$$

with their common Hamiltonian H and the Poisson bracket $\{,\}$. However, this assumption would be highly unrealistic for a many-particle system. Even if the *fundamental* dynamics were given, the *induced* Hamiltonian for the considered system would very sensitively depend on the state of the "environment", which cannot be assumed to be known better than that of the system itself. Borel (1914b) showed long ago that even the gravitational effect resulting from shifting a small rock at the distance of Sirius by a few centimeters would completely change the microscopic state of a gas in a vessel here on earth within seconds after the retarded field has arrived (see also Chap. 3). In a similar connection, Ernst Mach spoke of the "profound interconnectedness of things". Borel's surprising conclusion follows from the enormous amplification in subsequent collisions of those tiny differences in molecular trajectories that result from the slightly different forces. Conversely, small microscopic variations in the state of the gas must affect its environment, thus leading in turn to slightly different effective Hamiltonians for the gas. This will in general have strong ("chaotic") effects in the microscopic states which form the original ensemble, and thus induce *statistical* correlations of the gas with its environment. Their neglect leads to an increase of ensemble entropy.

increased ensemble entropy

This *effective local dynamical indeterminism* can be taken into account in this universe (when calculating *forward* in time) by means of stochastic forces (using a *Langevin equation*) for the individual states, or by means of a corresponding *master equation* for an ensemble of states, $\rho(p, q; t)$. The increase of the local ensemble entropy is thus attributed to an uncertain effective Hamiltonian. In this way, statistical correlations with the environment are regarded as dynamically irrelevant for the *future* evolution. An example is Boltzmann's collision equation (where the arising irrelevant correlations are *intrinsic* to the gas, however). The justification of this time-asymmetric procedure forms a basic problem of physics and cosmology (Zeh 2001).

When applying the conventional quantization rules to the Liouville equation (2.8) in a formal way, one obtains the *von Neumann equation* (or *quantum Liouville equation*),

V. Neumann
parallel with
Liouville eq^n

$$i\hbar\frac{\partial\rho}{\partial t} = [H, \rho] \quad , \tag{2.9}$$

for the dynamics of "statistical operators" or "density operators" ρ. Similarly, expectation values $\langle A \rangle = \mathrm{tr}(A\rho)/\mathrm{tr}(\rho)$ of observables A formally replace mean values $\bar{a} = \int dp\, dq\, a(p, q)\rho(p, q)$ of the state functions $a(p, q)$. Expectation values of a *restricted* set of observables would again represent a generalized coarse graining for the density operators. The von Neumann equation (2.9) is thus unrealistic for similar reasons as is the Liouville equation (2.8), although quantitative differences between these two cases may arise from the different energy spectra – mainly at low energies. Discrete spectra have relevant consequences for macroscopic systems in practice only in exceptional cases, while they often prevent mathematically rigorous proofs of ergodic or chaos theorems that are valid only in excellent approximation. However, whenever quantum correlations do form in analogy to classical correlations (as is the rule), they lead to far more profound consequences than their classical counterparts.

In order to understand these differences, the concept of a density matrix has to be derived from that of a (pure) quantum state instead of being postulated by means of quantization rules. According to Sect. 2.2, the probability for a state $|n\rangle$ to be "found" in a state $|\alpha\rangle$ in an appropriate measurement is given by $|\langle n \,|\, \alpha\rangle|^2$. Its *mean* probability in an ensemble of states $\{|\alpha\rangle\}$ with

Major difference
in QM

probabilities p_α, representing incomplete information about the initial state α, is, therefore, $p_n = \sum p_\alpha |\langle n \mid \alpha \rangle|^2 = \mathrm{tr}\{\rho P_n\}$, where $\rho = \sum |\alpha\rangle p_\alpha \langle\alpha|$ and $P_n = |n\rangle\langle n|$. This result remains true for general observables $A = \sum a_n P_n$ in place of P_n. The ensemble of wave functions $|\alpha\rangle$, which thus defines a density matrix as representing a *state of information* about the quantum state, need not consist of mutually orthogonal states, although the density matrix can always be diagonalized in terms of its eigenbasis. Its representation by a *general* ensemble of states is therefore far from unique – in contrast to a classical probability distribution. Nonetheless, the density matrix could still be shown to obey a von Neumann equation *if* all states contained in the ensemble were assumed to evolve unitarily according to one and the same Hamiltonian.

However, the fundamental nonlocality of quantum states means that in general the state of a local system does *not even exist in principle:* it cannot be merely unknown. As a consequence, there *is no* induced local Hamiltonian that would allow (2.9) to apply (see Kübler and Zeh 1973). In particular, a time-dependent Hamiltonian would in general require a (quasi-)classical source for its potential energy term. This specific quantum aspect is easily overlooked when the density matrix is introduced axiomatically by "quantizing" a classical probability distribution on phase space.

Quantum nonlocality means that the generic state of a composite system ("system" and environment, say),

$$|\Psi\rangle = \sum_{m,n} c_{mn} \left|\phi_m^{\text{system}}\right\rangle \left|\phi_n^{\text{environment}}\right\rangle \quad , \tag{2.10}$$

does not factorize in any basis. The expectation values of all *local* observables,

$$A = A^{\text{system}} \otimes \mathbb{1}^{\text{environment}} \quad , \tag{2.11}$$

are calculated by "tracing out" the environment,

$$\langle\Psi| A |\Psi\rangle \equiv \mathrm{tr}\left\{ A |\Psi\rangle\langle\Psi| \right\} = \mathrm{tr}_{\text{system}}\left\{ A^{\text{system}} \rho^{\text{system}} \right\} \quad . \tag{2.12}$$

Here, the density matrix ρ^{system}, which in general has nonzero entropy even for a pure (completely defined) global state $|\Psi\rangle$, is defined by a partial trace,

$$\rho^{\text{system}} \equiv \sum_{m,m'} \left|\phi_m^{\text{system}}\right\rangle \rho_{mm'}^{\text{system}} \left\langle\phi_{m'}^{\text{system}}\right|$$
$$:= \mathrm{tr}_{\text{env}}\left\{ |\Psi\rangle\langle\Psi| \right\} \equiv \sum_{m,m'} \left|\phi_m^{\text{system}}\right\rangle \sum_n c_{mn} c_{m'n}^* \left\langle\phi_{m'}^{\text{system}}\right| , \tag{2.13}$$

where $\mathrm{tr} \equiv \mathrm{tr}_{\text{system}} \otimes \mathrm{tr}_{\text{env}}$. It represents a specific coarse-graining, *viz.* the restriction to all subsystem observables. This "reduced density matrix" can be *formally* represented by various ensembles of subsystem states (including its eigenrepresentation or diagonal form), while according to its derivation it describes here *one* pure though entangled *global* state. A density matrix thus based on entanglement has been called an "improper mixture" by d'Espagnat (1966). It can evidently *not* explain genuine ensembles of definite measurement outcomes. If proper and improper mixtures were identified for operationalist reasons, then decoherence would indeed completely solve the measurement problem. Unfortunately, this argument is circular, since the identification of proper and improper mixtures is itself based on the measurement postulate.

Regardless of its origin and interpretation, the density matrix can be replaced by its partial Fourier transform, known as the *Wigner function* (see also Sect. 3.2.3):

$$
W(p, q) := \frac{1}{\pi} \int e^{2ipx} \rho(q + x, q - x)\, dx
$$

$$
\equiv \frac{1}{2\pi} \int \int \delta\left(q - \frac{z + z'}{2}\right) e^{ip(z - z')} \rho(z, z')\, dz dz'
$$

$$
=: \mathrm{trace}\{\Sigma_{p,q}\rho\} \quad . \tag{2.14}
$$

The third line is here written in analogy to the Bloch vector, $\pi_i = \mathrm{trace}\{\sigma_i \rho\}$, since

$$
\Sigma_{p,q}(z, z') := \frac{1}{2\pi} e^{ip(z - z')} \delta\left(q - \frac{z + z'}{2}\right) \tag{2.15}
$$

is a generalization of the Pauli matrices (with an index pair p, q instead of the vector index $i = 1, 2, 3$ – see also Sect. 4.4 of Zeh 2001). Although the Wigner function is *formally* analogous to a phase space distribution, it does, according to its derivation, by no means represent an ensemble of classical states (phase space points). This fact is reflected by its potentially negative values, while even a Gaussian wave packet, which leads to a non-negative Wigner function, is nonetheless *one* (pure) quantum state.

The degree of entanglement represented by an improper mixture (2.13) is conveniently measured by the latter's formal entropy, such as the linear entropy $S_{lin} - \mathrm{trace}(\rho - \rho^2)$. In a "bipartite system", the *mutual* entanglement of its two parts is often controllable, and thus may be *used* for specific applications (EPR-Bell type experiments, quantum cryptography, quantum "teleportation", etc.). This is possible as far as the entanglement is not obscured by a mixed state of the total system. Therefore, effective measures have been proposed to characterize the *operationally available* entanglement in mixtures (Peres 1996, Vedral *et al.* 1997). However, these measures do not represent the true and complete entanglement, since a mixed state, which reduces this measure, is either based on entanglement itself (of the bipartite system as a whole with its environment), or the consequence of *averaging* over an ensemble of unknown (but nonetheless entangled) states.

The eigenbasis of the reduced density matrix can be used, by orthogonalizing the correlated "relative states" of the environment, in order to write the total state as a single sum,

$$
|\Psi\rangle = \sum_k \sqrt{p_k} \left|\hat{\phi}_k^{\mathrm{system}}\right\rangle \left|\hat{\phi}_k^{\mathrm{environment}}\right\rangle \quad . \tag{2.16}
$$

While Erhard Schmidt (1907a) first introduced this representation as a mathematical theorem, Schrödinger (1935b) used it for discussing entanglement as representing what he called "probability relations between separated systems". It was later shown to be useful for describing quantum nonlocality and decoherence by means of a universal wave function (Zeh 1971, 1973).

The two orthogonal systems $\hat{\phi}_k$ of the Schmidt form (2.16) are *determined* (up to degeneracy of the p_k's) by the total state $|\Psi\rangle$ itself. A time dependence of $|\Psi\rangle$ must therefore in general affect both the Schmidt states *and* their (formal) probabilities p_k – hence also the subsystem entropy, such as $\sum p_k(1 - p_k)$ – see Kübler and Zeh (1973), Pearle (1979), and Albrecht (1993). The induced subsystem dynamics is thus not "autonomous". Similar to the motion of a shadow that merely reflects the regular motion of a physical object, the reduced information content of the subsystem density matrix by itself is insufficient to determine its own change. Likewise, Boltzmann had to introduce his *Stoßzahlansatz*, based on statistical assumptions, when neglecting *statistical* correlations between particles (instead of quantum entanglement). The exact dynamics of any local "system" would in general require the whole Universe to be taken into account; no effective time-dependent Hamiltonian would suffice.

Effective "open systems quantum dynamics" has indeed been postulated *as an approximation for calculating forward in time* in analogy to the Boltzmann equation in the form of semigroups or master equations. An equivalent formalism was introduced by Feynman and Vernon (1963) in terms of path integrals (see Chap. 5). This approach can not explain the appearance of proper mixtures unless it is meant to *postulate* a fundamental correction to the Schrödinger equation. The formal theory of master equations will be discussed in Chap. 7 (see also Zeh 2001). Their approximate foundation in terms of a global unitary Schrödinger equation requires very specific (statistically improbable) cosmic initial conditions.

According to a universal Schrödinger equation, quantum correlations with the environment are permanently created with great efficiency for all macroscopic systems, thus leading to decoherence, defined as the irreversible dislocalization of phase relations (see Chap. 3 for many examples). The apparent (or "improper") ensembles, obtained for subsystems in this way, often led to claims that decoherence be able (or meant) to solve the measurement problem.[13] The apparent nature of these ensembles has then in turn been used to

---

[13] In the Schmidt basis, interference terms are *exactly* absent by definition. Hepp (1972) used the formal limit $N \to \infty$ to obtain this result in a *given* basis (though this may require infinite time). However, the global state always remains one pure superposition. The Schmidt representation has therefore been used instead to specify the Everett branches, that is, to define the ultimate "pointer basis" $|\chi_n^{\text{observer}}\rangle$ for each observer (cf. Zeh 1973, 1979, Albrecht 1992, 1993, Barvinsky and Kamenshchik 1995). It is also used in the "modal interpretation" of quantum mechanics (cf. Dieks 1995).

declare the program of decoherence a failure. As explained in Sect. 2.3, both claims miss the point. However, decoherence represents a *crucial dynamical step* in the measurement process. The rest may remain a pure epistemological problem (requiring only a reformulation of the psycho-physical parallelism in consistent quantum mechanical terms). If the Schrödinger equation is exact, the observed quantum indeterminism can only reflect that of the observer's identity – not one to be found in objective dynamics.

The process of decoherence leads to a new type of (generalized) dynamical coarse graining. If phase relations between certain Hilbert subspaces of a system permanently disappear by means of decoherence, the latter's reduced density matrix may be approximated in the form

$$\rho = \sum_{m,n} P_m \rho P_n \approx \sum_n P_n \rho P_n \quad , \tag{2.17}$$

where P_n projects on to the n-th decohered subspace, while $\sum P_n = \mathbb{1}$. The dynamics of formal probabilities, $p_n(t) := \mathrm{tr}\{P_n\rho(t)\}$, may then be written as a master equation, similar to the Pauli equation,

$$\dot{p}_n = \sum_m A_{nm}(p_m - p_n) \quad , \tag{2.18}$$

and valid in the time direction defined by decoherence, as was shown by Joos (1984). Since (2.18) describes an ensemble of stochastic evolutions $n(t)$, it defines probabilities for coarse-grained "histories", corresponding to time-ordered sequences of projections $P_{n_1}(t_1) \ldots P_{n_k}(t_k)$. Probabilities for such histories in discrete time steps can be written as

$$p(n_1, \ldots, n_k) = \mathrm{tr}\{P_{n_k}(t_k) \ldots P_{n_1}(t_1)\rho(t_0)\} \tag{2.19}$$

(using the property $P^2 = P$ of projection operators). States n_k dynamically arising according to a stochastic equation may contain "consistent memories" (or "time capsules" in Barbour's terminology – Barbour 1999). Individual stochastic histories obey a *quantum Langevin* equation (an indeterministic generalization of the Schrödinger equation). Models have been proposed by Diósi (1986), Belavkin (1989a), Gisin and Percival (1992), and others – see also Diósi and Kiefer (2001). They are often assumed to hold *exactly*, thus incorrectly interpreting the evolving mixture as a "proper" one.

In the theory of "consistent histories" (Griffiths 1984, Omnès 1992, 1995), *formal* projections P_n are called "events", regardless of any dynamics. These events are thus *not* dynamically described within the theory – in accord with the Copenhagen interpretation, where events are assumed to occur "out of the blue" or "outside the laws of nature". However, only those histories $n_1, \ldots n_k$ are then *admitted by postulate* (that is, assumed to "occur" in reality) which possess "consistent" probabilities in the sense of being compatible with a stochastic evolution. This condition *requires* (a weak form of) decoherence that is *not* necessarily based on entanglement (cf. Omnès 1999). It is sometimes described in terms of a "new logic", reflected by Omnès' (1995) surprising conclusion that "the formalism of logic is not time-reversal invariant, as can be seen in the time ordering of the (projection) operators". However, logic has nothing to do with the concept of time. A property a at time t_1 that is said to "imply" a property b at time $t_2 > t_1$ would describe a *causal* (that is, dynamical) rather than *logical* relationship. This confusion of the concepts of causes and reasons seems to have a long tradition in philosophy, while even in mathematics the truth of logical theorems is often inappropriately *defined* by means of logical operations that have to be performed in time (thus mimicking a causal relationship).

In the theory of decoherence, *apparent* events in the detector are described dynamically by a universal Schrödinger equation, using certain initial conditions for the environment, as a very fast but smooth formation of entanglement. Similarly, "particles" *appear* in the form of narrow wave packets in space as a consequence of decoherence. The identification of observable events with a decoherence process applies regardless of any conceivable subsequent genuine collapse. Decoherence is not only responsible for the classical aspects of quantum theory, but also for its "quantum" aspects (see Sect. 3.4.1). All fundamental physical concepts seem to be continuous and in accordance with a Schrödinger equation (generalized to relativistic field functionals).

This description of quantum events as fast but continuous evolutions also avoids any superluminal effects that have been shown (with mathematical rigor – see Hegerfeldt 1994) to arise (not very surprisingly) from explicitly or tacitly assumed instantaneous quantum jumps between exact energy or particle number eigenstates.[14] The latter require infinite exponential tails that can never *form* in a relativistic world. Supporters of explicit collapse mechanisms are quite aware of this problem, and try to avoid it (cf. Diósi and Lukácz 1994 and Chap. 8). In the nonlocal quantum formalism, *dynamical locality* is achieved by using Hamiltonian operators that are spatial integrals over a Hamiltonian operator density. This form prevents superluminal signalling and the like.

---

[14] Such superluminal "phenomena" are reminiscent of the story of *Der Hase und der Igel* (the race between *The Hedgehog and the Rabbit*), narrated by the Grimm brothers. Here, the hedgehog, as a competitor in the race, does not run at all, while his wife is waiting at the end of the furrow, shouting in low German "Ick bin all hier!" ("I'm already here!"). Similar arguments hold for "quantum teleportation", where an appropriate nonlocal state that contains the state to be ported as a component has to be well *prepared* (cf. also Vaidman 1998). Experiments clearly support this view of a continuous evolution instead of quantum jumps (cf. Fearn, Cock, and Milonni 1995). Teleportation *would* be required if reality were local and physical properties entered existence "out of the blue" in fundamental *quantum events*. Therefore, this whole concept is another artifact of the Copenhagen interpretation.

2.5 Conclusions

Let me recall the interpretation of quantum theory that has emerged in accordance with the concept of decoherence:

(1) General quantum superpositions (including nonlocal wave functions) represent *individual* physical states (Sect. 2.1.1).

(2) According to a universal Schrödinger equation, most superpositions are almost immediately, and in practice irreversibly, *dislocalized* by interaction with the environment. Although the resulting nonlocal superpositions still *exist*, we do not know, in general, what they *mean* (or how they could be observed). However, if dynamics is local (described by a Hamiltonian density in space, $H = \int h(\mathbf{r})d^3\mathbf{r}$), certain approximately factorizing components of the global wave function may be dynamically autonomous after this decoherence has occurred, and nonlocal superpositions cannot return into local ones if statistical arguments apply to the future (Zeh 2001).

(3) Any observer (assumed to be local for empirical and dynamical reasons) who observes a subsystem of the nonlocal superposition becomes part of this entanglement. Those of his component states which are then related only by *nonlocal* phase relations must describe "different observers" (novel psychophysical parallelism).

(4) Because of this dynamical autonomy of decohered world components ("branches"), there is no other reason than conventionalism to deny the existence of "the other" components which result from the Schrödinger equation ("many minds interpretation" – Sect. 2.3).

(5) Probabilities are meaningful only as frequencies in series of repeated measurements. In order to derive the observed Born probabilities in terms of frequencies, we have to *postulate* merely that we are living in an "Everett branch" with not extremely small norm.

is that kosher?

3 Decoherence Through Interaction with the Environment

E. Joos

The observation that all objects in our universe interact with each other to a greater or lesser degree seems rather trivial at first sight. The strength, and thereby the consequences, of interactions vary enormously, depending on the situation considered. Sometimes, even in the framework of classical physics, surprising results have been found. A prominent example, already mentioned in Sect. 2.3, is the influence of a small mass of a few grams, as far away as the star Sirius, on the trajectories of air molecules here on earth (Borel 1914a, Brillouin 1964). This example demonstrates that even a coupling which is considered as weak (as gravity certainly is over such distances) can have a great influence on systems that are sensitive enough to this kind of interaction,

> "The representation of gaseous matter... composed of molecules with positions and velocities which are rigorously determined at a given instant is therefore a *pure abstract fiction*;... as soon as one supposes the indeterminacy of the external forces, the effect of collisions will very rapidly *disperse* the trajectory bundles which are supposed to be infinitely narrow, and the problem of the subsequent movement of the molecules becomes, within a few seconds, very indeterminate, in the sense that an enormously large number of different possibilities are *a priori* equally probable."[1] (Borel 1914b)

This Sirius problem is a rather serious problem, even in classical physics. If events here on earth depend on what does (or does not) happen in arbitrarily distant systems, how can we explain all the regularities which are so important for our experienced "classical" world? In quantum theory we encounter an additional severe problem: Situations of this kind lead dynamically to a kinematically holistic, entangled wave function for the entire universe. How can we even find (or define) any subsystems (objects) such as trees or cats?

---

[1] "La représentation d'une masse gazeuse ... formé de molécules dont les positions et les vitesses à un instant donné sont rigoureusement déterminées, est donc une *pure fiction abstraite*; ... aussi que l'on suppose l'indétermination des forces extérieures, l'effet des chocs *disperse* très rapidement les faisceaux de trajectoires supposés infiniment déliés et le problème du mouvement ultérieur des molécules devient, en très peu de secondes, très indéterminé, en ce sens qu'un nombre colossalement grand des possibilités différentes sont *a priori* également probables."

In classical physics, where Hamilton's equations are considered as fundamental, the dynamics of a certain object will in general depend on the state of all other, even very distant, systems. This means that the Hamiltonian $H_{\text{subsystem}}(q, p, t)$ which determines the evolution of a local system described by canonical coordinates p and q, depends on the actual state $\{q_{\text{env}}(t), p_{\text{env}}(t)\}$ of the environment, that is

$$H_{\text{subsystem}}(q, p, t) = H_{\text{total}}\big(q, p; q_{\text{env}}(t), p_{\text{env}}(t)\big). \qquad (3.1)$$

The definition of a "subsystem" is, already at this classical level, as ambiguous as it is in quantum theory. Certain "relevant" or "accessible" degrees of freedom are singled out and called "the system". In a statistical description, ensembles of Hamiltonians may be more appropriate, since the effect of the environment is usually treated as an "uncontrollable perturbation" (see also Sect. 3.4.3). For this purpose, the Liouville equation is replaced by some master equation which is equivalent to a (time-asymmetric) stochastic dynamics (master equations are discussed in detail in Chap. 7). Nevertheless, a unique state can still be assumed to *exist*. Each part of the entire system has its own state, described by the coordinates q and momenta p, even if a statistical description by distributions in phase space is appropriate (since the state is not known or not dynamically determined in practice).

The importance of including the environment, finally the whole universe, for a proper understanding of irreversibility has occasionally been expressed in studies of the ergodic problem, e. g.

> "... the approach to true equilibrium is governed by interactions between the system and the outside world, not by interactions within the system itself. ... Either we make a causal, Hamiltonian description of the whole Universe, or else we must allow for an essential element of randomness in the description of the motion of the limited system under study. ... The random element here is not due to accidental shortcomings of the observer, but rather to the fact that the observer restricts his observations to a finite part of the Universe. ... Statistical mechanics is not the mechanics of large, complicated systems; rather it is the mechanics of limited, not completely isolated systems." (Blatt (1959) in a spirited discussion of irreversibility soon after Hahn's spin-echo experiments).

The situation is completely different in quantum theory, where usually there are no states of the subsystems defined anymore. Interactions generally lead to a non-separating global state for the whole system, even if it did factorize initially. As explained in detail in Chap. 2, this *kinematical nonlocality* cannot be simply be viewed as an artefact of the formalism, or perhaps as a sign of its incompleteness (Einstein, Podolsky and Rosen 1935), since the global properties of an entangled quantum state cannot be explained as purely statistical correlations between local systems. The violation of Bell's inequalities (Bell 1964, Aspect 1976, Clauser and Shimony 1978, Aspect, Grangier and Roger 1981, Aspect, Dalibard and Roger 1982, Greenberger *et al.* 1990, Kwiat *et al.* 1995) demonstrates that quantum correlations are not simply statistical correlations between parts of a larger system (see also d'Espagnat 1979, Mermin 1990a, 1990b, 1994, and App. A3).

Since a system that is interacting with its environment cannot have a state on its own, the only possible description in the standard quantum formalism is by means of its density matrix. The local density matrix, which contains the complete information for the (probabilities of) the results of all measurements that can be performed at this system, will therefore be the central subject throughout this section. We will often use the term "local" as a synonym for "subsystem". In classical physics objects are considered distinct if they occupy different positions in space. But even in classical mechanics such a property is not invariant under canonical transformations. When we speak of "local systems" and "local density matrices" we always call attention to certain degrees of freedom from the totality of all coordinates needed to describe the whole system. Often the singled-out degree of freedom will have a *spatial* meaning, in particular in Sect. 3.2. The definition of "system" nevertheless remains problematic, both in classical and in quantum physics. As was already discussed in Sect. 2.2, one should always keep in mind that the concept of the density matrix relies ultimately on the assumption that at some stage a collapse (or an equivalent branching in the Everett interpretation) occurs. The formal ensemble of states characterizing the local density matrix is not a "real" ensemble in the sense of statistical mechanics [2] . Sometimes these "mixed states" are therefore called "improper mixtures" (d'Espagnat 1966, 1976, 1995). In analogy to the classical case, no Schrödinger or von Neumann equation can be expected to hold for them.

Even if a complete set of density matrices for all subsystems were given, such a description would remain incomplete in an essential way, in contrast to classical physics, where the specification of the state for each degree of free-

---

[2] The formal operation of calculating a local density matrix from an entangled global state is often called "coarse graining". This nomenclature is misleading, since it suggests that the change from a global to a local description is equivalent to a neglect of information as in statistical mechanics, where the exact state is replaced by an appropriate ensemble of "possible states". Since a subsystem does not have a state at all in the case of entanglement, the neglect of quantum correlations cannot be interpreted in purely statistical terms. A similar nonsensical wording is the widely used "averaging over the states of the environment", motivated (and misled) by the fact that the formal procedure of calculating the subsystem density matrix *looks like* the analogous operation in probability theory. There simply *is* nothing to average over – before an observation has been made. (Similar remarks apply to the Wigner "distribution", which does not distribute anything.) "Paying attention" to only one subsystem always contains the axiom of measurement in an essential way.

dom implies a complete characterization of the global state. Moreover, even a statistical description, including correlations between subsystems, cannot fully encompass quantum correlations.

In order to compare classical and quantum descriptions, consider the Schmidt decomposition (2.16) of a generic pure state,[3]

$$|\Psi\rangle = \sum_n \sqrt{p_n} |\varphi_n\rangle |\Phi_n\rangle , \qquad (3.2)$$

describing, for example, the resulting global state after a measurement interaction (compare (2.5) on page 27). A collapse with respect to this basis would lead to a ("classical") ensemble of correlated states described by

$$\rho_{\text{class.}} = \sum_n p_n |\varphi_n\rangle \langle\varphi_n| \otimes |\Phi_n\rangle \langle\Phi_n| , \qquad (3.3)$$

while the original pure state (3.2) still contains quantum correlations,

$$\rho = |\Psi\rangle \langle\Psi|$$
$$= \rho_{\text{class.}} + \sum_{n \neq m} \sqrt{p_n p_m} |\varphi_n\rangle \langle\varphi_m| \otimes |\Phi_n\rangle \langle\Phi_m| . \qquad (3.4)$$

As can be seen from (3.4), the additional interference terms can never be considered as "small", even if there are many contributions.[4] Interference does not simply vanish for "complicated" states (such as cats). The increase

---

[3] In the usual Schmidt decomposition the whole system is divided into *two* parts (Schmidt 1907a,b). A simple derivation can be found in Appendix A of Albrecht (1992), see also Ekert and Knight (1995) or Sect. 2.5 of Nielsen and Chuang (2000); for a more mathematically oriented treatment see App. A3. It has been shown that the more general situation of three or more parts allows a similar expansion with a single sum (Elby and Bub 1994, see also Peres 1995). The essence of the extended theorem is that a generalized Schmidt decomposition is unique, *if* it exists (as it does only for a subset of states). The well-known degeneracy in (3.2) no longer appears for three or more subsystems. See also next footnote.

[4] In (3.2)–(3.4) we used the Schmidt representation for simplicity. If a fixed set of basis states is employed instead, a double sum with complex coefficients occurs in (3.2). Equation (3.4) would then contain a fourfold sum. If the two subspaces have dimension N_1 and N_2, respectively, the number of contributions in an expansion of a general pure state is $N_1 N_2$. In the Schmidt expansion (3.2) this number is reduced to the *minimum* of dimensions of the two Hilbert spaces, $\min(N_1, N_2)$, the so-called *Schmidt rank*. This corresponds to the fact that the Schmidt states diagonalize the respective subsystem density matrices. The density matrix of the higher-dimensional system has therefore at least $\max(N_1, N_2) - \min(N_1, N_2)$ zero eigenvalues. This degeneracy is a consequence of the assumed purity of the global state. Another consequence is the fact that entropies for both systems are always the same.

of information when an observer "reads" the measurement result corresponds to the transition from the ensemble described by $\rho_{\text{class.}}$ to one of its members (compare the discussion in Sect. 2.2).

When treating the interaction of a quantum system with its environment, one usually assumes that the state factorizes at some "initial" time $t = 0$, i. e.,

$$|\Psi(0)\rangle = |\varphi\rangle |\Phi\rangle \tag{3.5}$$

or, if both systems are described by ensembles,

$$\rho(0) = \rho_\varphi \otimes \rho_\Phi. \tag{3.6}$$

Initial states of this kind are often motivated by a "preparation" of the system of interest, an assumption which implicitly contains a collapse of the wave function. Obviously, $\rho(0)$ in (3.6) is related to $\rho_{\text{class.}}$ in (3.3); now even classical correlations (that is, statistical correlations describing missing information about the system states) between the two parts have been omitted.

As discussed at length in Chap. 2, measurement-like interactions cause a strong quantum entanglement of macroscopic objects with their natural environment. The accompanying delocalization of phases then effectively "destroys" superpositions between "macroscopically different" states with respect to a local observer, so that the object *appears* to be in one or the other of those states. The relevant mechanisms will be described in the next subsection. Whenever we use the term "destroyed" in the following, this expression is always meant as a synonym for "unavailable with respect to a restricted set of observations" (usually the local ones). As long as no collapse is assumed, phase relations are of course still present in the whole system and never "destroyed".

The destruction of coherence by strong coupling to the environment does not mean, however, that the so-called "macroscopic quantum effects", which also include many atoms (for example circulation quantization in superfluid helium, the Josephson effect, lasers etc.) are affected in the same way. There is an important structural difference between these two kinds of states. If we express these states in a very crude and schematic way by means of one- or two-particle wave functions $|\psi_i\rangle$, the N-particle superposition in the case of "macroscopic quantum effects" (such as superconductivity, the quantum Hall effect and others) is of a form which we may write down as

$$|\Psi\rangle = (a |\psi_1\rangle + b |\psi_2\rangle)^N, \tag{3.7}$$

while superpositions of macrostates (like that of the moon being at different places, or Schrödinger's cat being dead or alive, or even a cat and a dog (Joos 2000)) would look like

$$|\Psi\rangle = a |\psi_1\rangle^N + b |\psi_2\rangle^N \tag{3.8}$$

(Leggett 1980, 1987, Schrödinger 1935a). This demonstrates again that the number of particles *alone* does not explain the quantum or classical behavior of a system.[5] Instead, various kinds of coupling to the environment have to be taken into account in an essential way.

The entanglement between "systems" may depend on their definition in a delicate way. For example, two independent particles with opposite spin $|\uparrow\rangle$ and $|\downarrow\rangle$, respectively, would be described by the non-entangled state

$$|\uparrow\rangle_1 |\alpha\rangle_1 |\downarrow\rangle_2 |\beta\rangle_2 , \qquad (3.9)$$

where $|\alpha\rangle$ and $|\beta\rangle$ describe the spatial state of each particle. If these are fermions, this state has to be replaced by its antisymmetrized version

$$\frac{1}{\sqrt{2}}(|\uparrow\rangle_1 |\alpha\rangle_1 |\downarrow\rangle_2 |\beta\rangle_2 - |\downarrow\rangle_1 |\beta\rangle_1 |\uparrow\rangle_2 |\alpha\rangle_2). \qquad (3.10)$$

This seems to be an entangled EPR-state, but this is not the case, since it still has the local interpretatation "there is a spin-up particle with wave function $|\alpha\rangle$" (Ghirardi and Marinatto 2002). The often-discussed EPR-singlet state, on the other hand would in this notation be given by

$$\frac{1}{\sqrt{2}}(|\uparrow\rangle_1 |\downarrow\rangle_2 - |\downarrow\rangle_1 |\uparrow\rangle_2)(|\alpha\rangle_1 |\beta\rangle_2 + |\beta\rangle_1 |\alpha\rangle_2). \qquad (3.11)$$

The latter state describes quantum nonlocality, whereas the former appears entangled only in the particle representation. (The artificial entanglement disappears in the more appropriate mode-number representation.)

Similar remarks apply to nonlocal states of the electromagnetic field, where entanglement is properly described as entanglement between field amplitudes rather than entanglement between "photons". Even then the degree of entanglement depends on the definition of field modes. A "one-photon state" of the form

$$\frac{1}{\sqrt{2}}(|0\rangle_A |1\rangle_B + |1\rangle_A |0\rangle_B) \qquad (3.12)$$

is usually considered nonlocal, if modes A and B are located in different spatial regions. On the other hand, this state can be rewritten in the product form

$$|0\rangle_{A'} |1\rangle_{B'} \qquad (3.13)$$

by a redefinition of modes (Bogoliubov transformation). The choice of an "appropriate" representation seems always to contain some kind of locality condition (van Enk 2002).

---

[5] A Weber bar used for the detection of gravitational waves may weigh several tons. Nevertheless, at low temperatures it is treated as a quantum mechanical oscillator (Braginski, Vorontsov and Thorne 1980). Superpositions of macroscopically different states can be observed in superconducting loops containing Josephson junctions. The two classical states correspond to currents generated by millions of Cooper pairs (van der Wal et al. 2000, Friedman et al. 2000).

In the next section we will explore the different mechanisms which play a role in the interaction of a system with its environment. The irreversible formation of quantum correlations will be discussed in detail for various situations. A particularly important consequence is the *appearance* of localized macroscopic objects, to be discussed in Sect. 3.2. Two important limiting cases for the dynamical behavior under the influence of the environment (master equations and the freezing of any motion, the so-called "Zeno effect") will be analyzed in Sect. 3.3. The stability of states with respect to environmental influence is analyzed for rather general situations. Some questions related to the interpretation of quantum states and decoherence are the topic of Sect. 3.4. The subsequent Chap. 4 extends the analysis given in this chapter to the case of quantum electrodynamics (decoherence of electromagnetic fields in the presence of charges) and quantum gravity (emergence of classical spacetime). We usually set $\hbar = 1$, except for some special cases where it appears sensible to include $\hbar$ (for example, when discussing the Wigner function).

3.1 The General Mechanisms of Decoherence

In this section we discuss the typical behavior of a system interacting in an irreversible manner with its environment. An initially factorizing state will in general not keep this property if some kind of interaction is present, but will evolve into an entangled one. This evolution leads to a behavior of a subsystem's density matrix which may be drastically different from the properties the system would show in isolation.

3.1.1 Dynamics of Quantum Correlations

In many situations a system acts on its environment in a certain way, while the back-reaction of the surroundings on the considered system is negligibly small. This is the canonical situation of a "measurement-like process". It corresponds to (and can be considered as the definition of) the unitary part of a "measurement of the first kind" (compare (2.4, 2.5) on page 27; see also Sect. 3.4.3). For this case, many essential aspects of decoherence can be discussed in a particularly simple way.[6] In this chapter, whenever we use a phrase like "system B is measuring property α of system A", we only refer to the *dynamical* evolution of the joint system according to a Schrödinger equation with appropriate coupling between the two systems. No collapse is assumed when we use the term "measurement" in this informal sense.

---

[6] Even if "ideal measurement" may only be an approximation (for example, conservation laws may preclude appropriate Hamiltonians (Wigner 1952, Araki and Yanase 1960), it is an important ingredient when the "movability of the cut" between observer and observed system is discussed; compare the chain of states in (2.6) and (2.7).

In the following we will repeatedly discuss the interaction between a system of interest (which we will designate alternately "local system", "object", "system 1", or just "the system") with a second system (which is called "system 2" or "environment"). Since this interaction is in many cases "measurement-like", it is appropriate to start our discussion with a review of von Neumann's treatment of (ideal) measurements (von Neumann 1932). The role of the environment is first taken by the apparatus, but, since macroscopic, the apparatus is itself strongly coupled to the rest of the universe. Finally, a chain of states as in (2.6) will ensue.

If the interaction Hamiltonian is of von Neumann's form[7]

$$H_{\text{int}} = \sum_n |n\rangle \langle n| \otimes \hat{A}_n, \tag{3.14}$$

where $\hat{A}_n$ are (rather arbitrary, but n-dependent) operators acting only in the Hilbert space of system 2, an eigenstate $|n\rangle$ of the "observable" measured by this interaction will not be changed, while system 2 acquires "information" about the state $|n\rangle$ in the sense that its state changes in an n-dependent way,

$$|n\rangle |\Phi_0\rangle \xrightarrow{t} \exp(-iH_{\text{int}}t) |n\rangle |\Phi_0\rangle = |n\rangle \exp(-i\hat{A}_n t) |\Phi_0\rangle$$
$$= : |n\rangle |\Phi_n(t)\rangle . \tag{3.15}$$

The resulting environmental (apparatus) states $|\Phi_n(t)\rangle$ are generally called "pointer positions", although they do not need to correspond to any states of actually present measurement devices. They are simply the states of the "rest of the world". For a general initial state of system 1, the linearity of the Schrödinger equation then immediately yields (cf. (2.5) on page 27)

$$\left(\sum_n c_n |n\rangle \right) |\Phi_0\rangle \xrightarrow{t} \sum_n c_n |n\rangle |\Phi_n(t)\rangle , \tag{3.16}$$

i. e. a correlated state representing a superposition of all "measurement results". The local (subsystem) density matrix changes accordingly,

$$\rho_S = \sum_{n,m} c_m^* c_n |n\rangle \langle m| \xrightarrow{t} \sum_{n,m} c_m^* c_n \langle \Phi_m|\Phi_n\rangle |n\rangle \langle m| . \tag{3.17}$$

Non-diagonal elements (in the basis defined by the interaction) are thereby multiplied by a factor which is given by the overlap of the pointer states

---

[7] Of course not every interaction can be brought into this form. With the general interaction Hamiltonian given as $H_{\text{int}} = \sum_{n,m,\alpha,\beta} \sigma_{nm\alpha\beta} |n\rangle \langle m| \otimes |\alpha\rangle \langle \beta|$, one may wish to perform a partial diagonalization of the matrix σ (for fixed α, β) to achieve a form like that shown in (3.14). This would be possible if σ were Hermitean in this subspace, i. e., $\sigma_{nm\alpha\beta}^* = \sigma_{mn\alpha\beta}$. Hermiticity of H_{int} only requires $\sigma_{nm\alpha\beta}^* = \sigma_{mn\beta\alpha}$, however. Another characterization is that the Hamiltonian (3.14) is special in having factorizing eigenstates.

corresponding to the respective quantum numbers. Diagonal elements are unchanged (hence the attribute "ideal"). If the environmental states are orthogonal,

$$\langle \Phi_m | \Phi_n \rangle = \delta_{nm} \,, \tag{3.18}$$

that is, if apparatus states discriminate system states (otherwise system 2 would not be called "apparatus"), the system density matrix becomes diagonal in this basis,

$$\rho_{\mathrm{S}} \to \sum_n |c_n|^2 \, |n\rangle \, \langle n| \,. \tag{3.19}$$

During this evolution, interference terms (nondiagonal elements) are thus destroyed locally (or rather delocalized) in this basis, which is defined by the interaction Hamiltonian. This means that the phase relations characterizing the superposition become inaccessible for *local* observations. The system now *appears* classical with respect to the property given by the quantum number n. In other words, no interference effects between different n can be observed *at this system* anymore, if the above process is assumed to be irreversible.[8] Then the (classical) assumption that the system *is* in one of the states $|n\rangle$ cannot be proven wrong by observations at this system (although no collapse was assumed so far). The improper mixture defined by (3.19) behaves in the same way as a proper mixture with respect to observations restricted to the system. This "apparent collapse" should not be confused, however, with the collapse in von Neumann's theory of measurement (cf. the discussion in Sect. 2.3).

The basis selected by a process of the form (3.16) is defined entirely by the properties of the interaction Hamiltonian (its diagonalizing representation) and at this stage is independent of the properties of the local system (see also Sect. 3.1.3). The selected basis will, however, not be "stable" under the system's evolution, in general, that is, the Hamiltonians of system and/or environment will not commute with H_{int}. The interplay between the system's intrinsic dynamics and the interaction with the environment will play an important role later in this chapter.

If the evolution described by (3.16) is viewed as a model of system-apparatus coupling, the apparatus itself, since macroscopic, will interact

---

[8] There are exceptional situations where the evolution described by (3.16) can be reversed, although this may often be difficult to achieve technically. For a discussion of a reversible Stern–Gerlach apparatus see Wigner (1963), Englert, Schwinger and Scully (1988), Schwinger, Scully and Englert (1988). For an experiment showing coherent recombination of polarized neutrons see Summhammer *et al.* (1983). In quantum optical experiments a shielding from the environment can be achieved by enclosing both systems in a reflecting cavity (Haroche and Kleppner 1989). Another well-known example is given by the spin-echo experiments performed by Hahn (1950), see also Rhim, Pines and Waugh (1971). The main interest of decoherence studies is in those normal situations, where correlations spread out irreversibly (perhaps with the sole exception of a recollapsing universe, see Sect. 4.2).

strongly with its environment. By the very same mechanism, correlations are then delocalized again, leading to a diagonal density matrix for system+ apparatus: After establishing the system-apparatus correlations (3.16), information about the measurement result is rapidly transferred to the environment E,

$$\left(\sum_n c_n \, |n\rangle \, |\Phi_n\rangle\right) |E_0\rangle \xrightarrow{t} \sum_n c_n \, |n\rangle \, |\Phi_n\rangle \, |E_n\rangle \qquad (3.20)$$

with $\langle E_n|E_m\rangle \approx 0$, thus

$$\rho_{\mathrm{SA}} = \sum_n |c_n|^2 \, |n\rangle \, \langle n| \otimes |\Phi_n\rangle \, \langle \Phi_n| \qquad (3.21)$$

for the combined system-apparatus subsystem. Again, the interaction coupling the apparatus to its environment is assumed to have the form (3.14), thereby *defining* the "pointer states" $|\Phi_n\rangle$ dynamically (Zurek 1981). The density operator ρ_{SA} now describes a classical correlation between system and apparatus states (see (3.3)), as expected after a genuine measurement (cf. Sect. 2.3). Some explicit models represent the apparatus by a harmonic oscillator, while the interaction with the environment is modelled by coupling to an ensemble of oscillators (see Haake and Walls 1987, Haake and Zukowski 1993, Braun, Haake and Strunz 2001). It should be emphasized again that (3.21) does not solve the measurement problem. In particular, an ignorance (i.e. ensemble) interpretation of (3.21), as suggested by some authors, would wrongly identify an improper mixture with a proper one.

If a collapse of the wave function after completion of the unitary part of the measurement interaction is assumed, that is, if the entangled state (3.16) or (3.20) is replaced by one of its components, the local description changes accordingly, because the measurement result is assumed to be known (e. g., in the form of a photodetection "event"). The state of the "observed" system is thereby replaced stochastically by a pure state corresponding to the assumed measurement result. Such collapse models are very popular, particularly in quantum optics applications[9] (the time-dependent states are then termed "quantum trajectories", see Carmichael 1993). The local description in these phenomenological models is given by a stochastic evolution of pure states (the detailed form depends on the assumed measurement/collapse process). A comparison of some of these models was given by Wiseman and Milburn (1993b),[10] see also Chaps. 7 and 8, a simplified model is described

---

[9] See, for example, Dalibard, Castin and Mølmer (1992), Dum, Zoller and Ritsch (1992), Teich and Mahler (1992), Goetsch and Graham (1993), Hegerfeldt (1993), Mølmer, Castin and Dalibard (1993), Wiseman and Milburn (1993a,b), Garraway and Knight (1994), Goetsch and Graham (1994), Keller and Mahler (1994), Goetsch, Graham and Haake (1995), Garraway, Knight and Steinbach (1995); recent monographs are Gardiner (1991) and Carmichael (1993). Compare also Gisin and Percival (1992).

[10] While some authors use stochastic models only for calculational convenience (a vector requires less computer memory than a (high-dimensional) density matrix;

in Sect. 3.3.1.2. Of course, such a collapse must *not* be interpreted simply as only representing increased knowledge.

The assumed collapse does not change the (dynamics of) the local density matrix, if the result is not observed. Otherwise, quantum correlations could be used for (perhaps superluminal) signalling. Therefore, the ensemble of all "quantum trajectories" must exactly reproduce the conventionally calculated density matrix (which in most cases follows a Lindblad master equation) by construction. None of these stochastic models can be distinguished from standard quantum theory by local measurements. On the other hand, they suffer from the problem that the assignment of states (even if these remain unknown) to a subsystem is *incompatible* with standard quantum theory, in the same sense as assigning a certain spin state to a particle in a two-particle singlet state – an assumption proved wrong in EPR-experiments.

The measurement scheme presented so far represents a "controllable" measurement, i. e. information about the state of system 1 is transferred to system 2 and can there in principle be extracted (by appropriate measurements which discriminate between the states $|\Phi_n\rangle$). This corresponds, in particular, to experimental setups devised to measure a certain "observable". A standard example is a Stern–Gerlach device, measuring the spin of an atom by establishing a well-controlled correlation between spin and position. However, under more general circumstances the environment may also be described by an equilibrium ensemble of all "pointer positions", for example motivated by a quasi-ergodic rapid movement ("fluctuation") of the pointer in a thermal environment,

$$\rho_2 = \frac{1}{N} \sum_{n=1}^{N} |\Phi_n\rangle \langle \Phi_n| . \tag{3.22}$$

Assume for simplicity that the interaction with $|n\rangle$ shifts the pointer among its N possible states (cyclically) by n . Then the evolution corresponding to (3.16) with initial state of the form (3.6) yields

$$\rho_{\text{tot}}(0) = \sum_{n,m} c_m^* c_n |n\rangle \langle m| \otimes \frac{1}{N} \sum_{k=1}^{N} |\Phi_k\rangle \langle \Phi_k|$$

$$\longrightarrow \quad \rho_{\text{tot}}(T) = \frac{1}{N} \sum_{m,n,k} c_m^* c_n |n\rangle |\Phi_{k+n}\rangle \langle \Phi_{k+m}| \langle m| . \tag{3.23}$$

---

the price paid is that many stochastic realizations have to be computed to reliably reconstruct the density matrix results), some stochastic equations can be derived by assuming certain measurement mechanisms: "The relevant model for a given experimental situation depends on the method by which information is to be extracted from the light leaving the system. That is to say, the state of a quantum system is always conditioned on (and in fact can be identified with) our knowledge of the system obtained from a measuring apparatus which effectively behaves classically." (Wiseman and Milburn 1993b). Clearly, the "knowledge obtained from the system" finally implies a collapse – or Everett branching – , cf. Sect. 2.3)

This "measurement" is "uncontrollable", since no information can be extracted from the pointers,

$$\rho_2(T) = \text{tr}_1\rho_{\text{tot}}$$

$$= \frac{1}{N} \sum_{n,k=1}^{N} |c_n|^2 |\Phi_{n+k}\rangle \langle\Phi_{n+k}|$$

$$= \frac{1}{N} \sum_{k=1}^{N} |\Phi_k\rangle \langle\Phi_k| = \rho_2(0) \tag{3.24}$$

i. e. the density matrix does not change. Nevertheless, interference between different n's is still destroyed, precisely as for the pure pointer state in (3.17),

$$\rho_1(T) = \text{tr}_2\rho_{\text{tot}}(T) = \sum_n |c_n|^2 |n\rangle \langle n|. \tag{3.25}$$

An important example for this mechanism is the interaction of a system with thermal radiation. Photons may "measure" the position of an object by being scattered from it (see the next section), while the heat bath of radiation is not altered appreciably. Decoherence is still possible since initially the system did not participate in the equilibrium. Again, the term "measurement" here only means a certain kind of interaction (leading to (3.23)) without assuming any collapse of the wave function.

So far we have considered an idealized model for the interaction with the environment, in which only the changes caused by the interaction are taken into account. For a complete treatment, the system's intrinsic dynamics has to be included. For example, system 1 may rapidly re-establish coherence due to its own internal dynamics, regardless of the form of the interaction. A typical example is given by resonance fluorescence in atomic physics, where loss of coherence between two atomic states due to radiation damping is compensated by a driving laser field (considered as a part of the atomic Hamiltonian) leading to damped oscillations between the two atomic states ("Rabi oscillations"). The interplay between de-coherence by the environment and re-coherence due to the internal dynamics will play a major role in all subsequent discussions. A particularly explicit example can be found in Sect. 3.3.1.2, where we will discuss an experimental realization of the quantum Zeno effect. Another, completely different, kind of re-coherence would be a (conspirative) back-flow of spread-out quantum correlations, the time-reverse of (3.16). Seemingly such processes play no role in our *present* universe, aside from very special experimental situations (sufficiently closed systems) as in the famous spin-echo experiment.

The formation of correlations as in (3.16) is the fundamental mechanism for the destruction of local coherence. How fast and under what circumstances does an initially uncorrelated (separating) state evolve into a correlated one? This question can be investigated by again using the Schmidt decomposition

(Kübler and Zeh 1973, Joos and Zeh 1985, Nemes and de Toledo Piza 1986). If, at an initial instant $t = 0$, the total state is a product state $|\varphi_0\rangle |\Phi_0\rangle$, the Schmidt representation (3.2) has only one component, i. e., $p_0(0) = 1$ and $p_{n\neq 0}(0) = 0$. The formation of quantum correlations is revealed by the appearance of more than one component in (3.2). Since the coefficients p_i have formal properties of probabilities (they sum up to one), the initial value of p_0 will decrease with increasing distance in time (in both time directions) according to

$$p_0(t) = 1 - At^2 \tag{3.26}$$

(to lowest order in t). The coefficient A thereby represents a measure of how fast the systems become entangled. It can be expressed[11] as

$$A := \sum_{j\neq 0, k\neq 0} |\langle \varphi_j \Phi_k |H| \varphi_0 \Phi_0 \rangle|^2$$
$$= \langle \varphi_0 \Phi_0 |H (\mathbb{1} - |\varphi_0\rangle \langle\varphi_0|) (\mathbb{1} - |\Phi_0\rangle \langle\Phi_0|) H| \varphi_0 \Phi_0 \rangle. \tag{3.27}$$

A measures the "rate of de-separation" or of entanglement (or *Verschränkung*, as it was called by Erwin Schrödinger) between the two systems. If the interaction H_{int} is given by a product of two operators acting in the respective subspaces,

$$H_{\text{int}} = W_\varphi \otimes W_\Phi, \tag{3.28}$$

A simplifies to a product of two contributions

$$A = A_\varphi \cdot A_\Phi. \tag{3.29}$$

Each factor is given by the variance of the initial states with respect to the interaction operators,

$$A_\varphi = \langle \varphi_0 |W_\varphi^2| \varphi_0 \rangle - \langle \varphi_0 |W_\varphi| \varphi_0 \rangle^2 \tag{3.30}$$

(similarly for A_Φ). A straightforward generalization can be written down for a general interaction. Note that A is independent of the Hamiltonians of the two systems, hence it describes directly the entangling effect of the interaction[12]. It depends on the initial state (of both systems), which in general will change according to the intrinsic dynamics. Therefore, the de-separation parameter A gives important information of how fast a system *becomes* entangled with its environment for a given initial state. For example, oscillator eigenstates with high quantum number are very sensitive to coupling to other

---

[11] If the number of Schmidt components is less than the dimension of the larger of the two Hilbert spaces (as is usually the case), the basis may be completed by adding linearly independent (e. g. orthogonal) vectors. In the first line of (3.27) φ_j and Φ_k are arbitrary basis vectors orthogonal to φ_0 and Φ_0.

[12] In "quantum information theory" this fact is usually expressed as a locality condition: local manipulations (and "classical communication") do not create entanglement.

oscillators. They de-factorize very rapidly. This shows that these states, which are often associated with the "classical limit" of quantum mechanics, are dynamically unstable and therefore do *not* behave classically. A discussion of the parameter A for various oscillator states is given in Sect. 3.3.3.1, see also the oscillator states in Figs. 3.2 and 3.14. We will see that coherent states are most stable against coupling to the environment.

The expressions given above can be intuitively understood if one considers the interaction (3.28) as a model of measurement along the lines of the von Neumann theory. Since W_φ plays the role of the "measured observable", only a superposition of eigenstates (i.e., a non-zero variance of eigenvalues) leads to a correlated state (compare (3.16)). On the other hand, measurement (entanglement) only occurs, if the "pointer moves". If the initial state of the pointer were an eigenstate of $W_\varPhi$ it would only acquire a phase factor – only superpositions can act as a measurement device (In the original von Neumann model $W_\varPhi = \hat{p}$ induces a shift in x – so wave packets small in x (but broad in p) rapidly become orthogonal).

For a von-Neumann type model (measurement of the first kind, compare (3.14) and (3.15)), the de-separation can be calculated as follows. Consider (as a special case of (3.14)) the interaction Hamiltonian

$$H = \gamma \sum_n \alpha_n |n\rangle \langle n| \otimes \hat{p}, \qquad (3.31)$$

where the momentum operator $\hat{p}$ leads to a shift of a "pointer" depending on the measured value n. (If n is a continuous variable, this is exactly the interaction discussed by von Neumann.) From (3.29) and (3.30) one finds

$$A \geq \frac{\gamma^2 \overline{(\alpha_n - \overline{\alpha}_n)^2}}{4b^2}, \qquad (3.32)$$

where the expectation values are calculated from the initial pointer wave function of width b. The ratio γ/b describes the efficiency of the measurement by giving the time scale on which pointer wave packets corresponding to different results no longer overlap. The larger the spread in "pointer values" α_n, the stronger the correlation. This demonstrates that, particularly in measurement-like situations, separating states cannot be dynamically stable (unless the initial state is an eigenstate).

As another example consider interaction through a long-range potential, e. g., a Coulomb-like electric or gravitational force. If both systems are particles assumed to be initially in narrow wave packets of width b_i, centered at r_1^0 and r_2^0,

$$\Psi(r_1, r_2) = \varphi(r_1)\Phi(r_2), \qquad (3.33)$$

an interaction potential $V(r_1 - r_2)$ leads to the formation of correlations with a rate of de-separation given by

$$A = b_1^2 b_2^2 \sum_{i,j} V_{ij} \qquad (3.34)$$

where
$$V_{ij} = \partial_i \partial_j V(r_1^0 - r_2^0). \tag{3.35}$$

This can be interpreted as a measurement of the position of particle 1 by particle 2 and vice versa. The efficiency is proportional to $1/b_i$, since up to order t^2 an interaction potential transfers only momentum (without shifting the wave packet spatially). Therefore, in this case the width of the *momentum* wave function is relevant. For example, the gravitational interaction between the earth and the moon leads rapidly to mutual decoherence. Similarly, a charge can show only a limited coherence range in the presence of other charges (see also Sect. 4.1 and Chap. 6). Note that any interaction described by a potential $V(r)$ is diagonal in position, hence *position* is always the distinguished "observable" measured by the interaction (although the interplay with the intrinsic dynamics of the system may lead to modifications).

In (3.26) the behavior of the Schmidt expansion was considered only for short times. For simple models the time dependence of the component wave functions and eigenvalues can be studied beyond the short-time expansion, see Albrecht (1992, 1993). A description of entanglement in terms of generalized Bloch vectors has been given by Schlienz and Mahler (1995), see also Barnett and Phoenix (1989).

If the coupling strength depends on a (non-dynamical) parameter, superpositions of states differing in the value of the latter decohere. This is the case for charged particles and also for gravitational interaction between two masses, thereby leading to a mass-superselection rule (compare also Sects. 3.1.3, 4.2, and Chap. 6).

3.1.2 Scattering Processes

The interplay between the intrinsic dynamics of a system and the coupling to its environment in general gives rise to a complicated time dependence of the local density matrix. This can be expressed as an integro-differential equation (see Sect. 7.3), which reduces to an autonomous dynamics of the system only under special conditions. The typical scenario for such a simplification is given by the case of scattering processes. If the duration of a single scattering process is short compared to the typical time scales of evolution of the system by itself, the total evolution can be approximated by a combined dynamics described by an evolution equation

$$i\frac{\partial \rho}{\partial t} = [H_{\text{internal}}, \rho] + i\frac{\partial \rho}{\partial t}\bigg|_{\text{scatt.}} \tag{3.36}$$

where $\frac{\partial \rho}{\partial t}\big|_{\text{scatt.}}$ may be expressed by means of an appropriate S-matrix. This means that the change of ρ is given by the sum of contributions due to the internal dynamics and the changes caused by the scattering processes. The additional contribution to the von Neumann equation will in general lead to an increase in local entropy, since it implicitly contains the time-asymmetric assumption that system and scattered particles are initially uncorrelated (cf. the Sommerfeld radiation condition). The density matrix in (3.36) is a reduced density matrix (improper mixture), since the degrees of freedom of the scattered particles have been traced out. Equations of this kind were discussed for the classical case by Boltzmann, when he introduced his "Stoßzahlansatz" for deriving the Second Law. They are now usually called "master equations". For a general discussion see Chap. 7; many examples can be found in the following subsections. The technical advantage offered by (3.36) is the fact that the contributions from internal dynamics and scattering processes can be calculated separately (they do not "interfere"). This would not be the case for the exact subsystem dynamics (see Sect. 7.3). In the language of system-reservoir theories this is expressed by the statement that correlations decay very fast compared to the timescales of the system dynamics (Markov approximation). Usually the second term in (3.36) will describe the effect of many independent scattering processes (see below). There are some interesting situations where such a combined (alternating) dynamics can be controlled by applying measurements at well-defined instants. As an example, an experiment devised to test the quantum Zeno effect by a sequence of scattering processes will be described in Sect. 3.3.1.2.

If the scattering processes are frequent but individually inefficient, a further simple result can be derived. If recoil can be neglected, the matrix elements of the local density matrix are multiplied at each scattering event by a factor which is the overlap of the two "pointer positions" $|\Phi_n\rangle$ and $|\Phi_m\rangle$ scattering off the states $|n\rangle$ and $|m\rangle$, respectively, see (3.17). These may conveniently be expressed by using the S-matrix, as indicated in (3.15), thus

$$\rho_{nm} \xrightarrow{\text{scatt.}} \rho_{nm} \langle \Phi_m | \Phi_n \rangle$$
$$= \rho_{nm} \langle \Phi_0 | S_m^\dagger S_n | \Phi_0 \rangle. \tag{3.37}$$

If this multiplying factor is close to unity (that is, if a single scattering does not "resolve" different system states), we can write

$$\langle \Phi_0 | S_m^\dagger S_n | \Phi_0 \rangle = 1 - \varepsilon. \tag{3.38}$$

During many collision events, occurring with a rate Γ, non-diagonal terms will eventually be destroyed exponentially, since

$$\rho_{nm} \longrightarrow \rho_{nm}(1 - \varepsilon)^{\Gamma t}$$
$$\approx \rho_{nm} \exp(-\Gamma \varepsilon t). \tag{3.39}$$

Hence we find

$$\left. \frac{\partial \rho_{nm}}{\partial t} \right|_{\text{scatt.}} = -\lambda \rho_{nm}(t) \tag{3.40}$$

with

$$\lambda = \Gamma \left(1 - \left\langle \Phi_0 \left| S_m^\dagger S_n \right| \Phi_0 \right\rangle \right) \tag{3.41}$$

(see also Harris and Stodolski 1981). In this case the additional term in the equation of motion is proportional to ρ_{nm} itself, a special case of the more general form discussed by Lindblad, Zwanzig and others (see also Sects. 3.1.3, 3.3.2 and Chap. 7). Equation (3.40) describes damping of coherence in the given basis. This process is often called "dephasing", although it originates from nonlocal quantum correlations, instead of "randomized phases" (the physical interpretation of phase averaging and stochastic forces in quantum theory will also be discussed in Sects. 3.4.3 and 9.1). If decay processes are included (e. g. in optical Bloch equations), an additional contribution from decay appears, thereby generalizing (3.40), see Sect. 3.3.2.2 on page 133.

3.1.3 Environment-Induced Superselection Rules

If a system occurs in states $|\Psi_1\rangle$ and $|\Psi_2\rangle$, but never in a superposition

$$|\Psi\rangle = a\,|\Psi_1\rangle + b\,|\Psi_2\rangle \tag{3.42}$$

one says that a superselection rule is separating the Hilbert space into two or more subspaces (often characterized by unitary inequivalent representations of a certain algebra of observables). Historically, superselection rules were first formulated for conserved, discrete quantities such as electric or baryonic charge (Wick, Wightman and Wigner 1952, Wightman 1995). In this way *selection rules* (conservation laws) were elevated to *superselection rules*. As discussed in Sect. 2.1.2, there are many other situations where superselection rules are found empirically, e. g., for large molecules, or, quite generally, for all "macroscopically different" states.

What does it mean to say that states of the form (3.42) do not occur? It means that the system appears always to be *either* in the state $|\Psi_1\rangle$ *or* in the state $|\Psi_2\rangle$, i. e., all observations can successfully be described by a density matrix of the form

$$\rho = p_1\,|\Psi_1\rangle\,\langle\Psi_1| + p_2\,|\Psi_2\rangle\,\langle\Psi_2| \tag{3.43}$$

(This is sometimes expressed by the statement that the relative phases in (3.42) are unobservable; these would be revealed by measuring an appropriate "observable" such as $A = |\Psi_1\rangle\,\langle\Psi_2| + |\Psi_2\rangle\,\langle\Psi_1|$ giving $\langle A \rangle = a^*b + ab^*$ for (3.42) and $\langle A \rangle = 0$ for (3.43)).

As should be obvious from the preceding sections, a measurement-like interaction with the environment can lead to precisely such a behavior. This is indeed the explanation offered by decoherence. If all superselection rules can be reduced to decoherence mechanisms (as we assume in this chapter), then strictly isolated systems can never show superselection rules. While axiomatic superselection rules are usually considered to be exact, the diagonal form

(3.43) is in the framework of decoherence generally an approximation, and the validity of a superselection rule depends on the concrete situation.[13]

Let us consider a simple model (following Zurek 1982) which is useful for demonstrating the essential mechanism of how decoherence can lead to superselection rules. A system S is assumed to interact with its environment via a von Neumann interaction of the form (3.14). The environment is thereby assumed to discriminate between groups of energy eigenstates of system S. In the system Hamiltonian

$$H_S = \sum_i E_i \left| i \right\rangle \left\langle i \right| \tag{3.44}$$

one may group the states according to a "quantum number" α,

$$H_S = \sum_\alpha \sum_{i_\alpha=1}^{N_\alpha} E_{i_\alpha} \left| i_\alpha \right\rangle \left\langle i_\alpha \right|, \tag{3.45}$$

where the states within each group α are indexed by i_α. There are N_α states belonging to the quantum number α, which describes a conserved quantity in this model. With respect to the system Hamiltonian this is so far just a relabelling of states. The division of the system Hilbert space into subspaces with fixed quantum number α may be conveniently described by introducing a complete set of projection operators

$$P^{(\alpha)} = \sum_{i_\alpha=1}^{N_\alpha} \left| i_\alpha \right\rangle \left\langle i_\alpha \right|. \tag{3.46}$$

---

[13] Usually superselection rules are related to symmetries. This will be discussed in detail in Chap. 6. Here we only mention that some of the innocent-looking symmetry arguments may appear rather questionable. For example, the univalence superselection rule (forbidding superpositions of integer and half-integer angular momentum states) is believed to result from rotational invariance of all physical systems under rotations by an angle of 2π around a fixed axis (Wick, Wightman and Wigner 1952, Hegerfeldt, Kraus and Wigner 1968). However, this plausible requirement may well be viewed as contradictory to the very assumption of the existence of spinors which do not display a 2π-symmetry (but instead, as has been confirmed experimentally, a 4π-symmetry – see Aharonov and Susskind 1967b, Rauch et al. 1975, Werner et al. 1975, Rauch et al. 1978, Greenberger 1983). Formally, this inconsistency is related to the use of non-single-valued group representations for physical states ("ray representations"). These non-unique group representations obviously lead to a violation of the superposition principle. If the latter is to be given priority, such representations should be avoided altogether. In the case of rotations, empirical evidence then shows directly that SO(3) is the *wrong* group. The covering group SU(2) has instead to be viewed as the proper quantum rotation group. Similar arguments may be put forward in the case of the Galilei group, concerning its relation to a mass superselection rule (see Chap. 6). All these "proofs" of superselection rules seem to be based on a classical prejudice.

Let the environment Hamiltonian be given as

$$H_{\mathrm{E}} = \sum_j \varepsilon_j \, |e_j\rangle \, \langle e_j| \, , \qquad (3.47)$$

while the coupling between S and E is assumed to have the form

$$H_{\mathrm{SE}} = \sum_{\alpha,j} \gamma(\alpha,j) P^{(\alpha)} \otimes |e_j\rangle \, \langle e_j| \, . \qquad (3.48)$$

The structure of this interaction defines the quantum number α: The coupling constants γ are assumed to depend only on α (and j), and to have the same value for each state belonging to the same group. Since H_{SE} commutes with H_{S}, this simple model applies only to situations where α is a conserved quantity. The Schrödinger equation can then be solved exactly, because all three parts of the Hamiltonian commute with each other. The general factorizing initial state

$$|\Psi(0)\rangle = \left(\sum_\alpha \sum_{i_\alpha=1}^{N_\alpha} c_{i_\alpha} \, |i_\alpha\rangle \right) \otimes \sum_j d_j \, |e_j\rangle \qquad (3.49)$$

will then evolve into

$$|\Psi(t)\rangle = \sum_\alpha \sum_{i_\alpha} \sum_j c_{i_\alpha} d_j \exp \left\{ -\mathrm{i}t \left[E_{i_\alpha} + \varepsilon_j + \gamma(\alpha,j) \right] \right\} |i_\alpha\rangle \, |e_j\rangle \, , \qquad (3.50)$$

and the density matrix describing system S reads

$$\rho_S - \sum_\alpha \sum_{i_\alpha=1}^{N_\alpha} \sum_{j_\beta=1}^{N_\beta} c_{i_\alpha} c_{j_\beta}^* \, |i_\alpha\rangle \, \langle j_\beta| \exp\{-\mathrm{i}t[E_{i_\alpha} - E_{j_\beta}]\}$$
$$\times \sum_k |d_k|^2 \exp\{-\mathrm{i}t[\gamma(\alpha,k) - \gamma(\beta,k)]\}. \qquad (3.51)$$

The decoherence factor

$$z_{\alpha\beta} = \sum_k |d_k|^2 \exp\{-\mathrm{i}t[\gamma(\alpha,k) - \gamma(\beta,k)]\} \qquad (3.52)$$

is the relevant quantity in this model. It corresponds to the overlap of pointer wave functions as described at the beginning of Sect. 3.1.1. If the initial state has only nonvanishing components for a single value of α, that is, $c_{i_\alpha} = \delta_{\alpha\alpha_0} \tilde{c}_{i_\alpha}$, this factor equals unity, and the dynamics of ρ is unitary – despite the coupling to the environment. This dynamical property defines a *coherent subspace* or *superselection sector*. On the other hand, if we initially have coherence between different values of the quantum number α, which may be viewed to define what is sometimes called a "superselection charge", interference between different α will be suppressed, since $|z_{\alpha\beta}| < 1$ for $\alpha \neq \beta$ and $t > 0$. If there are many contributions in the sum (3.51), coherence will rapidly disappear (for estimates see Zurek (1982) and Sect. 7.6, where we will discuss this model from a different viewpoint). If the system S interacts with several independent systems, (3.51) is replaced by a product of many factors, all of them smaller than one. This will generally lead to a very strong suppression of coherence. The above model is similar to the measurement model studied in Sect. 3.3.2.1. on page 126.

For charged particles, the associated Coulomb fields are – because of the *discrete* nature of electric charge – always macroscopically different, hence orthogonal, when represented quantum-mechanically.

The model outlined above should give the general idea of how a superselection rule can be understood to emerge from dynamical decoherence. We will discuss more realistic situations in the following sections. In the above model, the interaction (3.48) is diagonal in the system (energy) eigenstates (as well as those of the environment). This leads to superselection rules for conserved quantities, commuting with the system Hamiltonian. These quantities are therefore sometimes called "superconserved", since only eigenstates of the Hamiltonian (not their superpositions) occur (Epstein 1960). If this restriction is relaxed, we enter the wide field of general "macroscopic properties", where transitions are possible even though interference is absent. Typically such situations are described by rate equations. Even quantities which are not conserved can acquire stability through the quantum Zeno effect (Sect. 3.3.1)[14].

For each coherent subspace spanned by the set of states $|i_\alpha\rangle$ we can build superpositions for a fixed value of α,

$$|\Psi_\alpha\rangle = \sum_{i_\alpha} c_{i_\alpha} |i_\alpha\rangle . \tag{3.53}$$

If states $|\Psi_\alpha\rangle$ are defined dynamically in this way, it may be useful to introduce a "supercharge observable" Q, which has the states $|\Psi_\alpha\rangle$ as (highly degenerate) eigenstates, according to

$$Q |\Psi_\alpha\rangle = q_\alpha |\Psi_\alpha\rangle . \tag{3.54}$$

This allows us to briefly give a comparison with some other common formulations of superselection rules. These fall into two categories. Some require that certain states do not occur (as we have done at the beginning of this section), while others claim that any state is allowed, but certain observables are not available. For the first category, all "physically allowed" density operators commute with Q (Wick, Wightman and Wigner 1970, Houtappel, Dam and Wigner 1965),

$$[\rho, Q] = 0 \tag{3.55}$$

---

[14] One may speculate that superconserved quantities, such as electric charge, are not conserved *per se*, but attain this property only through intense coupling to the environment. In the case of electric charge, the Coulomb field, usually treated as a non-dynamical entity, may be the carrier of this coupling. Needless to say, pursuing such a program would require a complete reformulation of quantum electrodynamics.

since they are assumed diagonal in α,

$$\rho = \sum_{\alpha} \rho^{(\alpha)}, \tag{3.56}$$

where each component density matrix is an arbitrary mixture of states confined to one subspace,

$$\rho^{(\alpha)} = \sum_k p_k \left| \Psi_\alpha^k \right\rangle \left\langle \Psi_\alpha^k \right|. \tag{3.57}$$

This is sometimes expressed in the form that the Hilbert space of the system decomposes into a direct sum of coherent subspaces,

$$\mathcal{H} = \bigoplus_{\alpha} \mathcal{H}_\alpha \tag{3.58}$$

in such a way that pure states are restricted to lie in the union of coherent subspaces,

$$\bigcup_{\alpha} \mathcal{H}_\alpha \tag{3.59}$$

(Bogolubov *et al.* 1990).

As is well known, at the level of expectation values, emphasis may be shifted from states to observables (although in our opinion such an approach is misleading, since observables have to be derived from the interaction with appropriate measurement instruments). This yields another common way to state a superselection rule. For the general superposition

$$|\Psi\rangle = \sum_{\alpha} c_\alpha |\Psi_\alpha\rangle \tag{3.60}$$

expectation values of arbitrary observables A contain interference terms ($\alpha \neq \beta$)

$$\langle \Psi | A | \Psi \rangle = \sum_{\alpha,\beta} c_\beta^* c_\alpha \langle \Psi_\beta | A | \Psi_\alpha \rangle. \tag{3.61}$$

If a superselection rule holds, only diagonal terms in the sum survive (the relative phases are unobservable),

$$\langle \Psi | A | \Psi \rangle = \sum_{\alpha} |c_\alpha|^2 \langle \Psi_\alpha | A | \Psi_\alpha \rangle \tag{3.62}$$

(compare the step from (3.42) to (3.43)). This is expressed by the statement, that for all *available* observables matrix elements connecting different values of α are zero, that is, all observables commute with the "superselection charge" Q,

$$[Q, A] = 0. \tag{3.63}$$

Then Q is called a "classical observable". The validity of (3.63) is achieved by excluding precisely those (non-diagonal) observables which would allow prove of coherence between different values of α.

In the framework of decoherence, none of these artificial restrictions of the superposition principle has to be postulated. They arise dynamically in a natural way.[15] Strictly speaking, there are no fundamental superselection rules at all, since the phase relations in (3.50) could *in principle* be made evident by appropriate measurements. If correlations spread out irreversibly, any local observation (in a very weak sense) can no longer reveal these phase relations, however, and thus seems to show a superselection rule.

3.2 Localization of Objects

As a first – and most important – application of the previous considerations we will in this section discuss the problem of spatial localization of macroscopic bodies. The resulting equations of motion are examples of master equations which we will explore in a more general context in Sect. 3.3.2 and Chap. 7.

Macroscopic objects are always observed in spatially well-localized states, quite in contrast to micro-objects, which are usually found in energy eigenstates. If the center-of-mass coordinate of a macroscopic body were described by a Hamiltonian $H = p^2/2m$, energy eigenstates would be represented by plane waves, in sheer contradiction to what is observed! Obviously, macroscopic objects never occur in energy eigenstates, but only in time-dependent wave packets. This means that only a small subset of all possible states of macro-objects is realized in Nature. This fact cannot be understood from the quantum mechanics of mass points alone, as was repeatedly stressed by Einstein in his correspondence with Max Born.

Born argued that the spreading of an initially well-localized wave packet is completely negligible in the limit of large mass, and this should suffice to ensure the "classical limit". However, the superposition principle allows far

---

[15] If there exists more than one superselection charge for a certain system, the corresponding observables are assumed to commute with each other. This fact is sometimes referred to as "Hypothesis of Commuting Superselection Rules" (Streater and Wightman 1964). From the viewpoint of the above model, the validity of this conjecture is almost trivial. In the case of more than one superselection rule, the interaction must have the form $H_{\mathrm{SE}} = H_{\mathrm{SE}}^{(\alpha)} + H_{\mathrm{SE}}^{(\beta)} + ...$, where each term has the same structure as given in (3.48). If two projectors $P^{(\alpha)}$, $P^{(\beta)}$ did not commute, an eigenstate of $P^{(\alpha)}$, say, would not be dynamically stable under the decohering influence of the interaction diagonal in $P^{(\beta)}$. A quite similar situation is encountered in models where position and momentum are measured "simultaneously" by means of an interaction which is given as a sum of two von Neumann terms, diagonal in position and momentum, respectively. No dispersion-free states can persist under such circumstances.

more states than can be called "classical". Einstein expressed this fact clearly in a letter to Max Born (Jan 1, 1954),

> "Your opinion is quite untenable. It is in conflict with the principles of quantum theory to require that the Ψ-function of a "macro"-system be "narrow" with respect to the macro-coordinates and momenta. Such a demand is at variance with the superposition principle for Ψ-functions. ... Let Ψ_1 and Ψ_2 be two solutions of the same Schrödinger equation. Then $\Psi = \Psi_1 + \Psi_2$ also represents a solution of the Schrödinger equation, with equal claim to describing a possible real state. When the system is a macrosystem, and when Ψ_1 and Ψ_2 are "narrow" with respect to the macro-coordinates, then in by far the greater number of cases, this is no longer true for Ψ. Narrowness in regard to macro-coordinates is a requirement which is not only *independent* of the principles of quantum mechanics, but, moreover, *incompatible* with them. ... On the other hand, the following objection is valid in nearly all cases, but only of secondary significance: that the Schrödinger equation will lead to a dispersion of "narrowness" in the course of time." (Born 1969) [16]

Even the mentioned slow spreading of massive wave packets does not suffice, unless such a packet describes a real state of affairs from the very beginning,

> "One would then have to be very surprised if a star or a fly, seen for the first time, appeared somehow quasi-localised." [17]

On the other hand, macroscopic bodies cannot avoid scattering photons and other particles. Since the scattering process depends in an essential way

---

[16] "Deine Auffassung ist ganz unhaltbar. Es ist mit den Prinzipien der Quantentheorie unvereinbar zu fordern, daß die Ψ-Funktion eines "Makro"-Systems bezüglich der Makrokoordinaten und -impulse "eng" sein soll. Eine solche Forderung ist unvereinbar mit dem Superpositionsprinzip für Ψ-Funktionen. ... Ψ_1 und Ψ_2 seien zwei Lösungen derselben Schrödinger-Gleichung. Dann ist $\Psi = \Psi_1 + \Psi_2$ ebenfalls eine Lösung der Schrödinger-Gleichung mit gleichem Anspruch darauf, einen möglichen Realzustand zu beschreiben. Wenn das System ein Makro-System ist, und wenn Ψ_1 und Ψ_2 "eng" sind in Bezug auf die Makro-Koordinaten, so ist dies in der weitaus überwiegenden Zahl der möglichen Fälle für Ψ nicht mehr der Fall. Enge bezüglich der Makro-Koordinaten ist eine Forderung, die nicht nur *unabhängig* ist von den Prinzipien der Quantenmechanik, sondern auch *unvereinbar* mit diesen Prinzipien. ... Demgegenüber ist der ebenfalls in fast allen Fällen gültige Einwand nur von sekundärer Bedeutung: daß die Schrödinger-Gleichung zu einer Zerstreuung der "Enge" mit der Zeit führt."

[17] "Dann muß man sich aber sehr wundern, daß ein Stern oder eine Fliege, die man zum ersten Mal sieht, so etwas wie quasilokalisiert erscheinen..."

on the position of the object, as in a microscope, interference terms between different positions in the density matrix of the scattering center are destroyed[18]. Even microscopic objects under continuous measurement (that is, strong coupling to their environment) are subject to this effect (e. g. an α-particle leaving a track in a bubble chamber).

Within extremely good approximation, scattering of photons or atoms off a macroscopic object (even very small dust particles or large molecules) causes no recoil, so that this process follows the scheme of an (ideal) measurement (compare Sect. 3.1.1). Clearly, a single photon cannot resolve arbitrarily small distances. Therefore its "pointer states", which are correlated with two slightly different positions of a macroscopic object, are usually far from orthogonal. This means that this "measurement" is necessarily incomplete.[19] Under usual circumstances this inefficiency is overcompensated by the vast number of scattering processes which occur in realistic situations even in small time intervals.

3.2.1 Localization Through Ideal Measurements

For reasons stated in the last paragraph, in the main part of this section we will neglect any back-action. Therefore, the scheme of ideal measurements (see (3.16) – (3.17)) can be applied in a rather direct manner. This limit is a central feature of decoherence and therefore essential for the understanding of classical properties of certain objects. For this reason we will discuss the resulting dynamics in considerable detail. The more general case of "quantum Brownian motion" (including recoil) will be only briefly reviewed and compared in the subsequent sections. We emphasize again that the term "measurement" is used throughout this chapter only in the sense of establishing a particular kind of quantum correlations (usually uncontrollable) between two systems in the sense discussed in Sects. 2.3 and 3.1.1, even though finally some collapse (or equivalent branching) has to be assumed in order to establish the link to subjective perception (compare Sect. 3.2.5).

3.2.1.1 Spatial Decoherence. Let $|x\rangle$ represent position eigenstates of a mass point representing a macro-object, $|\chi\rangle$ the state of an incoming particle. During the scattering $|x\rangle$ will be assumed to remain unchanged (no recoil) in the approximation to be used here, while the scattered state will carry away

---

[18] We use the expression "destroyed" in the same sense as in previous sections. It means that certain interference terms are unobservable for "local" observations. In this section the term "local" is used in its common meaning, discriminating objects in space.

[19] Some authors use the formalism of "operations and effects" to treat incomplete measurements. For an analysis of continuous position measurements along these lines see Caves and Milburn (1987) and references therein.

information about the position of the scattering center. This can be written as

$$|x\rangle |\chi\rangle \xrightarrow{t} |x\rangle |\chi_x\rangle = |x\rangle S_x |\chi\rangle, \qquad (3.64)$$

where the resulting state of the scattered particle may be conveniently expressed by means of a scattering matrix S for times t larger than the scattering time, as indicated in the equation.

If the position of the scattering center is described by a wave function $\varphi(x)$, one obtains

$$\int d^3x\, \varphi(x) |x\rangle |\chi\rangle \xrightarrow{t} \int d^3x\, \varphi(x) |x\rangle S_x |\chi\rangle. \qquad (3.65)$$

The (reduced) density matrix of the scattering center is thereby multiplied by a factor representing the overlap of the corresponding scattered states (compare (3.17) on page 48),

$$\rho(x, x') = \varphi(x)\varphi^*(x') \xrightarrow{t} \varphi(x)\varphi^*(x') \left\langle \chi \left| S_{x'}^\dagger S_x \right| \chi \right\rangle. \qquad (3.66)$$

In the following we will mainly consider the case where a *single* scattering event does *not* resolve the distance $|x - x'|$, i. e. we assume the wavelength of the scattered particle to be larger than this distance, $\lambda \gg |x - x'|$. Otherwise interference terms are destroyed in a single scattering event. Both limiting cases are governed by a few parameters describing the interaction process. A detailed calculation is given in App. A1, compare also the extensions described in Sect. 3.2.1.4; for a review of treatments including recoil see Sects. 3.2.2 and 5.1.

If the scattering interaction is translationally invariant, the dependence on position of the S-matrix in momentum representation is given by a phase factor,

$$S_x(k, k') = S(k, k')e^{-i(k-k')x}, \qquad (3.67)$$

where $S(k, k')$ is the usual scattering matrix with scattering center at the origin. The expression $\left\langle \chi \left| S_{x'}^\dagger S_x \right| \chi \right\rangle$ can then be calculated for incoming plane waves $|\chi\rangle = |k\rangle$ (perhaps contained in a thermal ensemble). Adding the contributions from many individually ineffective scattering events yields an exponential damping of interference in the position representation of the density matrix of the scattering center (Wigner 1983, Joos and Zeh 1985),

$$\rho(x, x', t) = \rho(x, x', 0) \exp\{-\Lambda t(x - x')^2\}. \qquad (3.68)$$

Here the quantity

$$\Lambda = \frac{k^2 N v \sigma_{\text{eff}}}{V} \qquad (3.69)$$

measures the "localization rate" in the short-distance limit $k|x - x'| \ll 1$. In (3.69) k is the wave number, Nv/V the incoming flux, while σ_{eff} (to be calculated from the S-matrix) is of the order of the total cross section. In the limit of large distances, $k|x - x'| \gg 1$, destruction of coherence is simply governed by scattering rates, independent of $|x - x'|$ (see App. A1).

There is in general a vast number of scattering processes contributing to the destruction of interference. They will depend on the properties of the interaction of the considered object with its actual surroundings. In Table 3.1 we list some "localization rates" Λ resulting from various scattering processes.

Table 3.1. Localization rate Λ in $cm^{-2}s^{-1}$ for three sizes of "dust particles" and various types of scattering processes according to (3.68) (from Joos and Zeh 1985). This quantity measures how fast interference between different positions disappears for distances smaller than the wavelength of the scattered particles. For larger distances, decoherence rates are just given by the scattering rates, and thus independent of $x - x'$.

	$a = 10^{-3}\,cm$ dust particle	$a = 10^{-5}\,cm$ dust particle	$a = 10^{-6}\,cm$ large molecule
Cosmic background radiation	10^6	10^{-6}	10^{-12}
300 K photons	10^{19}	10^{12}	10^6
Sunlight (on earth)	10^{21}	10^{17}	10^{13}
Air molecules	10^{36}	10^{32}	10^{30}
Laboratory vacuum (10^6 particles/cm^3)	10^{23}	10^{19}	10^{17}

Some further values were calculated by Tegmark (1993). We quote some of his results in Table 3.2. In addition to well-known scattering processes this table contains some estimates concerning "nonstandard" theories such as quantum gravity (following suggestions of Hawking (1982) and Ellis, Mohanty and Nanopoulos (1990)[20]) and the theory proposed by Ghirardi, Rimini and Weber (1986). Obviously the effects of "normal decoherence" render it rather difficult to test these alternatives by experiment. An assessment of alternative theories has been given by Pearle and Squires (1994), see also Chap. 8.

Here on earth, scattering of air molecules is most important under normal conditions. The small value of the thermal de Broglie wavelength of massive particles together with the large number of scattering events leads to rapid decoherence. Even in intergalactic space, scattering of photons from the cosmic background radiation cannot be considered negligible for macroscopic objects.

Equations (3.68) and (3.69) describe the effect of scattering for small distances $k|x - x'| \ll 1$, leading to a typical decoherence timescale $t_{dec} \approx$

---

[20] In these models the interference-damping exponential is *quadratic* in t, in contrast to (3.68). This is to be expected in models where a *single* pointer is coupled to the measured system.

Table 3.2. Some values of the localization rate Λ given by Tegmark (1993). The values given for the case of a dust particle agree well with those of earlier work (Table 3.1).

	Free electron	10^{-3} cm dust particle	Bowling ball
300 K air at 1 atm pressure	10^{31}	10^{37}	10^{45}
300 K air in lab vacuum	10^{18}	10^{23}	10^{31}
Sunlight (on earth)	10^{1}	10^{20}	10^{28}
300 K photons	10^{0}	10^{19}	10^{27}
Background radioactivity	10^{-4}	10^{15}	10^{23}
Quantum gravity	10^{-25}	10^{10}	10^{22}
GRW effect	10^{-7}	10^{9}	10^{21}
Cosmic background radiation	10^{-10}	10^{6}	10^{17}
Solar neutrinos	10^{-15}	10^{1}	10^{13}

$\frac{1}{\Lambda|x-x'|^2}$. In the opposite limit $k|x - x'| \gg 1$, where a single scattering event destroys coherence, the decoherence timescale is just given by the scattering rate, that is $t_{\mathrm{dec}} \approx \frac{V}{N v \sigma_{\mathrm{tot}}} \approx \frac{k^2}{\Lambda}$ (see App. A1).

As an example consider a tiny dust particle of the size of a virus (10^{-5}cm). Under normal conditions, scattering of air molecules leads to a decoherence timescale of the order of 10^{-13} s. Since this value scales with the particle density, in a laboratory situation radiation effects may become dominant. In this example, the 300K thermal background (Rayleigh scattering) yields a decoherence time of about 1 s for a distance of 10^{-6}cm, 10^{-4} s for 10^{-4} cm, and 10^{-5} s for $|x - x'| \gtrsim 10^{-2}$ cm, respectively.

Indeed, recent experiments with C_{60} and C_{70} fullerene molecules (Arndt et al. 1999, Nairz et al. 2001, Brezger et al. 2002) showed the expected interference patterns, in line with the above estimates[21] (cf. also Alicki 2001). The most important decoherence mechanism for C_{60} is emission of (thermal) radiation from the internally hot molecule. These and other examples demonstrate the strong dependence of decoherence effects on the actual situation. It should be emphasized, however, that in all cases decoherence is based on the arising entanglement with the environment – not on an "increase of information".

As an illustration of the dynamical influence of the environment let us consider the destruction of coherence between spatially separated components of a wave function. As an example take a superposition of two Gaussian wave packets, which may arise in a double-slit experiment,

$$\Psi(x) = \Psi_1(x) + \Psi_2(x)$$
$$= N_1 e^{-(x-a_1)^2} + N_2 e^{-(x-a_2)^2}. \tag{3.70}$$

---

[21] We wish to thank Werner Wetzel for valuable exchange.

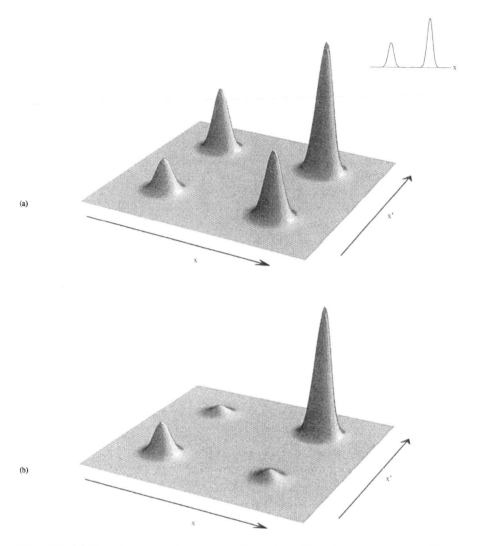

Fig. 3.1. (**a**) Density matrix of a superposition of two Gaussian wave packets. The wave function is shown in the inset. Even if the two packets do not overlap, such a state must not be identified with an ensemble of independent packets. The superposition may be distinguished from the ensemble by observing interference effects, for example. Coherence between the two parts of the wave function is represented by the off-diagonal contributions in the picture. (**b**) The density matrix after interference terms have been partially destroyed by decoherence (see also Zurek 1991).

The corresponding density matrix $\rho(x, x') = \Psi^*(x')\Psi(x)$ (shown in Fig. 3.1) contains *four* peaks, $\Psi_1^*(x')\Psi_1(x)$ and $\Psi_2^*(x')\Psi_2(x)$ along the main diagonal $x = x'$, plus two "interference terms" $\Psi_2^*(x')\Psi_1(x)$ and $\Psi_1^*(x')\Psi_2(x)$.

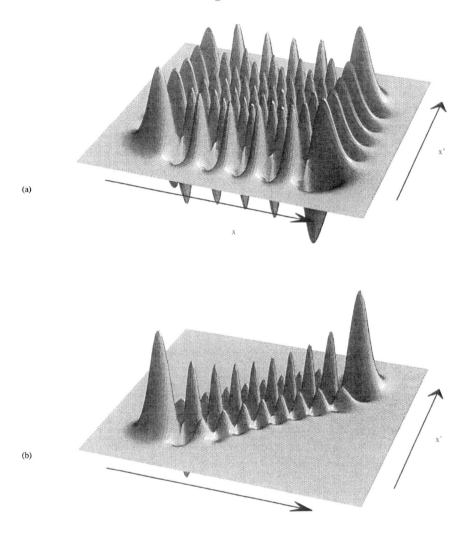

Fig. 3.2. (**a**) Density matrix of an energy eigenstate of a harmonic oscillator for $n = 9$ in the position representation. (**b**) the density matrix at a later time, showing non-diagonal terms suppressed by decoherence according to (3.68). Note that the oscillations in position probability (along the diagonal) are preserved. This would not be the case if the system would approach thermal equilibrium.

Suppression of coherence according to (3.68) damps the off-diagonal terms, so that the density matrix approaches that of a mixture of Ψ_1 and Ψ_2.

Another interesting example is the effect of (spatial) decoherence on energy eigenstates of an harmonic oscillator. These are often – motivated by the "correspondence principle for high quantum numbers" – associated with the classical limit of quantum mechanics. However, these energy eigenstates are very sensitive to decoherence, especially for large quantum numbers, as we will analyze in more detail in Sect. 3.3.3.1. Figure 3.2 shows the suppression of spatial decoherence for the oscillator eigenstate with quantum number $n = 9$.

For the general case, the density matrix $\rho(x, x')$ is complex, so it is not as easy to visualize as in the examples displayed in the above figures. A commonly used real description is given by the Wigner function which we will use in Sect. 3.2.3 (compare Figs. 3.13 and 3.14).

3.2.1.2 Equation of Motion. The complete dynamics for the center-of-mass motion of a particle (including decoherence) can now be written as a Boltzmann-type equation which combines the free dynamics with the effect of the scattering processes (compare (3.36)). From (3.68) we see that the equation of motion in one space dimension then reads

$$i\frac{\partial \rho}{\partial t} = [H, \rho] - i\Lambda[x, [x, \rho]], \tag{3.71}$$

or, in the position representation for a "free" particle with $H = p^2/2m$,

$$i\frac{\partial \rho(x, x', t)}{\partial t} = \frac{1}{2m}\left(\frac{\partial^2}{\partial x'^2} - \frac{\partial^2}{\partial x^2}\right)\rho - i\Lambda(x - x')^2\rho. \tag{3.72}$$

Solutions of this equation can be constructed in various ways. We will here employ a Gaussian density matrix by using the ansatz

$$\rho(x, x, t) = \exp\left\{-\left[A(t)(x - x')^2 + iB(t)(x - x')(x + x') + C(t)(x + x')^2\right.\right.$$
$$\left.\left. + iK(t)(x - x') + L(t)(x + x') + D(t)\right]\right\} \tag{3.73}$$

D ensures conservation of the trace, $A(t)$ describes the range of coherence, while $C(t)$ specifies the extension of the ensemble in space. For a hermitean density matrix ρ, all functions $A, B, \ldots$ are real. Other representations are, of course, possible and may sometimes be more useful, for example the momentum representation $\rho(p, p')$, or the Wigner function $W(x, p)$ (see Sect. 3.2.3, Eq. (3.137)). In App. A2 we will use the so-called characteristic function (the Fourier transform of the Wigner function), which has the advantage that the equation of motion is only of first order.

The ansatz (3.73) leads to a system of coupled ordinary differential equations. Their general solution is constructed in App. A2. Note, that the solutions of (3.72) depend only on t/m and the product Λm as dynamical parameters. Although these are sometimes useful, in this section we prefer to express all quantities in terms of the physical parameters. For the initial (pure) state

$$\varphi(x, 0) = \left(2\pi b^2\right)^{-1/4}\exp\left(-x^2/4b^2\right), \tag{3.74}$$

a Gaussian wave packet of width b, the density matrix is given by $A(0) = C(0) = 1/8b^2$. The time dependence of some relevant quantities for this initial condition is shown in various diagrams below for three values of the localization rate Λ (more general expressions can be found in App. A2). $\Lambda = 0$ corresponds to no interaction with the environment (pure Schrödinger dynamics). The curves labelled 1,2,3 show the increasing influence of decoherence. The values chosen in the pictures do not correspond to any particular physical situation (realistic values are given in the tables), but only serve to illustrate the behavior of the solutions.

Figure 3.3 shows the time-dependence of linear entropy (3.75). Since the equation of motion contains the influence of time-directed scattering processes, local entropy is expected to increase. This can be easily proven directly from (3.71). The logarithmic (von Neumann) entropy increases as well as the linear entropy

$$S_{\text{lin}}(t) = \text{tr}\left(\rho - \rho^2\right) = 1 - \sqrt{\frac{C(t)}{A(t)}}$$

$$= 1 - \frac{mb}{\sqrt{\frac{2}{3}\Lambda t^3 + m^2 b^2 + \frac{4}{3}\Lambda^2 b^2 t^4 + 8\Lambda m^2 b^4 t}} \quad . \tag{3.75}$$

(The last equalities in this and the following equations refer to the solution deriving from the initial state (3.74).) For pure states the entropy is zero. In contrast to the von Neumann entropy, S_{lin} cannot grow beyond unity. Similar to the coherence length (which we will discuss below), entropy measures how strongly the environment destroys coherence between different positions by delocalizing phases, as can be seen from expanding (3.75) in powers of t,

$$S_{\text{lin}} = 4\Lambda b^2 t + O(t^2). \tag{3.76}$$

The departure from the pure-state value zero is proportional to the coupling Λ and larger for broad wave packets, which are particularly sensitive to position measurements. The linear time dependence is a typical feature of master equations (Markov approximation). Quite generally, the time derivative of entropy gives a measure of how fast a system becomes entangled with its environment. Linear entropy is thus directly related to the de-separation parameter A introduced in (3.26). For further discussion of the behavior of linear entropy see Sect. 3.3.3.1.

The width of the ensemble described by ρ increases even for unitary evolution (the well-known "spreading of the wave packet"). Here an additional term appears, since

$$\Delta x = \frac{1}{\sqrt{8C(t)}}$$

$$= \frac{1}{2}\sqrt{\frac{t^2}{m^2 b^2} + \frac{8}{3}\frac{\Lambda t^3}{m^2} + 4b^2} \quad . \tag{3.77}$$

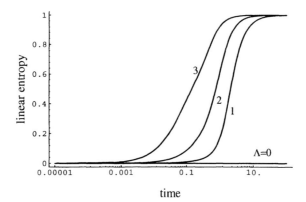

Fig. 3.3. Time dependence of the linear entropy (3.75). Without interaction ($\Lambda = 0$), the entropy is constant. The three curves labelled 1,2,3 show the behavior of entropy with increasing coupling to the environment.

This means that the ensemble spreads more rapidly than without decoherence ($\Lambda = 0$). This is related to the increase in mean energy (see below).

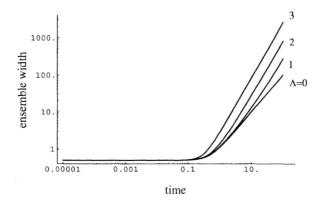

Fig. 3.4. Extension in space of the probability distribution described by ρ. The figure shows the spreading of the wave packet according to the unitary Schrödinger dynamics, as well as the effect of the coupling to the environment. The curve corresponding to $\Lambda = 0$ represents only the coherent spreading. In all other cases ρ describes an improper mixture.

A quantity of particular interest is the coherence length $l_x(t)$, which is a measure of the distance over which quantum interference can still be found (the width of ρ in the x–x' direction). It is defined by

$$l_x(t) = \frac{1}{\sqrt{8A(t)}}$$

$$= \frac{1}{2}\sqrt{\frac{3t^2 + 8\Lambda b^2 t^3 + 12m^2 b^4}{2\Lambda t^3 + 3m^2 b^2 + 4\Lambda^2 b^2 t^4 + 24\Lambda m^2 b^4 t}} \quad . \tag{3.78}$$

For short times it shows a behavior similar to that of linear entropy,

$$l_x(t) = b\left(1 - 4\Lambda b^2 t\right) + O(t^2). \tag{3.79}$$

This means that initially the damping of coherence is proportional to the coupling and the sensitivity of the wave packet to position measurements (i. e., its spatial width). For large times the coherence length *decreases* as

$$l_x(t) \overset{t\to\infty}{\longrightarrow} \frac{1}{\sqrt{2\Lambda t}}, \tag{3.80}$$

independently of the initial width b. One should keep in mind, that "large" can here easily be as small as 10^{-12} sec for macroscopic objects in view of the large values of Λ (see tables). For intermediate times the increase in coherence length (the spreading of the wave function according to the Schrödinger equation) competes with the opposite effect of the position measurements (i. e. decoherence through scattering processes).[22] A dimensionless measure of decoherence is given by the ratio of $l(t)$ and $\Delta x(t)$ instead of $l(t)$ alone. Interestingly, this measure of decoherence,

$$\delta(t) := \frac{l_x(t)}{\Delta x(t)}, \tag{3.81}$$

(Morikawa 1990) is directly related to the linear entropy for Gaussian states, since

$$\frac{l_x(t)}{\Delta x(t)} = \sqrt{\frac{C(t)}{A(t)}} = \mathrm{tr}\rho^2 = 1 - S_{\mathrm{lin}}(t). \tag{3.82}$$

$\delta(t)$ is therefore a monotonically decreasing quantity.

The width of the momentum distribution (proportional to the expectation value of $H = \frac{p^2}{2m}$, since $\langle p \rangle = 0$ in this example) is given by

$$\Delta p = \sqrt{2\left(A(t) + \frac{B(t)^2}{4C(t)}\right)}$$

$$= \sqrt{\frac{1}{4b^2} + 2\Lambda t} \quad . \tag{3.83}$$

---

[22] Closer inspection shows that $l(t)$ reaches a minimum at $t_1 = \frac{1}{4\Lambda b^2}[(1 + 192\Lambda^2 m^2 b^8)^{1/3} - 1]$ (with $l(t_1) = b/\left(1 + 192\Lambda^2 m^2 b^8\right)^{1/6}$) and a maximum at $t_2 = \left(\frac{3m^2 b^2}{\Lambda}\right)^{1/3} > t_1$ with $l(t_2) = 1/\left(2\left(\sqrt{3}\Lambda mb\right)^{1/3}\right)$.

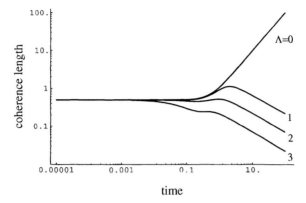

Fig. 3.5. Coherence length $l_x(t)$. It is a measure of the spatial extension over which the described object can show interference effects. Except for zero coupling to the environment ($\Lambda = 0$), the coherence length decreases for large times proportional to $1/\sqrt{\Lambda t}$ and asymptotically vanishes in this model.

It displays a linear increase in energy, independent of the initial state. The reason for this phenomenon is the neglect of recoil in the present model (see (3.64)). Although only valid far from thermal equilibrium, this is an excellent approximation for macroscopic objects (see Sect. 3.2.2.3).

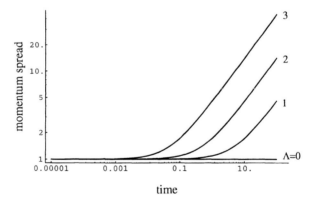

Fig. 3.6. Width of the momentum distribution. For a free particle it is a constant of motion, since energy is conserved. Otherwise the mean energy increases forever as the environment has formally infinite temperature in the present approximation.

Because the equation of motion (3.71) contains the effect of both decoherence and internal dynamics, not only the coherence in position, but also phase relations of momentum are affected. After performing a Fourier transformation to the momentum representation $\rho(p, p')$, one finds for the coherence range in momentum

$$l_p(t) = \sqrt{\frac{B^2 + 4AC}{2A}}$$

$$= \frac{m}{2} \left(m^2 b^2 + \frac{2}{3} \Lambda t \frac{1 + 2\Lambda b^2 t}{1 + 8\Lambda b^2 t} \right)^{-1/2}. \tag{3.84}$$

In the limit of infinite mass, l_p is constant. Otherwise it also decreases with increasing time because of the interplay between internal dynamics and decoherence. For short times it is given by

$$l_p(t) = \frac{1}{2b} - \frac{\Lambda}{6b^2 m^2} t^3 + O(t^4). \tag{3.85}$$

The decay of momentum coherence is initially "slow" (only in third order of time) since decoherence in the momentum representation is a secondary, indirect effect caused by the intrinsic system dynamics. Interestingly, the *ratio* $l_p/\Delta p$ also equals $\mathrm{tr}(\rho^2) = l_x/\Delta x$ (compare (3.75), (3.82)) for Gaussian states.

The fact that coherence in momentum is also damped, although the environment only discriminates different positions, is obvious from the following example. Consider a superposition of two well-localized wave packets at the same position but with very different (mean) momenta. Although initially this state will not be affected appreciably, the internal dynamics (i. e., just momentum conservation in our case of a "free" particle) will cause a spatial splitting of the initial state into two packets, allowing decoherence to set in. On the other hand, if we had only decoherence in momentum, described by a term $\propto [p, [p, \rho]]$ in the equation of motion, an analogous superposition would not be affected in the same way. Formally, the "pointer observable" (position) does not commute with the system Hamiltonian.

3.2.1.3 Decohering Wave Packets. Superpositions of states can, in contrast to mixtures, display characteristic interference effects in appropriate situations. A simple example is a superposition of two plane waves with wave vectors k_1 and k_2,

$$\Psi(x) = \frac{1}{\sqrt{2}} \left(e^{ik_1 x} + e^{ik_2 x} \right). \tag{3.86}$$

This state yields an interference pattern in the position distribution according to

$$\rho(x, x, t) = 1 + \cos \left[(k_1 - k_2) x + \frac{k_1^2 - k_2^2}{2m} t \right]. \tag{3.87}$$

In the presence of the environment, the fringe contrast will be reduced. This example was discussed by Savage and Walls (1985b) for the more general case with friction (see Sect. 3.2.2.1, Eq. (3.101)). They obtained a position distribution of the form (in our notation)

$$\rho(x,x,t) = 1 + \exp[-\eta] \cos\left[(k_1 - k_2)x - \frac{1 - \mathrm{e}^{-\gamma t}}{2m\gamma}\left(k_1^2 - k_2^2\right)\right], \qquad (3.88)$$

where

$$\eta = \frac{2\Lambda}{m^2\gamma^3}\left[\gamma t/2 - \frac{3}{4} + \mathrm{e}^{-\gamma t} - \frac{1}{4}\mathrm{e}^{-2\gamma t}\right](k_1 - k_2)^2. \qquad (3.89)$$

(γ is the friction coefficient defined in Sect. 3.2.2.1). For negligible friction, the above result reads

$$\rho(x,x,t) = 1 + \exp\left[-\frac{\Lambda t^3\left(k_1^2 - k_2^2\right)}{3m^2}\right]\cos\left[(k_1 - k_2)x - \frac{\left(k_1^2 - k_2^2\right)t}{2m}\right].$$
$$(3.90)$$

The visibility of the interference fringes is strongly reduced, while the spatial structure remains unaffected.

When the two components do not permanently overlap (as in the above example), a superposition of two spatially distinct wave packets (as shown in Fig. 3.1) can still be distinguished from a mixture when the two packets are brought to interference. This is the often-discussed situation of a double-observed on a screen, where the packets originating coherently from two slits overlap. If the "path" of the particle is observed, interference vanishes (Feynman, Leighton and Sands 1965).

A situation of this kind can be modelled by using the equation of motion discussed in this section. If we start with two wave packets moving upon each other, the probability distribution $P(x,t) = \langle x\,|\rho(t)|\,x\rangle$ will show a typical interference pattern when the two packets overlap, see Fig. 3.7a.

As soon as the interaction with the environment is included, the pattern diminishes (b), finally it disappears entirely (c). (The figures were generated by superposing solutions of (3.71), see App. A2.)

A similar situation was explored by Caldeira and Leggett. In their treatment the two packets are confined in an oscillator potential, one initially in the ground state, the other one displaced sufficiently. Due to the oscillations they interfere periodically, while the environment continually destroys the coherence between the two packets (Caldeira and Leggett 1983a, 1985). For an analysis in terms of the Wigner function see Paz, Habib and Zurek (1993).

The double-slit and similar experiments have been common ground for debates on the meaning of quantum theory from the early days (Bohr 1949) until today. Thought experiments with particles were realized in experiments in neutron interferometry (Rauch 1984, Badurek, Rauch and Tuppinger 1986, Zeilinger et al. 1988) and more recently in the rapidly developing field of atom optics. Coherence properties of a beam of He atoms, diffracted by a standing light wave, were studied by Pfau et al. (1994). For an experiment with iodine molecules see Bordé et al. (1994). Superpositions of still larger objects

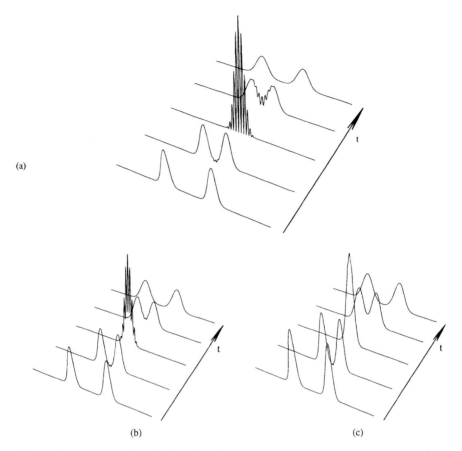

Fig. 3.7. (**a**) According to the Schrödinger equation, two moving wave packets (representing *one* particle) display an interference pattern in the position probability distribution when they meet in space. (**b**) The same situation with a small decohering influence of the environment. The interference pattern is still visible, but damped to some extent. (**c**) Two wave packets moving under the influence of strong decoherence. Even when they share the same region of space, no interference fringes are visible.

are possible, if decoherence is weak enough. Interference of C_{60} (fullerene) molecules was observed although these molecules were (internally) in highly excited states (Arndt *et al.* (1999), for a theoretical analysis see Facchi, Mariano and Pascazio (2002) and Alicki (2001)).

The disappearance of interference fringes when the "path" of the particles is measured is still interpreted in various ways, either (historically motivated) as resulting from the uncertainty relations, as a consequence of quantum entanglement with the detector/environment, or as a stochastic force exerted on the particle. Some relevant references are Wheeler (1978), Wootters and

Zurek (1979), Stern, Aharonov and Imry (1990), Scully, Englert and Walther (1991), Tan and Walls (1993), Dürr, Nonn and Rempe (1998). In the framework adopted here the disappearance of interference has its origin in the nonlocal properties of quantum correlations, which are incompatible with any statistical local model. For a controversial exchange of ideas see Storey *et al.* (1994) and Englert, Scully and Walther (1995), compare also Schulman (1996). As long as quantum correlations are not spread out of reach, coherence can be regained in appropriate ("quantum eraser") experiments (Hillery and Scully 1983, Kwiat, Steinberg and Chiao 1992, Herzog *et al.* 1995, see also Chapman *et al.* 1995). An overview of particle interferometry concepts and experiments can be found in Greenberger, Horne and Zeilinger (1993).

3.2.1.4 More General Recoil-Free Decoherence. The coherence-damping term in (3.71) was originally derived for short distances $|x - x'|$, where individual scattering processes are ineffective (Joos and Zeh 1985). The treatment can be improved to give the more general result (Gallis and Fleming 1990)

$$i\frac{\partial \rho(\boldsymbol{x}, \boldsymbol{x}')}{\partial t} = \langle \boldsymbol{x} \,|[H, \rho]|\, \boldsymbol{x}' \rangle - iF(\boldsymbol{x} - \boldsymbol{x}')\rho(\boldsymbol{x}, \boldsymbol{x}'), \qquad (3.91)$$

with

$$F(\mathbf{r}) = \int dq\, n(q)v(q) \int \frac{d\Omega\, d\Omega'}{2} \left(1 - e^{i(\boldsymbol{q} - \boldsymbol{q}')\boldsymbol{r}}\right) |f(\boldsymbol{q}, \boldsymbol{q}')|^2 \qquad (3.92)$$

(f is the scattering amplitude, nv the particle flux). The short-scale expression can be recovered from (3.92) by expanding the exponential, while for large distances F now approaches a constant value (see App. A1).

An equation of the *form* (3.91) was also suggested by Ghirardi, Rimini and Weber (1986) as a new fundamental (irreversible) dynamics replacing the Schrödinger equation. They *postulated* a master equation of the form

$$i\frac{\partial \rho}{\partial t} = [H, \rho] - i\lambda(\rho - T[\rho]), \qquad (3.93)$$

where

$$T[\rho] = \left[\frac{\alpha}{\pi}\right]^{1/2} \int_{-\infty}^{\infty} dx\, e^{-(\alpha/2)(q-x)^2} \rho\, e^{-(\alpha/2)(q-x)^2}. \qquad (3.94)$$

Here q is the position operator of the considered system. This equation also leads to a destruction of coherence between different positions of the object, as can be seen by writing it in the position representation,

$$i\frac{\partial \rho(x, x', t)}{\partial t} = \langle x \,|[H, \rho]|\, x' \rangle - i\lambda \left(1 - e^{-(\alpha/4)(x-x')^2}\right) \rho(x, x', t). \qquad (3.95)$$

We mention this model here only because of the formal similarity of the equations of motion. The interpretation is rather different since (3.93) is intended to describe a *proper* mixture, quite in contrast to the interpretation in the context of decoherence (Joos 1987a, Ghirardi, Rimini and Weber 1987). Equation (3.93) and related models for a stochastic evolution of wave functions are beyond the framework of standard quantum theory and will be discussed in Chap. 8.

3.2.2 Decoherence and Dissipation

If the wave packet describing the center-of-mass of a macroscopic object were to follow (reversible) Newtonian dynamics, any interference over macroscopic distances must be effectively absent (i. e., destroyed very quickly), while friction should remain irrelevant on the same timescale. This seems to contradict the view that the relaxation rate is a measure for the strength of the interaction of a system with its environment. If there is no effective relaxation, then the interaction should be negligible altogether. This argument is fallacious, however, since the relevant timescales may be extremely different for macroscopic objects (a numerical example is given in Table 3.3).

3.2.2.1 Quantum Brownian Motion. While scattering of a photon off a dust particle drastically changes the state of the photon, it hardly influences the state of the dust particle, even though recoil cannot be neglected completely. From thermodynamic considerations one would indeed expect an approach to equilibrium for sufficiently long times. This cannot be described by a model which neglects momentum conservation. The corresponding terms in the equation of motion for a massive object coupled to a "heat bath" have been derived and discussed by many authors. The resulting more general equations, usually discussed as "quantum Brownian motion", complete the picture presented so far. As can easily be shown, the "minimal decoherence" discussed in the previous section is by far the dominating effect for macroscopic objects (see Sect. 3.2.2.3 below). The approach to thermal equilibrium described by the generalized models is not relevant at all for most objects which we usually regard as "macroscopic". After all, the planet Jupiter has been moving around the sun perfectly decohered (i. e., in a well-defined position) for eons, without its orbit being affected by friction caused by photons. In this section we will therefore only give a brief overview of this vast field. We will discuss some important extensions, concentrating on their relation to the simpler equations derived above. More material on quantum Brownian motion can be found in Sect. 5.1.

In classical physics, the motion of a Brownian particle is described phenomenologically by a Langevin equation of the form

$$m\ddot{x} + \eta\dot{x} + V'(x) = F(t), \tag{3.96}$$

where m is the mass of the particle, η the damping constant describing the viscosity of the medium, V a potential and F the so-called "fluctuating force".

The "random" force F is assumed to have zero average,

$$\langle F(t)\rangle = 0. \tag{3.97}$$

In the standard model it also has no memory, that is, its correlation function can be written (fluctuation-dissipation relation) as

$$\langle F(t)F(t')\rangle = 2\eta k_B T\delta(t - t')$$
$$= 2m^2 D\delta(t - t'), \tag{3.98}$$

where we have introduced the diffusion constant D. Physically, the δ-function in (3.98) hides a fast but finite timescale. In this simplified treatment it is also assumed that higher order terms factorize into products of second order correlation functions. In the case of a δ-correlation, the spectral distribution of the fluctuating force is constant. Therefore it is called "white noise" (see, for example, Gardiner (1983) and App. A7). If the ensemble of all possible trajectories is considered, well-known procedures lead from the Langevin equation (3.96) to the Fokker–Planck equation (Risken 1984) for the probability distribution P of particle positions and momenta,

$$\frac{\partial P(x,p,t)}{\partial t} = -\frac{\partial}{\partial x}(pP) + \frac{\partial}{\partial p}(V'P) + \frac{\eta}{m}\frac{\partial}{\partial p}(pP) + D\frac{\partial^2}{\partial p^2}P. \tag{3.99}$$

(This form of the Fokker–Planck equation is sometimes called "Kramers equation".)

The standard quantization procedures cannot be applied to systems with friction, since no time-independent Hamiltonian or Lagrangian description exists. Essentially, two lines of approach can be found in the literature. The first one employs new quantization schemes (Dekker 1977), while the second one (the so-called system-reservoir approach) treats the particle as part of a larger quantum system which is quantized according to the usual rules (Ford, Kac and Mazur 1965). Of course, only the second approach can be considered as fundamental. For discussions on the relation of the two views see Senitzky (1960), Caldeira and Leggett (1983a) and Ford, Lewis and O'Connell (1988).

Most models consider a particle linearly coupled to a set of harmonic oscillators (or a linear field). The particle itself is usually also assumed to be bound by a harmonic potential (then it is called the "distinguished oscillator"). These linear models can (in principle) be solved exactly. For example, the integrals occurring in the mostly used Feynman–Vernon influence-function technique are purely Gaussian. An exposition of this calculational method (Feynman and Vernon 1963) is given in Sect. 5.1. The various models differ mainly in their assumptions about the distribution of energies and couplings to the oscillator bath, and the initial conditions.

Caldeira and Leggett considered coupling of an oscillator to a heat bath consisting of non-interacting oscillators described by the Hamiltonian

$$H = \frac{p^2}{2m} + V(x) + x\sum_k C_k R_k + \sum_k \left[\frac{P_k^2}{2M} + \frac{1}{2}M\omega_k^2 R_k^2\right]. \tag{3.100}$$

The first two terms describe the system of interest, the third one a linear coupling, while the last sum represents the Hamiltonian of the ensemble of oscillators. Since the coupling is proportional to the position operator of the particle, it has von Neumann's form (3.14) and therefore represents an ideal measurement of the *position* of the distinguished oscillator by the other oscillators. Position is only an exact "pointer basis" in the limit of infinite mass m, however (we will come back to this point in Sect. 3.3.3). The environment is assumed to be in a thermal state with temperature T. Under appropriate approximations the authors arrive at the equation of motion

$$i\frac{\partial\rho}{\partial t} = [H, \rho] + \frac{\eta}{2m}[x, \{p, \rho\}] - i\eta k_B T[x, [x, \rho]] \tag{3.101}$$

(Caldeira and Leggett 1983a, 1985, see also Agarwal 1971, Dattagupta 1984). Here we have written the coefficients in terms of the classical friction constant η appearing in (3.96). The last term obviously corresponds to the interference-reducing term in (3.71), while the second one on the rhs describes the classical friction force, which leads to damping of spatial motion. The simpler equation (3.71) can be obtained by taking the limit $T \to \infty$, $\eta \to 0$ while keeping the product $\Lambda = \eta k_B T$ fixed. This limit is appropriate if the localization timescale is much shorter than the relaxation timescale on which the system comes to equilibrium with its environment. In other words, this limit is reasonable if friction can be ignored. We will mainly discuss (3.101) in the position representation where it can be written for a harmonically bound particle as

$$i\frac{\partial\rho(x, x', t)}{\partial t} = \left[\frac{1}{2m}\left(\frac{\partial^2}{\partial x'^2} - \frac{\partial^2}{\partial x^2}\right) + \frac{m\omega^2}{2}\left(x^2 - x'^2\right) - i\Lambda(x - x')^2\right.$$
$$\left. + i\frac{\gamma}{2}(x - x')\left(\frac{\partial}{\partial x'} - \frac{\partial}{\partial x}\right)\right]\rho(x, x', t) \quad . \tag{3.102}$$

The relaxation constant γ is given by $\gamma = \eta/m$.[23] Solutions for (3.102) are constructed in App. A2.

The ensemble of oscillators coupled to the particle may be replaced by a scalar field. Then a similar model is obtained, described by the action

$$I = \int dt\, dx \left\{\frac{1}{2}\left[\dot{\Phi}^2 - \left(\frac{\partial\Phi}{\partial x}\right)^2\right] + \delta(x)\left[\frac{\dot{q}^2}{2} - \varepsilon q\dot{\Phi} - \frac{\Omega_0^2}{2}q^2\right]\right\} \tag{3.103}$$

(Unruh and Zurek 1989). Here the oscillator with bare frequency ω_0 is represented by the coordinate q, while the field Φ is assumed to be in thermal equilibrium. ε is the coupling constant. The field Φ acts like a measuring

---

[23] There are several different definitions of the "damping constant γ" in the literature, so some care has to be used in comparing equations from different sources. We will adhere in this section to the definitions given above, where γ is the relaxation rate for the spatial motion; hence the relaxation rate for energy is 2γ.

apparatus for the position of the oscillator. Due to the structure of the coupling, the field irreversibly carries away information about the position of the particle. This is equivalent to a scattering process as discussed earlier. The field therefore "remembers" the result of the position measurements.

The resulting master equation may be written in the form

$$i\frac{\partial\rho}{\partial t} = [H,\rho] + 2\gamma_{UZ}[q,\{p,\rho\}] - 4i\gamma_{UZ}mh(t,T)[q,[q,\rho]] - 4i\gamma_{UZ}f(t,T)[q,[p,\rho]]$$
(3.104)

(Gallis and Fleming 1990). Here the damping constant is $\gamma_{UZ} = \varepsilon^2/4$. The coefficients h and f depend on further details of the model, such as the cutoff-frequency, but within appropriate approximations the equation given by Caldeira and Leggett can be recovered. Unruh and Zurek discussed solutions of (3.104) in considerable detail. The destruction of coherence in the position representation and the corresponding entropy increase are similar to that shown in our figures below. Their model also clearly displays the approach to equilibrium for $t \to \infty$.

It is instructive to compare some solutions without friction with one of the more general models. For this purpose we will use the equation of motion (3.96), derived by Caldeira and Leggett and others. Quite generally, we expect an approach to equilibrium values for all physical quantities at large times. This can indeed be confirmed for the same situations which we already considered previously. We will thus again follow the evolution of an initially pure Gaussian wave packet of width b (see (3.74)).

Let us first consider the behavior of the (spatial) coherence length. As already pointed out, it decreases generally, because the environment effectively measures position. On the other hand, it should approach the thermal de Broglie wavelength when the system comes to equilibrium. In Fig. 3.8 we compare the coherence length for "pure decoherence" with its behavior when the system is also weakly damped. The following figures show on the left the results of Sect. 3.2.1.2 for comparison. The plots to the right follow from solutions of (3.102) with $\omega = 0$ and the same values for Λ as used before and an additional damping constant of $\gamma = 10^{-3}\text{s}^{-1}$. Of course, in the case $\Lambda = 0$ we have also chosen $\gamma = 0$ (no interaction with the environment).

As expected, for $t \to \infty$ the coherence length approaches a limit which is given by

$$l(t = \infty) = \frac{1}{2}\sqrt{\frac{\gamma}{\Lambda}}.$$
(3.105)

This value is of the order of the thermal de Broglie wavelength, since

$$l(t = \infty) = \frac{\lambda_{\text{th}}}{\sqrt{8\pi}}$$
(3.106)

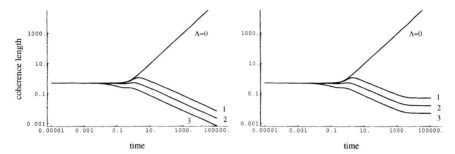

Fig. 3.8. Time dependence of coherence length. The left picture shows decoherence only (while friction can be ignored). On the right, the corresponding solutions including a small friction constant show the approach to equilibrium. The coherence length finally settles to a value of the order of the thermal de Broglie wavelength of the measured object. The curves labelled 1,2,3 correspond to the same values of Λ as in Figs. 3.3 to 3.6. In the plots on the rhs with $\Lambda \neq 0$ we have chosen $\gamma = 10^{-3}\text{s}^{-1}$.

with[24]

$$\lambda_{\text{th}} = \sqrt{\frac{2\pi}{mk_{\text{B}}T}}. \tag{3.107}$$

The ensemble width also shows a different behavior when friction is included. Asymptotically it scales as

$$\Delta x \overset{t \to \infty}{\longrightarrow} \frac{\sqrt{2\Lambda}}{m\gamma} \cdot \sqrt{t} \tag{3.108}$$

in this model. The behavior at long times depends very sensitively on the chosen model; some authors find a logarithmic time dependence, for example. The result (3.108) is the typical behavior of simple diffusion processes. It coincides with the result of classical Brownian motion with (spatial) diffusion constant $D_x = \frac{2\Lambda}{m^2\gamma^2} = \frac{2k_{\text{B}}T}{m\gamma}$.

The momentum coherence length obviously must approach zero (as is also the case for pure decoherence), since the density matrix for a canonical ensemble is diagonal in energy (momentum). This is displayed in Fig. 3.10.

For large times it vanishes according to

$$l_p \overset{t \to \infty}{\longrightarrow} \frac{m\gamma}{\sqrt{8\Lambda t}} \tag{3.109}$$

This behavior must not be misinterpreted (Venugopalan 1994, 1998) to mean that the pointer basis is now given by momentum instead of position. As already emphasized, decoherence in position is by far the dominating effect for macroscopic objects far from equilibrium (see also Sect. 3.2.2.3 below).

---

[24] This common definition of the thermal de Broglie wavelength differs from the de Broglie wavelength $\lambda_{\text{dB}} = h/p$ of a particle with energy $E = \frac{1}{2}k_{\text{B}}T$ by a factor $\sqrt{2\pi}$.

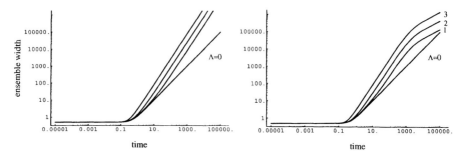

Fig. 3.9. Comparison of the time dependence of the ensemble width.

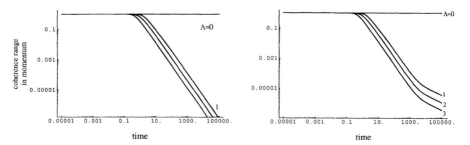

Fig. 3.10. Comparison of the time dependence of coherence range in the momentum representation. Since in equilibrium the system is described by a canonical density matrix which is diagonal in energy, hence also momentum, this quantity approaches zero in both cases.

Entropy increases monotonically in general (except for zero coupling)[25]. The analytical expression for the logarithmic entropy looks cumbersome, so we do not give it here (cf. (A2.38) in App. A2), but only display its behavior in Fig. 3.11.

The *linear* entropy approaches its maximum value of unity according to

$$S_{\text{lin}} \xrightarrow{t \to \infty} 1 - \frac{m\gamma^2}{\sqrt{8\Lambda^3 t}}. \tag{3.110}$$

---

[25] The equilibrium state is not always the state of maximal entropy. For example, a harmonic oscillator immersed in a heat bath of very low temperature will always relax to the ground state (with zero entropy), although for finite times the local entropy can assume large values, depending on the initial state (which may also be pure, for example, a "Schrödinger cat" state, cf. (3.278)). Phase relations in the initial state are then completely transferred into a complicated (but also pure) state of the environment. For a comparison of the oscillator model with black hole evaporation see Anglin *et al.* (1995).

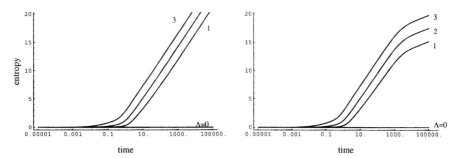

Fig. 3.11. Comparison of the time dependence of the logarithmic (von Neumann) entropy. The behavior of linear entropy looks very similar for the two situations compared here (see Fig. 3.3).

There exist, however, unphysical solutions of (3.102). If we express the linear entropy in terms of the Gaussian ansatz (3.73),

$$S_{\text{lin}} = \text{tr}\left(\rho - \rho^2\right) = 1 - \sqrt{\frac{C}{A}}, \tag{3.111}$$

its time derivative can be found as (see App. A2)

$$\frac{d}{dt}S_{\text{lin}} = \frac{1}{2}\sqrt{\frac{C}{A^3}}(\Lambda - 2\gamma A). \tag{3.112}$$

If the initial state is described by a wave packet of width b, we have $A = C = \frac{1}{8b^2}$, hence the entropy can assume negative values for small times if γ is too large or if b is too small (more precisely, if the width b is smaller than the thermal de Broglie wavelength of the object). This pathological behavior was observed by several authors (see, for example, Ambegaokar 1991, Hu, Paz and Zhang 1992). It does not occur in more sophisticated treatments, where the friction coefficient is time-dependent and initially small (clearly the very notion of friction makes only sense after many scattering events have occurred so that the system has already approached equilibrium to some extent). In some treatments a double commutator proportional to $[p, [p, \rho]]$ is added to (3.101), thereby describing decoherence in momentum. The resulting master equation is then of the Lindblad type (3.241) (hence assures positivity) if this additional term is of comparable order of magnitude (Diósi 1993a), see also Munro and Gardiner (1996).

Finally, the mean energy, for a free particle given by the spread in momentum, does not increase indefinitely as in the case of recoil-free decoherence, but approaches the value demanded by the equipartition principle.

It is given by

$$(\Delta p)^2 = \langle 2mE \rangle = \frac{e^{-2\gamma t}}{4b^2} + \frac{\Lambda}{\gamma}\left(1 - e^{-2\gamma t}\right) \overset{t \to \infty}{\longrightarrow} \frac{\Lambda}{\gamma} = 2m \times \frac{1}{2}k_B T. \tag{3.113}$$

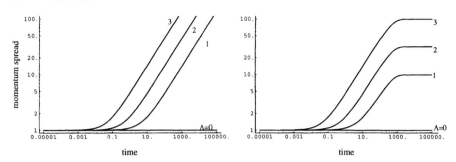

Fig. 3.12. Comparison of the time dependence of the dispersion of momentum which is proportional to the mean energy in the considered case of a "free" particle. Without recoil it increases linearly in time, since then the environment corresponds to a reservoir of infinite temperature. With damping included, the mean energy approaches its equilibrium value for large times.

Most treatments of quantum Brownian motion and other master equations assume a factorizing initial state of the form

$$\rho = \rho_{\text{system}} \otimes \rho_{\text{environment}}, \tag{3.114}$$

as we have also done in our previous discussions. This does not always correspond to the experimental situation, however. For example, through an incomplete position measurement with outcome $q_0 - \delta/2 < q < q_0 + \delta/2$ the system density matrix is changed according to the von Neumann/Lüders rules,

$$\rho \to P_q \rho P_q, \tag{3.115}$$

where

$$P_q = \frac{1}{\sqrt{N}} \int_{q_0 - \delta/2}^{q_0 + \delta/2} dq \, |q\rangle \langle q| \tag{3.116}$$

projects on the measured interval δ (N is a normalization factor). The new density matrix may then still contain correlations between the two systems. The dynamics resulting from such more general initial states has also been studied extensively. For a review see Grabert, Schramm and Ingold (1988).

In this overview we confined ourselves to the most fundamental and basic models for quantum Brownian motion, which allowed us to discuss the relation to ideal decoherence in the simplest possible way. There exists a vast literature on this subject, extending the basic models in various directions. For important work see for example Hu, Paz and Zhang (1992), Gallis (1992), Hu, Paz and Zhang (1993), Gallis (1993), Krzywicki (1993), Paz, Habib and Zurek (1993), Hu and Matacz (1994), Vacchini (2001). For an extension of the scalar field model (including a spin variable) to estimate the sensitivity of quantum computers to the coupling to their environment see Unruh (1995). An application to a Stern-Gerlach measurement model has been discussed by Venugopalan, Kumar and Ghosh (1995a).

3.2.2.2 Ehrenfest Theorems. The Ehrenfest relations express a formal connection between the time dependence of mean values of canonically conjugate variables and the Hamilton equations of classical mechanics. Although mean values are of course insufficient to derive classical behavior from quantum mechanics, the validity of the Ehrenfest relations is an important requirement for a partial derivation of classical physics. If the system is governed by a Schrödinger equation with Hamiltonian

$$H = \frac{p^2}{2m} + V(x),\qquad(3.117)$$

the mean values for position, momentum and energy obey the relations

$$\frac{d}{dt}\langle x\rangle = \frac{\langle p\rangle}{m},\qquad(3.118)$$

$$\frac{d}{dt}\langle p\rangle = -\left\langle \frac{d}{dx}V(x)\right\rangle,\quad\text{and}\qquad(3.119)$$

$$\frac{d}{dt}\langle H\rangle = 0.\qquad(3.120)$$

Note that in (3.119) $\left\langle \frac{d}{dx}V(x)\right\rangle$ is different from $\frac{d}{dx}V(\langle x\rangle)$, except for the case of a harmonic oscillator. Smoothness of the potential, more precisely, a weak position dependence of the de Broglie wavelength, is an important ingredient in the WKB approximation. The applicability of this method, corresponding to the use of rays in optics, does of course not explain classical motion. Since in realistic situations the system no longer obeys a Schrödinger equation, the Ehrenfest relations are not necessarily valid anymore. Even in the case of an ideal (passive) measurement by the environment, the average behavior of the system can change drastically – it may even become frozen (see the section on the quantum Zeno effect below). If the system is immersed in a heat bath, as is assumed in the models of quantum Brownian motion, energy will not be conserved and the motion will be damped by friction. All this can rather easily be seen by re-evaluating the time dependence of expectation values of physically interesting quantities according to

$$\frac{d\langle O\rangle}{dt} = \frac{d}{dt}\mathrm{tr}(O\rho) = \mathrm{tr}(O\frac{\partial\rho}{\partial t})\qquad(3.121)$$

for a time-independent observable O.

For the minimal case (3.71),

$$i\frac{\partial\rho}{\partial t} = \left[\frac{p^2}{2m} + V, \rho\right] - i\Lambda[x,[x,\rho]],\qquad(3.122)$$

the result is (Joos and Zeh 1985, Gallis and Fleming 1991)

$$\frac{d}{dt}\langle x \rangle = \frac{\langle p \rangle}{m}, \tag{3.123}$$

$$\frac{d}{dt}\langle p \rangle = -\left\langle \frac{d}{dx}V(x) \right\rangle, \quad \text{and} \tag{3.124}$$

$$\frac{d}{dt}\langle H \rangle = \frac{\Lambda}{m} \tag{3.125}$$

The Ehrenfest relations are thus maintained, except for an increase of mean energy at a constant rate, independent of the state of the system.

Generalizations of (3.122) with a non-unitary term of the form

$$\frac{\partial}{\partial t}\rho(x, x', t)\bigg|_{\text{non-unitary}} = -F(x - x')\rho(x, x', t) \tag{3.126}$$

with a positive and even function F lead to the same equations, with the energy gain now given by

$$\frac{d}{dt}\langle H \rangle = \frac{F''(0)}{m}. \tag{3.127}$$

Hence all models describing a passive recording (ideal measurement) of position by the environment share this increase in mean energy (Ballentine 1991a). The role of this process can clearly be seen by considering the more general model of Caldeira and Leggett which includes recoil. The equation of motion (3.101),

$$i\frac{\partial}{\partial t}\rho = [H, \rho] + \frac{\eta}{2m}[x, \{p, \rho\}] - i\eta k_B T[x, [x, \rho]] \tag{3.128}$$

yields the new Ehrenfest relations

$$\frac{d}{dt}\langle x \rangle = \frac{\langle p \rangle}{m}, \tag{3.129}$$

$$\frac{d}{dt}\langle p \rangle = -\left\langle \frac{d}{dx}V(x) \right\rangle - \frac{\eta}{m}\langle p \rangle, \quad \text{and} \tag{3.130}$$

$$\frac{d}{dt}\langle H \rangle = \frac{2\eta}{m}\left[\frac{k_B T}{2} - \left\langle \frac{p^2}{2m} \right\rangle\right]. \tag{3.131}$$

Equation (3.130) demonstrates clearly the role of the additional friction term, while energy is approaching thermal equilibrium in (3.131).

As mentioned above, the replacement of $\left\langle \frac{d}{dx}V(x) \right\rangle$ by $\frac{d}{dx}V(\langle x \rangle)$ is substantial for the validity of semiclassical approximations. For chaotic systems, this approximation turns into a poor one very rapidly, even when starting with a small wave packet. We will discuss chaotic motion in Sect. 3.3.4.

3.2.2.3 Decoherence Versus Friction. For nearly all macroscopic objects, friction does not play any role on timescales on which decoherence is already important. How are these two effects related to each other? It is straightforward to compare the two relevant timescales from the equation of motion (3.101) given by Caldeira and Leggett, where the non-unitary contributions are ($\gamma = \eta/m$)

$$i\frac{\partial}{\partial t}\rho\bigg|_{non-unitary} = \frac{\gamma}{2}[x, \{p, \rho\}] - im\gamma k_B T[x, [x, p]]. \tag{3.132}$$

The second term, describing decoherence without recoil, leads to the suppression of coherence over a distance $\Delta x = x - x'$ according to (3.68),

$$\rho(x, x', t) = \rho(x, x', 0)\exp\left[-m\gamma k_B T (x - x')^2 t\right]. \tag{3.133}$$

This defines the *decoherence timescale*

$$t_{dec} = \frac{1}{m\gamma k_B T(\Delta x)^2} = \frac{1}{\Lambda(\Delta x)^2}, \tag{3.134}$$

while effects of friction are characterized by

$$t_{friction} = \frac{1}{\gamma} \tag{3.135}$$

(see (3.130) above). The ratio of the two rates is therefore

$$\frac{\text{decoherence rate}}{\text{relaxation rate}} = mk_B T(\Delta x)^2 \sim \left(\frac{\Delta x}{\lambda_{th}}\right)^2. \tag{3.136}$$

Under usual circumstances, this ratio is enormous, because it relates the range of coherence with the value of the thermal de Broglie wavelength of the *object*. For a typical macroscopic situation, e. g. $m = 1\,g$, $T = 300\,K$, $\Delta x = 1\,cm$, a value of $\sim 10^{40}$ results (Zurek 1986 , Caldeira and Leggett 1985). This shows that classical *reversible* dynamics is quite compatible with strong *irreversible* interaction with the environment.

As an example where friction plays a role, consider a particle track in a bubble chamber. If the particle's momentum is damped by a small amount, corresponding to $\gamma t = 0.05$, say, the coherence length for this situation turns out to be about $10^{-9}\,m$ for a neutron at $T = 100\,K$. This is much less than the width of a bubble track (Mott 1929, Savage and Walls 1985b).

The relation between the timescales of decoherence and relaxation is of a quite general nature. The decoherence timescale is always given by the relaxation timescale *times* a factor which measures the "distance" between the two macroscopically distinct states. This distance is here given by $\left(\frac{\Delta x}{\lambda_{th}}\right)^2$, in the case of quantum optical "Schrödinger cat" states it is proportional to the square of the mean photon number (Milburn and Walls 1988). An example is given at the end of Sect. 3.3.3.1. This quadratic parameter-dependence of the ratio of decoherence to dissipation timescales follows from rather general considerations in the limit of slow (negligible) system dynamics (Braun, Haake and Strunz 2001). Compare also the discussion of the damped oscillator in Sect. 3.3.2.2.

An interesting analogy is found in the case of the kicked rotor, governed by a Hamiltonian of the form $H(p, q, t) = \frac{p^2}{2} - \frac{K}{4\pi^2} \cos(2\pi q) \sum\limits_{n=-\infty}^{\infty} \delta(t - n)$. This is one of the standard examples for investigations of "quantum chaos". The classical model shows diffusive spreading of momentum in the chaotic regime, i. e. $\langle p^2(t) \rangle = \langle p^2(0) \rangle + D(K) t$, while in the quantum case the initial increase in $\langle p^2 \rangle$ is limited to a finite range, then replaced by a quasiperiodic motion. The latter is understood as a coherence effect similar to Anderson localization in disordered solids. If quantum coherence is destroyed, it is found that this localization breaks down even for very weak noise or weak coupling to a heat bath or measurement device. On the other hand, the diffusion constant remains largely unaffected, if the coupling is not too strong. The kicked rotator is discussed in more detail in Sect. 3.3.4.

We close our review of quantum Brownian motion with a comparison of the physical constants which appeared in the various equations we discussed so far. Table 3.3 summarizes the basic parameters in the dynamical equations together with their relationships. Again the importance of decoherence, as well as the relative unimportance of friction can be seen from the numerical values given here for a dust particle interacting only with the 3 K background radiation. The interaction with the 3 K photons rapidly transfers enough information about the position of the dust particle to the radiation field to assure destruction of any coherence between macroscopically different positions. On the other hand, friction would slow down the motion of the dust grain only on timescales greater than the age of the universe.

3.2.3 Wigner Function Description of Decoherence

The Wigner function is a commonly used representation for pure or mixed states of a quantum system described by a continuous degree of freedom.[26] In the standard case of the quantum mechanics of a particle it is related to the position representation of the density matrix ρ by the definition

$$W(x, p) = \frac{1}{\pi \hbar} \int_{-\infty}^{\infty} dy \, e^{2ipy/\hbar} \rho(x - y, x + y) \qquad (3.137)$$

---

[26] For an extension to discrete systems see Wootters (1987), compare also (2.14) on page 36. For the case of a harmonic oscillator many other representations are found useful, particularly in quantum optics applications (see, for example, Walls and Milburn 1994, Scully and Zubairy 1997).

Table 3.3. Some constants appearing in the simplest forms of the equation of motion of a massive object interacting with its environment.

Quantity	Physical meaning	Important relations	Numerical example: 10μm dust particle in intergalactic space
Localization rate Λ	Destruction of coherence between different positions	$\rho(x, x', t) =$ $\rho(0)e^{-\Lambda t(x-x')^2}$ $\Lambda = \eta k_B T$	$\Lambda = 10^6 \frac{1}{\text{cm}^2 \text{s}}$
Relaxation rate γ	Relaxation of kinetic energy to equilibrium	$\gamma = \frac{\eta}{m}$ see also (3.131) $\frac{\gamma}{\Lambda} = \frac{1}{mk_B T} = \frac{\lambda_{th}^2}{2\pi}$	$\gamma = 10^{-25} 1/s$ $\approx 1/(10^{16} \text{years})$
Viscosity η	Damping of motion	$\eta = m\gamma$, see (3.96)	
Thermal de Broglie wavelength λ_{th}	Wavelength related to thermal energy	$\lambda_{\text{th}} = \sqrt{\frac{2\pi}{mk_B T}}$	$\lambda = 10^{-15}$ cm for $m = 10^{-8}$ g and $T = 3K$
Diffusion constant D	Definition depends on situation. Used here: diffusion in momentum	$D = \frac{\eta k_B T}{m^2} = \frac{\Lambda}{m^2}$	$D = 10^{22} \frac{1}{\text{cm}^2 \text{s} \text{g}^2}$

(Wigner 1932, Hillery *et al.* 1984).[27] In the special case of a pure state this relation reads

$$W_{\text{pure}}(x, p) = \frac{1}{\pi\hbar} \int_{-\infty}^{\infty} dy\, e^{2ipy/\hbar} \psi^*(x + y)\psi(x - y) \tag{3.138}$$

(we inserted Planck's constant in the last two equations.). The generalization of these definitions to many degrees of freedom is straightforward.

We introduce the Wigner function with some hesitation, since it is often misinterpreted as representing a "phase space distribution", similar to a probability distribution of classical particles. Clearly such an interpretation is wrong. The simplest argument is that the Wigner function can assume negative values, hence it cannot represent a probability distribution. There is simply *no phase space in quantum mechanics.*[28] Nevertheless, the Wigner function is often a useful tool[29] in decoherence studies for investigating *correlations* between position and momentum if a system is continuously measured

---

[27] Analogously, one can associate a phase space function $A(x, p)$ with any (Weyl-ordered) operator $\hat{A}(\hat{x}, \hat{p})$.

[28] The term *phase space* (as parameter space of the Wigner function) is here meant as it is commonly used in classical mechanics. A more general notion is used in Chap. 6.

[29] For a discussion of squeezed state properties employing the concept of "interference in phase space" see Schleich and Wheeler (1987), Schleich, Walls and

by its environment. There exist many other (pseudo-)distribution functions in quantum mechanics[30], but one can show that the Wigner function follows uniquely if certain appropriate assumptions are made. Some of these are

(1) $W(x,p)$ is real. As mentioned above, it can, however, become negative. A theorem due to Hudson states that the only pure state which leads to a strictly positive Wigner function is a Gaussian. As we shall show below, however, decoherence can smoothen the oscillations in W. An extreme example is a canonical ensemble of oscillators, where W is manifestly positive.

(2) The probability distributions $\rho(x,x)$ for position and $\tilde{\rho}(p,p)$ for momentum can easily be recovered from W according to

$$\int dp\, W(x,p) = \rho(x,x), \qquad (3.139)$$

$$\int dx\, W(x,p) = \tilde{\rho}(p,p), \qquad (3.140)$$

hence the normalization of W reads

$$\int dx\, dp\, W(x,p) = \operatorname{tr} \rho = 1. \qquad (3.141)$$

(3) The overlap of two Wigner functions corresponding to two pure states Φ and Ψ, respectively, is given by

$$2\pi \int dx\, dp\, W_\Psi(x,p) W_\Phi(x,p) = \left| \int dx\, \Psi^*(x)\Phi(x) \right|^2. \qquad (3.142)$$

If the two states are orthogonal, the right hand side of (3.142) is zero, and one can see immediately that W cannot be positive definite.

(4) If part of the degrees of freedom are integrated over in the Wigner function, one finds the Wigner function which is associated with the reduced density matrix corresponding to the remaining degrees of freedom. Again, the reader should be aware of the interpretational problems involved in this operation, cf. footnote 2 on page 43.

If the system follows a Schrödinger or von Neumann equation with potential $V(x)$, the time-dependence of the Wigner function is given by the so-called "Moyal bracket" $\{H,W\}_{MB}$,

$$
\begin{aligned}
\frac{\partial}{\partial t} W(x,p,t) &= \{H,W\}_{MB} \\
&= -i \sin(i\hbar\{H,W\}_{PB})/\hbar \qquad (3.143) \\
&= \{H,W\}_{PB} + \sum_{n\geq 1} \frac{\hbar^{2n}(-1)^n}{2^{2n}(2n+1)!} \partial_x^{2n+1} V(x) \partial_p^{2n+1} W(x,p),
\end{aligned}
$$

Wheeler (1988). An extensive review of phase space methods is provided by Schleich (2001).

[30] The Husimi distribution, defined as expectation value for coherent states, is even more misleading since it is manifestly positive.

The Moyal bracket coincides with the Poisson bracket $\{H, W\}_{PB}$ only for the special case of at most quadratic potentials. Consider the simple case of a one-dimensional harmonic oscillator. The Wigner function corresponding to the energy eigenstates reads

$$W_{\text{osc}}(x, p) = \frac{(-1)^n}{\pi} e^{-2H/\omega} L_n(4H/\omega), \qquad (3.144)$$

where L_n are the Laguerre polynomials and $H = p^2/2m + m\omega^2 x^2/2$ is the Hamiltonian. Note that the functions (3.144) are *not* peaked around a "classical orbit" in phase space (see also Fig. 3.14). The most localized Wigner function is obtained for a coherent state

$$\psi_{\text{coh}}(x) = \left(\frac{\omega}{\pi}\right)^{1/4} e^{-\omega(x-x_0)^2/2 + ip_0 x}, \qquad (3.145)$$

and reads

$$W_{\text{coh}}(x, p) = \frac{1}{\pi} e^{-\omega(x-x_0)^2 - (p-p_0)^2/\omega}. \qquad (3.146)$$

In the following we will discuss the effect of decoherence in terms of the Wigner function for two situations. The first one is the suppression of coherence for two spatially extended wave packets, the second one is decoherence for wave functions approximated by WKB states.

Let us first consider the same states that we have already used for illustration in Sect. 3.2.1.1, a superposition of two Gaussian wave packets described by

$$\psi(x) = c_1 e^{-(x-a)^2/(4b^2)} + c_2 e^{-(x+a)^2/(4b^2)}. \qquad (3.147)$$

(cf. (3.70 on page 67)). If $a \gg b$ we have the typical non-classical situation displayed in Fig. 3.1. Instead of the four peaks in the position representation, we now find in the Wigner function description two peaks at the classically expected phase space positions (in our example at $p = 0$ and $x = \pm a$) plus an interference term centered around $x = 0$, see Fig. 3.13a. The latter is typical for non-Gaussian states (more precisely, for all states which are not a *single* Gaussian). Note that the ripples displayed in Fig. 3.13a have a frequency proportional to the distance a between the two peaks of the wavefunction. A similar pattern would arise from two wave packets separated in momentum space. A general superposition could then show a very fine-grained interference pattern with structures on a sub-Planck scale $\Delta x \Delta p < \hbar$. Such states would be very sensitive to small shifts (Zurek 2001).

The effect of decoherence, the suppression of *spatial* interference according to (3.68),

$$\rho(x, x', t) = \rho(x, x', 0) \exp\left\{-\Lambda t(x - x')^2\right\}, \qquad (3.148)$$

leads for the Wigner function corresponding to the initial state (3.147) to the time dependence

$$
W(x,p,t) = \sqrt{\frac{2b^2}{\pi\,(1+8b^2\Lambda t)}} \left\{ |c_1|^2 \exp\left(-\frac{(x-a)^2}{2b^2} - \frac{2b^2p^2}{1+8b^2\Lambda t} \right) \right.
$$

$$
+ |c_2|^2 \exp\left(-\frac{(x+a)^2}{2b^2} - \frac{2b^2p^2}{1+8b^2\Lambda t} \right)
$$

$$
+ 2\exp\left(\frac{-x^2}{2b^2} - \frac{2b^2p^2 + 4a^2\Lambda t}{1+8b^2\Lambda t} \right)
$$

$$
\left. \times \left(\mathrm{Re}\,(c_1 c_2^*) \cos\frac{2ap}{1+8b^2\Lambda t} + \mathrm{Im}\,(c_1 c_2^*) \sin\frac{2ap}{1+8b^2\Lambda t} \right) \right\} \tag{3.149}
$$

The oscillating terms in the last line of (3.149) display the non-classical character of this state. Figure 3.13 shows the same situation as Fig. 3.1: Interference between different superpositions is destroyed by decoherence. In the Wigner function picture, the oscillating terms are damped. If the measurement by the environment destroys coherence between the two well-separated ($a \gg b$) packets, we have $a^2 \gg 1/\Lambda t \gg b^2$ (this is the case displayed in the figures), hence the size of the oscillating terms in (3.149) is given by the expression $\exp\left\{ - \left(\frac{x^2}{2b^2} + 2b^2p^2 \right) \right\} \times \exp\left[-4a^2\Lambda t \right]$.

The damping of oscillations in the Wigner function is a general consequence of the non-unitary dynamics, as can be seen by writing the equation of motion in the Wigner function representation. The time dependence of (3.148)

$$
\frac{\partial}{\partial t}\rho(x,x',t) = -\Lambda(x-x')^2\rho \tag{3.150}
$$

translates into

$$
\frac{\partial}{\partial t}W(x,p,t) = \Lambda\frac{\partial^2 W}{\partial p^2}. \tag{3.151}
$$

This equation describes a *diffusion* in momentum. We emphasize again, that an interpretation of this dynamics as resulting from random kicks (as in classical theory) would mistakenly intermingle classical and quantum concepts (cf. also our discussion of stochastic forces in Sect. 3.4.3). The time-dependence (3.148) corresponds to the well-known solution of the diffusion equation (3.151),

$$
W(x,p,t) = \frac{1}{\sqrt{4\pi\Lambda t}}\int_{-\infty}^{\infty} dp'\, W(x,p',0)\exp\left[-\frac{(p-p')^2}{4\Lambda t} \right]. \tag{3.152}
$$

(This equation is analogous to (3.400) on page 179). It can be shown that these solutions are strictly positive after a typical decoherence timescale of the order $\sqrt{\frac{m}{\Lambda}}$ (Diósi and Kiefer 2002). In the more general case (3.101) the

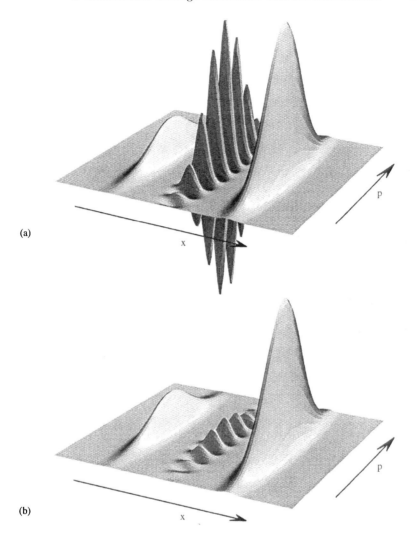

Fig. 3.13. The Wigner function equivalent to the density matrices shown in Fig. 3.1. (a) represents the superposition of two Gaussian wave packets. Strong oscillations together with negative values indicate the coherence between the two wave packets. In (b) these oscillations are partially damped by decoherence (compare also Fig. 2 in Paz, Habib and Zurek 1993). A similar behavior can be found for field amplitudes (cf. Sect. 3.3.3.1) in quantum optical experiments, compare Fig. 1 in Haroche *et al.* (1993).

equation of motion assumes the Fokker–Planck *form*

$$\frac{\partial}{\partial t} W(x, p, t) = \left(-\frac{p}{m} \frac{\partial}{\partial x} + m\omega^2 x \frac{\partial}{\partial p} + \Lambda \frac{\partial^2}{\partial p^2} + \gamma \frac{\partial}{\partial p} p \right) W \qquad (3.153)$$

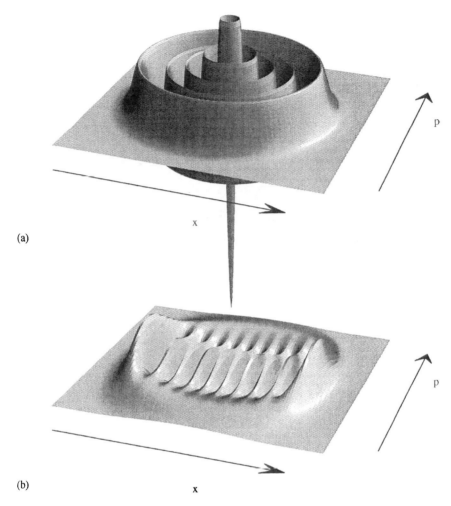

Fig. 3.14. Wigner function for a harmonic oscillator eigenstate with $n = 9$. (**a**) The pure state shows strong oscillations indicating its non-classical character as well as the symmetry between position and momentum. (**b**) Decoherence acts like diffusion in the momentum variable. Note, that decoherence does *not* lead to thermal equilibrium. Such a state would be described by a single Gaussian centered at the origin.

(see also (3.337) on page 158).

Figure 3.14 shows the effect of decoherence for an energy eigenstate of a harmonic oscillator, analogous to Fig. 3.2. The symmetry between position and momentum, typical for a harmonic oscillator, is broken by the coupling to the environment (compare, however, the discussion in Sect. 3.3.3.1).

As already mentioned, in some models the evolution of the density matrix governed by master equations is replaced by a stochastic process. Correspondingly, the time dependence of the density matrix or Wigner function displays a different behavior; only after averaging over an ensemble of stochastic realizations is the usual description reproduced. (Compare, for example, our Figs. 3.1, 3.2, 3.13, and 3.14 with the figures shown in Garraway, Knight and Steinbach (1995), in the latter case, however, calculated for a *damped* oscillator obeying (3.251)).

For a detailed discussion of solutions of (3.153) and generalizations thereof, see Paz, Habib and Zurek (1993).

Another interesting case is the Wigner function corresponding to a WKB state, i. e. a state of the form

$$\psi(x) = C(x)e^{iS(x)}, \tag{3.154}$$

where $C(x)$ is assumed to be slowly varying compared to $S(x)$. This separation of scales in $\psi(x)$ is achieved by expansion in a power series with respect to a small parameter. For the latter, one often uses Planck's constant, although a rigorous analysis of course would require a dimensionless quantity. For simplicity we will use the common language in the following.

Naively, one might expect a result like $|C|^2\delta\left(p - \frac{dS}{dx}\right)$ for the Wigner function. This has in fact been claimed in the literature, see e. g. Halliwell (1987) and the criticism in Anderson (1990). This is, however, not correct (apart from simple cases, such as the free particle where $W \sim \delta(p-p_0)$). The reason is that one must take care when interchanging the integration in (3.138) with the limit $\hbar \to 0$ (as the wave function W is not analytical at $\hbar = 0$). The correct expression was given by Berry (1977), see also the discussion in Habib (1990). It contains the Airy function, i. e., it is an oscillating function, with its peak shifted from the classical trajectory by an amount which vanishes for $\hbar \to 0$. The non-positivity of the WKB Wigner function is easily interpreted: the usual WKB state is not localized in configuration space and the WKB form is therefore not yet sufficient to ensure classical behavior.

As already shown above, the coupling to other degrees of freedom can lead to a smoothing of the Wigner function. Thus, a "decohered WKB state is localized".

Consider the simple two-dimensional example

$$\psi(x,y) = C(x)e^{iS(x)}\chi(x,y) \tag{3.155}$$

where χ is a Gaussian,

$$\chi(x,y) = \left(\frac{\mathrm{Re}\big(\Omega(x)\big)}{\pi}\right)^{1/4} \exp\left[-\frac{\Omega(x)}{2}y^2 + if(x)y\right] \tag{3.156}$$

with f real. A state of this form can result, for example, from a Born–Oppenheimer approximation where x is a "slow variable", and y is a "fast" variable. Such states have been frequently employed in quantum cosmology, and we shall also use them in Sect. 4.2.

The reduced density matrix corresponding to the variable x is obtained in the usual way by tracing out the y degree of freedom,

$$
\rho(x, x') = \int_{-\infty}^{\infty} dy \, \psi^*(x', y)\psi(x, y)
$$

$$
= \rho_0(x, x') \int_{-\infty}^{\infty} dy \, \chi^*(x', y)\chi(x, y), \qquad (3.157)
$$

where $\rho_0(x, x') = C(x')C(x)\exp\left[i\left(S(x) - S(x')\right)\right]$. Using the explicit expression above, one finds

$$
\rho(x, x') = \rho_0(x, x') \left(\frac{4\mathrm{Re}\left(\Omega(x')\right)\mathrm{Re}\left(\Omega(x)\right)}{[\Omega^*(x') + \Omega(x)]^2} \right)^{1/4} \exp\left(-\frac{[f(x) - f(x')]^2}{2\left[\Omega^*(x') + \Omega(x)\right]} \right).
$$
$$(3.158)$$

Note that the probabilities are not changed, i. e. $\rho(x, x) = \rho_0(x, x)$, that is, the y degree of freedom plays the role of a measuring device performing an ideal measurement of x (cf. (3.64)). The second factor can be written in an exponential form by expanding x and x' around $\bar{x} = (x+x')/2$ and neglecting terms of higher order than $(x - x')^2$. The result reads

$$
\left(\frac{4\mathrm{Re}\left(\Omega(x')\right)\mathrm{Re}\left(\Omega(x)\right)}{[\Omega^*(x') + \Omega(x)]^2} \right)^{1/4} \approx \exp\left(\frac{-i\,\mathrm{Im}(\Omega')}{4\mathrm{Re}(\Omega)}(x - x') - \frac{|\Omega'|^2}{16\mathrm{Re}(\Omega)^2}(x - x')^2 \right),
$$
$$(3.159)$$

where $\Omega' \equiv d\Omega/dx$. Equation (3.159) is exact up to second derivatives in Ω (which have cancelled in this expression). The effect of decoherence is manifest in the remaining coherence length

$$
\sigma_{\mathrm{dec}}^2 = \frac{8\mathrm{Re}(\Omega)^2}{|\Omega'|^2}. \qquad (3.160)
$$

In the following we put $f = 0$ for simplicity.

The Wigner function corresponding to $\rho(x, x')$ reads

$$
W(x, p) = \frac{1}{\pi} \int dy \, C(x + y)C(x - y)e^{i[S(x-y) - S(x+y) + 2py]}
$$

$$
\times \exp\left(\frac{i\,\mathrm{Im}(\Omega')}{2\mathrm{Re}(\Omega)}y - \frac{|\Omega'|^2}{4\mathrm{Re}(\Omega)^2}y^2 \right). \qquad (3.161)
$$

If we evaluate this integral in a saddle point approximation we find

$$
W(x, p) \approx \frac{2C^2}{\sqrt{\pi}} \left| \frac{\mathrm{Re}(\Omega)}{\Omega'} \right| \exp\left(-\frac{4\mathrm{Re}(\Omega)^2}{|\Omega'|^2} \left[p - S' + \frac{\mathrm{Im}(\Omega')}{4\,\mathrm{Re}(\Omega)} \right]^2 \right). \qquad (3.162)
$$

Several observations can be made here.

(1) The oscillations in W have been damped. The Wigner function (3.162) is manifestly positive (cf. Diósi and Kiefer 2002).

(2) One recognizes from (3.161) a *correlation* between p and x:

$$p \approx S' - \frac{\mathrm{Im}(\Omega')}{4\,\mathrm{Re}(\Omega)},\tag{3.163}$$

i. e. the mean momentum is shifted from its WKB expression $S' \equiv dS/dx$ by a term which expresses the *back reaction* of the "traced out" degree of freedom, y. Such a term was averaged over in our first treatment of localization in Sect. 3.2.1.1, where we assumed an *isotropic* environment (no "radiation pressure", see also App. A1).

(3) The width of the Gaussian in (3.162), expressing the degree of correlation between x and p, is given by

$$\sigma^2_{\mathrm{corr}} = \frac{|\Omega'|^2}{8\,\mathrm{Re}(\Omega)^2}.\tag{3.164}$$

Upon comparing this with (3.160) one finds the "decoherence–correlation uncertainty relation"

$$\sigma_{\mathrm{dec}}\sigma_{\mathrm{corr}} = 1.\tag{3.165}$$

To ensure classical behavior, therefore, a certain compromise between decoherence and correlations must be achieved (see also the discussion in Chap. 5).

The above expression (3.159) for the decoherence factor is valid only in regions of small $x - x'$. However, if many degrees of freedom are traced out, the corresponding overlap of environmental states will become small for larger $x - x'$ and the expression (3.158) is realistic for all arguments. This is similar to the cumulative effect of many individually negligible scattering processes in Sect. 3.2.1.1.

3.2.4 Molecular Structure

Most molecules are considered by chemists to exist in certain well-defined "configurations", that is, with their atomic nuclei localized "classically" in certain relative positions. These positions change slowly, while the electron wave functions follow adiabatically. Only molecules consisting of only a few atoms are known to occur in energy and angular momentum eigenstates. The classical configurations are usually derived by using the Born–Oppenheimer approximation, that is, by an expansion in terms of the electron-to-nucleon mass ratio m/M.

The molecular Hamiltonian has the form

$$H = -\sum_i \frac{\Delta_i}{2M_i} - \sum_k \frac{\Delta_k}{2m} + V(x, X), \qquad (3.166)$$

where x collectively denotes the electron, X the nucleon coordinates, respectively. In the Born–Oppenheimer approximation the nuclear kinetic term is neglected to lowest order. All terms appearing in the time-independent Schrödinger equation for the *isolated* molecule are then expanded with respect to the electron/nucleon mass ratio $\varepsilon = \left(\frac{m}{M}\right)^{1/4}$ in the form

$$E = \sum_{n=0} \varepsilon^n E^{(n)}, \quad \Psi = \sum_{n=0} \varepsilon^n \Psi^{(n)}, \quad H = \sum_{n=0} \varepsilon^n H^{(n)}. \qquad (3.167)$$

Using this ansatz the nuclear coordinates are expanded around their equilibrium positions. Since $H^{(0)}$ contains no kinetic term for the nucleons, one obtains

$$\Psi^{(0)} = \varphi(x, X)\delta(X - X_0). \qquad (3.168)$$

In lowest order, the nuclear positions X_0 enter only parametrically, and $E^{(0)}$ depends only on the *relative* positions of the nuclei.[31]

This means in particular that to lowest order rotational states of the entire molecule are completely degenerate, and therefore very sensitive to perturbations. Nuclear positions can then easily be "measured" by low energy particles. The measurement of electron positions would require much higher energy. Correspondingly, rotational levels are observed only for small molecules, while larger molecules are always found in oriented states.

The non-occurrence of eigenstates of the molecular Hamiltonian is particularly striking for molecules showing optical activity. We will therefore limit our discussion to this important case in the remaining part of this section. Other important examples are isomeric molecules (described by the *same* Hamiltonian) or the knot structure of DNA molecules with the same monomer sequence.

The two relevant configurations of an optically active molecule[32] are viewed as mirror images of each other, so such molecules cannot be described

---

[31] Terms of the order ε^2 contain the oscillation of the nuclei relative to their equilibrium positions, only to order ε^4 rotational motion is coupled to the other electronic and nuclear degrees of freedom.

[32] The study of substances changing the polarization of light passing through goes back to Biot and Pasteur in the 19th century. Any optically active substance was found to exist in two versions, one called dextrorotatory, the other levorotatory, characterizing their influence on the plane of linearly polarized light. It was soon attributed to properties of molecules dissolved in the liquids showing this phenomenon by Pasteur in 1848. The two kinds of, chemically identical, molecules were called optical isomers, enantiomorphs, or optical antipodes. A mixture of equal parts ("racemate") does not show optical activity. Optical activity is also found in some solids, notably quartz (SiO_2), where this phenomenon was first

by parity eigenstates for symmetry reasons. The problem of chiral molecules was first treated in the framework of quantum mechanics by Hund in 1927. In organic chemistry only substances with an "asymmetric C-atom", that is, a C-atom with *three* different ligands are optically active. If the four ligands are different, the molecule and its mirror-image are distinct, that is, they are not related by a (proper) rotation in space. If we replace the C-atom by a three-valence object (such as nitrogen), the simplest possible configuration is a tetrahedron consisting of four different building blocks (either single atoms or groups of atoms), as shown in Fig. 3.15.

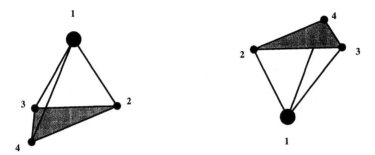

Fig. 3.15. Simplest possible structure of an optically active molecule. If all four elements (which can be single atoms or groups of atoms) are different, the molecule (*left*) and its mirror image (*right*) cannot be transformed into each other by a proper rotation.

The two classical configurations are related by space-inversion, not by a proper rotation, hence they are now usually called chiral. A simple example would be NHDT, a compound similar to ammonia (NH_3), but with two hydrogen atoms replaced by the heavier isotopes deuterium (D) and tritium (T) (Pfeifer 1980). Unfortunately, this substance is difficult to access experimentally. (For a discussion of other similar molecules, such as monodeuteroaniline and naphtazarin, see Amann (1995) and references therein.) Let us consider as the relevant degree of freedom the relative position of atom #1 with respect to the plane spanned by the other atoms in Fig. 3.15. The effective potential for this inversion coordinate has the form of a double well, as shown in Fig. 3.16.

Since the interaction is invariant under space inversion, every non-degenerate eigenstate has a definite parity and therefore cannot be chiral. The ground state $|1\rangle$ is symmetric, the first excited state $|2\rangle$ antisymmetric in this

discovered by Arago in 1811. In this case optical activity originates from the crystal structure (it disappears after melting), while in the case of solid sugar, crystal structure and molecular structure are both partially responsible for optical activity.

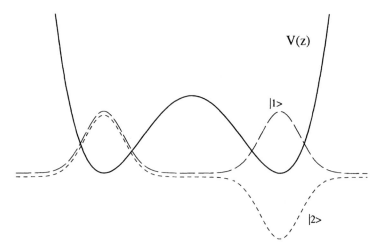

Fig. 3.16. Schematic picture of the effective potential for the inversion coordinate in a model for a chiral molecule and the two lowest-lying eigenstates. The ground state is symmetrically distributed over the two wells. Only linear combinations of the two lowest-lying states are localized in one well, corresponding to a classical configuration.

model[33]. Only the linear combinations

$$|L\rangle = \frac{1}{\sqrt{2}}(|1\rangle + |2\rangle)$$

$$|R\rangle = \frac{1}{\sqrt{2}}(|1\rangle - |2\rangle) \tag{3.169}$$

are concentrated in one of the minima of the potential.[34] These are not stationary states, however, but will oscillate with a frequency given by the energy difference between ground state and first excited state

$$\omega = \frac{(E_2 - E_1)}{\hbar}. \tag{3.170}$$

---

[33] In general, wave functions can only be calculated approximately in realistic models. An interesting method to construct exact solutions for toy models is to suggest a wave function and then to construct the corresponding potential from the time-independent Schrödinger equation, see Caticha (1995).

[34] Another alternative would be given by the combinations $|1\rangle \pm i|2\rangle$ which have very interesting properties. Upon spatial inversion, these two states switch roles, so they are chiral. On the other hand, they would show no optical activity under usual circumstances where the difference of probabilities for right- and lefthanded states is measured. Therefore these states would behave like a racemic mixtures in conventional measurements (Cina and Harris 1995).

This frequency is actually observed in small molecules like ammonia (here 24 GHz).[35] In these cases, oscillations are not observed directly, however, but are inferred from spectroscopic data. A more direct observation of oscillations in a two-state system is achieved in the strangeness-oscillations of neutral K-mesons (Perkins 1982).

Molecules should therefore not occur as objects with a definite shape, quite in contrast to what is believed in chemistry. Why are these molecules not at least *sometimes* produced in energy eigenstates, as would be natural for microobjects? Why is chirality such a stable property, that it can even survive chemical reactions (recall that nearly all molecules in living organisms show a one-sided handedness!)? This problem was clearly seen by Hund as early as in 1927:

> "Thus if a molecule admits two different nuclear configurations being the mirror images of each other, the stationary states do not correspond to a motion around one of these two equilibrium configurations. Rather, each stationary state is composed of left-handed and right-handed configurations in equal shares. ... The fact that the right-handed or left-handed configuration of a molecule is not a quantum state might appear to contradict the existence of optical isomers."[36] (Hund 1927)

His way out of this dilemma was the observation, that a position measurement would localize the system in one of the two wells and it would remain

---

[35] A very similar situation occurs in the theory of superconducting rings containing a Josephson junction (the so-called "superconducting quantum interference devices (SQUIDs)"). The dynamics of the flux Φ through the ring is described in an idealized case by a Schrödinger equation for a potential like that shown in Fig. 3.16. If dissipation is included, an equation similar to those for quantum Brownian motion is used. Transitions between states localized in one well are referred to as "macroscopic quantum tunnelling", oscillations similar to those in the ammonia molecule are termed "macroscopic quantum coherence" (see e. g., Caldeira and Leggett (1983b), Leggett (1984, 1987), Leggett and Garg (1985), Tesche (1986).) A general analysis of two-state systems coupled to a thermal environment has been given by Leggett *et al.* (1987), see also Dekker (1991). SQUIDs can be used not only to test quantum mechanics, but also alternative theories, see e. g., Benatti, Ghirardi and Grassi (1995).

Superpositions of macroscopically different states of a current in a superconducting ring (in the range of microamperes) have been observed (van der Wal *et al.* 2000, Friedman *et al.* 2000).

[36] "Wenn bei einer Molekel zwei einander spiegelbildlich entsprechende und verschiedene Anordnungen der Kerne möglich sind, so sind die stationären Zustände nicht Bewegungen in der Nähe einer der beiden Gleichgewichtsanordnungen. Jeder stationäre Zustand ist vielmehr aus Rechts- und Linksanordnungen gleichmäßig zusammengesetzt. ... Es könnte scheinen, als stünde die Tatsache, daß die Rechts- oder Linksanordnung einer Molekel nicht Quantenzustand ist, im Widerspruch mit der Existenz der optischen Isomeren."

there for a long time because the transition frequency ω is extremely small for large molecules. Hund's explanation was apparently considered as sufficient for a long time. (It was shown, for example, that Hund's model is very stable with respect to perturbation of the double-well potential, see Fischer and Mittelstaedt 1990). This corresponds precisely to the common argument that the spreading of the wave function of a billiard ball is negligible because of its large mass. As we have repeatedly emphasized, arguments based on dynamical stability alone cannot bypass the superposition principle.

Currently, there are essentially three types of explanation (for an overview see Wightman and Glance 1989, Wightman 1995, cf. also Woolley 1976, 1982, 1986, Claveric and Jona-Lasinio 1986).

The first is the suggestion that the coupling to the radiation field causes some sort of phase transition for nearly degenerate states. A model supporting this view was studied in great detail by Pfeifer (1980), see also Amann (1991a,b). In this so-called "spin-boson model", the (infinitely many) soft photons of the transverse radiation field lead to a "phase transition" for chiral states. For certain values of the model parameters (such as the coupling strength), chiral states acquire the role of eigenstates of the "dressed" molecule (i. e., the molecule plus radiation field). This alone, however, does not explain the non-ocurrence of superpositions of these dressed states, as would be required by the superposition principle. A superselection rule excluding these superpositions is motivated[37] by mathematical idealizations (leading to "inequivalent representations of the commutation relations"). They rely decisively on the fact that the number of degrees of freedom is strictly *infinite* (for models including the radiation field). Similar mathematical procedures were invoked to "derive" other superselection rules (see Chap. 6). However, it appears questionable that mathematical subleties of this kind should have any physical relevance. After all, the assumption, that our universe can be described by only a finite (but very large) number of degrees of freedom, can never be disproved by experiment. Even in the restricted field of spin-boson models it seems to be an open question whether

---

[37] "...the Fock representation [for a spin-Boson model] does *not* admit superselection rules! Hence, the *exclusive* use of the Fock representation leads to empirically incorrect results." (Amann 1991a). The absence of strict superselection rules in the usual formulation of quantum theory is thus given as a motivation for constructing new theories without apparent empirical basis, since these superselection rules can be derived through decoherence. Compare, however, the change of attitude expressed in Amann (1995). In this paper, the decohering influence of the environment is modelled by a stochastic dynamics of pure states (thereby sidestepping quantum nonlocality). Similar problems were repeatedly encountered in the quest for a "phase operator" for the harmonic oscillator. Many intricacies disappear if one starts with a finite number of states, without taking the limit of infinitely many states too early (see for example, the description of oscillator phase by Barnett and Pegg (1990)).

the desired superselection rules induced by the radiation field really exist.[38] Note also that these models are completely time-symmetric, in contrast to decoherence, where superselection rules arise from time-directed processes.

The parity violation found in weak interactions has also been made responsible for the dominance of chiral states. In the presence of parity-violating interactions, the Hamiltonian loses its space-inversion symmetry, so that chiral states could become eigenstates. However, this would not explain why superpositions do not occur.

Finally, the influence of the environment of the molecule, for example scattering of other molecules, could lead to the sought-for classical behavior. Indeed, this is the explanation offered by decoherence. If the resulting states of scattering other particles off a molecule depend on chirality,

$$|R, L\rangle \, |\Phi_0\rangle \xrightarrow{t} |R, L\rangle \, |\Phi_{R,L}\rangle \,, \tag{3.171}$$

phase relations are delocalized, hence a superselection rule is induced (cf. Sect. 3.1.3). In addition, transitions between $|R\rangle$ and $|L\rangle$ no longer occur at the rate indicated by (3.170), but are strongly suppressed by the quantum Zeno effect (Simonius 1978, Harris and Stodolski 1981, 1982, Bixon 1982). This additional stabilization of chirality through continuous measurement will be discussed in Sect. 3.3.1.

Obviously, this mechanism is not very effective for small molecules such as ammonia, which can be prepared in any state[39]. Where is the border line? For a crude estimate we can use the results of the previous section, by considering the fact that scattering of other molecules effectively amounts to a position measurement, precisely in the sense of (3.171) above, that is, the scattered states depend on the *orientation* of the molecule. A considerable effect can be expected, since the resolution of this "measurement" is given by the thermal de Broglie wavelength of the scattering particles, which are usually of the order of atomic sizes.

According to (3.68), coherence over a distance Δx is destroyed on a timescale given by (3.134)

$$\Delta t|_{\text{decoherence}} \sim \frac{1}{\Lambda (\Delta x)^2}, \tag{3.172}$$

with the decoherence rate Λ (3.69). Here the relevant length scale Δx is represented by the size d of the molecule. This is to be compared with the

---

[38] "The discussion of ground states of the full Hamiltonian for the joint system (molecule and environment) is difficult and unsettled up to now. Hence, the question ... [does the radiation field generate molecular superselection rules?] ... remains open." (Amann 1991a).

[39] Quantitative estimates for possible methods to produce non-classical ("Schrödinger cat") superposition states in the vibrational coordinate of a diatomic molecule, along with methods of detection by time-dependent spectroscopy, have been given by Walmsley and Raymer (1995).

transition frequency between chiral states (3.170),

$$\Delta t|_{\text{molecule}} = \frac{2\pi}{\omega} = \frac{h}{\Delta E}, \tag{3.173}$$

where ΔE is the energy difference between the two lowest-lying states. If the decoherence timescale is much shorter than the transition time, we can expect stabilization of chiral states. Using results from Sect. 3.2.1.1, a simple estimate can be made. If we consider scattering of air molecules, for example, Λ can be calculated from (3.69) by inserting the geometric cross section $\sigma = \pi d^2$, the thermal de Broglie wavelength of the scattering particles ($k = \frac{2\pi}{\lambda_{\text{dB}}}$, $\lambda_{\text{dB}} = \hbar\left(\frac{2\pi}{mk_{\text{B}}T}\right)^{1/2}$) and the mean thermal velocity $v = \left(\frac{k_{\text{B}}T}{m}\right)^{1/2}$, to yield

$$\Delta t|_{\text{decoherence}} \sim \frac{\hbar^2}{d^4 m^{1/2}(k_{\text{B}}T)^{3/2}N/V} \tag{3.174}$$

(here we inserted Planck's constant and omitted all numerical factors of order one; N/V is the particle density). For the example of a very small molecule such as ammonia, the result is $\Delta t \sim 10^4\,\text{s}$ for laboratory vaccum (10^6 particle/cm^3), or $\Delta t \sim 10^{-3}\,\text{s}$ for atmospheric pressure. This is to be compared with the transition frequency. Obviously even a small molecule such as ammonia is *near* the border line. This is also obvious from experiment: the inversion spectra at frequencies of 24 GHz and 1.6 GHz disappear at pressures of approximately 0.5 and 0.04 atm for NH_3 and ND_3, respectively. For large molecules the transition times are so tremendously large that decoherence works faster by many orders of magnitude.

Even if the transition probability between chiral states is extremely small this does not imply that symmetric states cannot be prepared. Such states would necessarily emerge from a symmetric ground state in a potential with only *one* minimum. If such a potential is adiabatically changed into the double-well form shown in Fig. 3.16, a symmetric state (such as $|1\rangle$) must result. Suggestions for how to produce symmetric molecular states by mechanisms of this kind have been put forward by Quack (1986), cf. also Cina and Harris (1995) and references therein.

A transition from a potential with one minimum to a double-well potential (in one or more dimensions) is often considered in theories describing symmetry breaking in phase transitions (e. g. in the Ginzburg–Landau theory, for an experiment with superfluid helium see Hendry *et al.* 1994). As long as only unitary dynamics is assumed, a symmetric superposition of asymmetric states results. Only a collapse of the wave function would change such a state into a symmetry- breaking asymmetric state (Zeh 1975, Joos 1987b); such a symmetry-breaking collapse could have important cosmological implications, see also Sect. 9.4.

3.2.5 Decoherence in the Brain

The question of the applicability of quantum theory and the significance of decoherence for brain processes is of fundamental importance for at least two reasons. First, there is an epistemological aspect concerning the very formulation of physical theories. Second, if we treat the brain as governed by the usual quantum laws, one may ask whether or not quantum coherence effects can play any significant role. Surprisingly, quantitative answers to the second question surfaced only recently.

Why should one extend quantum theory up to (or near to) the level of conscious perception? One possible answer is, that a physical theory may be viewed as devoid of any meaning if it does not contain a part saying "and this is perceived by the observer." This was clearly expressed by von Neumann in his theory of the measurement process – and his justification to introduce the collapse of the wave function as a second dynamics (cf. also Sect. 2.3). He reasoned that measurements have to be described as a chain of physical interactions and that this chain has to terminate somewhere, since (von Neumann 1932, Chap. VI)

> "Experience only asserts something like: an observer has made a certain (subjective) perception, but never such as: a certain physical quantity has a certain value".[40]

Indeed, these fundamental questions emerge again and again whenever the very nature of physical theories is scrutinized. An interesting discussion of this problem can be found in Heisenberg's autobiographical notes (Heisenberg 1973), where he reminisces about talking with Einstein on how to construct new physical theories,

> " 'One cannot observe the orbits of electrons in an atom..', [I said]. 'Seriously, you can't believe', Einstein replied, 'that a physical theory can be based only on observable quantities.'... 'I thought,' I asked astounded, ' that exactly this idea was essential for your theory of relativity?' 'Perhaps I used this kind of philosophy,' Einstein answered, 'but it is nonsense nevertheless. ... It may be of heuristic value to recall what one really observes. But from a principal point of view it is quite wrong to insist on founding a theory on observed quantities alone. In reality just the opposite is true. Only the theory decides about what can be observed. ...On [the] entire long path from a process up to our conscious perception we need to know how Nature is working in order to claim that we have observed anything at all .' "[41]

---

[40] Denn die Erfahrung macht nur Aussagen von diesem Typus: ein Beobachter hat eine bestimmte (subjektive) Wahrnehmung gemacht, und nie eine solche: eine physikalische Größe hat einen bestimmten Wert."

[41] " 'Die Bahnen der Elektronen im Atom kann man nicht beobachten..', [sagte ich]. 'Aber Sie glauben doch nicht im Ernst', entgegnete Einstein, 'daß man

Nearly all current models for "neural networks", "pattern recognition" and related fields rely on classical pictures (see for example Haken 2002). On the other hand, it has been suggested repeatedly, that quantum effects may play *some* role in the functioning of the brain. The most radical proposal was Wigner's linkage of conscious perception with the collapse of the wave function (Wigner 1962, see also Squires 1990, Stapp 1993, Whitaker 2000). Later work includes Penrose's suggestion that quantum coherence may be of importance in the working of microtubules (Penrose 1994), supposedly acting like a "quantum computer". Concrete results for realistic scenarios were achieved by Tegmark (2000).

The key question is whether the relevant degrees of freedom are sufficiently isolated from their environment to allow for quantum coherence effects. Even without any detailed calculations, this looks rather implausible, since the brain constitutes a fairly "hot" environment.

Considering the central process of neuron firing, one may study a superposition of a neuron in "firing" and "non-firing" state. These two states differ in the location of millions of Na^+ and K^+ ions separated by a membrane by a distance of several nm. Possible decoherence mechanisms include collisions with other ions or water molecules and long-range Coulomb interaction with distant ions. Ion-ion collisions can be treated similar to the models described in Sect. 3.2.1. The relevant deBroglie wavelengths are about 0.03 nm, much smaller than the membrane thickness over which coherence would have to be maintained. Obviously an extremely short decoherence time will result. For details see Tegmark (2000), a summary is given in Table 3.4.

Table 3.4. Decoherence timescales in the brain (Tegmark 2000).

Object	Environment	Decoherence timescale
Neuron	Colliding ion	10^{-20} s
Neuron	Colliding H_2O	10^{-20} s
Neuron	Nearby ion	10^{-19} s
Microtubule	Distant ion	10^{-13} s

in eine physikalische Theorie nur beobachtbare Größen aufnehmen kann.'... 'Ich dachte,' fragte ich erstaunt, 'daß gerade Sie diesen Gedanken zur Grundlage Ihrer Relativitätstheorie gemacht hätten?' 'Vielleicht habe ich diese Art von Philosophie benutzt,' antwortete Einstein, 'aber sie ist trotzdem Unsinn. ... Es mag heuristisch von Wert sein, sich daran zu erinnern, was man wirklich beobachtet. Aber vom prinzipiellen Standpunkt aus ist es ganz falsch, eine Theorie nur auf beobachtbare Größen gründen zu wollen. Denn es ist ja in Wirklichkeit genau umgekehrt. Erst die Theorie entscheidet darüber, was man beobachten kann. ...Auf [dem] ganzen langen Weg vom Vorgang bis zur Fixierung in unserem Bewußtsein müssen wir wissen, wie die Natur funktioniert, wenn wir behaupten wollen, dass wir etwas beobachtet haben.' "

Calculations of this kind strongly support the view that (1) quantum theory can be successfully extended to regions "close" to subjective perception and (2) the workings of the brain such as neuron firing and related processes are decohered so strongly that the classical models employed so often appear well founded in principle. Typically, dynamical timescales for neuron firing and cognitive processes are in the range of 10^{-4} to 10^0 seconds, whereas decoherence timescales are many orders of magnitude shorter.

3.3 Dynamical Consequences

If the unitary Schrödinger dynamics of a system is "disturbed" by interaction with other systems, the most important effect studied in the previous sections is the destruction of coherence with respect to certain system states. This represents the essential consequence of decoherence. Without coupling to the environment the system displays a certain time dependence, governed by its own Hamiltonian, which will now be altered in various ways. In the last section we have already discussed the interplay between the spreading of the wave packet (free evolution) and the influence of scattering processes (decoherence) for a mass point. Here we will investigate two important limiting cases which display quite different behavior of a system subjected to the decohering influence of its environment.

The first effect is the – at first sight quite astonishing – consequence that the motion of a system may become frozen by repeated measurements, even if these are performed "ideally" (recoil-free). This phenomenon is now usually termed the "quantum Zeno effect". We will compare the phenomenological description, where the measurement is represented by the collapse of the state vector, with unitary models that include the interaction with the measurement device. This analysis will demonstrate that the quantum Zeno effect can be understood as a consequence of the unitary Schrödinger dynamics and does not require a collapse of the wave function.

In contrast to objects which are frozen by continuous measurement, the behavior of systems which are successfully described by master equations is obviously very different. These systems seem to be quite inert with respect to measurements. The prototypical example is exponential decay (typical for radioactive nuclei, unstable elementary particles, virtual bound states, etc.). For these objects it does not matter whether, or how often, the system is monitored. In their natural environment, decaying systems are continuously observed in many ways. In α-decay, the charge of the emitted particle is immediately recognized by the environment, for example. It is therefore necessary to understand why decay processes nevertheless occur independently of the presence of the environment, that is, why the Zeno effect does not play any role. Decoherence will in fact turn out to be indispensable for the proper understanding of such quasiclassical behavior. As an example of a rate equation which describes a classical stochastic process, we will reconsider the derivation of Pauli's master equation.

3.3.1 The Quantum Zeno Effect

The inhibition of transitions by frequent measurements as a prediction of quantum theory was discovered independently by several authors. The now popular name "Zeno effect" (or "Zeno paradox") was suggested by Misra and Sudarshan (1977).[42] The question of how frequent observations influence a quantum system arises naturally when one tries to understand the action of a continuously active measurement device, such as a Geiger counter. In contrast to the axiomatic measurement theory (instantaneous collapse of the wave function) such a device does not measure the state of a system at a certain instant, but monitors the system continually (in a very delicate sense).

The quantum Zeno effect was mentioned even earlier, e. g. by Yourgrau (1968), who called it "Turing's paradox". It was indeed already used by von Neumann in 1932 as a means to transform any state from the Hilbert space of a system into any other one.[43] Other names were proposed, such as "watchdog effect" (suggested by John Wheeler, see Kraus 1981), "watched pot behavior" (since "a watched pot never boils", see Wolsky 1976), and others[44]. The effect has often been considered as counterintuitive, absurd or even as a defect of

---

[42] Zeno of Elea (c. 495 BC–c. 430 BC), a greek philosopher and mathematician, is especially known for his four paradoxes whose purpose was to support the philosophy of Parmenides. He intended to prove that motion is impossible in all his paradoxes. These are the Achilles paradox (Achilles can never pass the tortoise in a race), the dichotomy paradox, the arrow paradox (a flying arrow is actually at rest), and the stadium paradox. In particular the dichotomy paradox resembles the situation now discussed as "quantum Zeno paradox": Before an object can travel a certain distance, it must have been (been found?) at half the distance. But this interval can also be divided in half, and so on. So the object had to pass through an infinite number of intermediate points. Therefore the object can never reach its goal, Zeno concluded. From a modern point of view, the paradoxes arise from an insufficient mathematical understanding of the continuum.

[43] In order to show that all pure states have the same entropy, von Neumann considered transformations of a state φ into another state ψ. This could be accomplished by the Schrödinger equation with an appropriate Hamiltonian. Instead he employed a sequence of measurements (von Neumann 1932, p. 195) to transform φ into ψ with probability arbitrarily close to one, a scheme which is more or less identical to that used much later by Aharonov and Vardi (1980). This procedure reduces to the popular description of the quantum Zeno effect, if φ and ψ are identical.

[44] Unfortunately, the nomenclature is marred by ambiguous definitions. Except for the term "Zeno effect", all other names are used in various, sometimes contradictory, ways. Special care is indicated when the subject includes terms like "inverse Zeno effect" or "anti-Zeno effect".

quantum theory. A crucial point in the early phenomenological discussions was certainly the assumed collapse of the wave function during measurement. Is there a limit to how fast a measurement can be performed? Is the collapse of the wavefunction really needed? We will see that the Zeno effect can be understood as a dynamical effect of unitary evolution, resulting from strong coupling of a system to its environment.

3.3.1.1 Phenomenological Description.

Let us consider a simple model for the measurement of a decaying system. Suppose the system is prepared in its "undecayed" state $|u\rangle$ at some initial instant $t = 0$. Evolution according to the Schrödinger equation will lead to a superposition of this undecayed state with some orthogonal ("decayed") states $|d_k\rangle$, with amplitudes a_u and a_{d_k}, respectively,

$$
\begin{aligned}
|\Psi(t)\rangle &= \exp(-\mathrm{i}Ht)\,|u\rangle \\
&= a_u(t)\,|u\rangle + \sum_{d_k \neq u} a_{d_k}(t)\,|d_k\rangle .
\end{aligned}
\tag{3.175}
$$

The probability of finding the system still "undecayed" (i.e., in the state $|u\rangle$) at a later time $t > 0$ is then

$$
\begin{aligned}
P(t) &= |a_u(t)|^2 \\
&= |\langle u\,|\exp(-\mathrm{i}Ht)|\,u\rangle\,|^2.
\end{aligned}
\tag{3.176}
$$

Expanding the exponential in powers of t gives

$$
P(t) = 1 - (\Delta H)^2 t^2 + O(t^4)
\tag{3.177}
$$

with[45]

$$
(\Delta H)^2 = \langle u\,|H^2|\,u\rangle - \langle u\,|H|\,u\rangle^2 .
\tag{3.178}
$$

The essential feature of this result is the *quadratic* time dependence of the survival probability. There is no linear term in $P(t)$, in contrast to exponential decay (Khalfin 1968).

Let us now suppose that the measurement performed on the same unstable system is carried out not just once, but is repeated N times in the interval $[0, t]$. The probability that it will be found undecayed in all N measurements is then given by

$$
P_N(t) \approx \left[1 - (\Delta H)^2 \left(\frac{t}{N}\right)^2\right]^N > 1 - (\Delta H)^2 t^2 = P(t).
\tag{3.179}
$$

The non-decay probability is always increased, that is, the decay is suppressed; in the limit of arbitrary dense measurements it comes to a complete halt,

$$
P_N(t) = 1 - (\Delta H)^2 \frac{t^2}{N} + \dots \overset{N \to \infty}{\longrightarrow} 1.
\tag{3.180}
$$

---

[45] The inverse of the energy width is sometimes called "Zeno time", $\tau_Z = 1/\Delta H$.

Thus under continuous measurement the system will not decay at all!

Obviously this result cannot be of any relevance for most systems, especially in the macroscopic domain. Staring at my pencil will not keep it from falling from my desk. We have already seen in Sect. 3.2.2.2 that the Ehrenfest theorems remain valid for a mass point even under continuous observation. Radioactive decay does not depend on the properties or the presence of the (chemical) environment. Indeed, exponentially decaying systems are not influenced at all by repeated measurements. If $P(t)$ *were* exponential,

$$P(t) = \exp(-\Gamma t), \tag{3.181}$$

observation at intermediate times trivially would not change P, since

$$P_N(t) = \left(\exp\left(-\Gamma \frac{t}{N} \right) \right)^N = \exp(-\Gamma t), \tag{3.182}$$

quite in contrast to (3.179). Equivalently, we may say that the naive expectation of a decaying system being *either* in the undecayed *or* in a decayed state at all times (leading to the classical prejudice, that the semigroup law $P(t_1)P(t_2) = P(t_1 + t_2)$ should hold) is incompatible with quantum theory. This assumption, which amounts to a complete absence of coherence between certain states (and replacing the smooth Schrödinger evolution by "quantum jumps"), leads directly to the Zeno effect (Ekstein and Siegert 1971, Bunge and Kalnay 1983). All this is related to the well-known fact that an exponential decay law cannot be exact in quantum theory (the "Markov limit" can never be reached, see Sect. 7.4), because energy is bounded from below (Khalfin 1958).

If we express the initial undecayed state $|u\rangle$ in terms of energy eigenstates $|E\rangle$ (for simplicity we assume a continuous energy spectrum and neglect degeneracy),

$$|u\rangle = \int_{E_{\min}}^{\infty} dE\, a_u(E)\, |E\rangle\,, \tag{3.183}$$

the non-decay probability is given by the Fourier transform of the energy distribution $\omega(E) = |a_u(E)|^2$ of the initial state,

$$P(t) = \left| \int_{E_{\min}}^{\infty} dE\, \omega(E) \exp(-\mathrm{i}Et) \right|^2. \tag{3.184}$$

It can be shown (by using the Paley–Wiener theorem on Fourier transforms) that for large times $P(t)$ decreases less rapidly than any exponential function, if $E_{\min}$ is finite. Combining these results we see that exponential decay can never be exact, although the *range* where it is approximately valid can be extremely large[46]

---

[46] For the decay of the charged pion, the times T_1 and T_2 separating the quadratic (short-time), exponential (intermediate), and non-exponential (large-time) re-

(Khalfin 1958, Petzold 1959, Winter 1961). The usual Lorentzian ansatz (Weisskopf and Wigner 1930)

$$\omega(E) \sim \frac{1}{(E - E_{\text{res}})^2 + \left(\frac{\Gamma}{2}\right)^2} \tag{3.185}$$

would lead to an exact exponential law, but requires that energy is unbounded (for a detailed review see Fonda, Ghirardi and Rimini 1978).

If we take into account the fact that not only the experimenter, but also the natural environment is monitoring the decay, then the long-time deviations from the exponential law become unobservable not only in practice, but in principle (as long as these measurements are irreversible). If we assume an exponential law at "moderate" times,

$$P(t) \approx 1 - \Gamma t, \tag{3.186}$$

we see that repeated observations will *enforce* an exponential law, since

$$P_N(t) \approx \left(1 - \Gamma \frac{t}{N}\right)^N \to \exp(-\Gamma t). \tag{3.187}$$

This limit (which amounts to an integration of the corresponding rate equation) is valid only if there is no backflow of the decay-products, i. e., no interference of $|d_n\rangle$ states with $|u\rangle$ (Nakazato, Namiki and Pascazio 1994). If the system, together with its decay products, is enclosed in a reflecting cavity, decay properties may change drastically, see Lewenstein, Mossberg and Glauber (1987), Haroche and Kleppner (1989).

The quadratic time-dependence, as in (3.177), is typical for oscillating systems. In view of the foregoing discussions it is natural to look for the Zeno effect in the simplest possible quantum system, a two-state system.

Consider a spin-half particle precessing in a constant magnetic field oriented in the z-direction. Let the system at $t = 0$ have its spin in the $+x$-direction (denoted by $|\rightarrow\rangle$), which is a superposition of the energy eigenstates

---

gions were estimated to be of the order $T_1 = 10^{-14}/\Gamma$ and $T_2 = 190/\Gamma$, where Γ is the pion decay rate (Chiu, Sudarshan and Misra 1977). Obviously it is hopeless to look for a non-exponential decay law by counting pions (see also Home and Whitaker 1986, Levitan 1988). Of course, there are many (at first sight similar) situations where an exponential decay law is only a poor approximation (see, e. g. Fu and Ramaswami 1991). For an analysis of the dynamics of optical transitions see Sudbery (1988).

Exploring the transition region between the Zeno limit and exponential decay may lead to some interesting effects, for example, an *increased* decay rate (relative to the Markov region). This "anti-Zeno effect" is possibly the only observable consequence for *some* common unstable systems (usually showing exponential decay), since overly intense (or frequent) measurements, trying to reach the Zeno region, would destroy these systems, for example by particle creation (Kofman and Kurizki 2000). See also Sect. 3.3.2.1.

$|\uparrow\rangle$ and $|\downarrow\rangle$,

$$|\Psi(0)\rangle = |\rightarrow\rangle = \frac{1}{\sqrt{2}}\left(|\uparrow\rangle + |\downarrow\rangle\right). \tag{3.188}$$

The interaction with the magnetic field B will lead to a precession in the x–y plane with the Larmor frequency $\omega = \frac{|e|B}{2mc}$, hence

$$|\Psi(t)\rangle = \cos(\omega t/2)\,|\rightarrow\rangle - \sin(\omega t/2)\,|\leftarrow\rangle. \tag{3.189}$$

If we measure the spin in the x-direction at $t = \delta t$, the probability for the system to collapse into $|\rightarrow\rangle$ is

$$P_1 = \cos^2\frac{\omega\delta t}{2} \approx 1 - \frac{(\omega\delta t)^2}{4}. \tag{3.190}$$

If the measurement is repeated N times in the interval $[0, T]$, i. e., $\delta t = T/N$, we find, as in (3.179),

$$P_N \approx \left(1 - \left(\frac{\omega T}{2N}\right)^2\right)^N \overset{N\to\infty}{\longrightarrow} 1, \tag{3.191}$$

that is, the repeated measurements freeze the "natural motion", the precession in the x–y plane.

The sequence of measurements could in principle be realized by a succession of Stern–Gerlach devices, each one oriented in the x-direction. A formally identical setup was suggested by Peres. If we send a beam of linearly polarized light through a vessel containing an optically active liquid, the plane of polarization will rotate in the same manner as the spin in the above example. Measurement of (linear) polarization can now be accomplished by using a sequence of polarization filters, all oriented in the same direction as the incoming light. Inserting one more *absorbing* filter will *increase* the light intensity at the end of the container. In the limit of arbitrarily many filters the light will just pass through the liquid without ever changing its polarization (Peres 1980). This example provides a classical (wave) analogue to the spin experiment.[47]

Freezing the internal motion is not the only option. If, instead of always measuring the spin along the same axis, one were to perform a sequence of measurements along a "path" defined by

$$|\sigma_n\rangle = \cos\alpha_n\,|\rightarrow\rangle - \sin\alpha_n\,|\leftarrow\rangle \tag{3.192}$$

---

[47] The analogy can even be driven a step further. A filter absorbs the "unwanted" components orthogonal to the "allowed" ones. This is exactly what the collapse of the wave function does. Since the dismissed components can no longer interfere with the remaining ones, the "natural" evolution is thrown back to start again with the leftover component alone (Home and Whitaker 1993).

with $\alpha \propto n\delta t$, then, if the path followed the natural motion, i. e. $\alpha_n = n\omega\delta t/2$, we would trivially find $P \equiv 1$. If we now switch off the magnetic field, but perform the same sequence of measurements, the conditional probability of finding $|\sigma_n\rangle$, provided that $|\sigma_{n-1}\rangle$ was found before, is

$$P_c(n) = |\langle \sigma_n|\sigma_{n-1}\rangle|^2 = \cos^2 \frac{\omega\delta t}{2}. \qquad (3.193)$$

The probability of finding always "spin up" along the prescribed path is then

$$P_N = \prod_{n=0}^{N} P_c(n) = \left(\cos \frac{\omega T}{2N} \right)^{2N}. \qquad (3.194)$$

For large N this yields

$$P_N \to \exp\left(-\frac{1}{N} \left(\frac{\omega T}{2} \right)^2 \right) \stackrel{N\to\infty}{\longrightarrow} 1, \qquad (3.195)$$

that is, the spin is found to precess with frequency ω even without the presence of the magnetic field. Therefore, even ideal measurements can completely take control of the system's dynamics:

"..., the only reasons for the spin rotation are those measurements. This result is quite surprising because of the accepted assumption that if the outcome of a measurement of some dynamical variable is certain (i. e., with probability one), then the state of the system was not disturbed." (Aharonov and Vardi 1980)

A measurement device for (optical) polarization could, for example, be realized by using a beam splitter. If one of the beams is absorbed by an opaque object, only one component can propagate further; otherwise, if there is no obstacle, the beams can be coherently recombined (unitary evolution). In a properly adjusted Mach-Zehnder interferometer, a previously "dark" (because of destructive interference) exit channel could show a signal if one arm is blocked, even though the photon is *not* absorbed (only its destructive component is – in conflict with a particle picture). Early ideas about such "negative result" or "interaction-free" measurements (Elitzur and Vaidman 1993, Penrose 1994, Krenn, Summhammer and Svozil 1996) can be combined with the Zeno mechanism (Bennett 1994, Kwiat *et al.* 1995, Kwiat, Weinfurter and Zeilinger 1996). One of these schemes is exemplified in Fig. 3.17.

If a horizontally polarized photon is sent through N polarization rotators (or repeatedly through the same one) each of which rotates the polarization by an angle $\Delta\Theta = \frac{\pi}{2N}$, the photon ends up with vertical polarization, that is the probability to find horizontal polarization is zero,

$$P_H = 0. \qquad (3.196)$$

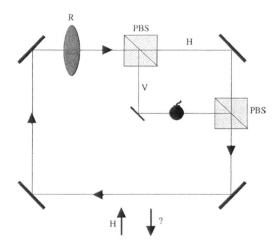

Fig. 3.17. Scheme of "interaction-free interrogation" as a variant of the Zeno effect. Without the absorbing object (the bomb), the polarization of the injected photon (initially horizontal) is rotated by the rotator R by a small angle on every passage. The two polarizing beam splitters PBS have no effect, if properly adjusted, since horizontal and vertical components are recombined coherently. If an absorbing object is present, the vertical polarization is removed at every passage. Inspecting the photon after many cycles allows one to infer the presence of the object with high probability, while the photon is only very infrequently absorbed.

If this evolution is interrupted by a horizontal polarizer (absorber) the probability of transmission is (similar to the previous examples) given by

$$P'_H = \cos^{2N} \Delta\Theta = \cos^{2N} \frac{\pi}{2N} \approx 1 - \frac{\pi^2}{4N} \longrightarrow 1. \qquad (3.197)$$

To implement this idea, a photon is injected into the setup shown in Fig. 3.17 and goes N times through the rectangular path, as indicated. The initial polarization is rotated at R by an angle $\Delta\Theta = \frac{\pi}{2N}$ on each passage. In the absence of the absorbing object, the polarizing beam splitters, making up a Mach-Zehnder interferometer, are adjusted to have no effect. That is, the vertical component V is coherently recombined with the horizontal one (H) at the second beamsplitter to reproduce the rotated state of polarization. If, however, the "bomb" is present, the vertical component is absorbed at each step. After N cycles, the photon is now still horizontally polarized, thereby indicating the presence of the object with probability near one, or has been absorbed (with arbitrarily small probability). For details of experimental setups see Kwiat *et al.* (1999). Possible practical applications include imaging of photosensitive objects (White *et al.* 1998).

One should be aware of the fact, that the term "interaction-free" is seriously misleading, since the Zeno mechanism is a consequence of *strong* interaction. Part of this conceptual confusion is related to the classical particle pictures often used in the interpretation of interference experiments, in particular "negative-result measurements". We will come back to this topic in Sect. 3.4.1.

If the system under consideration is described by a density matrix instead of a pure state, the phenomenological description we have used above can be easily generalized: The initial "undecayed" state $|\Psi\rangle$ in (3.175) is replaced by a density matrix constrained to the respective subspace of the total Hilbert space. If this subspace is defined by the projection operator P, we have initially

$$\rho(0) = P\rho P \tag{3.198}$$

The sequence of N measurements during the interval $[0,t]$ is then described by an alternating sequence of unitary evolutions lasting $\tau = \frac{t}{N}$ and projections P for the (unnormalized) density matrix, thus

$$\rho_N(t) = PU(\frac{t}{N})PU(\frac{t}{N})P...\rho_0 PU^\dagger(\frac{t}{N})P...$$
$$= T^N \rho_0 (T^\dagger)^N \tag{3.199}$$

with the non-unitary effective evolution operator

$$T = P \exp(-iHt/N)P. \tag{3.200}$$

The "survival" probability after N steps is then given by the trace of ρ_N,

$$P_N(t) = \mathrm{tr}\rho_N(t) = \mathrm{tr}(T^N \rho_0 T^{\dagger N})$$
$$\overset{N\to\infty}{\longrightarrow} 1 \tag{3.201}$$

(Misra and Sudarshan 1977). If the projections vary with time, situations analogous to those described in (3.192)–(3.195) can be contemplated (Balachandran and Roy 2002). Note that the very same expressions are utilized in the "consistent histories" approach (compare Chap. 5), where projections are used to establish "consistency conditions", while here they correspond to measurements.

For discrete systems this description is trivially equivalent to the one given at the beginning of this section. If the projection reduces the state to a multidimensional subspace, the dynamics of the system is confined to that subspace, while a dynamical superselection rule forbids transitions between these subspaces (Facchi and Pascazio 2002), compare also Sect. 3.1.3. In the case of continuous degrees of freedom, the result may possibly depend on the precise definition of the projections used as well as on how the limit $N \to \infty$ is taken. These mathematical issues are related to the *physical* question whether there is a Zeno effect for continuous degrees of freedom, for example, whether the motion of free particle is affected by repeated measurements. Various answers have been given to this question. In Sect 3.2.2.2 we pointed out that continuous measurement does *not* slow down the motion of a mass point (the

Ehrenfest theorems are still valid). This may be understood as a consequence of the fact that, for a continuous degree of freedom, any measurement with finite resolution necessarily is too coarse to invoke the Zeno effect. On the contrary, using projections such as $P = \int dx\, \chi(x)\, |x\rangle \langle x|$ where χ is the characteristic function on a finite interval $[0, L]$, there are claims that under such measurements the particle is confined to this interval in the Zeno limit, providing an effective infinite potential wall (Facchi *et al.* 2001). This issue can only be discussed properly by treating the measurement not by a projection but more realistically by an appropriate interaction, as done below (compare also Sect. 3.3.2.1 on Pauli's master equation).

3.3.1.2 An Experimental Test. Cook (1988) proposed an experiment with a single trapped ion to demonstrate the quantum Zeno effect (see also Sect. 3.3.2.3). It was later realized not with a single atom, but instead with an ensemble of about $5\,000$ Be$^+$ ions confined in a Penning trap (Itano *et al.* 1990).

In this experiment the population of two levels is measured by coupling them to a third atomic level which decays rapidly by emitting fluorescence light. The first two levels represent the "measured object", the third level together with the emitted photons play the role of the measurement device. The relevant part of the level structure is shown schematically in Fig. 3.18. Level 2 is an excited metastable state, such that spontaneous decay from level 2 to level 1 can be neglected. If the system is in its ground state at $t = 0$ and a driving field at resonance frequency $(E_1 - E_2)/\hbar$ is applied, the probability for the ground state oscillates according to $P_1(t) = \cos^2(\Omega t/2)$, where Ω is the Rabi frequency. If a measurement of the state of the atom is made after a short time $t \ll 1/\Omega$, the probability to find it in the ground state is $P_1(t) \approx 1$, while $P_2(t) \approx \frac{1}{4}\Omega^2 t^2$ for the excited level 2.

Level 3 is assumed to be connected only to level 1 by a strongly allowed (dipole) transition, i. e., level 3 decays rapidly.

The measurement is carried out by driving the $1 \leftrightarrow 3$ transition with a short optical pulse. In the phenomenological description, this pulse is equivalent to a measurement, and thus causes a collapse of the system state to the measured state, i. e. level 1 or 2, since information about the population in the 1-2-system is available in the environment: If level 1 is found, the ion emits some fluorescence photons by oscillating between level 1 and level 3 until the measuring pulse is turned off. Alternatively, if it is found in level 2, the system does not emit any photons (a so-called "null measurement"[48]).

---

[48] The concept of a "null-measurement" or "negative-result-measurement", where, for example, a detector does *not* fire, relates to a classical picture in that it presupposes a collapse of the state vector. If a measurement device is (sufficiently strongly) coupled to a system, all possible states of this device which discriminate states of the measured system have to be treated on an equal footing, even if among these there is a pointer state where "nothing happens" (the pointer does

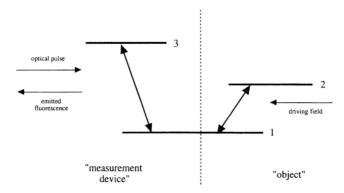

Fig. 3.18. Level structure for an experiment demonstrating the quantum Zeno paradox. The Rabi oscillations of the transition between levels 1 and 2 (driven by a resonant radiofrequency field) are monitored by exciting the optical transition $1 \rightarrow 3$ resulting in light emission from level 3 through spontaneous emission. In this way the $1 \leftrightarrow 3$ transition together with the emitted light acts like a (nearly ideal) measurement device discriminating between levels 1 and 2.

In this way the electromagnetic field carries away the information about the state of the ion. In addition, the measurement is ideal (non-destructive): If it is followed immediately by a second one, the result will almost always be the same.

The theoretical description at this phenomenological level is quite simple. In the usual Bloch vector description for a *two*-state system, the three components of the real Bloch vector[49] $\boldsymbol{R}$ are connected with the density matrix elements in the rotating frame (where the Bohr frequency is eliminated) according to

$$\boldsymbol{R} = \begin{pmatrix} \rho_{12} + \rho_{21} \\ \mathrm{i}(\rho_{12} - \rho_{21}) \\ \rho_{22} - \rho_{11} = P_2 - P_1 \end{pmatrix}. \tag{3.202}$$

---

not move, $|\Phi_0\rangle = |\Phi_n\rangle$ in (3.15), although an entangled state still results). In particular, "null results" must not be interpreted as an absence of interaction. See Sect. 3.4.1, compare also Fig. 3.17.

[49] The Bloch vector description is a common tool in atomic physics and other fields. It is nothing more than a convenient representation of the density matrix of a two-state system (an expansion in terms of the Pauli matrices σ_i, $R_i = \mathrm{tr}\rho\sigma_i$). Since the Bloch vector is real, it can be easily visualized. In some sense, the Bloch vector is a finite-dimensional analogue of the Wigner function, compare (2.14) on page 36 (for an analysis of the discrete Wigner function see Wootters 1987). For an N-level system, the Bloch vector has $N^2 - 1$ real components (Hioe and Eberly 1981, Joos 1989, Schlienz and Mahler 1995).

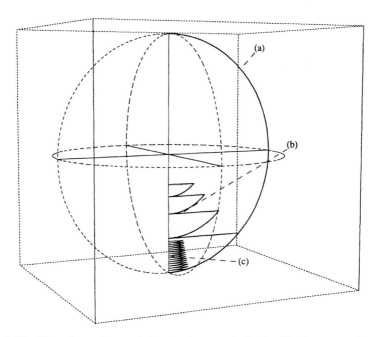

Fig. 3.19. Time dependence of the density matrix in the Bloch vector picture for three different runs of the quantum Zeno experiment. In case (a), without measurement pulse, unitary evolution leads the Bloch vector from the south pole (ground state) to the north pole (complete inversion) of the Bloch sphere during a half-period of the Rabi oscillation. In case (b) the smooth time-dependence is interrupted by four measurements, each projecting the state back onto the z-axis on a very fast timescale. After each projection, unitary dynamics starts again, but with a new (shorter) initial vector. Case (c) clearly displays the Zeno effect, since here (with 16 measurements) the system moves only slightly away from the initial state.

The equation of motion for $\boldsymbol{R}$ is

$$\frac{d}{dt}\boldsymbol{R} = \boldsymbol{\Omega} \times \boldsymbol{R} \tag{3.203}$$

where $\boldsymbol{\Omega} = (\Omega, 0, 0)$. In this geometrical picture $\boldsymbol{R}$ precesses about $\boldsymbol{\Omega}$ with fixed magnitude and angular velocity Ω as long as (3.203) is valid. (Pure states correspond to points on the surface of the "Bloch sphere", mixed states are represented by points inside the unit sphere.) Let the system be in state $|1\rangle$ at $t = 0$, i. e. $\boldsymbol{R} = (0, 0, -1)$. If the $1 \rightarrow 2$ transition is driven by an on-resonance π-pulse (of duration $T = \pi/\Omega$) the probability of finding the system in level 2 at time $t = T$ would be unity without the measurements. Now this unitary evolution is interrupted by n measurement pulses. Just before the first measurement pulse at $t = \pi/(n\Omega)$,

$$\boldsymbol{R} = \big(0, \sin(\pi/n), -\cos(\pi/n)\big). \tag{3.204}$$

According to the collapse rules to be used here, the measurement projects the system to either level 1 or level 2. For the density matrix this means that the interference terms ρ_{12} and ρ_{21} between these states are reset to zero, while the populations ρ_{11} and ρ_{22} remain unchanged. Hence, after the first measurement pulse the Bloch vector is projected onto the z-axis (see Fig. 3.19),

$$\boldsymbol{R} = \big(0, 0, -\cos(\pi/n)\big). \tag{3.205}$$

This vector is the same as it was at $t = 0$, except that its magnitude has been reduced by a factor $|\cos(\pi/n)|$. If n measurements are performed in the time interval $[0, T]$, the Bloch vector after the last π-pulse is finally

$$\boldsymbol{R}(T) = \big(0, 0, -\cos^n(\pi/n)\big) \tag{3.206}$$

(instead of $\boldsymbol{R} = (0, 0, -1)$ without measurements). Hence the probability for level 2 at time $t = T$ is

$$P_2(T) = \frac{1}{2}\big(1 - \cos^n(\pi/n)\big) \approx \frac{1}{2}\left(1 - \exp\left(-\frac{1}{2}\pi^2/n\right)\right) \xrightarrow{n \to \infty} 0. \tag{3.207}$$

$P_2(T)$ is a monotonically decreasing function of n, going to zero as n goes to infinity. This means that in this limit only $|1\rangle$ will be found, and the transition to $|2\rangle$ is completely inhibited.

In the actual experiment (Itano $et\ al.$ 1990) levels 1 and 2 were realized by hyperfine sublevels of the $2s^2S_{1/2}$ ground state of $^9Be^+$-ions, level 3 was a $2p^2P_{3/2}$ state, decaying only to level 1. The number of emitted photons per pulse was about 72 per ion, only a small fraction was actually detected, however. After necessary corrections, the measured data closely followed (3.207).

Further experimental tests were proposed using neutrons (Pascazio $et\ al.$ 1993, Inagaki, Namiki and Tajiri 1992, for an analysis of experimental limitations in reaching the Zeno limit see Nakazato $et\ al.$ 1995); an optical realization with beam-splitters was suggested by Agarwal and Tewari (1994). For an application to the theory of cellular automata see Grössing and Zeilinger (1991).

Evidence for non-exponential decay at short times was found in observations of tunneling of atoms from potential well (Wilkinson $et\ al.$ 1997). In this experiment a sample of about 10^5 ultra-cold sodium atoms is trapped by a magnetic field and intersecting laser beams. The observed initial survival probability was flat for short times and, after a sharp fall-off, turned into exponential decay. In this way the initial quadratic Zeno region, exponential decay, as well as the transition region (sometimes called "anti-Zeno effect") were observed in a single experiment (Fisher, Gutiérrez-Medina and Raizen 2001).

The destruction of coherence by the environment may also lead to changes in the dynamics of chemical reactions. Solvation effects are usually treated by classical models. However, quantum models, for example for the photodissociation of the HF-Na van der Waals complex, show dynamical changes similar to those discussed above, cf. Prezhdo (2000) and references therein.

Schemes similar to Cook's proposal were suggested for inhibiting unwanted transitions, see for example Agarwal, Scully and Walther (2001)[50]. For an experiment similar to the Cook/Itano setup, but testing the Zeno effect on a *single* ion, see Balzer *et al.* (2000).

3.3.1.3 Models for the Quantum Zeno Effect. So far we have applied the rules for calculating probabilities of observations, i. e. the collapse of the wave function, directly to the system of interest. In the example of the previous subsection the "system" consisted of only two states of a multi-level atom. As already discussed in the introductory sections, the cut between observer and observed system can be shifted into the macroscopic range by including more and more of the environment to the observed system. This is useful for understanding the dynamical processes involved. A measurement certainly takes a finite time (in the above example, level 3 must be excited and a photon be emitted). Clearly, all relevant interactions can be treated more completely by using the Schrödinger equation (even though this does not solve the collapse problem).

A simple extension of the above two-state model already takes into account the essential physical elements: the third atomic level as well as the emitted photons are also described quantum-mechanically (Ballentine 1991b).

If we start at $t = 0$ with the ion in level 1, then without further interactions the driving of the $1 \leftrightarrow 2$ transition would lead to the usual Rabi oscillations, while the relevant mode of the radiation field is not changed,

$$|\Psi(t)\rangle = [\cos(\Omega t/2)\,|1\rangle + \sin(\Omega t/2)\,|2\rangle]\,|\text{vac}\rangle. \qquad (3.208)$$

The measurement pulse at $t = \tau$ couples levels 1 and 3, for simplicity here described by a π-pulse interaction (leading to a complete inversion [51], $|1\rangle \leftrightarrow |3\rangle$), hence

$$|\Psi(\tau)\rangle = [\cos(\Omega\tau/2)\,|3\rangle + \sin(\Omega\tau/2)\,|2\rangle]\,|\text{vac}\rangle. \qquad (3.209)$$

Now level 3 decays rapidly back to level 1, leaving as a trace a photon in the respective field modes (we denote this state by $|\gamma\rangle$),

$$|\Psi(\tau^+)\rangle = \cos(\Omega\tau/2)\,|1\rangle\,|\gamma\rangle + \sin(\Omega\tau/2)\,|2\rangle\,|\text{vac}\rangle. \qquad (3.210)$$

This is a *correlated* state of the same structure as in a unitary description of an ideal measurement (compare (3.16)). At this stage coherence between levels

---

[50] Note that these authors emphasize that their work had nothing to do with the Zeno effect.

[51] If the laser field is so weak that the probability of emitting a fluorescence photon is small, (3.209) contains a superposition of $|1\rangle$ and $|3\rangle$. Correspondingly, in (3.210) the state of the radiation field correlated with $|1\rangle$ contains a superposition of $|\text{vac}\rangle$ and $|\gamma\rangle$. Since now the "pointer states" in (3.210) are no longer orthogonal (incomplete measurement), coherence is destroyed only partially and the dynamical Zeno effect is reduced (Peres and Ron 1990).

1 and 2 is already irreversibly delocalized, since $\langle\gamma|\mathrm{vac}\rangle = 0$ (This process would still be reversible, however, if the atom were confined to a reflecting cavity, where the emitted photon could not escape!). The two above steps are now repeated, so that the action of the driving field during $\tau < t < 2\tau$ leads to

$$|\Psi(2\tau)\rangle = \cos(\Omega\tau/2)[\cos(\Omega\tau/2)\,|1\rangle + \sin(\Omega\tau/2)\,|2\rangle]\,|\gamma\rangle$$
$$+ \sin(\Omega\tau/2)[\cos(\Omega\tau/2)\,|2\rangle - \sin(\Omega\tau/2)\,|1\rangle]\,|\mathrm{vac}\rangle\,, \qquad (3.211)$$

where the second line originates from the unitary evolution of state $|2\rangle$. Applying the measuring pulse with subsequent decay of level 3 yields

$$|\Psi(2\tau^+)\rangle = \cos^2(\Omega\tau/2)\,|1\rangle\,|2\gamma\rangle$$
$$+ \cos(\Omega\tau/2)\sin(\Omega\tau/2)\,|2\rangle\,|\gamma\rangle$$
$$+ \cos(\Omega\tau/2)\sin(\Omega\tau/2)\,|2\rangle\,|\mathrm{vac}\rangle$$
$$- \sin^2(\Omega\tau/2)\,|1\rangle\,|\gamma'\rangle\,. \qquad (3.212)$$

(The two one-photon states appearing in the last equation are orthogonal.) Every step adds new components corresponding to different "alternatives" (see also Section 3.4.1 for a discussion of various approaches using "histories" see Chap. 5), the final probability of finding level 1 at $t = \pi/\Omega$ is again given by (3.207).

Alternatively one could consider the time dependence of the atomic 3-state-system, either derived from the pure-state description outlined in the previous subsection (Pearle 1989, Petrosky, Tasaki and Prigogine 1990, 1991, Pascazio and Namiki 1994, Mihokova, Pascazio and Schulman 1997) or directly use the appropriate 3-level Bloch equations (Block and Berman 1991, Frerichs and Schenzle 1991, Altenmüller and Schenzle 1993). Especially the naive use of optical Bloch equations, though formally correct, hides the conceptual problems, since these equations already implicitly contain the collapse of the wave function.[52] For stochastic models, using the concept of "quantum trajectories" see Beige and Hegerfeldt (1996) and Power and Knight (1996). One should keep in mind the questionable status of such approaches (cf. Chap. 8 and Sect. 3.4.3).

In order to analyze the freezing process that apparently occurs because of repeated measurements it is useful to consider a quite general model, where a two-state system is continuously coupled to a "pointer", following the discussion given by von Neumann in his quantum theory of measurement (Kraus 1981, Joos 1984).

---

[52] We strongly disagree with statements made by some of these authors, suggesting that the use of "irreversible quantum mechanics" more or less eliminates the measurement problem.

A two-state system is assumed to interact with a measurement device via the Hamiltonian

$$
\begin{aligned}
H &= H_0 + H_{\text{int}} \\
&= V(|1\rangle\langle 2| + |2\rangle\langle 1|) + E|2\rangle\langle 2| + \gamma\hat{p}(|1\rangle\langle 1| - |2\rangle\langle 2|), \quad (3.213)
\end{aligned}
$$

where γ is a coupling constant and $\hat{p}$ shifts the "pointer" wave function to the "left" or to the "right", depending on the state of the measured system. The measurement is complete when the pointer wave functions for the two possible outcomes no longer overlap, that is, for times $t > B/\gamma$, where B is the width of the pointer wave packet. Since we are interested in the limit of strong coupling, V can be treated as a perturbation (although this model can be solved exactly, see Kraus 1981, Joos 1983, Chumakov, Hellwig and Rivera 1995). With the initial state $|1\rangle|\Phi\rangle$ the probability of finding the two-state system in state $|2\rangle$ is in Born approximation given by

$$
\rho_{22}(t) = \begin{cases} V^2 t^2 + O(t^3), & t \ll B/\gamma \\ \dfrac{\pi V^2}{\gamma}\left|\tilde{\Phi}\left(\dfrac{E}{2\gamma}\right)\right|^2 t, & t \gg B/\gamma \end{cases} \quad (3.214)
$$

where $\tilde{\Phi}(p)$ is the momentum representation of the pointer state. An example is shown in Fig. 3.20. For short times, the transition probability grows quadratically in time, in accordance with the general result (3.177). Interestingly, the system shows constant transition rates for $t \gg B/\gamma$, similar to a "decaying" state described by a master equation. This will become more transparent in the extended model of Sect. 3.3.2.1. These rates, however, depend on the details of the interaction, and are suppressed for strong coupling, when the Zeno effect dominates. Figure 3.21 shows an example of how the transition probability depends on the coupling constant γ.

This and similar models show that the quantum Zeno effect can be explained by means of the dynamical (unitary) coupling to a measurement device. Other models have been considered by Milburn (1988), Greenland and Lane (1989), Damnjanovic (1990); for a comparison of the above model with the Itano experiment see Chumakov, Hellwig and Rivera (1995); for an analysis in the framework of algebraic quantum theory see Blanchard and Jadczyk (1993), a general review is given by Nakazato, Namiki and Pascazio (1996). A derivation of the Zeno effect via a variant of the adiabatic theorem was proposed by Facchi (2002). With regard to the saying that "a watched pot never boils", one could put it the way, that "it is not so much that one watches the pot, but that the combined system and observer no longer can boil" (Schulman 1998).

These models show, that the reason for the Zeno effect lies in the dislocalization of phases needed for the natural (isolated) evolution of a system. Even recoil-free measurement-like interactions lead to a slowing down.

Systems with only a few states are typical "quantum systems". It is therefore not surprising that they are very sensitive to being monitored by other

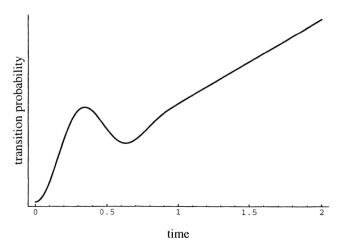

Fig. 3.20. Time dependence of the probability of finding state $|2\rangle$ if the system was prepared in $|1\rangle$ at $t = 0$, under continuous (strong) coupling to a meter. For very small times this function grows quadratically as is generally required (see (3.177)). As soon as the pointer can resolve the two states (at $t \approx 1$ in this example) it grows linearly in time (constant transition rates), see Joos (1984).

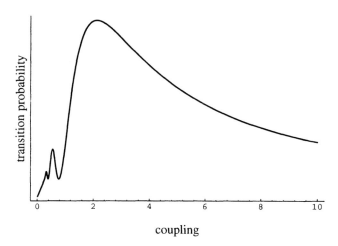

Fig. 3.21. Probability of finding state $|2\rangle$ at a fixed time as a function of the coupling constant γ. For strong coupling the transitions are always suppressed by a factor $1/\gamma$.

systems. The question of why macrosystems are not "frozen" by the Zeno effect will be adressed in the next section.

If the time-independent coupling to the "pointer" in (3.213) is replaced by an impulsive coupling (leading to a completion of measurement in a short interval), a sequence of such interactions produces an entangled wave function containing a sum over all possible "histories" (sequences of measurement results). Even with a one-dimensional pointer, all measurement results can be registered if the coupling constant γ is doubled each time, i. e., the binary representation of the final position of the pointer contains the information about each single result. Such a model (and its realization via photon polarization measurements) was considered by Dicke (1989). The relation between pulsed and continuous coupling was investigated by Schulman (1998). See also Sect. 3.4.1.

As mentioned in Sect. 3.2.3, for chiral molecules, the Zeno effect may drastically increase the stability of spatially well-defined configurations, since the spatial structure of large molecules is continuously registered by the environment (Simonius 1978, for models of position measurements of an atom confined in a double-well potential see Gagen, Wiseman and Milburn 1993, Altenmüller and Schenzle 1994). This enhancement of stability may be especially important for biological systems. For a numerical study of the influence of collisions on transitions in a multilevel molecule see Bruno *et al.* (2002).

The Zeno effect can possibly also be used in quantum computers in order to "stabilize" results and reduce errors (Chuang *et al.* 1995), see also Facchi and Pascazio (2002).

3.3.2 Master Equations

We have already encountered master equations when considering the interaction of a dust particle with its natural surroundings. Here we will discuss other master equations from the viewpoint of decoherence. From the host of special equations we will pick out only a few examples which display most clearly the role of decoherence. A more general framework, in particular the formalism developed by Zwanzig, will be presented in Chap. 7.

3.3.2.1 Pauli Equation. In many situations, transitions between "properties" α of a certain system are successfully described by rate equations of the form

$$\frac{d}{dt}P_\alpha = \sum_\beta A_{\alpha\beta}P_\beta = \sum_{\beta\neq\alpha}(A_{\alpha\beta}P_\beta - A_{\beta\alpha}P_\alpha) \tag{3.215}$$

where P_α is the probability for the occurrence of property α, while $A_{\alpha\beta}$ are transition rates. This is a typical equation for a classical random process. In quantum theory, "properties" are represented by states (or groups of states) and P_α are diagonal elements of the density matrix describing the system. Thus, (3.215) is replaced by

$$\frac{d}{dt}\rho_{\alpha\alpha} = \sum_\beta A_{\alpha\beta}\rho_{\beta\beta}. \tag{3.216}$$

Such equations describe an autonomous dynamics for the *diagonal* part of the density matrix $\rho_{\alpha\beta}$ in a *given* basis. Equations of this kind are often derived by starting from a von Neumann equation for the system and applying suitable approximations (Pauli 1928, van Hove 1957, see also Kreuzer 1981). In particular, a repeated and time-directed application of the so-called "random-phase approximation" is necessary to obtain the desired results. Obviously, in (3.216) the interference terms $\rho_{\alpha\beta}$ ($\alpha \neq \beta$) seem to play no dynamical role.[53] Their complete absence, however, would be at variance with the assumption of a von Neumann equation, since

$$i\frac{d}{dt}\rho_{\alpha\alpha} = \sum_{\beta}(H_{\alpha\beta}\rho_{\beta\alpha} - \rho_{\alpha\beta}H_{\beta\alpha}) \equiv 0 \tag{3.217}$$

if

$$\rho_{\alpha\beta} = \rho_{\alpha\alpha}\delta_{\alpha\beta}. \tag{3.218}$$

This extreme result, the freezing of the motion of a system by total destruction of interference, is equivalent to the quantum Zeno paradox.

If the properties α are "macroscopic", they have to be considered as being continuously measured by their environment, as was the case for the dust particle discussed in Sect. 3.2.1. On the other hand, we have seen in the preceding section that intense coupling to the environment strongly influences transition rates and in extreme cases can again lead to a complete freezing of any transitions. This does not seem to be the case in situations where rate equations are found useful. The decay rate of a nucleus, for example, is independent of its chemical environment and decay is never slowed down by the presence of a detector. This can be understood by extending the continuous measurement model (3.213) discussed above to the case of macroscopic states, each represented by *many* microstates (Joos 1984).

Following Pauli's work one may consider a many-state system with Hamiltonian

$$H = \sum_{\alpha E} E\,|\alpha E\rangle\,\langle\alpha E| + \sum_{\alpha E \neq \alpha' E'} V_{\alpha E, \alpha' E'}\,|\alpha E\rangle\,\langle\alpha' E'| \tag{3.219}$$

where α is assumed to distinguish between macroscopically different properties. We focus on transitions caused by the "perturbation" V between the α's. One should be aware of the fact that the notion of "perturbation" is only defined with respect to a given basis; if H were diagonalized from the beginning, there would be no perturbation term! Often the "unperturbed"

---

[53] The absence of non-diagonal terms in (3.216) cannot be adequately explained by assuming random phases for the contributing transition amplitudes, an argument still found in many textbooks; if the transition rate is calculated in Born approximation (as was first done by Pauli), then the total rate contains, for an N state system, contributions from N diagonal terms and from $N(N-1)$ non-diagonal elements. Even under the assumption of random phases, the latter would give a contribution of the same order $\sqrt{N(N-1)} \approx N$ as the diagonal terms (Joos 1983). Hence a further mechanism to suppress interference terms is needed.

Hamiltonian is defined by dividing the system into certain subsystems. For example, a few eigenstates of an atom may be coupled to a "reservoir" of many states of the radiation field. The "perturbation" terms then represent the coupling between these "local" systems, leading to the decay of excited states, for example. On other occasions it may be appropriate to include more degrees of freedom into the "undisturbed" set of states ("dressing").

When we start with the most general initial state with property α_0,

$$|\Psi(0)\rangle = \sum_{E_0} c_{\alpha_0 E_0}(0) |\alpha_0 E_0\rangle, \qquad (3.220)$$

the probability of finding another property α is in Born approximation given by

$$\rho(\alpha E, \alpha E, t) \approx 2\pi t \sigma_{\alpha_0}(E_0)|V(\alpha E, \alpha_0 E_0)|^2 \rho(\alpha_0 E_0, \alpha_0 E_0, 0) \qquad (3.221)$$

for times $t \gg 1/\delta$, where $\sigma|V|^2\rho$ is approximately constant over the resonance range δ ($\sigma_\alpha(E)$ are level densities). This is essentially "Fermi's Golden Rule". Integration of the resulting master equation

$$\dot{\rho}(\alpha E, \alpha E, t) \approx \frac{\Delta\rho}{\Delta t} = \sum_{\alpha'} A_{\alpha\alpha'}(E)\rho(\alpha' E, \alpha' E, t) \qquad (3.222)$$

requires that interference terms between different properties α play no dynamical role. On the other hand, they would have to occur according to the von Neumann equation if there are transitions at all.

Motivated by Pauli's approach we will now look in some detail at this rather general model for transitions of a system while it has some its properties continually monitored by the environment. This will allow us to find in general terms a wide range of behavior from Zeno freezing to transitions governed by Fermi's Golden rule, unaffected (but motivated) by coupling to a pointer. Also intermediate deviations from the Golden Rule, where measurement increases transition rates (sometimes termed "anti-Zeno effect" or "inverse Zeno effect") can easily be found in this model (Joos 2003).

Let again states be characterized by their enery E with respect to an "undisturbed" Hamiltonian, while the label α is supposed to be monitored by the environment, here modelled by a simple pointer wave function (see Fig. 3.22). The total Hamiltonian of our model then consists of three parts, H_0 characterizing the basis states, V causing transition between these states, and H_{meas} containing the interaction with a measuring device (or the natural environment). Hence we are led to consider the dynamics governed by

$$\begin{aligned} H &= H_0 + V + H_{\mathrm{meas}} \\ &= \sum_{\alpha E} E |\alpha E\rangle \langle \alpha E| + \sum_{\alpha E \neq \alpha' E'} V_{\alpha E, \alpha' E'} |\alpha E\rangle \langle \alpha' E'| \\ &\quad + \sum_{\alpha E} \gamma(\alpha)\hat{p} |\alpha E\rangle \langle \alpha E|. \end{aligned} \qquad (3.223)$$

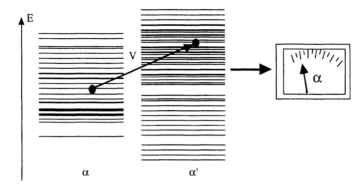

Fig. 3.22. Transitions between groups of states are monitored by a pointer. The symbolic measurement device in the figure represents the interaction with the environment (which may or may not contain an experimental setup). Transition probabilities often follow Fermi's Golden rule (rates governed by transition matrix elements V and level densities at resonance energy), but may be influenced by the presence of the environment monitoring certain features α of initial or final states.

Starting with some initial state with property α_0

$$|\Psi(0)\rangle = |\alpha_0 E_0\rangle \, |\Phi\rangle \,, \tag{3.224}$$

where $|\Phi\rangle$ is the state of the environment, we use the interaction picture with respect to V,

$$|\Psi^I\rangle = e^{i(H_0 + H_{\mathrm{meas}})t} \, |\Psi\rangle \,. \tag{3.225}$$

Since both H_0 and H_{meas} are diagonal in the same basis we immediately get for the transformed Hamiltonian

$$V^I(\tau) = e^{i(H_0 + H_{\mathrm{meas}})t} V e^{-i(H_0 + H_{\mathrm{meas}})t} \tag{3.226}$$

$$= \int dp \sum_{\alpha E \neq \alpha' E'} V_{\alpha E, \alpha' E'} \, e^{i(E - E' + [\gamma(\alpha) - \gamma(\alpha')]p)t} \, |\alpha E\rangle \, |p\rangle \, \langle p| \, \langle \alpha' E'| \,.$$

The initial state (3.224) evolves in lowest order perturbation theory (Joos 1983, Joos 1984, Ruseckas and Kaulakys 2001) into

$$|\Psi(t)\rangle = \int dp \, |p\rangle \, \Phi(p) \left\{ e^{i(E_0 + \gamma(\alpha_0))t} |\alpha_0 E_0\rangle \right. \tag{3.227}$$

$$\left. - \sum_{\alpha E \neq \alpha' E'} V_{\alpha E, \alpha' E'} \frac{e^{-i(E + \gamma(\alpha)p)t} \big(e^{i(E - E_0 + [\gamma(\alpha) - \gamma(\alpha_0)]p)t} - 1 \big)}{(E - E_0) + [\gamma(\alpha) - \gamma(\alpha_0)]p} |\alpha E\rangle \right\} \,.$$

The density matrix $\rho_{\alpha E, \alpha' E'}$ can be obtained from the relative states $|\Phi_{\alpha E}\rangle$, defined by

$$|\Psi\rangle = \sum_{\alpha E} |\alpha E\rangle \, |\Phi_{\alpha E}\rangle \tag{3.228}$$

as

$$\rho_{\alpha E, \alpha' E'} = \langle \Phi_{\alpha' E'} | \Phi_{\alpha E} \rangle. \tag{3.229}$$

The transition probability into a *single* state which is different from the initial one is then

$$P_{\alpha E} = \langle \Phi_{\alpha E} | \Phi_{\alpha E} \rangle \tag{3.230}$$

$$= 4 \int dp \, |V_{\alpha E, \alpha_0 E_0}|^2 |\Phi(p)|^2 \frac{\sin^2(E - E_0 + [\gamma(\alpha) - \gamma(\alpha_0)]p)t/2}{(E - E_0 + [\gamma(\alpha) - \gamma(\alpha_0)]p)^2}.$$

If there are many final states with the same property α, occurring with level density $\sigma_\alpha(E)$, the final result is,

$$P_\alpha = 4 \int dp \, dE \, \sigma_\alpha(E) |V_{\alpha E, \alpha_0 E_0}|^2 |\Phi(p)|^2 \frac{\sin^2(E - E_0 + [\gamma(\alpha) - \gamma(\alpha_0)]p)t/2}{(E - E_0 + [\gamma(\alpha) - \gamma(\alpha_0)]p)^2}. \tag{3.231}$$

If we turn off the interaction with the pointer ($\gamma = 0$), the usual Golden Rule result is recovered, with transition rates proportional to matrix elements and level densities at the resonance peak, as described above.

From (3.230) we can see that the coupling to a monitoring environment leads to the following changes:

(1) the center of resonance is shifted from its original value $E = E_0$ to the new value $E = E_0 - (\gamma(\alpha) - \gamma(\alpha_0))p$. This means, transition rates can be expected to change in various ways. In particular, new transitions, previously forbidden by energy conservation may come into range (this appears to be the background of the "anti-Zeno effect" discussed by some authors, see below). For large coupling the shift must eventually lead to an off-resonance condition, hence to a suppression of all transitions; this is again the Zeno effect.

(2) Every value of p produces a new resonance, with weight $|\Phi(p)|^2$. So even if there is only a discrete and possibly small set of target states, the broadening induced by the coupling to a multi-state (in our model continuous) pointer leads to the same effect as is well-known from the Golden Rule: the transition probability increases linearly in time, leading to constant transition rates (although these are generally dependent on the details of the interaction).

Except for very small times we may thus in (3.230) use the replacement

$$\frac{\sin^2(xt/2)}{x^2} \longrightarrow \frac{\pi}{2} t \, \delta(x). \tag{3.232}$$

We obtain as our modified probability for the appearance of the final state $|\alpha E\rangle$

$$P_{\alpha E} = 2\pi t \frac{|V_{\alpha E, \alpha_0 E_0}|^2}{\gamma(\alpha) - \gamma(\alpha_0)} \left| \Phi \left(\frac{E - E_0}{\gamma(\alpha) - \gamma(\alpha_0)} \right) \right|^2. \tag{3.233}$$

To simplify notation we set $\gamma(\alpha_0) = 0$ in the following (only the difference between measurement scales does matter). As mentioned above, for each value of p we have a contribution $2\pi t |V(E_{\text{res}})|^2$ at $E_{\text{res}} = E_0 - \gamma(\alpha)p$ with weight $|\Phi(p)|^2$. The width of this dynamically created resonance is

$$< (E - E_0)^2 >= \gamma(\alpha) \langle \Phi | p^2 | \Phi \rangle. \qquad (3.234)$$

The total transition probability into the new property α can then be calculated by summing over final states with level density $\sigma_\alpha(E)$,

$$P_\alpha = \int dE \, \sigma_\alpha(E) \, P_{\alpha E} \qquad (3.235)$$

$$= 2\pi t \int dE \, \sigma_\alpha(E) \, |V_{\alpha E, \alpha_0 E_0}|^2 \frac{\left| \Phi\left(\frac{E-E_0}{\gamma(\alpha)}\right) \right|^2}{\gamma(\alpha)}. \qquad (3.236)$$

This is the central result of our model. The most interesting situations are the following.

(1): Zeno limit. For large $\gamma(\alpha)$ we have

$$P_\alpha \approx \frac{2\pi t}{\gamma(\alpha)} \int dE \, \sigma |V|^2(E_0) |\Phi(0)|^2 \sim \frac{1}{\gamma(\alpha)}, \qquad (3.237)$$

which is always decreasing with $\gamma(\alpha)$, except under the unrealistic assumption that $\sigma |V|^2$ is constant over an arbitrarily large energy interval (compare the discussion on the validity of the exponential decay law at the beginning of Sect. 3.3.1.1).

(2) Golden Rule limit. For small $\gamma(\alpha)$ we can replace

$$\frac{\Phi\left(\frac{E-E_0}{\gamma(\alpha)}\right)}{\gamma(\alpha)} \longrightarrow \delta(E - E_0) \qquad (3.238)$$

and recover Pauli's result

$$P_\alpha = 2\pi t \, \sigma_\alpha(E_0) |V(E_0)|^2. \qquad (3.239)$$

At the same time, interference terms between different α are damped to a small value (of the order $\frac{B V_{\alpha\alpha'}}{\gamma(\alpha)}$), where B is the width of the pointer wave function.

(3) Anti-Zeno effect. In realistic settings, the available energies are limited to a certain range $[E_{\min}, E_{\max}]$. If we assume σV^2 to be nearly constant in this range, we find from (3.236)

$$P_\alpha = \frac{2\pi t}{\gamma(\alpha)} \sigma |V|^2 \int_{E_{\min}}^{E_{\max}} dE \, |\Phi|^2 = 2\pi t \sigma |V|^2 \int_{\frac{E_{\min} - E_0}{\gamma}}^{\frac{E_{\max} - E_0}{\gamma}} dz \, |\Phi(z)|^2 \qquad (3.240)$$

This expression changes smoothly from the small Zeno value (3.237) $2\pi t(E_{\max} - E_{\min}) \sigma |V|^2(E_0) \frac{|\Phi(0)|^2}{\gamma(\alpha)}$ to the Golden rule result proportional to the norm of

transition rate

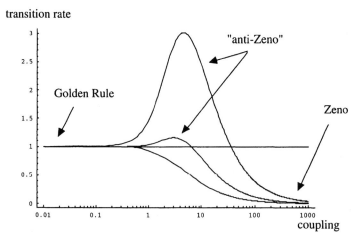

Fig. 3.23. Continuous coupling to a pointer changes the transition rate from an initial state $|\alpha_0 E_0\rangle$ to a group of final states in various ways. For small coupling we find the standard Golden rule result (here normalized to unity). Increasing the coupling to the measuring agent may in some cases increase the transition probability by shifting the effective resonance frequency to regions with higher level density or larger transition matrix elements (anti-Zeno effect). Strong interaction always leads to decreasing transition rates (Zeno effect).

Φ. To have some form of anti-Zeno effect, that is, transition rates increasing with measurement intensity, obviously requires a non-monotonous behavior of $\sigma|V|^2$, whatever the form of the pointer wavefunction.

An example is shown in Fig. 3.23, where the transition rate for weak coupling is constant (and equals the standard Golden Rule value at resonance energy), shortly increases (we took $\sigma|V|^2$ proportional to some power of E in this example) and finally decreases as required by the Zeno argument. Similar results were found in a model with repeated projection on energy eigenstates (Kofman and Kurizki 2002), or in a model of electron tunneling from a quantum dot (Elattari and Gurvitz 2000), see also Facchi, Nakazato and Pascazio (2001), where the term "inverse Zeno effect" was used.

The model shows that coupling to the environment leads to constant transition rates which are unaffected by the measurement if the coupling is "coarse enough" to discriminate only between macroscopic properties. This may in turn be used to *define* what qualifies a property as macroscopic: it must be robust against monitoring by the environment (see also Sect. 3.3.3). Since the transition rates are now independent of the interference terms, we have effectively an autonomous dynamics for the probabilites $\rho_{\alpha\alpha}$. This dynamics can be described by the master equation (3.215) to a very good approximation. The exponential decay law can thus be viewed as *enforced* by continuous measurement of the decay status by the environment (compare the argument following (3.186)). In contrast, if the decaying system is placed in a reflecting cavity, decay is no longer exponential (Haroche and Kleppner 1989).

Rate equations provide an excellent model for many (open) systems. Of course, as soon as quantum coherence plays a dominant role, e.g., in driven quantum systems, the neglect of non-diagonal terms as in the Pauli equation is no longer adequate. This was found in the early descriptions of nuclear spin resonance (Bloch 1946), later in laser theory (Lax 1966, Gordon 1967, Haken 1984), and atom-laser interactions (where optical Bloch equations are a common tool, see Loudon 1983, Gardiner 1991, Walls and Milburn 1994, Scully and Zubairy 1997). We will discuss some simple examples in the next subsection. The monography of Weiss (1999) presents a detailed account of open-system theory.

3.3.2.2 Lindblad's Form of Master Equations. Lindblad has derived a general expression for quantum-mechanical master equations from certain axioms (in particular the existence of a semigroup, which presupposes time-directed irreversibility and an idealized Markov approximation). The general status of such equations will be discussed in Chap. 7 in more detail. In this subsection we are only interested in their relation to decoherence.

The general form of the dynamics as given by Lindblad can be written as

$$i\frac{\partial}{\partial t}\rho = [H, \rho] - \frac{i}{2}\sum_k \left(L_k^\dagger L_k \rho + \rho L_k^\dagger L_k - 2L_k \rho L_k^\dagger\right) \tag{3.241}$$

where L_k are operators acting in the Hilbert space of states (Gorini, Kossakowski and Sudarshan 1976, Lindblad 1976). Since this equation is local in time (in contrast to the exact pre-master equation (7.31)), it is obviously Markovian. If the operators L_k are hermitean, the non-unitary part in (3.241) can be written as a double commutator,

$$i\frac{\partial}{\partial t}\rho = [H, \rho] - \frac{i}{2}\sum_k [L_k, [L_k, \rho]], \tag{3.242}$$

a structure we already encountered in 3.71).

In Sect. 3.1.3 we discussed the dynamical consequences of scattering processes: A measurement-like interaction will in this case exponentially damp interference in a certain basis which is defined by the properties of the interaction. Each scattering process of the form

$$|n\rangle |\Phi_0\rangle \xrightarrow{\text{scatt.}} |n\rangle S_n |\Phi_0\rangle \tag{3.243}$$

implies for the reduced density matrix

$$\rho_{nm} \xrightarrow{\text{scatt.}} \rho_{nm}\langle\Phi_0 | S_m^\dagger S_n | \Phi_0\rangle, \tag{3.244}$$

hence

$$\frac{\partial}{\partial t}\rho_{nm}\bigg|_{\text{scatt.}} = -\lambda\rho_{nm}(t). \qquad (3.245)$$

Such a damping of coherence without change of the diagonal elements is represented by *hermitean* operators L in the Lindblad equation, as can easily be seen as follows.

Let

$$L = \sum_n f(n)\,|n\rangle\,\langle n| \qquad (3.246)$$

be hermitean (i. e. $f(n)$ real). Then the non-unitary part of (3.241) reads

$$\frac{\partial\rho}{\partial t}\bigg|_{\text{non-unitary}} = -\frac{1}{2}\left(L^\dagger L\rho + \rho L^\dagger L - 2L\rho L^\dagger\right) \qquad (3.247)$$

$$= -\frac{1}{2}\sum_n f^2(n)\left(|n\rangle\,\langle n|\,\rho + \rho\,|n\rangle\,\langle n|\right) \qquad (3.248)$$

$$+ \sum_{n,m} f(n)f(m)\,|n\rangle\,\rho_{nm}\,\langle m| \qquad (3.249)$$

and the kl matrix element $\rho_{kl} = \langle k\,|\rho|\,l\rangle$ obeys

$$\frac{\partial\rho_{kl}}{\partial t}\bigg|_{\text{non-unitary}} = -\frac{1}{2}\big(f(k) - f(l)\big)^2\rho_{kl}(t), \qquad (3.250)$$

containing (3.245) as a special case. Entropy always increases for hermitean operators L. For the motion of a dust particle (3.71), L is proportional to the position operator, $L = \sqrt{2\Lambda}\hat{x}$. Note that (3.101), which includes friction, does not have the Lindblad form.

Equation (3.250) implies an exponential damping of coherence for super-positions of eigenstates of the Lindblad generator L. For a harmonic oscillator, two cases are worth considering. If $L \propto a^\dagger a$ (a "phase-damped oscillator"), we have

$$\dot{\rho} = -i\omega\left[a^\dagger a, \rho\right] - \frac{\gamma}{2}\left[a^\dagger a, \left[a^\dagger a, \rho\right]\right], \qquad (3.251)$$

corresponding to a non-destructive measurement of photon number (see also Sect. 3.3.3.3). Coherence between *number* states is then damped exponentially according to

$$\langle n\,|\rho(t)|\,m\rangle = e^{-i\omega(n-m)t}e^{-\gamma t(n-m)^2/2}\,\langle n\,|\rho(0)|\,m\rangle, \qquad (3.252)$$

while diagonal terms are unaffected[54]. On the other hand, in the standard case of a damped oscillator the Lindblad generator is proportional to the

---

[54] Decoherence in the *energy* basis has been suggested as *modification* of the fundamental Schrödinger equation by Milburn (1991). An equation of the form (3.251) appears as an approximate dynamics in this theory. For discussions of this model see Finkelstein (1993), Milburn (1993), Moya-Cessa *et al.* (1993), Flores (1995), Chen, Kuang and Ge (1995).

destruction operator, $L \propto a$. Since a has coherent states as eigenstates, one finds for the dynamics,

$$\dot{\rho} = -i\omega \left[a^\dagger a, \rho \right] + \frac{\gamma_\downarrow}{2} \left(2a\rho a^\dagger - a^\dagger a\rho - \rho a^\dagger a \right) \tag{3.253}$$

approximately (since L is not hermitean) an exponential damping of interference for a superposition of different *coherent* states (while a *single* coherent state remains pure). If the environment has non-zero temperature an additional contribution with $L \propto a^\dagger$ appears,

$$\dot{\rho} = -i\omega \left[a^\dagger a, \rho \right] + \frac{\gamma_\downarrow}{2} \left(2a\rho a^\dagger - a^\dagger a\rho - \rho a^\dagger a \right) + \frac{\gamma_\uparrow}{2} \left(2a^\dagger \rho a - aa^\dagger \rho - \rho aa^\dagger \right), \tag{3.254}$$

where the ratio $\gamma_\uparrow/\gamma_\downarrow$ is given by the Boltzmann factor $\exp(-\omega/k_B T)$. The probability $P(n) = \langle n \,|\rho|\, n \rangle$ of finding n quanta follows a Pauli master equation,

$$\partial_t P(n) = \gamma_\downarrow [(n+1)P(n+1) - nP(n)] + \gamma_\uparrow [nP(n-1) - (n+1)P(n)]. \tag{3.255}$$

In contrast to the situation considered in the last subsection, non-diagonal terms are not small, but their dynamics decouples from that of the occupation probabilities.

The behavior of the damped oscillator is discussed in many quantum optics textbooks, see also Milburn and Walls (1988) and Phoenix (1990).

The decomposition of the Lindblad generator L in (3.246) is obviously connected to the structure of the von Neumann interaction term in (3.14). If the system is already in an eigenstate of this interaction, it is stationary with respect to this influence (see also Sect. 3.3.3). If we look at the non-unitary part (3.242) of the Lindblad equation, this means that ρ is stationary if it commutes with the generator L (in some stochastic models which "unravel" the master equation, the approach to a such a stationary state is called "localization" (Garraway and Knight 1994, Percival 1994), see also Chap. 8).

Usually the time-scales and the form of the system–environment coupling are fixed for a given system. On the other hand, for an ion confined in an electromagnetic trap the interaction can easily be controlled, e.g. by changing the frequencies or intensities of laser beams. In this way it is possible to study how decoherence, described by Lindblad equations with various generators L, affects an experimentally accessible system (Poyatos, Cirac and Zoller 1996).

3.3.3 Dynamical Stability of States

One of the most characteristic features distinguishing quantum from classical states is their sensitivity to measurements. While we could imagine many kinds of measurements on classical systems, at least in principle, without disturbing the system in an essential way, this is no longer true for genuinely quantum systems. On the other hand, our everyday experience of a classical world relies heavily on predictable, stable behavior of classical objects. One could go even further and *define* classical systems by the property of being immune (to a certain extent) to the influence of their environment.

3.3.3.1 Sensitivity to the Presence of an Environment.

Which states are robust and which are most sensitive to the presence of an environment? The quantum theory of measurement tells us that there exists a special set of states which are not changed during measurement, namely eigenstates of the observable to be measured. The corresponding observable is defined by the interaction Hamiltonian coupling the system to the measurement device (or environment), as described in Sect. 3.1.1. If the density matrix of the system before measurement commutes with the observable, the system is completely insensitive to this measurement. Thus, if $[\rho, H_{\text{int}}] = 0$, all states diagonalizing ρ are obvious candidates for this sort of dynamical stability. In general, however, the state of the system will be altered, either by the system's own dynamics, or during measurement according to von Neumann's projection postulate (describing *phenomenologically* the dynamics of measurement),

$$\rho_{\text{after}} = \sum_i P_i \rho_{\text{before}} P_i, \qquad (3.256)$$

where P_i are projectors corresponding to the different possible outcomes of the measurement. The above equation may readily be generalized to a sequence of measurements, that is

$$\rho_{\text{after}} = \sum_{i,j,\dots,n} P_n \cdots P_j P_i \rho_{\text{before}} P_i P_j \cdots P_n. \qquad (3.257)$$

If P_i are Heisenberg operators which take into account the dynamics of ρ, this leads to the concept of histories (see also Chap. 5). Generalizations of (3.257) are sometimes used in stochastic models for describing continuous measurements (see Chap. 8).

The sensitivity of a given state may be assessed by comparing ρ_{before} and ρ_{after}. Obviously, if $[\rho, H_{\text{total}}] = 0$, the state will not change at all. Eigenstates of conserved quantities are robust by definition, but generally such states will not be available for local subsystems. For example, in quantum Browian motion the interaction favors localized states, while position is not conserved since the system Hamiltonian $H = \frac{p^2}{2m} + V(x)$ is not diagonal in the pointer basis. Hence one cannot expect to find states which remain completely unaffected by the environment.

Nevertheless, one can ask which states among all possible states of a system are least sensitive to "disturbance" by the environment (Zurek 1993a, Zurek, Habib and Paz 1993). Such a "predictability sieve" would single out those states which are most robust and therefore "classical".

There are several methods to accomplish this task. They are all concerned with the question of how strongly (and how fast) a system *becomes* entangled with its environment. For this purpose, one could, for example, study the behavior of local entropy. Alternatively the de-separation parameter A introduced in Sect. 3.1.1 gives a measure of how fast an initially factorizing state evolves into an entangled one. Of course, these two quantities are intimately related to each other. We will first discuss the time dependence of entropy.

In the case of quantum Brownian motion, decoherence is efficient on a much shorter timescale than friction (compare Sect. 3.2.2.3 on page 89), so the influence of the environment can approximatively be described by the equation of motion (3.71)

$$i\frac{\partial}{\partial t}\rho = [H_{\text{system}}, \rho] - iA[x, [x, \rho]]. \tag{3.258}$$

In the (unrealistic) case of "perfect robustness" a pure initial state would remain pure and evolve along a certain trajectory in Hilbert space. On the other hand, a typical quantum state would rapidly become quantum-correlated with the environment, that is, local entropy will increase. In the following we use the increment of linear entropy,

$$S_{\text{lin}} = \text{tr}\left(\rho - \rho^2\right), \tag{3.259}$$

as a measure of robustness. Pure states correspond to $S_{\text{lin}} = 0$, while $0 \leq S_{\text{lin}} \leq 1$. Its rate of change is given by

$$\frac{d}{dt}S_{\text{lin}} = -2\,\text{tr}\left(\rho\dot{\rho}\right)$$
$$= 2\,A\,\text{tr}\left(\rho^2 x^2 - \rho x \rho x\right) \tag{3.260}$$

for the evolution described by (3.258).

If the system is initially in the pure state $|\varphi\rangle$, this leads to an entropy increase proportional to the *spatial* dispersion of the initial state,

$$\frac{d}{dt}S_{\text{lin}} = 4A\left(\langle x^2\rangle - \langle x\rangle^2\right), \tag{3.261}$$

with $\langle x\rangle = \langle\varphi\,|x|\,\varphi\rangle$ and $\langle x^2\rangle = \langle\varphi\,|\,x^2\,|\,\varphi\rangle$. As expected, localized states are singled out (compare (3.30) on page 53). The smaller the spread in position, the less the state of the system will be affected through the effective position measurement. In particular, superpositions of macroscopically different positions are quickly "destroyed".

It is not sufficient, however, to consider only the time derivative of S, since position is not a conserved quantity. This is particularly obvious for the case of the harmonic oscillator, where position and momentum interchange their roles from every quarter period to the next. Therefore, a more sensible measure is the entropy production during one period of oscillation. For an oscillator with period $\tau = 2\pi/\omega$ it is given by

$$S(\tau) = 2\Lambda \int_0^\tau dt \left\langle \varphi \left| \left(x - \langle x \rangle \cos \omega t + (m\omega)^{-1}(p - \langle p \rangle) \sin \omega t \right)^2 \right| \varphi \right\rangle$$
$$= \Lambda \left(\Delta x^2 + \frac{\Delta p^2}{m^2 \omega^2} \right) \tag{3.262}$$

(Zurek, Habib and Paz 1993). This quantity is minimized by wave packets with $\Delta x \Delta p = \frac{1}{2}$ and with position and momentum widths given by the ground-state wave function of the harmonic oscillator,

$$\Delta x^2 = \frac{1}{2m\omega}, \quad \Delta p^2 = \frac{m\omega}{2}. \tag{3.263}$$

Thus, for harmonic oscillators the well-known *coherent* states (Schrödinger 1926, Glauber 1963) are least sensitive to a coupling to an environment which "measures" position (For a generalization to harmonic chains see Tegmark and Shapiro 1994; cf. also Gallis 1996, Venugopalan 1999). This means that the formation of quantum correlations with the environment is rather slow for these states (in contrast to squeezed states (Milburn and Walls 1983a)). Similar mechanisms can be expected to explain the dominance of classical fields (described by coherent states for field modes) in situations where a strong coupling to matter is important, e. g. in lasers (see also Sect. 4.1). An example will be given below.

The robustness of certain states should, however, not be taken as an indication that such systems are truly classical in possessing a state by themselves at all times. Quantum entanglement is unavoidable (and implicitly contained in (3.258)), hence quantum nonlocality forbids "truly classical" states in a strict sense.

The method of the "predictability sieve" as outlined above may also be applied to a "free" particle. If we use Gaussian wave packets as test functions, the solution of the equation of motion (3.258), now with $H = \frac{p^2}{2m}$, can be used to calculate the linear entropy as a function of time for an initial state of width b (see also (3.75) and Fig. 3.3),

$$S_{\text{lin}} = 1 - \left(\frac{3b^2}{4(\Lambda m)^2 b^2 \tau^4 + 2\Lambda m \tau^3 + 24\Lambda m b^4 \tau + 3b^2} \right)^{\frac{1}{2}}. \tag{3.264}$$

As a function of $\tau = \frac{t}{m}$ it increases monotonically from 0 (pure state) to 1, while it shows a local minimum at $b = \left(\frac{\tau}{2\sqrt{3}} \right)^{\frac{1}{2}}$ when viewed as a function of the initial width b. Since this minimum depends on time, there is no unique optimal width, although the entropy clearly displays a "valley of robustness" (see Fig. 3.24). As in the case of the oscillator, it is appropriate to consider the time-averaged (or time-integrated) entropy production, that is, to minimize the function

$$I(b) = \int\limits_0^\infty \left(S_{lin}(\tau) - 1 \right) \, d\tau. \qquad (3.265)$$

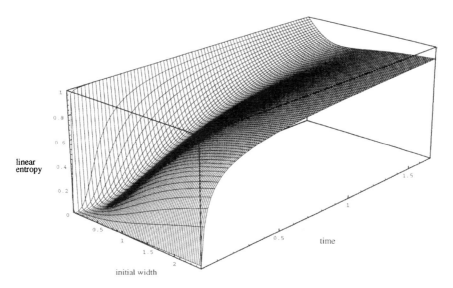

linear
entropy

initial width

time

Fig. 3.24. Linear entropy as a function of the width of the initial wavepacket and rescaled time $\tau = \frac{t}{m}$ for $\Lambda m = 1$. The effectiveness of entanglement ($S > 0$) depends strongly on the initial value of the width b.

The time-integrated linear entropy is shown in Fig. 3.25. As numerical analysis shows, its minimum is obtained at $b_{\mathrm{opt}} \approx 0.5(\Lambda m)^{-\frac{1}{4}}$. This length also follows from a dimensional analysis of (3.72), which contains a length scale $l_0 = (\Lambda m)^{-\frac{1}{4}}$ and a time scale $t_0 = \sqrt{\frac{m}{\Lambda}}$, see also Diósi and Kiefer (2002).

Similar results can be achieved by using the de-separation parameter A introduced in Sect. 3.1.1 (see (3.27) on page 53). The value of A yields a measure of short-time entanglement and thus is related to the treatment in (3.260) and (3.261) above.

For two coupled harmonic oscillators with Hamiltonian

$$H = \omega \left(a_1^\dagger a_1 + a_2^\dagger a_2 \right) + \varepsilon \left(a_1^\dagger + a_1 \right) \left(a_2^\dagger + a_2 \right) \qquad (3.266)$$

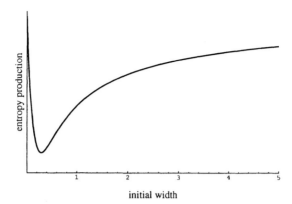

Fig. 3.25. Time-integrated entropy production $I(b)$ as a function of the width b of the initial wavepacket for $\Lambda m = 10$, showing the existence of an "optimal width" given by $b_{\mathrm{opt}} \approx 0.5(\Lambda m)^{-\frac{1}{4}}$.

one obtains (Kübler and Zeh 1973)

$$A = \langle \varphi_0 \Phi_0 | H(\mathbb{1} - |\varphi_0\rangle \langle \varphi_0|)(\mathbb{1} - |\Phi_0\rangle \langle \Phi_0|)H | \varphi_0 \Phi_0 \rangle$$
$$= \varepsilon^2(4Nn + 2N + 2n + 1), \tag{3.267}$$

if both oscillators are initially in number eigenstates, $|n\rangle_1$ and $|N\rangle_2$. For large quantum numbers (often identified with the "macroscopic limit" according to the correspondence principle) the systems become entangled very rapidly. This shows that the usual textbook arguments connecting large quantum numbers with classical physics are certainly wrong. These states are dynamically much too unstable to represent classical states. As shown above, coherent states are most robust.[55] This can also be seen by calculating the de-separation parameter for the interaction (3.266).

Let the initial state of system 1 be a squeezed state,

$$|\varphi_0\rangle = D(\alpha)S(\xi) |0\rangle, \tag{3.268}$$

where

$$D(\alpha) = \exp(\alpha a^\dagger - \alpha^* a) \tag{3.269}$$

---

[55] As repeatedly emphasized, the consequences of decoherence depend on the peculiarities of the coupling of a system to its environment. Clearly, then, the classical nature of coherent states is restricted to those "normal" situations which we have described extensively in this chapter. An interesting exchange of the role of coherent states and number eigenstates is found in the interaction of a two-level atom with a single-mode radiation field. If the radiation field is in a number state, single-frequency Rabi flopping occurs as in the semiclassical theory. On the other hand, for a coherent field state, a complicated dynamics ("collapse and revival") results (see Eberly, Narozhny and Sanchez–Modragon 1980, Rempe, Walther and Klein 1987, Gea–Banacloche 1990). This shows that "classical behavior" is very much dependent on the concrete situation.

generates a coherent state with complex amplitude α (defined by $a \left| \alpha \right\rangle = \alpha \left| \alpha \right\rangle$, see also (3.274)), and

$$S(\xi) = \exp \tfrac{1}{2} \left(\xi^* a^2 - \xi a^{\dagger 2} \right) \tag{3.270}$$

is the squeezing operator. Recall that for a harmonic oscillator with mass-frequency $\eta = m\omega$ the complex amplitude α is related to mean position and momentum by

$$\alpha = \bar{x} \sqrt{\frac{\eta}{2}} + \mathrm{i} \frac{\bar{p}}{\sqrt{2\eta}}, \tag{3.271}$$

so that the wave function in the position representation can be written as

$$\langle x | \alpha \rangle = \left(\frac{\eta}{\pi} \right)^{\frac{1}{4}} \exp \left[-\frac{\eta}{2} (x - \bar{x})^2 + \mathrm{i}\bar{p}x \right]. \tag{3.272}$$

This state obviously describes a Gaussian wave packet centered at $\bar{x}$ and with mean momentum $\bar{p}$.

It is convenient to express the squeezing parameter ξ in the form

$$\xi = r \mathrm{e}^{2\mathrm{i}\Phi} \tag{3.273}$$

($r = 0$ corresponds to a coherent state, Φ defines the direction of squeezing). If the second system is initially in the coherent state $\left| \beta \right\rangle$, for example, the de-separation parameter A has the form

$$A = \varepsilon^2 (\cosh 2r - \cos 2\Phi \sinh 2r). \tag{3.274}$$

A is *independent* of the coherent amplitudes α and β, and depends only on the squeezing parameters. The dependence on the phase of the squeezing operator can also easily be understood, see Fig. 3.26. Squeezing in the x-direction ($\Phi = 0$) decreases the width in x (while increasing it in p). The resulting state is less prone to the effect of position measurements (compared to the standard coherent state). On the other hand, squeezing in p ($\Phi = \pi/2$), broadening the wave function in x, leads to maximal entanglement. As discussed above, the time-averaged (or Φ-averaged) value (curve (2) in Fig. 3.26) is physically relevant in most environments.[56]

Another quite general example is represented by the action of "classical" oscillating fields on atomic states. In a semiclassical treatment this is often approximated by a Schrödinger equation (that is, unitary evolution) for the

---

[56] An important exception is of course given by quantum-optical experiments, where phase-sensitive interactions play a major role. Note that in the usual quantization schemes the role of position and momentum is played by quadrature components of the electric field (see also (3.297) below). For early tutorial reviews of squeezed states of light see Walls (1983) and Slusher and Yurke (1988).

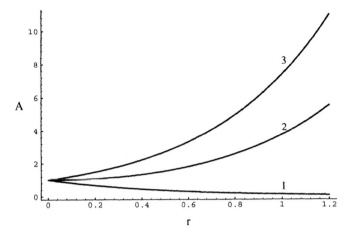

Fig. 3.26. Dependence of the de-separation parameter A on the squeezing param eter r for various values of the squeezing phase Φ for two interacting oscillato (H_{int} is given by (3.266) with $\varepsilon = 1$). (1) If the oscillator is squeezed in x-directic ($\Phi = 0$), sensitivity to the position-dependent coupling is reduced with increase squeezing. (2) For average phase any squeezing leads to increased decoherence con pared to a coherent state ($r = 0$). (3) Squeezing in momentum ($\Phi = \pi/2$) leads maximal broadening in space, hence to maximal decoherence.

atom alone, leading to well-known Rabi oscillations. In a full quantum trea ment of transitions between atomic states $|1\rangle$ and $|2\rangle$ one may start with Jaynes-Cummings Hamiltonian of the form

$$H = g(a^\dagger |1\rangle \langle 2| + a |2\rangle \langle 1|), \tag{3.27}$$

where g is an effective coupling constant, and $a, a^\dagger$ are mode operators for single-mode field. The calculation of the de-separation parameter

$$A = \langle \varphi_0 \Phi_0 | H (\mathbb{1} - |\varphi_0\rangle \langle \varphi_0|)(\mathbb{1} - |\Phi_0\rangle \langle \Phi_0|) H | \varphi_0 \Phi_0 \rangle \tag{3.27}$$

yields for an arbitrary initial state $|\varphi_0\rangle$ of the atom and the field in a cohere state $|\Phi_0\rangle = |\alpha\rangle$ the result

$$A = g^2 |\langle \varphi_0 | 2 \rangle|^4. \tag{3.27}$$

This means that the dynamics of the atom can remain unitary only as loi as it stays close to the ground state. As soon as there is a considerat contribution of the excited state, entanglement is unavoidable, although t degree of entanglement is *independent* of the field strength. In this sen the field acts similar to an external potential. Even if the driving field in a coherent state, however, the atomic dynamics necessarily acquires no unitary contributions from photon emission.

From the general expression for the de-separation parameter (cf. (3.2 – (3.30)) and the form of the interaction (3.266), it is obvious that (3.27

is proportional to the variance in x of the initial state. This resembles the behavior of linear entropy in (3.261). The short-time expansion of the Schmidt decomposition (3.2) always leads to a *quadratic* increase of linear entropy, in contrast to (3.260).

There is another interesting example which can be studied by this method. A superposition of field states with macroscopically different field strengths would also belong to the brand of "non-classical states". In quantum-optical discussions (Yurke and Stoler 1988, Phoenix 1990, Brune *et al.* 1992, Knight 1992, Davidovich *et al.* 1993, Garraway and Knight 1994, Schleich, Pernigo and Le Kien 1991, Goetsch, Graham and Haake 1995), superpositions of coherent states with different amplitude, for example

$$|\varphi_0\rangle = N(|\alpha\rangle + |-\alpha\rangle), \tag{3.278}$$

a superposition of coherent states with opposite phases, are called "Schrödinger cat states". Such states are similar to the example shown in Figs. 3.1 and 3.13. If this state is coupled to another oscillator by the interaction (3.266), the de-separation parameter becomes

$$A = \varepsilon^2 \left[1 + 2|\alpha|^2 \left(\cos 2\theta + \frac{1 - e^{-2|\alpha|^2}}{1 + e^{-2|\alpha|^2}} \right) \right] \tag{3.279}$$

(where $\alpha = |\alpha|e^{i\theta}$). For very non-classical states ($|\alpha| \gg 1$) one has

$$A = \varepsilon^2 \left(1 + 4|\alpha|^2 \cos^2 \theta \right). \tag{3.280}$$

The phase dependence has a similar origin as discussed above for squeezed states. Since Θ is usually a fast variable ($\Theta = \omega t$), averaging is appropriate in most situations, hence the sensitivity of "Schrödinger cat states" is given by

$$A = \varepsilon^2 \left(1 + 2|\alpha|^2 \right) \approx 2\varepsilon^2 |\alpha|^2. \tag{3.281}$$

Again, we find the result that non-classical states are extremely fragile when interacting with the environment (compare also Anglin and Zurek 1996).

An interesting way to realize such nonclassical states are cavity QED experiments. In particular, the interaction of highly excited Rydberg ($n \approx 50$) atoms with appropriate fields can be utilized to produce and detect mesoscopic quantum superpositions (Haroche 1995, Brune *et al.* 1996). If a superposition of two atomic states $|e\rangle$ and $|g\rangle$ is coupled off resonance to an (ideal, lossless) cavity containing n photons, the initial state has the form

$$|\Psi_1\rangle = (c_e |e\rangle + c_g |g\rangle) |n\rangle. \tag{3.282}$$

Because of detuning, the interaction leads only to a phase shift without exchange of energy,

$$|\Psi_2\rangle = \left(c_e e^{i\varepsilon(n+1)t} |e\rangle + c_g e^{-i\varepsilon n t} |g\rangle \right) |n\rangle. \tag{3.283}$$

Therefore, this scheme represents an ideal measurement of photon number (cf. (3.15)). The relative phases induced by the coupling to the photon field can be measured in a Ramsey interference experiment, a temporal analogue of Young's double-slit experiment. Note, that even for $n=0$, a phase shift exists ("vacuum Lamb shift"). If we replace the number eigenstate by a coherent state,

$$|\alpha\rangle = \sum_n d_n |n\rangle = \sum_n \frac{e^{-|\alpha|^2/2}\alpha^n}{\sqrt{n!}} |n\rangle \qquad (3.284)$$

the same interaction leads to an entangled state of the form

$$\begin{aligned}|\Psi_3\rangle &= c_e \sum_n d_n e^{i\varepsilon(n+1)t} |e\rangle |n\rangle + c_g \sum_n d_n e^{-i\varepsilon nt} |g\rangle |n\rangle \\ &= c_e e^{i\varepsilon t} |e\rangle |\alpha e^{i\varepsilon t}\rangle + c_g |g\rangle |\alpha e^{-i\varepsilon t}\rangle .\end{aligned} \qquad (3.285)$$

The internal state of the atom is now correlated to the phase of the field (the number of photons has not changed). As in EPR-experiments, it is important that one can freely choose the "observable" to be measured at the atom, i.e., energy eigenstates or linear combinations thereof. A measurement of the atomic energy (usually by field-ionization) would collapse the field state to either $|\alpha e^{-i\varepsilon t}\rangle$ or $|\alpha e^{+i\varepsilon t}\rangle$. In order to produce a superposition of such states, a $\pi/2$-pulse is applied *after* the atom has left the cavity (i.e. the atomic state is rotated according to $|e\rangle \to (|e\rangle + |g\rangle)/\sqrt{2}$ or $|g\rangle \to (|g\rangle - |e\rangle)/\sqrt{2}$, respectively). The (still entangled) state then reads

$$|\Psi_4\rangle = c_e e^{i\varepsilon t} \left(|e\rangle |\alpha e^{i\varepsilon t}\rangle + |g\rangle |\alpha e^{i\varepsilon t}\rangle\right)/\sqrt{2} + c_g \left(|g\rangle |\alpha e^{-i\varepsilon t}\rangle - |e\rangle |\alpha e^{-i\varepsilon t}\rangle\right)/\sqrt{2}. \qquad (3.286)$$

If now the atomic state is detected (i.e. we assume a collapse into $|e\rangle$ or $|g\rangle$), the field is prepared in one of the states

$$|\Phi^\pm\rangle = c_e e^{i\varepsilon t} |\alpha e^{i\varepsilon t}\rangle \pm c_g |\alpha e^{-i\varepsilon t}\rangle . \qquad (3.287)$$

What kind of state is produced depends on the experimental parameters, which may be chosen long after the interaction with the cavity has ceased ("delayed-choice experiment").

The *detection* of such a "phase cat" employs a similar mechanism. Such a detection must allow one to discriminate between a coherent superposition (3.287) and a mixture of component states. Since the field inside the cavity is difficult to measure directly, a second atom can be used as a detector. The correlation between the two atomic states shows interference terms, which are absent for a *mixture* of coherent states. The interference terms in (3.287) disappear on a time scale given by $T_{\text{cav}}/|\alpha|^2$ (T_{cav} is the characteristic damping time of the cavity), which can be interpreted as the average time for the absorption of the first photon in the cavity walls. This decoherence timescale shows the same structure that we already encountered in the discussion on quantum Brownian motion (cf. Sect. 3.2.2.3, Eq. (3.134)). Interferences disappear on a timescale which is given by the (dissipative) relaxation timescale times a factor which measures the "distance" between two states.

Instead of a "phase cat" as above, related experimental setups could be used to produce "amplitude cats" of the form $|\Phi^\pm\rangle = (|\alpha\rangle \pm |0\rangle)/\sqrt{2}$, or non-local (EPR) cat states of the form $|\Phi^\pm\rangle = (|\alpha\rangle |0\rangle \pm |0\rangle |\alpha\rangle)/\sqrt{2}$, by using two cavities. For detailed discussion of these schemes see Haroche *et al.* (1993), and Davidovich *et al.* (1996). A general overview is given by Haroche (1998).

Instead of cavity QED states one can also study formally quite similar situations for a single trapped ion. Here the motional states in an (effectively) harmonic potential play the role of photon states in QED. Simple superpositions of discrete energy eigenstates as well as coherent states can be produced by applying appropriate fields, manipulating the ions internal states as well as the confining potential. For experiments generating superpositions of localized states of motion in an ion trap see Monroe *et al.* (1996) and Myatt *et al.* (2000). These experiments generate decoherence in various conceptually quite different ways (compare Sect. 3.4.3).

3.3.3.2 Decoherence and Quantum Computation.

The realization that "information" must be considered as related to properties of objects in the physical world – and therefore be restricted by the laws governing these objects – can be traced back to discussions of Maxwell's demon, apparently able to violate the Second Law if he can gather information at no cost (Szilard 1929). Shannon's theory of information completely relied on classical concepts, which were found sufficient to exorcise Maxwell's demon (Landauer 1961, Bennett 1987, Leff and Rex 1990). In the last decade the extension of information theoretical concepts, incorporating the properties of quantum states, has led to a flourishing field of research (Nielsen 2002).

If information is fundamentally represented by material objects, then the laws of information theory should be governed by or derivable from the laws of physics – and not the other way round, as is sometimes suggested.

When the concept of a "bit" as representing two alternative states is replaced by a "qubit" (Schumacher 1995), the most general state for N qubits is a vector in 2^N dimensional Hilbert space,

$$|\Psi\rangle = \sum_x c_x |x\rangle, \tag{3.288}$$

where x runs over all 2^N values of a classical bit string. Quantum computations are implemented as reversible (unitary) operations, except for measurements (and "noise").

Since unitary operations depend on the precise definition of phases, any "disturbance" would lead to wrong results. Despite implementation difficulties, this field gained strong impetus when Shor discovered an algorithm (Shor 1994) for effectively factoring large numbers, a process which is believed to require exponential time for classical computers (current algorithms for factoring a number of L digits require a time $\propto \exp(L^{1/3})$, whereas quantum computers could solve this problem in $\propto L^2$ time.). The crucial feature is the possibility to put a quantum computer in any superposition state in a controlled way. In a sense, many input data are processed in parallel (Deutsch 1985). The desired result is extracted by an appropriate interference technique (Deutsch and Jozsa 1992, Chuang *et al.* 1995).

Error correction in classical communication theory is achieved by introducing some sort of redundancy, for example, by simply copying a message. This is not possible for qubits, since the linearity of quantum evolution does not allow "cloning" of arbitrary states.

Suppose, one wants to copy the state $|\psi\rangle$ to another degree of freedom, initially in some standard state $|0\rangle$ by a unitary operation U,

$$|\psi\rangle |0\rangle \longrightarrow U\left(|\psi\rangle |0\rangle\right) = |\psi\rangle |\psi\rangle. \tag{3.289}$$

Since this procedure should work for *any* target state, for example $|\psi\rangle$ and $|\varphi\rangle$, one would like to have

$$U\left(|\psi\rangle |0\rangle\right) = |\psi\rangle |\psi\rangle$$
$$U\left(|\varphi\rangle |0\rangle\right) = |\varphi\rangle |\varphi\rangle. \tag{3.290}$$

Taking the scalar product of these two equations yields

$$\langle\psi|\varphi\rangle = (\langle\psi|\varphi\rangle)^2 \tag{3.291}$$

hence $|\psi\rangle$ and $|\varphi\rangle$ must either be identical or orthogonal (Wootters and Zurek 1982, Dieks 1982). Arbitrary states cannot be cloned[57]. On the other hand, a "known state", that is, a state which is quantum-correlated with orthogonal environment states, can be copied at will.

There are other ways, however, to introduce the redundancy required for error correction. For example, a single qubit state may encoded in three qubits via

$$a|0\rangle + b|1\rangle \longrightarrow a|000\rangle + b|111\rangle. \tag{3.292}$$

This operation does not conflict with the no-cloning theorem. The idea of quantum error-correction codes (Shor 1995, Steane 1996, Bennett *et al.* 1996, Knill and Laflamme 1997) is to measure certain collective aspects of the encoded state or its "disturbed" version, which show what kind of error occurred and to correct the error accordingly. As an example, let a "bit-flip" occur in the first qubit,

$$a|000\rangle + b|111\rangle \longrightarrow a|100\rangle + b|011\rangle. \tag{3.293}$$

As in classical error correction some evidence about what kind of error occurred is needed. Evidently, a direct measurement of the bits would destroy

---

[57] The no-cloning theorem was an upshot of (erroneous) suggestions to use quantum correlations for signalling, thereby violating special relativity; for a historical account see Peres (2002).

the coherence required for further computation. One way out is to introduce "ancilla bits" and couple the above state via a measurement-like interaction to the ancilla bits to extract an "error syndrome", e. g., in the form ($\oplus$ means addition modulo 2, or "exclusive or")

$$|x_0 x_1 x_2\rangle |000\rangle \longrightarrow |x_0 x_1 x_2\rangle |x_0 \oplus x_1, x_0 \oplus x_2, x_1 \oplus x_2\rangle, \qquad (3.294)$$

resulting in

$$(a |100\rangle + b |011\rangle) |110\rangle. \qquad (3.295)$$

Measuring the ancilla state (with result 110) reveals that the first bit was flipped, hence a unitary operation can be applied to correct for this error. The ancillas act as an entropy sink: at every step, new "blank" ancillas have to be provided. Since the ancillas themselves are subject to "noise" it is a nontrivial question whether errors can be suppressed enough to achieve reliable calculations (Knill, Laflamme and Zurek 1998, Preskill 1998).

Even though there are intense efforts to implement quantum computing schemes experimentally, repeatedly severe objections have been raised (Landauer 1995, 1996, Haroche and Raimond 1996, Dyakonov 2001). For reviews of quantum computing and quantum communication see Bennett (1995), Nielsen and Chuang (2000), Bouwmeester, Ekert and Zeilinger (2000), Rieffel and Polak (2000), Galindo and Martín-Delgado (2002).

3.3.3.3 Quantum Nondemolition Measurements.

Predictable behavior of a system under the influence of repeated measurements is an important subject in the theory of "quantum nondemolition measurements". It originated from the problem of measuring gravitational waves, e. g. by using a Weber bar, where extremely small spatial displacements need to be detected (a typical number is 10^{-19} cm). Just measuring the relative position with the required accuracy would not suffice, since the subsequent spread of a prepared small wave packet will rapidly make a second position measurement (in order to test for the effect of a weak force) useless.[58] Position is therefore not an appropriate observable for this purpose. Momentum, on the other hand, is conserved for a free particle, so that a repeated observation would not be disturbed by the dynamics of the system. Such a quantity is called a "quantum nondemolition (QND) observable".

---

[58] Example: If we require for a gravitational wave detection an accuracy of $\Delta x = 10^{-19}$ cm, the uncertainty relation yields a momentum spread $\Delta p_0 \geq 1/(2\Delta x_0)$ for the initial preparation. During a time interval Δt, the free evolution will then lead to a position uncertainty of $\Delta x^2 = \Delta x_0^2 + \Delta p_0^2 \Delta t/m$ and therefore $\Delta x^2 \geq \Delta t/m$. This is called the "standard quantum limit". For a detector mass of 10 tons (Braginski, Vorontsov and Thorne 1980) and an expected period of $\Delta t = 10^{-3}$ s for the gravitational waves, this yields $\Delta x \geq 5 \cdot 10^{-19}$ cm, so that the sought effect may be completely hidden by "quantum noise" (Walls and Milburn 1994).

In the theory of QND measurements two steps are essential (Caves *et al.* 1980, Caves 1981, Meystre and Scully 1983, for reviews see Walls and Milburn 1994, Braginski and Khalili 1996, Bocko and Onofrio 1996). One must first find an appropriate observable that allows repeated measurements with great predictability, such as the momentum of a free particle in the above example. This requirement for an observable $A(t)$ is usually expressed in the form

$$\left[A^{\mathrm{I}}(t), A^{\mathrm{I}}(t')\right] = 0, \tag{3.296}$$

where A^{I} is the observable in the interaction picture. If the system is initially in an eigenstate of A, it will remain in that eigenstate, although the eigenvalue may change. Obviously, constants of the motion are always QND observables. For a harmonic oscillator, appropriate QND observables are the so-called quadrature phase operators

and
$$X_1(t) = a\mathrm{e}^{\mathrm{i}\omega t} + a^\dagger \mathrm{e}^{-\mathrm{i}\omega t} = \sqrt{2\omega}\left(x\cos\omega t - \frac{p}{\omega}\sin\omega t\right) \tag{3.297}$$

$$X_2(t) = -\mathrm{i}\left(a\mathrm{e}^{\mathrm{i}\omega t} - a^\dagger \mathrm{e}^{-\mathrm{i}\omega t}\right) = \sqrt{2\omega}\left(x\sin\omega t + \frac{p}{\omega}\cos\omega t\right). \tag{3.298}$$

They correspond to a rotating frame of reference with respect to x and p. These quantities are of particular importance in quantum-optical realizations of QND measurements.

The second condition to be fulfilled is the requirement that $A(t)$ be measured with some measurement device without "disturbing" the measured state. If the interaction Hamiltonian between the considered system and the meter is H_{int}, then

$$[A, H_{\mathrm{int}}] = 0 \tag{3.299}$$

guarantees a measurement without changing the state of the system. This amounts to an "ideal measurement" of A (measurement of the first kind), so A is a "pointer observable", as discussed in Sect. 3.1.1 (see also Walls, Collet and Milburn 1985). In the theory of QND measurements this condition is usually called "back action evasion". Clearly, H_{int} must then have the von Neumann form (3.14). For example, a QND scheme to measure the photon number $N = a^\dagger a$ (without absorbing the photon, as is usually the case in photodetection) could employ an interaction of the form

$$\begin{aligned} H_{\mathrm{int}} &= \chi a^\dagger a\, b^\dagger b \\ &= \sum_n |n\rangle\,\langle n| \otimes \left(n\chi b^\dagger b\right) \end{aligned} \tag{3.300}$$

where a refers to the to-be-measured signal field and b to a probe field. (Such a coupling can be realized in a four wave mixing process, see Milburn and Walls 1983b.) The second line in (3.300) shows that this interaction amounts to an ideal measurement of photon number. A detailed presentation of QND criteria in optical applications can be found in Walls and Milburn (1994). The relation between QND measurements and the Zeno effect has been discussed by Zurek (1984b).

3.3.4 Decoherence and Quantum Chaos

The motion of heavenly objects always stood for regularity and predictability. The notion of chaos, on the other hand, symbolizes unorderly, complicated and unpredictable motion. Strangely enough, chaos theory has its origin in investigations of the three-body Kepler problem by Poincaré. Poincaré also discussed the problem of weather forecast, a subject that helped to initiate modern chaos theories when the apparently simple Lorenz model displayed chaotic solutions (Poincaré 1908, Lorenz 1963; for a nontechnical account see Ruelle 1991). The difficulties associated with the solution of the three-body Kepler problem reappeared when the quantization of Helium failed in the early quantum theory (Einstein 1917). Periodic orbits are still used in certain approaches employing "semiclassical quantization" (Gutzwiller 1990). General references are Ford (1983), Ott(2002) (mainly classical chaos), Casati and Chirikov (1995a), Stöckmann (1999), and Haake (2001).

3.3.4.1 Classical Versus Quantum Chaos. One of the hallmarks of chaos is the extreme sensitivity of trajectories to initial conditions: In chaotic systems, small deviations can grow exponentially fast. A probability distribution in phase space tends to expand exponentially in certain directions (while it shrinks in other directions because phase space volume is preserved under deterministic evolutions). This exponential spreading is characterized by the largest Lyapunov exponent λ_L[59]

$$d(t) \approx d(0)e^{\lambda_L t}, \tag{3.301}$$

with $d(0)$ being some uncertainty in the initial position in phase space. This characterization of chaos cannot directly be applied to quantum states, since there are no classical trajectories, and consequently there is no obvious measure replacing the notion of distance in phase space. Quite to the contrary, the natural measure of distance between two quantum states, their scalar product, is invariant under Hamiltonian evolution (It will nonetheless prove useful in Sect. 3.3.4.4 below).

---

[59] Local stability of orbits can be investigated by comparing a solution $z(t)$ of the given equation of motion with an arbitrarily close trajectory $z + \Delta z(t)$. Hamiltonian systems with N degrees of freedom follow $\frac{dz_i}{dt} = \sum_j \eta_{ij} \frac{\partial H}{\partial z_j}$ with $\eta = \begin{pmatrix} 0 & 1 \\ -1 & 1 \end{pmatrix}$. Then the deviations obey to leading order $\frac{d}{dt}\Delta z_i = \sum_{j,k} \eta_{ij} \frac{\partial^2 H}{\partial z_j \partial z_k} \Delta z_k$. This linear relation has a solution of the form $\Delta z(t) = M(t)\Delta z(0)$. The 2N eigenvalues of the (symplectic) stability matrix M come in quartets, $\lambda, 1/\lambda, \lambda^*$ and $1/\lambda^*$; their product is unity for conservative systems (Liouville's theorem). The largest Lyapunov exponent may then be defined by the norm of M, $\lambda_L := \lim_{t\to\infty} \frac{\ln\|M(t,0)\|}{t}$. If the stability of a periodic orbit (with period T) is analyzed, the Hamiltonian flow is replaced by a discrete map; the limit is then replaced by $\lambda_L := \lim_{n\to\infty} \frac{\ln\|M(T,0)^n\|}{nT}$.

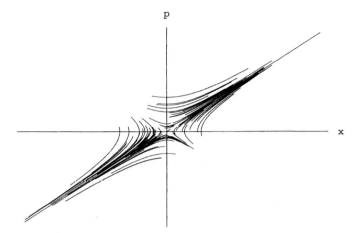

Fig. 3.27. An ensemble of trajectories starting near the origin of phase space for an inverted oscillator. Because of the local instability of the dynamics, the distribution expands exponentially in (x+p)-direction. The (semiclassical) evolution of *single* pure, initially localized, quantum state will then necessarily lead to a broad wavefunction, prone to strong decoherence.

The relevance of decoherence for the emergence of chaos has been over-looked for a long time. Nearly all studies were concerned with finding approximate ("quasiclassical") solutions of the Schrödinger equation, following the believe that "...all of the complicated classical behavior ... must be regarded as the limiting behavior of the corresponding quantal system when Planck's constant $\hbar$ is negligible. " (Berry 1978)

When it became obvious that unitary quantum theory can never lead to chaos in the classical sense (Ford and Ilg 1992), not only the "correspondence principle" (in one of its various forms) but also the completeness of quantum theory was put into question (Ford and Mantica 1992). Yet the notion of a classical orbit is a central concept in certain approximation methods which represent sophisticated extensions of the familiar WKB method (Gutzwiller 1990). The textbook treatments using a truncated version of the Ehrenfest theorem are insufficient for well-known reasons, in particular for nonlinear systems (cf. Sect. 3.2.2.2, see also Ballentine, Yang and Zibin 1994).

From our present point of view, the treatment of closed system evolving unitarily is clearly the wrong route from quantum to classical chaos. Indeed, it was found that the influence of "noise" drastically changes the behavior of these systems, for example by rapidly destroying localization (Ott, Antonsen and Hanson 1984). Historically, the instability typical for chaotic motion was found in discussion of the three-body Kepler problem. A classical analogue must include decoherence in the equation of motion while a naive application of the Schrödinger equation leads to results contradicting experience, such as a planetary wave function smeared out over the entire orbit (Zurek and Paz

1994). The resolution of these issues, offered by decoherence, should of course not be misinterpreted as influence of "noise" (compare Casati and Chirikov (1995b) and Zurek and Paz (1995b)). "Noise" is a classical concept and must not be identified with entanglement (see also Sect. 3.4.3).

Many studies of quantum systems with chaotic classical analogue have focused on the properties of energy spectra, but not so much on the transition from quantum to classical dynamics. It appears plausible, that the exponential divergence of trajectories should be reflected somehow in the energy level distribution. Indeed it was found that the nearest-neighbor distribution allows a classification. Typically (but exceptions do exist) it shows a Poissonian statistics for regular systems and resembles the Wigner distribution (level repulsion) of random matrix theory for chaotic systems (for details see Haake 2001).

A widely studied subfield involves systems driven by a periodically oscillating *classical* force. Such a Hamiltonian can only have phenomenological (and approximate) meaning for the special case of a driving field described by a coherent state. The latter can be justified only via decoherence (compare Sect. 3.3.3.1). The periodicity of time-periodic Hamiltonians allows one to introduce quasistationary Floquet eigenstates (cf. 3.315) and study the spectra of Floquet eigenvalues. Time periodic Hamiltonians are also of some experimental relevance (Stöckmann 1999, Friedrich 1998).

Many aspects of "quantum chaos", for example level distributions, parameter dependencies ("level dynamics") and even wave functions can be studied experimentally already in classical systems in view of the fact that the time-independent Schrödinger equation is often equivalent to a classical Helmholtz wave equation. This led to many experiments using sound waves in liquids or solids, microwaves in cavities of various shape ("quantum billards"), for an overview see Stöckmann (1999). One should, however, not be misled by these classical analogues. The very fact that these systems can be described by classical waves has its origin in decoherence.

Apparently, the common folklore is still that there must – according to the "correspondence principle" – exist some quantum states which should at least behave in a manner similar to their classical counterparts in an appropriate limit, such as large quantum numbers and/or small wavelengths. However, this hope faces serious problems in view of several examples where quantum systems typically show a behavior which is not even an approximation, but qualitatively different from of their classical analogues. One of the best understood effects is "quantum localization".

In the following we will briefly discuss the kicked rotator as a popular example, where the different behavior in the classical and quantum case can easily be studied. Then we turn to the important case of the motion of astronomical objects where a naive correspondence principle fails in a quite dramatic way. Inclusion of decoherence cures all these difficulties. The responsiveness to decoherence will turn out to yield an additional criterium to discriminate regular and chaotic systems.

3.3.4.2 Example: The Kicked Rotator.

One of the prototypical examples for quantization of chaotic systems is the one-dimensional kicked rotator. Here the difference in the behavior of the classical model, showing a diffusive spread in the chaotic regime, and the corresponding quantum model, showing dynamical localization, is most prominent. This model is not only of theoretical interest because of its simplicity, but is closely related to experimental situations, for example microwave ionization of hydrogen atoms, which may be described by a Hamiltonian of the form

$$H = \frac{p^2}{2m} - \frac{e}{r} + zF\cos(\omega t). \tag{3.302}$$

Here in addition to the Coulomb field the electron is subjected to an oscillating force in z-direction, given by the last term.

Replacing the oscillating force by a periodic kick[60] simplifies integration of the equation of motion. As a simple example consider the one-dimensional periodically kicked rotator which is defined by the Hamiltonian

$$H(p,\Theta,t) = \frac{p^2}{2I} + \frac{\lambda I}{\tau}V(\Theta)\sum_{n=-\infty}^{n=\infty}\delta(t-n\tau). \tag{3.303}$$

The coordinate Θ describes rotation around a fixed axis, p is the associated angular momentum. In the following we will usually set I and τ equal to unity.

Integrating the classical equations of motion from immediately after a kick to immediately after the next kick leads to the classical stroboscopic equations

$$p_{t+1} = p_t - \lambda V'(\Theta_{t+1}) \tag{3.304}$$
$$\Theta_{t+1} = \Theta_t + p_t \tag{3.305}$$

which are in this case formally identical to the Heisenberg equations of motion. The case

$$V(\Theta) = \cos(\Theta) \tag{3.306}$$

---

[60] These kicks are strictly periodic and should not be confused with any "random" force (which would generally lead to strong entanglement instead of a unitary evolution). Time-dependent Hamiltonians as in (3.302) are only an *approximation* valid under special conditions, e. g. when an atomic transition is driven by a coherent state of the electromagnetic field. The accuracy of this approximation is of some importance in quantum computing applications (van Enk and Kimble 2001). Compare also the de-separation parameter for the Jaynes-Cummings Hamiltonian (3.275) on page 142.

(which can be viewed as a periodically kicked pendulum) yields the so-called standard map (Casati *et al.* 1979)

$$p_{t+1} = p_t + \lambda \sin \Theta_{t+1} \tag{3.307}$$
$$\Theta_{t+1} = (\Theta_t + p_t) \bmod 2\pi \tag{3.308}$$

For sufficiently large λ, Θ_t describes a (pseudo-)random walk. If successive values of Θ are *assumed* to be uncorrelated, a diffusive behavior,

$$\overline{p_t^2} = \overline{p_0^2} + Dt \tag{3.309}$$

with diffusion constant

$$D = \frac{\lambda^2}{2\pi} \int_0^{2\pi} d\Theta\, V'(\Theta)^2$$
$$= \frac{\lambda^2}{2} \tag{3.310}$$

can be expected. This is confirmed by numerical results (see Fig. 3.28).

In the corresponding quantum model, the solution of the Schrödinger equation

$$i\frac{d}{dt}\Psi = H(t)\Psi(t) \tag{3.311}$$

with periodic Hamiltonian

$$H(t + n\tau) = H(t) \tag{3.312}$$

is characterized by the evolution operator over one period, again leading to a stroboscopic picture of the dynamics (a so-called "quantum map")

$$\Psi(n\tau) = F^n\Psi(0) \tag{3.313}$$

with the unitary Floquet operator $F = U(\tau)$.

For the special case of δ-kicks

$$H(t) = H_0 + \lambda V \sum_n \delta(t - n\tau) \tag{3.314}$$

the Floquet operator for evolution from immediately after one kick to immediately after the next one reads

$$F = U(\tau) = e^{-i\lambda V} e^{-iH_0\tau} \tag{3.315}$$

(We will set τ to unity in the following.) The time evolution of the quantum state may be discussed in the p-representation for the rotator

$$|\Psi(t)> = \sum_n a_n |n> \tag{3.316}$$

$$p|n> = n|n>, \tag{3.317}$$

while the kick term $V = \cos\Theta$ is diagonal in Θ representation

$$<\Theta|n> = \frac{e^{-in\Theta}}{\sqrt{2\pi}} \tag{3.318}$$

and

$$H_0|n> = \frac{n^2}{2}|n>. \tag{3.319}$$

The Floquet matrix F propagates the amplitude in momentum representation in the form

$$
\begin{aligned}
a_n(t+1) &= \sum_{n'} <n|F|n'> a_{n'}(t) \\
&= \sum_{n'} i^{(n'-n)} J_{n-n'}(\lambda) e^{-in'^2/2} a_{n'}(t)
\end{aligned} \tag{3.320}
$$

where J_k are Bessel functions of integer order.

Localization[61] means in this case that the spread in momentum does not increase indefinitely (and linearly in time) as in the classical model, but saturates after a certain number of kicks. This can be made plausible by expanding the time dependence in terms of the Floquet basis

$$F|\Phi_\nu> = e^{-i\Phi_\nu}|\Phi_\nu>. \tag{3.321}$$

Then the expectation value for the energy after N kicks reads

$$
\begin{aligned}
<p^2(N)> &= \sum_{\nu,\nu'} <\Phi_\nu|\Psi(0)><\Psi(0)|\Phi_{\nu'}> e^{-iN(\Phi_\nu - \Phi_{\nu'})} <\Phi_{\nu'}|p^2|\Phi_\nu> \\
&\approx \sum_\nu |<\Psi(0)|\Phi_\nu>|^2 <\Phi_\nu|p^2|\Phi_\nu>.
\end{aligned} \tag{3.322}
$$

If N is large, this expression becomes independent of the number of kicks and can therefore not grow indefinitely.

It has been shown (Grempel, Prange and Fishman 1984) that there is a strong analogy between the kicked rotator and Anderson localization in disordered solids (Anderson 1958, 1978). In both cases, the (quasi-)energy eigenstates are localized (momentum-localized Floquet states for the kicked rotator and spatially localized Bloch solutions for the Anderson model, respectively). In all these models, entanglement is ignored, however.

Localization is also found in other driven systems described by Hamiltonians similar to (3.302), but its presence or absence depends sensitively on

---

[61] This kind of "localization" is different from the localization induced by quantum correlations with the environment or the localization brought about by altering the Schrödinger equation (Chap. 8).

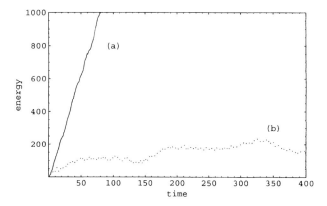

Fig. 3.28. Classical and quantum models of the kicked rotator lead to qualitatively different behavior. While the classical rotator shows diffusive increase in mean energy (a), the quantum rotator displays localization (b).

the precise structure of the dynamics (for a discussion of localization for the Hamiltonian $H(p, x) = p^2/2m - k \cos(x - l \sin t) + ax^2/2$ and its relation to decoherence see Karkuszewski, Zakrzeswki and Zurek 2001)

Localization requires coherence between states of different (angular) momentum. One may therefore expect that localization disappears as soon as some decoherence mechanism hinders coherent evolution. This was indeed found in various extensions of the above model.

One may choose, for example, a damping mechanism, so that the correspondence to the classical damped rotator (described by the so-called Zaslavsky's map) is maintained (Dittrich and Graham 1990a,b; for an alternative approach see Shiokawa and Hu (1995), for a review of dissipative quantum maps see Braun 2001). With the usual approximations, a Lindblad equation for the density matrix of the form

$$\dot{\rho} = \Lambda\rho = -i[p^2/2, \rho] + \gamma([I_-, \rho I_+] + [I_-\rho, I_+]) \tag{3.323}$$

with

$$I_- = I_+^\dagger = \sum_{n=0}^{\infty} \sqrt{n+1}(|n\rangle \langle n+1| + |-n\rangle \langle -n-1|) \tag{3.324}$$

may be derived. The corresponding quantum map for the damped kicked rotator then has the from

$$\rho(t+1) = e^{-i\lambda \cos \Theta} e^{\Lambda} \rho(t) e^{\lambda \cos \Theta} \tag{3.325}$$

The solutions to this dynamics show that even a weak damping, which does not noticeably disturb the classical motion, destroys localization and leads to a diffusive spread in momentum (as shown in curve (a) in Fig. 3.28). This effect resembles results discussed earlier for the free particle: Even weak coupling to the environment rapidly destroys coherence effects, while still leaving the motion of "robust" classical quantities unaffected. Again we have to draw the conclusion, that the disappearance of this quantum effect (localization) has much more to do with decoherence than with damping.

An alternative approach allows to revocer classical diffusion in the following way (Kaulakys and Gontis 1997). The time dependent dispersion in momentum (or energy) is given by (starting with $|p = 0\rangle$ for simplicity)

$$< p^2(t) >= \sum_n n^2 |a_n(t)|^2. \tag{3.326}$$

The change after one time step can be expressed as

$$< p^2(t+1) > = \sum_n n^2 |a_n(t+1)|^2$$

$$= \sum_{n,n'} n^2 J_{n-n'}^2 |a_n(t)|^2$$

$$+ \sum_n n^2 \sum_{n'<m'} J_{n-n'} J_{n-m'} \mathrm{Re}\,(\mathrm{b_{n'}} \mathrm{b_{m'}^*}) \tag{3.327}$$

with

$$b_n = \mathrm{i}^n \mathrm{e}^{-\frac{\mathrm{i}}{2}n^2} a_n. \tag{3.328}$$

The first term in the expression for $< p^2 >$ describes transition rates while the second term contains interference effects. If this second contribution is absent because of decoherence, the remaining time dependence is similar to the classical diffusion, since then

$$< p^2(t+1) > = \sum_{n,n'} n^2 J_{n-n'}^2 |a_n(t)|^2 \tag{3.329}$$

$$= < p^2(t) > + \frac{\lambda^2}{2}. \tag{3.330}$$

(using the identity $\sum_n n^2 J_{n-m}^2(\lambda) = m^2 + \frac{\lambda^2}{2}$)

3.3.4.3 Quantum (?) Chaos in the Solar System.
The strong dependence on initial conditions in nonlinear systems leads to the well-known exponential divergence in phase space, where an initially small compact volume is deformed locally in the way that it will shrink exponentially in one direction and expand exponentially in others, formally described by Lyapunov exponents. Since phase space volume is preserved, expansion is always accompanied by contraction (the sum of Lyapunov exponents is zero for Hamiltonian systems).

Most time scales for the solar system (given by the inverse of Lyapunov exponents) are estimated to be of the order of millions of years. For example, the obliquity of Mars (the angle between the spin axis of Mars and the normal to its orbit) is wildly chaotic, affecting the climate of Mars with a Lyapunov time of about 4 million yeras. There is strong evidence, that the orbit of all planets in the solar systems are chaotic (Sussman and Wisdom 1992, Laskar 1989, Murray and Holman 1999) in a similar time range. Some examples with much shorter times exist, for example the rotation of Hyperion, a moon of Saturn, with a Lyapuov exponent of $\lambda_L^{-1} \approx 20$ years.

Naive expectation would tell that planets or moons are so classical (because of their large mass) that quantum effects of chaos should never be of any relevance, even if the motion of these objects was (incorrectly) described by the Schrödinger equation for isolated systems. As with many expectations in the field of quantum chaos, this one turned out to be wrong, too.

In order to get an idea how instability in phase space translates into the quantum regime, consider a localized wave packet with an initial width $\Delta p(0)$ in the momentum distribution. If this momentum interval is squeezed exponentially with a (positive) Lyapunov exponent λ_L,

$$\Delta p(t) = \Delta p(0) \exp(-\lambda_L t) \tag{3.331}$$

then the uncertainty relations require an exponential divergence in position,

$$\Delta x(t) \geq \frac{\hbar}{\Delta p(0)} \exp(+\lambda_L t). \tag{3.332}$$

The classical description certainly breaks down as soon as Δx becomes so large that the nonlinearity of the potential becomes relevant. This scale may be estimated as (compare (3.335) below)

$$\chi \approx \sqrt{\frac{\partial_x V}{\partial_x^3 V}}. \tag{3.333}$$

The coherence length will reach the scale χ at the so-called Ehrenfest time

$$t_E = \frac{1}{\lambda_L} \log \frac{\Delta p(0)\chi}{\hbar}. \tag{3.334}$$

At this time a nonclassical situation will emerge: The wave is broad (in a sense like a Schrödinger cat state) and the notion of a trajectory becomes untenable. Even at the level of expectation values, classical behavior will be lost, since the Ehrenfest theorem ceases to be valid.

It is obvious that taking the naive limit $\hbar \to 0$ (leading to $t_E \to \infty$) is incorrect. Classicality does not follow from such as simple formal operation.

To put these heuristic arguments on a more solid basis one may consider the evolution of the system by means of the Wigner function (Habib, Shizume and Zurek 1998). Its equation of motion reads (cf. Sect. 3.2.3)

$$
\begin{aligned}
\dot{W}(x,p,t) &= \{H,W\}_{MB} \\
&= -i\sin(i\hbar\{H,W\}_{PB})/\hbar \tag{3.335} \\
&= \{H,W\}_{PB} + \sum_{n\geq 1} \frac{\hbar^{2n}(-1)^n}{2^{2n}(2n+1)!} \partial_x^{2n+1} V(x) \partial_p^{2n+1} W(x,p),
\end{aligned}
$$

where the last line requires the potential to be analytical. The so-called Moyal bracket $\{H, W\}_{MB}$ is equivalent to the von Neumann commutator $[H, \rho]$. The Poisson bracket term $\{H, W\}_{PB}$ is identical to the classical Liouville expression. Hence – ignoring problems of interpretation for the moment – one may reason that corrections to the classical motion are small, if the Wigner function is smooth enough, so that correction terms containing higher derivatives do not contribute significantly. But as argued before, this is impossible for times larger than the Ehrenfest time. At least at this time, quantum "corrections" can no longer be ignored – not even at the level of expectation values. An overestimate for the Ehrenfest time is (Berman and Zaslawsky 1978)

$$t \approx \frac{1}{\lambda_L} \log \frac{\text{action}}{\hbar}, \tag{3.336}$$

with "action" being some typical value characterizing the macroscopic system, certainly much larger than the Planck constant (and also much larger than initial phase space volume in (3.334)). These estimates are quite insensitive to the precise numerical values of the action (or $\hbar$, respectively), since the action enters only logarithmically (by contrast, the Ehrenfest time for non-chaotic systems scales according to some power law). After this time the planetary orbits should become completely nonclassical. This does not seem to be the case.

A particularly striking example is Hyperion's[62] chaotic rotational motion. An overestimate of the relevant action may be given by the product of Hyperion's orbital kinetic energy and its orbital period (Zurek and Paz 1995b, 1996). This yields an estimate of $t_E \approx 20$ years. Thus one would expect to find Hyperion in an extremely nonclassical state of rotation.

Taking into account the macroscopic nature of objects such as Hyperion, decoherence can now be included as described before. The equation of motion then reads

$$\dot{W} = \{H, W\}_{PB} + \sum_{n \geq 1} \frac{\hbar^{2n}(-1)^n}{2^{2n}(2n+1)!} \partial_x^{2n+1} V(x) \partial_p^{2n+1} W(x, p)$$
$$+ 2\gamma \partial_p(pW) + \Lambda \partial_p^2 W. \tag{3.337}$$

Here again the high-temperature model of Brownian motion (cf. Sect. 3.2.2.1) is employed. The contribution of the friction term $\gamma \partial_p(pW)$ can be neglected

---

[62] Hyperion was first discovered by W. Bond in 1848. Even though it moves in a regular elliptic orbit, its rate of rotation and the direction of its axis vary chaotically. Its unusal shape (roughly an ellipsoid with axes lengths of about 380, 290 and 230 km, respectively) was discovered in 1981 during the Voyager II mission. Later the irregular rotation was explained by gravitational influence of Saturn and of Titan, the largest moon of Saturn (Wisdom, Peale and Mignard 1984, Wisdom 1987). According to its tumbling rotation, the brightness of Hyperion varies chaotically (Klavetter 1989).

in most circumstances. Then two relevant time scales remain: The decoherence time (3.134)

$$t_{dec} = \frac{1}{\gamma} \left(\frac{\lambda_{dB}}{\Delta x} \right)^2,$$

(3.338)

limiting coherence, and the counter-acting effect of squeezing and expansion in phase space, which is characterized by a timescale $t_L \approx 1/\lambda_L$ given by the largest Lyapunov exponent. Both competing effects will compensate for typical scales

$$\Delta x \approx \lambda_{dB} \sqrt{\frac{\lambda_L}{\gamma}} \; ; \quad \Delta p \approx \sqrt{\frac{\Lambda_L}{\lambda}}.$$

(3.339)

For the planet Jupiter Δx turns out to be of the order $\Delta x \approx 10^{-29}$cm!

The interplay between decoherence and chaotic properties of the system also leads to a somewhat unusual behavior of entropy. The rate of entropy production is expected to be proportional to the coupling to the environment, that is, in this model, proportional to the localization rate Λ, as already shown in (3.76) and (3.261). This remains true for the initial stages. However, as soon as decoherence has effectively produced small wave packets, these will move quasi-classically along chaotic trajectories as in classical models. The rate of entropy production can then be expected to be given by the Lyapunov exponents alone, independently of the strength of decoherence.

Similar arguments were put forward for the Wigner function (Zurek and Paz 1994, 1995a,c) and found support in numerical studies (Monteoliva and Paz 2000).

3.3.4.4 Decoherence Through Chaotic Environments.

As already mentionend, it is difficult to find an appropriate measure for the distance between trajectories in order to give a quantum characterization of classical chaos. In classical physics there is an alternative way to describe chaotic dynamics: small changes in the Hamiltonian can also lead to large differences in the ensuing motion. This is not surprising, since different Hamiltonians automatically create different "initial" conditions. This latter characterization can be successfully translated into quantum theory. Since one and the same wavefunction evolves differently under (slightly) different Hamiltonians the overlap between the two versions may decrease rapidly if the dynamics is very sensitive to the structure of the Hamiltonian.

One may therefore consider the overlap

$$O(t) = | \langle \Psi(0) | U_{H_1}^{\dagger}(t) U_{H_2}(t) | \Psi(0) \rangle |^2$$

(3.340)

where H_1 and H_2 differ "slightly" as a measure of "chaoticity" (Peres 1993).

On the other hand, the overlap between two wave functions evolving under different Hamiltonians is important in measurement-like situations. For the case of the von Neumann interaction (3.14)

$$H_{\text{int}} = \sum_n |n\rangle \langle n| \otimes \hat{A}_n, \qquad (3.341)$$

interference terms in the density matrix are suppressed according to

$$\rho_{nm}(t) = \rho_{nm}(0) \langle \Phi_0| e^{i\hat{A}_m t} e^{-i\hat{A}_n t} |\Phi_0\rangle, \qquad (3.342)$$

$|\Phi_0\rangle$ being the initial state of the environmental degree of freedom. This mechanism was studied in the previous sections for many situations. A typicial and often used assumption is that the environment is described by many degrees of freedom, usually regular (like oscillators) and with weak coupling to a single degree of freedom. In particular, if the couplings $\hat{A}_n$ are "similar", only a small effect is to be expected. This may change drastically, if the environmental or pointer degree of freedom is chaotic, so that even a tiny n-dependence of the coupling leads to large changes in the evolution (Karkuszweski, Jarzynski and Zurek 2001).

To illustrate this idea we can use the kicked rotator model from above. Figure 3.29 shows the time dependence of the overlap $O(t)$ of wave functions evolving with nearby kick strengths λ. In the chaotic regime (large λ) the evolution of the wave functions depends in sensitive way on the kick strength, leading soon to orthogonality. Such an environment will therefore lead to very fast decoherence.

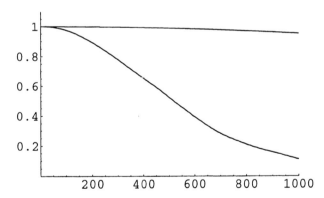

Fig. 3.29. The overlap between states evolving under slightly different ($\lambda_n = \lambda_m * (1 + 10^{-3})$) Hamiltonians for the kicked rotator. The upper line shows that usually similar Hamiltonians lead to similar states (here $\lambda = 0.5$) while when chaos reigns (lower curve, $\lambda = 5.1$), evolution from the same initial state easily leads to orthogonality and therefore decoherence.

3.4 Interpretational Issues

3.4.1 Null Measurements, Histories and Quantum Jumps

The concept of a null-measurement or negative result measurement was discussed long ago by Renninger (1960) and others (see also Sect. 9.1), and has again attracted attention in connection with experiments on single atoms. In contrast to earlier experiments, where an ensemble was employed to test quantum predictions – realized by working with many atoms at the same time – , these new experiments could implement the ensemble at most in the form of a time sequence of observations performed at the same object. It is therefore not surprising that this revived the age-old discussion about what quantum theory has to say about individual systems. Interestingly, the old dispute about the meaning of null measurements again played a role in the interpretation of quantum jump experiments with single atoms.

To understand the meaning of null measurements, it is important to realize that a null result is but one of several possible results of a measurement. It does not mean that nothing has happened to the system, or that the measurement device did not interact with it. Such interpretations would entertain a too classical picture. Consider, for example, a Stern–Gerlach device, set up to measure the z-component of spin with a (non-absorbing) detector positioned at the "upper" path (corresponding to $s_z = +\frac{1}{2}$) only. If the initial spin is in x-direction, for example, in half of all cases the detector will register the passage of a particle. A negative result means that the detector did *not* respond in a single run of the experiment. This ascertains that the particle leaving the apparatus has the spin $s_z = -\frac{1}{2}$. If a collapse is assumed as a consequence of the measurement, the state would change accordingly. If we extend the description by including the detector state (along the lines of the discussion in Sect. 3.1), a spin-up particle triggers the detector (changes its state from D_0 into $D_\uparrow$), while a spin-down particle does not, since such a wave packet does not interact with the detector at all, formally

$$|\uparrow\rangle \, |D_0\rangle \rightarrow |\uparrow\rangle \, |D_\uparrow\rangle \tag{3.343}$$

for spin up, but

$$|\downarrow\rangle \, |D_0\rangle \rightarrow |\downarrow\rangle \, |D_0\rangle \tag{3.344}$$

for the spin-down path which does not touch the detector. An x-polarized particle leads to an *entangled* state

$$(|\uparrow\rangle + |\downarrow\rangle) \, |D_0\rangle \rightarrow |\uparrow\rangle \, |D_\uparrow\rangle + |\downarrow\rangle \, |D_0\rangle \tag{3.345}$$

with $\langle D_\uparrow | D_0 | =\rangle 0$. Because the detector states are orthogonal, we find an apparent collapse of the spin states (provided the detector states are macroscopic, compare (3.21)). If the detector does not respond, one usually asserts that it is *known* that the particle took the lower path[63]. Clearly this interpretation is equivalent to assuming a collapse of the wave function. The claim that the particle did not interact at all with the detector in the case of a spin-down result must be wrong, since a spin-x state is different from an ensemble of z-up and z-down states (stated otherwise, the particle (wave) always takes *both* paths). For further discussions of null-measurements see Dicke (1986), Home and Whitaker (1992), and Elitzur and Vaidman (1993), compare also our description of so-called "interaction-free measurements" on page 116).

Repeated measurements, including null results, recently played a role in discussions about the observation of "Bohr's quantum jumps" in atoms. According to Bohr's 1913 theory, atoms only occur in energy eigenstates (certain classical orbits). The idea of their time-dependent superpositions was introduced much later. Upon emission or absorption of a photon they are assumed to "jump" between these stationary states. However, no sequence of such "jumps" can correctly predict coherence effects observed with superpositions of energy eigenstates. An operational definition in terms of repeated energy measurements was suggested by Cook (1988). Since an energy measurement projects the atom into an energy eigenstate, a sequence of such measurements should reveal the moment when the atom "jumps". If the measurements are made suffciently often, so that the individual jump probability is small compared to unity, one expects a "history" of measurement results where a sequence of identical results is from time to time interrupted by "quantum jumps", as displayed in Fig. 3.30. Note that these jumps are *not* identical with the collapse of the wavefunction upon measurement. A collapse (or something equivalent) happens at *every* measurement (every dot in the figure), whereas Bohr's jumps correspond only to *changes* of the measurement result. In a unitary description jumps do not happen at all (Zeh 1993), see also (3.367) below.

Following an earlier suggestion by Dehmelt (1975), Cook and Kimble (1985) proposed to observe the photon statistics of single-atom resonance fluorescence in a three-level system, as shown in Fig. 3.31. The ground state $|0\rangle$ is coupled to a metastable state $|2\rangle$, and the state of the two-level system $|0\rangle$, $|2\rangle$ can be measured by using another level $|1\rangle$ coupled by a strong optical transition to level $|0\rangle$. If the decay rates γ_1 and γ_2 of the two levels $|1\rangle$ and $|2\rangle$ are assumed to be very different, the following behavior may be expected (Cook and Kimble 1985): If the atomic electron is initially in state $|0\rangle$ it is quickly excited to level $|1\rangle$ which decays rapidly by spontaneous emission

---

[63] This is no longer true in Bohm's theory, where in a similar situation the detector could fire even if the "particle" took the other path, see e. g. Englert *et al.* (1992), Dürr *et al.* (1993), Dewdney, Hardy and Squires (1993). To assure consistency with quantum mechanics, the "ghost" particle travelling along the "wrong" path must then remain unobservable. Such results – discouraging the introduction of a particle trajectory as an additional concept – should, however, not be reason enough for a flight into an inconsistent Copenhagen phenomenalism (Englert *et al.* 1993).

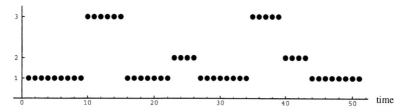

Fig. 3.30. A sequence of energy measurements as an operational definition of "quantum jumps" in an atom as proposed by Cook (1988). Note that in this example the first jump happens near $t = 10$, whereas a collapse of the wavefunction (or an equivalent branching) must be assumed to occur at least at every measurement (at $t = 1, 2, 3 \ldots$).

back to $|0\rangle$. Otherwise, if the electron is shelved in level $|2\rangle$, it is no longer available to scatter light through the strong transition, and the resonance fluorescence should be interrupted. Thus one expects a sequence of bright and dark periods in the fluorescence signal of the strong transition, a so-called random telegraph signal. Dehmelt's original suggestion aimed at using this disappearance of fluorescence to detect the presence of the metastable state.

Note that in an ensemble, where individual transitions could not be discriminated, fluorescence would not be interrupted by coupling to the metastable level, but only slightly reduced, according to the stationary solutions of the three-level Bloch equations. Therefore, the observation of this and related effects (photon-antibunching, for instance) requires the preparation of single quantum objects.

Traditionally, the statistical properties of a radiation field are characterized by intensity correlation functions. Fluctuations are described by functions such as

$$g^{(2)}(\tau) = \frac{\langle :I(t + \tau)I(t): \rangle}{\langle I(t) \rangle^2} \tag{3.346}$$

where

$$\langle :I(t + \tau)I(t): \rangle = \left\langle E^{(-)}(x, t)E^{(-)}(x, t + \tau)E^{(+)}(x, t + \tau)E^{(+)}(x, t) \right\rangle. \tag{3.347}$$

The intensity correlation function $g^{(2)}$ for the emission on the strong line of the three-level system shows two vastly different decay constants, indicating that for long times the emission of another photon becomes less likely. This provided a first hint for the existence of dark intervals in the fluorescence signal (Kimble, Cook and Wells 1986, Pegg, Loudon and Knight 1986, Schenzle and Brewer 1986, Schenzle, DeVoe and Brewer 1986). While the correlation function gives the probability that another photon is emitted at time τ on the strong transition, some authors considered instead the closely related probability $c(\tau)$ of finding the *next* photon (Cohen-Tannoudji and Dalibard 1986, Zoller, Marte and Walls 1987, Carmichael *et al.* 1989). $c(\tau)$ is found to

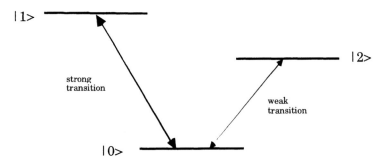

Fig. 3.31. Energy-level scheme for a single-atom experiment displaying "quantum jumps". The ground state $|0\rangle$ is weakly coupled to a metastable state $|2\rangle$. Transitions between $|0\rangle$ and $|2\rangle$ are monitored by driving the strong transition $|0\rangle \leftrightarrow |1\rangle$ and observing the emitted fluorescence photons. The same configuration can be used to explore the quantum Zeno effect (compare Fig. 3.18 on page 119). In the first experiment (Nagourney, Sandberg and Dehmelt 1986), the strong transition was realized by means of the $6^2S_{1/2} - 6^2P_{1/2} - 5^2D_{3/2}$ states of a laser-cooled trapped barium ion. The shelving level $|2\rangle$ was the $5^2D_{3/2}$ state with a lifetime of about 45 s.

be decaying again on two very different timescales. If the metastable state is not excited, c corresponds to a pure exponential decay with the fast decay rate γ_1. On the other hand, turning on a weak excitation of the metastable level $|2\rangle$ leads to the appearance of a slowly decaying second exponential. This is interpreted as indicating the occurrence of dark periods in the fluorescence of the strong transition, corresponding to a shelving of the electron in state $|2\rangle$. From the conditional probability c one builds the probability density $p_{[0,t)}(t_1, \ldots, t_n)$ which gives the probability that the atom emits (exactly) n photons at times $t_1, \ldots, t_n$ and no other photons in the time interval $[0, t)$. This probability can be written in the form

$$p_{[0,t)}(t_1, \ldots, t_n) = \left(1 - \int_{t_n}^{t} d\tau\, c(\tau)\right) c(t_n - t_{n-1}) \cdots c(t_1 - 0), \quad (3.348)$$

(Zoller, Marte and Walls 1987), that is, as a product of the conditional probability densities c describing photon emission at times $t_1, \ldots, t_n$ and the probability that no photon is emitted during the interval $[t_n, t]$. This expression shows the underlying Markov assumption: After emission of a photon, the atom can be assumed to be prepared in the ground state and the further evolution is independent of the previous history. A very readable exposition of theory and experiments has been given by Blatt and Zoller (1988).

The assumption that an atom is prepared in the ground state after emission of a photon gives an intuitive explanation of the phenomenon of *antibunching* (i. e. the short-time behavior of the correlation function, $g^{(2)}(0) = 0$). Usually "detection" and "emission" are identified in these treatments. (For an interesting discussion of the meaning of photon emission and photodetection see Carmichael *et al.* (1989), esp. Sect. 5.) Pegg and Knight argued that the calculations of the statistical properties of the observed fluorescence signal can be simplified considerably by using such collapse assumptions more explicitly (see also Porrati and Putterman 1987, Erber *et al.* 1989). They consider a measurement to be complete when photons are observed over a sampling time Δt which is short compared to the timescales of the weak transition, but large enough "that, if the atom is in one of the strong transition states $|1\rangle$ or $|0\rangle$ during that time, then there is a probability approaching unity that at least one photon will be detected" (Pegg and Knight 1988a). It is then assumed that a collapse occurs by either detection of at least one photon during Δt or by *nondetection* of a photon in this time, giving a definite value to ρ_{22} (either 1 or 0) at the end of each interval. From these considerations they estimate the average length of a dark period as

$$T_{\mathrm{D}}^{-1} = -\left.\frac{d}{dt}\rho_{22}\right|_{t=0} \quad \text{with} \quad \rho_{22}(0) = 1 \tag{3.349}$$

and the expected length of a bright period to be

$$T_{\mathrm{B}}^{-1} = \left.\frac{d}{dt}\rho_{22}\right|_{t=0} \quad \text{with} \quad \rho_{22}(0) = 0. \tag{3.350}$$

These expressions can be calculated from the optical Bloch equations appropriate for the considered situation (e. g., coherent or incoherent excitation). As usual for master equations, the time derivatives are here understood as average values over an interval larger than the decay times of the strong transition. The distribution of bright and dark periods allows measurement of the decay constants of very long-lived metastable states (of the order of seconds).

While the (axiomatic) collapse of the wave function is instantaneous (or nearly so), the sampling time Δt for collecting enough information to be certain whether the atom is in state $|2\rangle$ or somewhere in the $|0\rangle$–$|1\rangle$ plane is finite and can be related to parameters which can be inferred from experiments. If the (short) dark interval between emissions on the strong transition is t_d (if the transition is saturated this is given by the inverse of the decay rate γ_1), the "wave function collapse time" is estimated to be

$$\Delta t = t_d \ln \frac{T_{\mathrm{B}}}{t_d} \tag{3.351}$$

(Pegg and Knight 1988b, Pegg 1991). Apart from the (perhaps questionable) logarithmic factor, this is essentially the time in which a considerable amplitude for the emission of a photon on the strong transition develops (if the system starts in the ground state). This situation is analogous to the Stern–Gerlach experiment discussed at the beginning of this section: In (3.343) the state $|\uparrow\rangle$ is to be replaced by a superposition of the "active" states $|0\rangle$ and

$|1\rangle$, in (3.344) $|\downarrow\rangle$ corresponds to the "silent state" $|2\rangle$ and the role of the detector states is played by the modes of the radiation fields connected with the emitted fluorescence; $|D_0\rangle$ is the vacuum state, $|D_\uparrow\rangle$ a one-photon state. A superposition of atomic states then rapidly evolves into an entangled state as in (3.345). Therefore this "wavefunction collapse time" should rather be called "decoherence time": coherence between different alternatives no longer plays any role dynamically, as it is delocalized. The system dynamics can then be described by a rate equation (Joos 1984). Depending on the kind of excitation, the $|0\rangle$–$|2\rangle$ transition shows the Zeno effect (compare Sect. 3.3.1.2) as well as standard master behavior (for incoherent excitation). Since the emitted photons no longer interact with the atom – unless it is enclosed in perfectly reflecting cavity – we have effectively not only one but arbitrary many "detectors" coupled to the atom, each ready for irreversibly storing information about an emission "event".

We close this section with a simple schematic model for sequential measurements. Consider a two-state system which is monitored by coupling to a detector, which in turn, since macroscopic, is strongly coupled to its environment, i. e. finally to the whole universe. We have already seen a similar model in the discussions of the quantum Zeno effect (see also Sect. 3.1.1). As explained in Chap. 2, the description depends on whether and where the cut between system and (conscious) observer is located. In the following we compare three levels of description for a simple model. First we assume a collapse of the wave function of the system after each measurement interaction. The next step includes the measuring device (or, in a sense, the observer as a physical system). Finally, the interaction with the environment is taken into account. Only after this last step can we expect that the "pointer observable" and the sequence of values observed by the measurement device become well-defined (Zurek 1981).

Let the free evolution of the measured system be defined by the unitary operator

$$U_S(t) = e^{-iH_S t}. \tag{3.352}$$

We now suppose that a single measurement (discriminating system states $|k\rangle$) is performed on a timescale Δt. We assume the latter to be short enough to allow the system's evolution during Δt to be neglected. The measurement interaction

$$H_{SD} = \sum_k |k\rangle\langle k| \otimes A^{(k)} \tag{3.353}$$

then establishes a correlation between system and detector states (cf. (3.14)). In the phenomenological description a collapse into eigenstates of the observable defined by H_{SD} is assumed. In contrast, when maintaining a unitary evolution, we have to consider a sequence of interactions at times $t_1, \ldots, t_N$, starting with initial states $|S_0\rangle$ and $|D_0\rangle$ for system and detector. This evolution can then be written as

$$U_S(t-t_N)T_1 U_S(t_N-t_{N-1})T_1\cdots U_S(t_2-t_1)T_1 U_S(t_1-t_0)|S_0\rangle|D_0\rangle. \tag{3.354}$$

Here

$$T_1 = \exp(-iH_{SD}\Delta t) = \sum_k |k\rangle \langle k| \otimes \exp(-iA^{(k)}\Delta t) \tag{3.355}$$

describes the change induced by the measurement interaction during Δt. The coupling through H_{SD} must be so strong that it leads to orthogonal pointer states,

$$\langle D \,|\, \exp\!\big(iA^{(k)}t\big) \exp\!\big(-iA^{(l)}t\big) \,|\, D\rangle = \delta_{kl}, \quad t \gg \Delta t \tag{3.356}$$

where $|D\rangle$ is the initial state of the detector before any of the performed measurements.

The interaction of the macroscopic measurement device with its environment can be taken into account by adding another measurement-like process to this model in which the environment "recognizes" the macroscopically different detector readings $R = \{r_1, \ldots, r_N\}$ through

$$H_{DE} = \sum_R |R\rangle \langle R| \otimes B(R) \tag{3.357}$$

on a *very* short timescale δt (the decoherence timescale for the measurement device), in complete analogy to (3.353) – (3.356). Here the operators $B(R)$ are responsible for transferring information about the value R to the environment, as in (3.14). Although we assume $\delta t \ll \Delta t$, the change of the pointer state during measurement should not be affected by the Zeno effect, since it is macroscopic (Sect. 3.3.2.1). The complete time evolution can then be written in the form

$$U_S(t-t_N)T_2T_1U_S(t_N-t_{N-1})\cdots T_2T_1U_S(t_2-t_1)T_2T_1U_S(t_1-t_0)\,|S_0\rangle\,|D_0\rangle\,||E_0\rangle\rangle \tag{3.358}$$

with

$$T_2 = \exp(-iH_{DE}\delta t) = \sum_R |R\rangle \langle R| \otimes \exp\!\big(-iB(R)\delta t\big). \tag{3.359}$$

Here $||E_0\rangle\rangle$ denotes the initial state of the environment (in this subsection we write double brackets for environmental states). Again, states corresponding to different outcomes, i. e., different R, become orthogonal, now on the small timescale δt. More realistically, the state of the measurement device is monitored not only once, but continuously. The relevant mechanism is that any macroscopic change in the apparatus state is very rapidly recognized by the environment.

Each part in this unitary evolution is a three-step process: If the observed system were in a superposition $\sum_k c_k |k\rangle$ at time t_n, the following chain would ensue:

$$\left(\sum_k c_k \left|k\right\rangle\right) \left|\{r_1 r_2 \ldots r_{n-1} 0 \ldots 0\}\right\rangle \left|\left|\{r_1 r_2 \ldots r_{n-1} 0 \ldots 0\}\right\rangle\right\rangle \tag{3.360}$$

$$\overset{\Delta t}{\rightarrow} \left(\sum_k c_k \left|k\right\rangle \left|\{r_1 r_2 \ldots r_{n-1} r_n = k \ldots 0\}\right\rangle\right) \left|\left|\{r_1 r_2 \ldots r_{n-1} 0 \ldots 0\}\right\rangle\right\rangle$$

$$\overset{\delta t}{\rightarrow} \left(\sum_k c_k \left|k\right\rangle \left|\{r_1 r_2 \ldots r_{n-1} r_n = k \ldots 0\}\right\rangle\right) \left|\left|\{r_1 r_2 \ldots r_{n-1} r_n = k \ldots 0\}\right\rangle\right\rangle$$

For each system state $\left|k\right\rangle$, the corresponding measurement result $r_n = k$ is written to the detector's memory (second line). Shortly thereafter the environment "knows" the corresponding result, that is, the environment contains a "copy" of the pointer reading (third line). This is exactly the chain already discussed in Chap. 2 (cf. (2.6) and Fig. 2.3). The pointer basis is well-defined only after this copying process, compare (3.21 on page 50). For discussions on "cloning" an arbitrary quantum state see Wootters and Zurek (1982), Bennett et al. (1993) and Sect. 3.3.3.2.

Let us now consider a sequence of measurements performed at times $t = \tau, 2\tau, \ldots, N\tau$ and compare the above-mentioned three levels of description. Initially, the detector state can be assumed to contain only "zeros",

$$\left|D_0\right\rangle = \left|\underbrace{0, 0, \ldots 0}_{N}\right\rangle \tag{3.361}$$

(and similarly for the environment).

Level 1: The simplest description is obtained when we assume a collapse after each completed measurement. For ease of notation we consider a two-state system in the following. Let the system state at $t = 0$ be $\left|+\right\rangle$ and the system's own dynamics simply be given by

$$U_S(t)\left|+\right\rangle = \cos \Omega t \left|+\right\rangle + \sin \Omega t \left|-\right\rangle$$
$$U_S(t)\left|-\right\rangle = \cos \Omega t \left|-\right\rangle - \sin \Omega t \left|+\right\rangle. \tag{3.362}$$

The resulting (stochastic) sequence of states is shown in Fig. 3.32.

The probability for a sequence of measurement results is given by the product of the transition probabilities for each measurement, i. e.

$$P(\{a_1, a_2, \ldots, a_N\}) = R_{+a_1} R_{a_1 a_2} \ldots R_{a_{N-1} a_N} \tag{3.363}$$

with

$$R_{ab} = \begin{cases} \cos^2 \Omega \tau & \text{if } a = b, \\ \sin^2 \Omega \tau & \text{if } a \neq b. \end{cases} \tag{3.364}$$

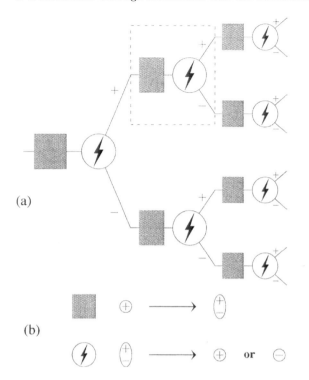

(a)

(b)

Fig. 3.32. (a) A sequence of measurements with a collapse of the wave function after every measurement event leads to a stochastic history of system states. Each step (indicated by the dashed square) consists of a unitary rotation of system states followed by a collapse of the wave function into an eigenstate of the "observable". (b) Explanation of the symbols used in (a). The shaded square stands for the (slow) rotation U_S of system states $|+\rangle$ and $|-\rangle$ between measurements (see (3.362)), leading to superposition of these states. In the circular regions the wave function is collapsed (indicated by the flash) with the respective probabilities to either $|+\rangle$ or $|-\rangle$. Only one component survives, hence only one of the emerging branches shown in the picture is actually realized.

This just expresses the well-known fact that a collapse induces a Markovian behavior. Equivalently, the (coarse grained) system's dynamics can be described by a rate equation for the density matrix, e. g.,

$$\dot{\rho}_{--}(t) \approx \frac{\rho_{--}(t+\tau) - \rho_{--}(t)}{\tau} = R(\rho_{++} - \rho_{--}) \tag{3.365}$$

with transition rate $R = R_{+-}/\tau$. (In this model the transition rate depends on the frequency of measurements; this is just the quantum Zeno effect, compare the experiment described in Sect. 3.3.1.2). Needless to say, the collapse assumption used here can in principle be proved wrong by coherently recombining the "split" components, as was, for example, demonstrated in neutron

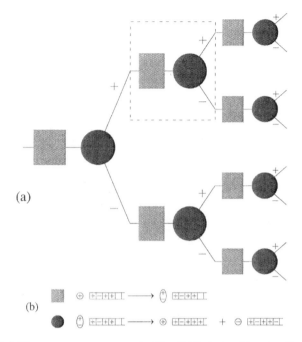

Fig. 3.33. (a) The same sequence as in Fig. 3.32, but without collapse and now including the measurement device in the unitary description. The resulting wave function is a superposition of all branches corresponding to possible measurement results. (b) Explanation of the symbols in (a). Between measurements the system states rotate via H_S (square), while the pointer state is not changed. The circle now describes a branching of the wave function (doubling the number of Schmidt components, cf. also Pascazio and Namiki 1994). Depending on the measured state, a new value is written to the detector's memory.

interferometers (Summhammer *et al.* 1983). Therefore, Fig. 3.32 does not give a valid description for such objects.

Level 2: The next level of description includes the measuring device. If we designate the pointer readings by $r = +$ and $r = -$ the sequence of states shown in Fig. 3.33 results.

It is important to realize that coherence between different "histories" is fully preserved in this state. "Inspecting" the detector, e. g. after the second measurement $(t > 2\tau + \Delta t)$, would reveal a superposition of different results. The system-apparatus state at this time reads

$$|\Psi(2\tau + \Delta t)\rangle = T_1 U_S(\tau) T_1 U_S(\tau) |+\rangle |0\ldots0\rangle \qquad (3.366)$$
$$= |+\rangle \left(\cos^2 \Omega\tau |+ + 0\ldots0\rangle - \sin^2 \Omega\tau |- + 0\ldots0\rangle\right)$$
$$+ \sin \Omega\tau \cos \Omega\tau |-\rangle \left(|+ - 0\ldots0\rangle + |- - 0\ldots0\rangle\right).$$

Obviously, the density matrix of the measuring apparatus is *not* diagonal in the pointer states, hence is does not even describe an improper mixture of measurement results. Furthermore, arbitrary superpositions of all four macroscopically different pointer readings could be produced by appropriate manipulations (for a related method to produce "Schrödinger cat" states see Savage, Braunstein and Walls (1990), cf. also Sect. 3.3.3.1).

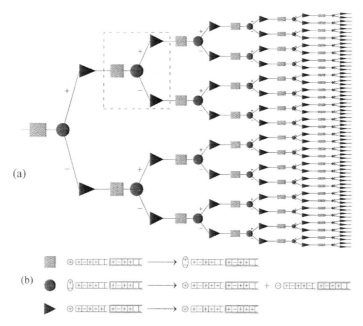

Fig. 3.34. (**a**) Unitary description including the environmental states. Each step (an example is marked by the dashed square) consists of three parts: a rotation of system states (indicated by a shaded square), branching via the measurement interaction (*circle*), and decoherence (*triangle*), where information about the measurement result is irreversibly transferred to the environment. (**b**) Pictorial representation of the three steps, corresponding to the unitary operators U_S, T_1 and T_2 defined in the text, showing examples of system, pointer and environment states. States of the environment are double-framed. See also Fig. 2.3.

Level 3: For a full unitary description the environment has to be included. The same sequence, now with additional correlated environmental states, is shown in Fig. 3.34.

Only after establishing quantum correlations with the environment is the "pointer basis" well defined, since phase relations between different memory states of the detector system are delocalized, cf. our discussion in Sect. 3.1.1.

After the second measurement (and rapid decoherence) the total state is

$$
\begin{aligned}
|\Psi(2\tau + \Delta t + \delta t)\rangle &= T_2 T_1 U_S(\tau) T_2 T_1 U_S(\tau) |+\rangle |D_0\rangle ||E_0\rangle\rangle \\
&= \cos^2 \Omega\tau |+\rangle |+ + 0 \ldots 0\rangle ||+ + 0 \ldots 0\rangle\rangle \\
&\quad + \cos \Omega\tau |-\rangle |+ - 0 \ldots 0\rangle ||+ - 0 \ldots 0\rangle\rangle \\
&\quad + \cos \Omega\tau \sin \Omega\tau |-\rangle |- - 0 \ldots 0\rangle ||- - 0 \ldots 0\rangle\rangle \\
&\quad - \sin^2 \Omega\tau |+\rangle |- + 0 \ldots 0\rangle ||- + 0 \ldots 0\rangle\rangle .
\end{aligned}
\qquad (3.367)
$$

This state contains four (generalized) Schmidt components. (For a less abstract example compare the chain of states (3.208) to (3.212) in our description of the quantum Zeno effect.) The density matrix describing the apparatus now represents an improper mixture of all four "histories" of measurement results (see (3.21), cf. also Paz and Zurek 1993). No manipulation of system and/or apparatus states can recover coherence, if the delocalization of phases into the environment is truly irreversible. This is usually the case in our present universe, with the exception of some special experimental situations. In a cosmological context, however, strict irreversibility is not guaranteed in a recollapsing universe. This situation is best discussed in a framework where space-time is also quantized, see Chap. 4.

3.4.2 Quantum "Information" and Teleportation

If one takes seriously the idea that information is without exception represented by states of physical objects, then there should be no fundamental difference between "classical" and "quantum" information. Indeed, the concept of classical information apparently is used exactly in those situations where decoherence plays a major role, for example during measurements or similar operations. It should therefore be possible to formulate all processes completely without ever using the concept of "classical information". As an example we look in the following at a reformulation of the process of so-called "quantum teleportation".

The scheme of quantum teleportation, where the state of a system is transferred (copied) to another degree of freedom (Bennett *et al.* 1993, Bouwmeester *et al.* 1997), is usually described by using the collapse of wavefunction onto certain states. Additionally, it is often claimed that the teleportation process *requires* transmission of "classical information". Clearly, all this must have a description in purely quantum terms. It can easily be shown, that the entire scheme can be represented by a sequence of well-chosen interactions and couched in a completely unitary quantum description.

Following the scheme suggested by Bennett *et al.*, the (arbitrary) state of particle 1 is to be transferred to particle 3, which is already entangled with particle 2. In addition to the standard description we now include states for the sender (Alice) and the receiver (Bob). If Alice and Bob are macroscopic, these states are necessarily part of an entangled state and should thus be understood as "Alice's state together with her environment" (the same for Bob). The initial state $|\Psi_0\rangle$ thus contains Alice, Bob, an entangled EPR pair, and the (possibly unknown) state of particle 1,

$$|\Psi_0\rangle = (\alpha\,|0\rangle_1 + \beta\,|1\rangle_1)\,|\text{Alice}_0\rangle\,\frac{(|0\rangle_2\,|1\rangle_3 - |1\rangle_2\,|0\rangle_3)}{\sqrt{2}}\,|\text{Bob}_0\rangle\,. \qquad (3.368)$$

This state can be rewritten in terms of the so-called Bell-states for particles 1 and 2,

$$|\text{Bell}_{1,2}\rangle = \frac{1}{\sqrt{2}}(|0\rangle_1\,|1\rangle_2 \pm |1\rangle_1\,|0\rangle_2)$$

$$|\text{Bell}_{3,4}\rangle = \frac{1}{\sqrt{2}}(|0\rangle_1\,|0\rangle_2 \pm |1\rangle_1\,|1\rangle_2) \qquad (3.369)$$

in the form

$$\begin{aligned}
|\Psi_0\rangle = |\text{Alice}_0\rangle\,\frac{1}{2}\,[\,&|\text{Bell}_1\rangle\,(\beta\,|1\rangle_3 - \alpha\,|0\rangle_3) - |\text{Bell}_2\rangle\,(\beta\,|1\rangle_3 + \alpha\,|0\rangle_3)\\
+\,&|\text{Bell}_3\rangle\,(\alpha\,|1\rangle_3 - \beta\,|0\rangle_3) + |\text{Bell}_4\rangle\,(\alpha\,|1\rangle_3 + \beta\,|0\rangle_3)]\,|\text{Bob}_0\rangle\,.
\end{aligned} \qquad (3.370)$$

The next step in the teleportation scheme is the "Bell-measurement" of the 1-2 system by Alice, leading in the conventional description to a collapse onto one of the four "Bell-states". This requires, of course, that objects 1 and 2 are located at Alice's position. In a unitary description, this measurement-like interaction changes the state of Alice (possibly including her environment if she is truly macroscopic) according to

$$|\text{Alice}_0\rangle\,|\text{Bell}_i\rangle \longrightarrow |\text{Alice}_i\rangle\,|\text{Bell}_i\rangle \quad (i = 1..4). \qquad (3.371)$$

Such a process may be modelled by an interaction of the form

$$H = \sum_{i=1}^{4} |\text{Bell}_i\rangle\,\langle\text{Bell}_i| \otimes A_i \qquad (3.372)$$

where

$$|\text{Alice}_i\rangle = \exp(-iA_iT)\,|\text{Alice}_0\rangle\,, \qquad (3.373)$$

where T is the measurement time (compare 3.14 on page 48). This interaction couples particles 1 and 2 (this coupling is contained in the operators A_i). Since particle 2 is already entangled with particle 3, this procedure affects the entire state. A unitary description leads therefore to a superposition of the form

$$\begin{aligned}
|\Psi_1\rangle = \frac{1}{2}\,[\,&|\text{Alice}_1\rangle\,|\text{Bell}_1\rangle\,(-\alpha\,|0\rangle_3 + \beta\,|1\rangle_3)\\
+\,&|\text{Alice}_2\rangle\,|\text{Bell}_2\rangle\,(-\alpha\,|0\rangle_3 - \beta\,|1\rangle_3)\\
+\,&|\text{Alice}_3\rangle\,|\text{Bell}_3\rangle\,(\alpha\,|1\rangle_3 - \beta\,|0\rangle_3)\\
+\,&|\text{Alice}_4\rangle\,|\text{Bell}_4\rangle\,(\alpha\,|1\rangle_3 + \beta\,|0\rangle_3)]\,|\text{Bob}_0\rangle
\end{aligned} \qquad (3.374)$$

The "classical information transmission" from Alice to Bob is equivalent to a measurement of Alice's state by Bob (there is no problem with the no-cloning theorem in this step, since all states $|\text{Alice}_i\rangle$ are orthogonal),

$$|\text{Alice}_i\rangle \, |\text{Bob}_0\rangle \longrightarrow |\text{Alice}_i\rangle \, |\text{Bob}_i\rangle \qquad (3.375)$$

leading to

$$
\begin{aligned}
|\Psi_2\rangle = \frac{1}{2}\big[&|\text{Alice}_1\rangle \, |\text{Bob}_1\rangle \, |\text{Bell}_1\rangle \, (-\alpha \, |0\rangle_3 + \beta \, |1\rangle_3) \\
+ &|\text{Alice}_2\rangle \, |\text{Bob}_2\rangle \, |\text{Bell}_2\rangle \, (-\alpha \, |0\rangle_3 - \beta \, |1\rangle_3) \\
+ &|\text{Alice}_3\rangle \, |\text{Bob}_3\rangle \, |\text{Bell}_3\rangle \, (\alpha \, |1\rangle_3 - \beta \, |0\rangle_3) \\
+ &|\text{Alice}_4\rangle \, |\text{Bob}_4\rangle \, |\text{Bell}_4\rangle \, (\alpha \, |1\rangle_3 + \beta \, |0\rangle_3)\big]
\end{aligned}
\qquad (3.376)
$$

The final step is Bob's application of a unitary transformation on his particle 3, depending on his information (*i.e., on his state*). This can be achieved by an Hamiltonian of the form

$$H = \sum_{i=1}^{4} |\text{Bob}_i\rangle \, \langle \text{Bob}_i| \otimes \Omega_i \qquad (3.377)$$

so that, for example,

$$\exp[-i\Omega_1 T](-\alpha \, |0\rangle_3 + \beta \, |1\rangle_3) = (\alpha \, |0\rangle_3 + \beta \, |1\rangle_3) \qquad (3.378)$$

The final state of the total system then reads

$$|\Psi_3\rangle = \frac{1}{2}\left[\sum_i |\text{Alice}_i\rangle \, |\text{Bob}_i\rangle\right] (\alpha \, |0\rangle_3 + \beta \, |1\rangle_3) \qquad (3.379)$$

If Alice and Bob are macroscopic, this state describes four decohered Everett branches. Note, that the final state $|\Psi_3\rangle$ factorizes again, but with respect to a different partition of the total Hilbert space.

Formally, Alice and Bob may be replaced by microscopic systems; no collapse is required for this scheme to work. Only if Alice and Bob are macroscopic, the transition from $|\Psi_0\rangle$ to $|\Psi_3\rangle$ is irreversible.

One may wonder, how it is possible to transfer the precise state $\alpha \, |0\rangle + \beta \, |1\rangle$ form particle 1 to particle 3. One could argue, that the specification of the values α and β already requires an infinite amount of information, whereas the "classical message" in the teleportation scheme contains only two bits. If this "classical message" is replaced by a (special) interaction – as described above – the entire process appears as a unitary rotation in Hilbert space.

Hence the appropriate characterization of this situation may well be that, "the teleportation procedure does not move the quantum state: the state was, in some sense, in the remote location from the beginning." (Vaidman 1998, cf. also Busch *et al.* 2001 , Braunstein 1996).

Similar remarks apply to the related scheme of "dense coding" (Bennett and Wiesner 1992). The four Bell states (3.369) form an orthonormal basis in the four-dimensional Hilbert space of the two qubits. All these states are nonlocal. *If* one has already prepared one of these states, one can accomplish transitions between all of them by unitary transformations on only *one* spin (spin-flip and/or phase change). Then it is claimed, that by manipulating one qbit one can encode *four* bits ("quantum dense coding"). This should not come as a surprise, since the initial state is nonlocal (holistic). "Touching" one particle necessarily affects the nonlocal quantum state as a whole. In contrast to decoherence, this entanglement is controllable.

3.4.3 True, False, and Fake Decoherence

The term "decoherence" is often identified with "disappearance of interference effects", formally described by damping of non-diagonal terms in a density matrix in a certain basis. In most parts of this book we are using this pragmatic representation. It is advisable, however, to clearly discriminate the physical mechanisms leading to the disappearance of coherence. These may be classified into three categories, from which perhaps only one deserves to be called "true" decoherence.

Pure states are coherent almost by definition. Parts, i. e. certain components, can act in a way which is observably different from the behavior of their superposition ("interference"). Let the parts be given by the Hilbert space vectors $|1\rangle$ and $|2\rangle$. The general superposition

$$|\Psi\rangle = a\,|1\rangle + b\,|2\rangle = e^{i\alpha}(\cos\frac{\Theta}{2}\,|1\rangle + e^{i\Phi}\sin\frac{\Theta}{2}\,|2\rangle) \qquad (3.380)$$

shows coherence between its components, conveniently expressed by the non-diagonal part ρ_{12} contained in the density matrix,

$$\rho = \begin{pmatrix} |a|^2 & ab^* \\ a^*b & |b|^2 \end{pmatrix} = \begin{pmatrix} \cos^2\frac{\Theta}{2} & \frac{1}{2}\sin\Theta\,e^{i\Phi} \\ \frac{1}{2}\sin\Theta\,e^{-i\Phi} & \sin^2\frac{\Theta}{2} \end{pmatrix} \qquad (3.381)$$

Decoherence in this basis means that the off-diagonal part ρ_{12} is reduced or completely eliminated,

$$\frac{1}{2} \geq |\rho_{12}| = |ab^*| \longrightarrow 0. \qquad (3.382)$$

True decoherence. The fundamental decoherence mechanism is "pure" entanglement with the environment – *without any dynamical change* of the component states, as already explained at the beginning of this Chapter,

$$\begin{aligned} |1\rangle\,|\Phi\rangle &\longrightarrow |1\rangle\,|\Phi_1\rangle \\ |2\rangle\,|\Phi\rangle &\longrightarrow |2\rangle\,|\Phi_2\rangle\,, \end{aligned} \qquad (3.383)$$

hence

$$(a\,|1\rangle + b\,|2\rangle)\,|\Phi\rangle \longrightarrow a\,|1\rangle\,|\Phi_1\rangle + b\,|2\rangle\,|\Phi_2\rangle\,. \qquad (3.384)$$

Locally, coherence is lost,

$$\rho_{12} \longrightarrow \rho_{12}\,\langle\Phi_2|\Phi_1\rangle \longrightarrow 0. \qquad (3.385)$$

The components *still exist*, but can no longer interfere, since the required phase relations are delocalized. This process has no analogue in classical physics. The damping of ρ_{12} often follows a Lindblad master equation with *hermitean* generators (cf. Sect. 3.3.2.2).

False decoherence. Coherence is trivially lost if one of the required components disappears. An important situation of this kind is represented by relaxation processes, for example, a decay of state $|2\rangle$ into state $|1\rangle$,

$$|1\rangle\,|\Phi\rangle \longrightarrow |1\rangle\,|\Phi'\rangle$$
$$|2\rangle\,|\Phi\rangle \longrightarrow |1\rangle\,|\Phi''\rangle\,. \qquad (3.386)$$

This is often called "amplitude damping". Clearly, interference must disappear together with the decay of component $|2\rangle$. Therefore the timescales T_1 for "longitudinal" and T_2 for "transversal" decay in the commonly used Bloch equations,

$$\dot{\rho}_{22} = -\frac{1}{T_1}\rho_{22}$$
$$\dot{\rho}_{12} = -\mathrm{i}\omega\rho_{12} - \frac{1}{T_2}\rho_{12} \qquad (3.387)$$

are connected by the well-known relation[64] $T_2 = 2\,T_1$. (In other words, processes such as "inelastic scattering" always imply dephasing.)

The role of the "environment" can also be played by other states of the *same* system, if the dynamics leads to disappearance of one component from the relevant subspace. For example, such "internal decoherence" may have the form

$$|1\rangle \longrightarrow |1\rangle$$
$$|2\rangle \longrightarrow \sum_{n>2} c_n\,|n\rangle\,, \qquad (3.388)$$

generated by an appropriate Hamiltonian.

Finally, the most direct way to remove a component is represented by a collapse of the wave function, for example during a measurement. If the

---

[64] T_1 and T_2 are called "longitudial" and "transversal" relaxation times in nuclear magnetic spectroscopy and related fields. In quantum optics they usually refer to population and polarization of atomic states. All two-level descriptions use Bloch equations of the same form.

measurement result is ignored (or unknown), one proceeds with

$$\rho = \begin{pmatrix} |a|^2 & 0 \\ 0 & |b|^2 \end{pmatrix} \tag{3.389}$$

which has exactly the same form as the density matrix resulting from complete entanglement. But interference is missing here for the trivial reason that one or the other component no longer exists.

Fake decoherence. Decoherence often arises from some averaging process. Two typical situations are noteworthy. The ensemble either consists of members undergoing the same unitary evolution but with different initial states or an ensemble of identically prepared states subjected to different Hamiltonians is employed. In both cases the fundamental dynamics of a *single* system is unitary, hence there is *no decoherence at all* from a microscopic point of view.

If instead of a single state (3.380) an ensemble $\{|\Psi_j\rangle\}$ of such states with different relative phases Φ_j is prepared in some way[65], ensemble averages for measurements are usually calculated from a density matrix (assuming all phases Φ_j equally likely) where

$$\rho = \frac{1}{N} \sum_{j=1}^{N} |\Psi_j\rangle \langle \Psi_j| = \begin{pmatrix} \cos^2 \frac{\Theta}{2} & \frac{1}{2} \sin \Theta \sum_{j=1}^{N} e^{i\Phi_j} \\ \frac{1}{2} \sin \Theta \sum_{j=1}^{N} e^{-i\Phi_j} & \sin^2 \frac{\Theta}{2} \end{pmatrix}$$

$$\approx \begin{pmatrix} |a|^2 & 0 \\ 0 & |b|^2 \end{pmatrix} \tag{3.390}$$

This sort of "dephasing" (cf. also Sect. 9.1) has its root only in the *incomplete* description by this averaged density matrix and leads always to a shorter dephasing time $T_2^* < T_2$ (compare (3.387)). Note also that the correct microscopic description would be given by a tensor product state

$$|\Psi\rangle = \otimes_j |\Psi_j\rangle \tag{3.391}$$

instead of (3.380). The state (3.391) is observably different as shown in many "echo"-experiments, since for each member of this ensemble the relative phases remain well-defined, even if they may be hard to access experimentally.

Phase randomization as a dynamical process appears as a special case of a system subjected to "stochastic forces" or "noise". One may derive pure phase damping induced by elastic collisions (i.e., without energy exchange)

---

[65] For example, consider an ensemble of spins rotating with different angular velocities (different ω in (3.387)) in an NMR experiment, leading to inhomogenuous broadening of a magnetic resonance signal.

by adding a random Stark shift term, so that the equation of motion for ρ_{12} reads

$$\dot{\rho_{12}} = -i[\omega + \delta\omega(t)]\rho_{12}. \tag{3.392}$$

Expanding the formal integral

$$\rho_{12}(t) = \rho_{12}(0)\exp(-i\omega t - i\int_0^t dt'\delta\omega(t')) \tag{3.393}$$

and assuming mean average $(<\delta\omega(t)>=0)$ and the Markow property

$$<\delta\omega(t)\delta\omega(t')>=2\gamma_c\delta(t-t'), \tag{3.394}$$

yields an exponential damping of phases,

$$\rho_{12}(t) = \rho_{12}(0)e^{-i\omega t}e^{-\gamma_c t}. \tag{3.395}$$

As another simple example consider a particle subjected to "random kicks" acting on an individual object. The corresponding *ensemble of unitary evolutions* represents again a non-unitary evolution of the density matrix. The result is similar to that of decoherence, while its interpretation is the same as in the case of dephasing by stochastic scattering. The assumption of *many* scattering events is entirely irrelevant.

Let a particle be subjected to "random kicks". This model is chosen in analogy to classical Brownian motion, although here it is applied to wave functions. If the original state of the particle is described by a wave packet $\varphi(x)$, a "kick", i.e., an instantaneous shift in momentum by p, introduces a phase factor,

$$\varphi(x) \to \varphi(x)e^{ipx}, \tag{3.396}$$

or for the density matrix,

$$\rho(x,x') \to \rho(x,x')e^{ip(x-x')}. \tag{3.397}$$

If we calculate the average action of kicks distributed according to a probability distribution of momentum transfers $P(p)$, we get

$$\rho(x,x') \to \int dp\, P(p)\rho(x,x')e^{ip(x-x')} =: f(x-x')\rho(x,x'). \tag{3.398}$$

Whatever the shape of $P(p)$, for such "kicks" we obtain a damping of *spatial* coherence, given by the Fourier transform of the momentum transfer distribution. For a Gaussian distribution $P(p) = \sqrt{\frac{\lambda}{\pi}}\exp\left(-\lambda p^2\right)$ we find the well-known result (cf. (3.68), (3.95) and App. A1)

$$\frac{\Delta\rho(x,x')}{\Delta t} \propto \left(1 - \exp\left[-\frac{(x-x')^2}{4\lambda}\right]\right)\rho(x,x'). \tag{3.399}$$

Since the distribution of kicks here does not depend on the state of the system, this treatment does not include any recoil. It corresponds to a Langevin equation with a stochastic force, but without a frictional term, and hence cannot describe approach to equilibrium.

The transition (3.398) can also be expressed as a transformation of the Wigner function, since (now writing q for the momentum transfer)

$$W(x,p) \longrightarrow \quad \frac{1}{\pi} \int dy \; e^{2ipy} \int dq \; P(q)\rho(x-y, x+y)e^{-2iqy}$$

$$= \int dq \; P(q)W(x, p-q) \tag{3.400}$$

This expression is the analogue of (3.152) on page 94.

It may thus appear that decoherence can also be obtained from "classical perturbations" (kicks) of the quantum system (Jayannavar 1992; for a model mimicking decoherence in EPR-situations by fluctuating external fields see Venugopalan, Kumar and Ghosh 1995b). This formal equivalence of density matrix equations hides once again the essential conceptual difference between the two types of interactions. For "classical noise" the system follows a *unitary* (even though uncontrollable in practice) dynamics; in each individual case it stays in a pure state (that may remain unknown because of an insufficiently known Hamiltonian). For a series of random successive kicks to the *same* particle, one would still obtain the result (3.397) with p simply representing the sum of all kicks. In contrast, decoherence leads deterministically to an entangled state that has quite different properties. "Noise" models are in fact only used in situations where this difference cannot be observed (for a comparison of noise and decoherence models see Allinger and Weiss 1995).

In recent experiments with single trapped ions various decoherence mechanisms were studied (Myatt *et al.* 2000). Here the case of "fake decoherence" is, for example, realized by applying noisy classical fields. Then the effects of decoherence can be undone by using suitably adjusted fields which reverse the "random" displacement of the ion.

The popular opinion that the measurement process causes an uncontrollable disturbance of the measured system seems to go back to Heisenberg's arguments in support of the uncertainty relations (Heisenberg 1927, 1958). This widespread interpretation is also used to explain the loss of coherence in current realizations of the double-slit (or similar) experiments with atomic beams (Pfau *et al.* 1994, Kurtsiefer *et al.* 1995). There the reduction of "visibility" of the diffraction pattern (cf. Fig. 3.7 on page 77) is attributed to a change in the momentum distribution of the scattered atoms, similar to (3.397) above. On the other hand, quantum correlations, which lead to the reduced fringe visibility, can (to a certain extent) be employed to regain coherence by appropriate correlation measurements that include the photons scattered off the atom (Chapman *et al.* 1995). EPR-type experiments showing a violation of Bell's inequality would then directly demonstrate that an interpretation in terms of "classical noise" cannot be maintained.

Note also that a stochastic model along the lines of (3.396) does not lead to ensembles of "classical trajectories". Since the wave function is only changed by a phase factor, wave packet dispersion and other non-classical features still persist for each member of the ensemble. This is in contrast to models with nonlinear stochastic dynamics (Diósi 1994b). Both lead, however, to the *same* equation of motion for the density matrix.

The notion of "noise" has also been used, albeit in a vague manner, for other – genuine quantum – processes, for example the decay of excited systems (another situation locally described by a dynamical semi-group – cf. Chap. 7). This is often (usually only verbally) attributed to "vacuum fluctuations". As long as interest is restricted to the behavior of the *local* system, such classical models cannot be disproved, while they are in observable conflict with the quantum nature of the total system that contains the electromagnetic field.

Even when the interaction of an atom with its radiation field is treated fully quantum mechanically, the use of the Heisenberg picture and the so-called "quantum Langevin equations" may lead to interpretation problems which partially have their origin in the classical pictures used. The interpretation of single terms in the equations often depends on the operator ordering (Milonni and Smith 1975, Meystre and Sargent 1991). For example, in normal ordering spontaneous emission is connected with the radiation reaction while the Langevin force is associated with vacuum fluctuations. When a symmetric ordering (motivated by the requirement of hermitian Heisenberg operators in the quantum Langevin equation) is employed instead (Dalibard, Dupont-Roc and Cohen-Tannoudji 1982), radiation reaction and vacuum fluctuations give equal contributions to spontaneous emission. (For an analysis of the Unruh effect using this approach see Audretsch and Müller 1994.) Such ambiguous arguments demonstrate that the Heisenberg picture is inappropriate for the interpretation of quantum states and quantum processes.

4 Decoherence in Quantum Field Theory and Quantum Gravity

C. Kiefer

The previous chapter was concerned with a detailed investigation of decoherence through interaction with the environment. While most examples were discussed in the framework of quantum mechanical systems, i.e. systems with finitely many degrees of freedom, it is the purpose of this chapter to extend these discussions to situations where quantum *fields* play an important role. Moreover, the inclusion of a quantised gravitational field will entail new conceptual issues such as the absence of a *fundamental* time parameter.

Why does the treatment of systems with infinitely many degrees of freedom justify a separate discussion? First, it is technically more involved, since issues like regularisation and renormalisation are expected to play a role. Second, the rapidly developing field of quantum optics is mainly concerned with the interaction of matter with the radiation field. We shall thus restrict our attention in the non-gravitational case to quantum electrodynamics (QED), although the extension of the principal ideas to other field theories should be straightforward. The present chapter also presents a preparation for the discussion of symmetries and superselection rules in Chap. 6, in particular for the discussion of the charge superselection rule in QED and the mass superselection rule in quantum gravity. Finally, the discussion enables one to understand, at least in principle, how a classical spacetime can emerge from quantum gravity as an approximate concept by decoherence. This is very important from a fundamental point of view, since the notion of a classical spacetime is a prerequisite for standard quantum field theory.

This chapter is divided into two parts. The first section deals with QED and addresses the emergence of classical behaviour for field strengths as well as for charges. The second section demonstrates how, in the framework of present approaches to quantum gravity, one can understand the emergence of a classical spacetime as well as the classicality of primordial fluctuations in the early Universe. It must be emphasised that – compared to the last chapter – the role of decoherence in quantum field theory has been much less explored, mainly since quantum field theory is usually applied in microscopic situations (scattering experiments), while electromagnetic and gravitational fields are usually macroscopic fields, and many problems remain to be investigated.

4.1 Decoherence in Quantum Electrodynamics

Depending on the given ("experimental") situation, one can study two aspects of the role that decoherence plays in QED. The first is concerned with the influence of the electromagnetic field on atoms or electrons. There, the role of the environment (the "irrelevant part") is played by the field, while the part of the relevant subsystem (the one which is studied in laboratory experiments) is played by atoms or electrons. This, in fact, constitutes an important part of the discussion in quantum optics, and we shall begin with a brief review of this aspect. The second, opposite, situation deals with the influence of matter on the electromagnetic field. The corresponding matter field degrees of freedom then serve as the "environmental reservoir" which is responsible for decoherence. The measurement of electromagnetic field quantities was first discussed in the pioneering work of Bohr and Rosenfeld (1933) in which they studied the operational meaning of field commutators. The particular aspect of decoherence for fields was, however, greatly neglected until recently. It will be the subject of the second part of this section.

4.1.1 "Measurement" of Charges by Fields

In this subsection the "environmental part" is played by the degrees of free-dom of the electromagnetic field. Its effect on the "relevant part" (atoms, electrons, . . .) is recognised from the reduced density matrix which is obtained by integrating out the electromagnetic field from the full quantum state (except for "dressing", see below). In the case of atoms, this was first studied by Landau (1927). At that time, a theory of quantum electrodynamics was still elusive. For this reason we shall briefly review his approach in the language of QED, see for example Diósi (1990a).

The full state vector of QED for matter plus electromagnetic field can be written, in the interaction representation, as

$$\Psi(t) = U(t)\Psi(t_{\text{in}}), \tag{4.1}$$

where $\Psi(t_{\text{in}})$ denotes an asymptotic in-state ($t_{\text{in}} \to -\infty$), and we indicate only the time-dependence of the wave functionals. The unitary evolution operator $U(t)$ is given by

$$U(t) = \mathrm{T} \exp\left(-i \int_{t'<t} dt' \; H_{\text{int}}(t')\right) = \mathrm{T} \exp\left(-i \int_{t'<t} d^4x \; \bar{\psi}\gamma^\mu\psi A_\mu\right), \tag{4.2}$$

where T denotes time-ordering, and a standard QED notation is employed.[1] Diósi (1990a) imposed the initial condition that at $t \to t_{\text{in}}$ the total system

---

[1] ψ and $\bar{\psi}$ are Dirac spinors, and A_μ is the vector potential. The integration runs over $d^4x \equiv dx^0 d^3x$.

is in a product state of charges and (real) photons. (This is where the arrow of time enters the calculation.) One thus has

$$\Psi(t_{\text{in}}) = |0_A\rangle \otimes |0_m\rangle, \tag{4.3}$$

where the initial vacuum states are those of the electromagnetic field ($|0_A\rangle$) and the matter field ($|0_m\rangle$), respectively. The role of matter may be played by an electron field as in (4.2) or by an atom (this is the case which is treated below).

The reduced density matrix for the matter fields is then given by tracing out the field degrees of freedom, i.e.

$$\rho(t) = \text{tr}_A |\Psi(t)\rangle\langle\Psi(t)|. \tag{4.4}$$

Using the "closed time path formalism" for Green functions[2] which is presented, for example, in Chou $et\ al.$ (1985), Diósi is able to present an exact formal expression for the reduced density matrix:

$$\rho(t) = \hat{T} \exp\left\{\frac{i}{2} \int_{x^0, y^0 < t} d^4x\, d^4y \left[J_+ D_F J_+ + J_- D_{\bar{F}} J_- \right.\right.$$
$$\left.\left. - J_+ D_+ J_- - J_- D_- J_+\right]\right\} \rho(t_{\text{in}}), \tag{4.5}$$

where the D - terms denote certain photon propagators. For example,

$$D_{\mu\nu}^F = i\langle 0_A \mid T A_\mu(x) A_\nu(y) \mid 0_A\rangle$$

is the standard Feynman propagator. Furthermore, J_+ (J_-) means that the field operator J is acting from the left (from the right), and $\hat{T}$ denotes time-ordering for J_+- fields and anti-time ordering for J_-- fields. The symbols J are a shorthand notation for the electromagnetic current operator $J^\mu = \bar{\psi}\gamma^\mu\psi$.

The result (4.5) gives a formal expression for the $exact$ reduced dynamics of the matter part. In the following, the matter part is supposed to be a system of atoms. Of course, the density matrix ρ obeys a highly non-Markovian equation. Evaluating (4.5) in perturbation theory up to order e^2 and assuming a Markov approximation, one finds in the energy representation the following master equation for the various components of the density matrix (cf. also Sect. 3.3.2):

$$\dot{\rho}_{nn} = -\Gamma_n \rho_{nn} + \sum_{r>n} \Gamma_{r\to n}\, \rho_{rr}, \tag{4.6}$$

$$\dot{\rho}_{nm} = \left[-i(\Delta E_n - \Delta E_m) - \frac{1}{2}(\Gamma_n + \Gamma_m)\right]\rho_{nm}, \quad n \neq m, \tag{4.7}$$

---

[2] This is a straightforward generalisation of standard path integral techniques, which is especially suited to treat density matrices, see also Chap. 5.

where ΔE_n denotes the energy shift of the nth atomic energy level (Lamb shift), and Γ_n denotes the full decay rate of the nth energy level, i.e.

$$\Gamma_n = \sum_{r<n} \Gamma_{n\to r} \ . \tag{4.8}$$

Because of the initial condition (no incoming photons), these equations describe only decay processes. In the dipole approximation one finds the well-known result for the spontaneous emission rate,

$$\Gamma_{r\to n} = \frac{4\omega_{rn}^3}{3\hbar c^3} |D_{rn}|^2, \tag{4.9}$$

where D_{rn} is the dipole matrix element between the levels r and n. The master equation (4.6), (4.7) can be derived by a variety of methods from models in quantum optics. We have chosen to sketch the above derivation, since it is closer to some of the approaches presented below and in Chap. 5.

Let us now focus our attention on the behaviour of the non-diagonal elements of the reduced density matrix, Eq. (4.7). As in our general discussion, the imaginary part in (4.7) expresses the back reaction effect of the field onto the atom, leading to a shift of the (unperturbed) energy eigenvalues (see our discussion of the Wigner function in Sect. 3.2.3). There is thus no contribution to decoherence from effects producing the Lamb shift. The real terms in (4.7) describe decoherence corresponding to a "non-ideal measurement" (decay into another state): the electromagnetic field carries away information about different energy eigenstates of the system. Consequently, interferences between these states cannot be observed at the system itself. This leads to the impression of a "quantum jump" after the decay state has been "measured" (cf. also Sect. 3.3.2). To cite Landau (1927) (we use our notation)

> Equation (4.7) shows that the quantities ρ_{nm} on the average always decrease in time. The reason for that apparently lies... in the increasing phase uncertainty which causes a decrease in the ability of the radiation to act coherently.[3]

He then estimates the decoherence timescale from (4.7) to be given by the decay time (in fact, the actual decoherence time is much shorter). He also emphasises that not only the radiation field connected with the system itself is responsible for decoherence, but also the coupling to other, external, fields which are part of the natural environment (he uses the word *Kohärenzstörung* = perturbation of coherence).

Since master equations of the above kind are extensively discussed in texts on quantum optics, see e.g. Cohen-Tannoudji (1992) or Walls and Milburn

---

[3] Die Gleichung (4.7) zeigt, daß die Größen ρ_{nm} mit der Zeit stets durchschnittlich abnehmen. Der Grund dafür liegt ... offenbar in der zunehmenden Phasenunbestimmtheit, welche eine Verminderung der Kohärenzfähigkeit der Strahlung herbeiführt.

(1994), we shall not go into more details here. Some examples have already been discussed in the preceding chapter. We also want to mention that decoherence in the atom-photon interaction is of relevance in order to understand properly Dirac's famous remark that a photon can only interfere with itself (Kiefer 1998a).

While the above example focused on decoherence of energy eigenstates, the question of the influence of the field on the *spatial* coherence properties (localisation) of charged particles (in particular, electrons) deserves some special attention: how far can a wave packet of a charged particle be extended in space ("split into two parts"), so that it can later be *coherently* recombined, i.e. so that the decohering effect of the radiation field is reversible? After all, we know from experience that there exists an upper limit for such a separation in realistic situations. Neutral particles, such as neutrons, can be coherently superposed over much larger distances than charged particles. Thermal radiation interacts with charged particles through Thomson scattering and leads to strong decoherence within a few seconds, see Joos and Zeh (1985) and Sect. 3.2.1.1, in particular Table 3.1.

For different charges no superpositions can be observed, because the charges are decohered by their own Coulomb field which is present at any distance. This aspect, leading to the charge superselection rule, will be discussed in detail in Chap. 6. For a spatially separated wave packet of one charge, decoherence can occur by emission of transversal photons which provide information about the "path" of the charge (*welcher Weg* information). The Coulomb field corresponds to photons of infinite wavelength, which cannot resolve the distance between the parts of the packet. Nicklaus and Hasselbach (1993) report about a coherent separation and recombination of an electron beam over at least a distance $d \approx 60$ µm, see also Hasselbach *et al.* (2000). The principal ingredients of the measurement are shown in Fig. 4.1. Use is made of a Wien filter which can shift the two packets longitudinally with respect to each other. Originally (top of Fig 4.1 right) interference was lost due to electrostatic fields of the deflection elements. Due to the shift caused by the Wien filter, interference fringes then become visible (middle) and eventually disappear after the shift becomes too big. The loss of coherence can be interpreted as being due to the fact that *welcher Weg* information is stored in the system (different flight times of the packets). As long as this information is not read out, recoherence can be achieved.

In order to obtain a quantitative estimate of environmental decoherence in this case, one can proceed as follows. Due to the acceleration of the electron, photons are emitted. In order that an emitted photon can discriminate between the two paths of the packet (and thus be able to lead to decoherence), its wavelength must be at least of the order of d, the lateral distance. Its energy E_γ must then be bigger than hc/d. Now, according to Larmor's formula the total instantaneous power for a nonrelativistic, accelerated electron is $P = 2e^2a^2/3c^3$. The acceleration a is at least of the order of $d/(\Delta t)^2$, where

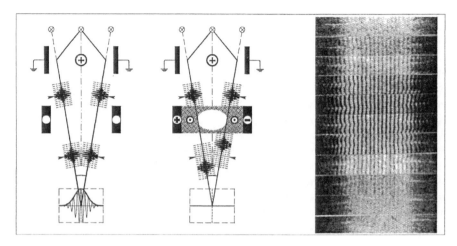

Fig. 4.1. Left: Electron biprism interferometer with Wien filter switched off (left) and Wien filter in its excited state (right): The wave packet shift exceeding the coherence length leads to the disappearance of the interference fringes. Middle: Restoration of contrast by a Wien filter: The longitudinal shift of the wave packets caused by electrostatic deflection elements (top) can be compensated (middle) and overcompensated (bottom) with the Wien filter (Nicklaus and Hasselbach (1993)). Right: The 11 micrographs correspond to 11 different excitations of the Wien filter, increasing from top to bottom. [From Hasselbach *et al.* (2000)].

Δt is the flight time. The probability to emit at least one photon should then be of the order

$$\frac{P\Delta t}{E_\gamma} \approx \frac{e^2}{hc} \left(\frac{d}{c\Delta t} \right)^3 . \tag{4.10}$$

In the experimental situation, this probability is indeed very small. Taking the values reported in Nicklaus and Hasselbach (1993), i.e. $d = 60$ μm and $\Delta t = 2 \times 10^{-8}$ s, it is only of the order of 10^{-17}. This is confirmed by the detailed discussion of emission through bremsstrahlung according to which coherence should be possible up to $d \approx 1$ cm (Breuer and Petruccione 2001), cf. also Dürr and Spohn (2000). A suggestion for an experimental test of decoherence for a spatially split electron was made by Anglin and Zurek (1996b): The two beams traverse a conducting plate, inducing image currents. These currents disturb the electron and phonon gas of the conductor. Because information about this disturbance remains in the conductor after the electron has left, decoherence for the electron beams arises (they are entangled with the electrons and phonons in the conductor) and they cannot be coherently recombined.

Another effect in this context has been suggested by Ford (1993), addressing the potential influence of vacuum fluctuations. In an electron interference experiment, the probability distribution of an electron which may coherently

"travel along two different classical spacetime paths" with respective amplitudes ψ_1 and ψ_2 is given by

$$n(x) = |\psi_1|^2 + |\psi_2|^2 + 2\text{Re}(\psi_1\psi_2^*). \tag{4.11}$$

Ford calculates the influence of the electromagnetic field on the interference
term in (4.11). He assumes a WKB approximation for the electron states
(non-spreading wave packets) and neglects spin effects. The above expression
for the probability distribution is then modified according to

$$n(x) = |\psi_1|^2 + |\psi_2|^2 + 2\text{Re}\left(\langle \varphi_2 \mid \varphi_1 \rangle \psi_1 \psi_2^*\right), \tag{4.12}$$

where the overlap factor between the respective photonic states φ_1 and φ_2
can be written as

$$\langle \varphi_2 \mid \varphi_1 \rangle = e^{W+i\phi}, \tag{4.13}$$

and an explicit expression is found for W,

$$W = -\frac{e^2}{2} \oint dx_\mu \oint dy_\nu \, D^{\mu\nu}(x,y). \tag{4.14}$$

Here, $D^{\mu\nu}$ denotes the Hadamard (anticommutator) Green function for the
electromagnetic field, $D^{\mu\nu}(x,y) = 1/2\langle 0_A \mid \{A^\mu(x), A^\nu(y)\} \mid 0_A \rangle$, and the
integration runs over the closed spacetime path formed by the respective
"paths" C_1 and C_2 in the superposition. Surprisingly, this expression is independent of the electron mass. It is also possible to rewrite this expression
in such a way that the separate contributions of "vacuum fluctuations" and
photon emission can be distinguished:

$$W_{\text{vac}} = -\frac{e^2}{2} \left(\int_{C_1} dx_\mu \int_{C_1} dy_\nu + \int_{C_2} dx_\mu \int_{C_2} dy_\nu \right) D^{\mu\nu}(x,y) \tag{4.15}$$

is the contribution from vacuum fluctuations, and

$$W_{\text{photon}} = W - W_{\text{vac}} = \frac{e^2}{2} \left(\int_{C_1} dx_\mu \int_{C_2} dy_\nu + \int_{C_2} dx_\mu \int_{C_1} dy_\nu \right) D^{\mu\nu}(x,y) \tag{4.16}$$

is the contribution from photon emission. Like in the expression (4.5) (from
which these expressions should also follow, cf. Diósi (1990b)), the vacuum contribution contains integrations over each "path" separately, while the photon emission contribution contains cross terms connecting the "paths". It
had been suggested that the "vacuum part" (4.15) also leads to decoherence.
Since, however, decoherence is an irreversible process, this cannot be true; the
vacuum part contributes only to the "dressing" of the electron, cf. Sect. 9.4
and the following discussion.

We want to add, therefore, some comments on the discussion about whether a decohering influence can be exerted by "real" photons only or also by

"virtual" ones. Santos (1994, 1995), for example, claimed that the electromagnetic vacuum can lead to decoherence even if the atom is in its ground state. This was criticised by Diósi (1995a) who argued that it is the "dressed" state which is experimentally accessible and that therefore virtual photons cannot cause decoherence. The system would thus be in a state where part of the electromagnetic degrees of freedom is automatically included. If no additional environmental interaction (which would act in an *irreversible* manner) were available, one would then expect arbitrary superpositions of such "dressed" states to occur, cf. also Sect. 9.4 and Sect. 3.2.4. From this point of view, one would expect that the contribution from vacuum polarisation, see (4.15) above, disappears after "dressing" is properly taken into account.

On the other hand, just such a "dressing" has been employed by various authors to *explain* classical properties such as a well-defined shape for chiral molecules (Pfeifer 1980; Amann 1991b, 1993), see also the review by Wightman (1995). Explicit calculations often use the *spin-boson model*, where a "system" with two degrees of freedom (mimicking, e.g., the two chirality states of a handed molecule) is coupled to infinitely many harmonic oscillators (the "electromagnetic field") in a dipole-type manner. The qualitative results indicate that two chiral states for the molecule emerge, which are exact eigenstates of a "classical operator", i.e. there is no operator available *among the chosen class of observables* possessing non-vanishing matrix elements with respect to these states. A *spatial* shape of the molecule, however, does not yet emerge, since the classical quantity in question is a two-valued quantity only (the "label" of the chirality state). The interpretation of this result seems at first glance to be analogous to a charge "dressed" by its Coulomb field, where phase relations cannot be observed by local observers, see Giulini *et al.* (1995) and Chap. 6. This influence by the Coulomb field is a *reversible* effect, and coherences can reappear if, for example, two spatially separated charges are brought together again. Note the difference to "real" decoherence, where irreversibility (e.g. through "escaping photons" as in the case of the electronic wave packets discussed above) plays a crucial role. In the case of a molecule, which as a whole is electrically neutral, the analogy with the far-reaching Coulomb field of a charge is, however, incomplete. For this reason, an infinite number of photons ("soft photons", "IR-singularity") is crucial to get the effect discussed by Amann (1991b, 1993). Such a "photon cloud" should, however, be attributed to the local system. Superpositions of such dressed states are thus expected to be observable unless they are decohered by an additional environment. Such an environment is, however, present, since it was shown in Chap. 3 that a definite shape of a large or medium-sized molecule emerges through the decohering influence of other molecules and light.

4.1.2 "Measurement" of Electromagnetic Fields by Charges

We now come to the second part of this section, which deals with the influence of charged matter on the quantum coherence of the electromagnetic field. Examples from quantum optics have already been discussed in Chap. 3. Here we shall briefly review a model discussed by Kiefer (1992) in the framework of scalar QED. We shall show in particular how superpositions between macroscopically different electromagnetic field strengths can be suppressed by interaction with charged matter fields. We consider the case where the electromagnetic part can be treated semiclassically and the matter part fully quantum mechanically. In particular, we consider the following superposition

$$\Psi \approx \sum_k C_k[\mathbf{A}]e^{iS_k[\mathbf{A}]}\chi_k[\phi, \phi^*, \mathbf{A}]. \tag{4.17}$$

The total state is thus a formal sum (k may also refer to a continuous variable) of many components, in each of which the electromagnetic field is in a WKB state ($\mathbf{A}$ denotes the vector potential and ϕ a complex scalar field); compare also the quantum-mechanical example presented in Sect. 3.2.3. Because of the WKB approximation, S_k thus obeys the electromagnetic Hamilton–Jacobi equation, and the matter states χ_k obey the approximate Schrödinger equation

$$i \int d^3x \frac{\delta S_k}{\delta \mathbf{A}} \frac{\delta \chi_k}{\delta \mathbf{A}} \equiv i \frac{\partial \chi_k}{\partial t_k} \approx H_\phi \chi_k, \tag{4.18}$$

where H_ϕ is the total Hamiltonian of QED without the pure electromagnetic part. Equation (4.18) corresponds to the limit of quantum field theory in "external" electromagnetic fields. It can be derived from full QED in a Born–Oppenheimer kind expansion scheme, see Kiefer *et al.* (1991) for details. The semiclassical degree of freedom (here given by the electromagnetic field) is sometimes called a mean field degree of freedom, see for example Habib *et al.* (1995). The discussion is performed, for simplicity, in the framework of scalar QED (the matter fields thus describe charged pions, for example), and the functional Schrödinger picture is used. The use of this framework enables our discussion to be as close as possible to the quantum mechanical case discussed in Chap. 3. The process of decoherence can be intuitively best understood by using the language of wave functions.

We shall now be more specific and consider the example where only one WKB component in (4.17) is present, and where the corresponding phase S is given by

$$S \approx -V\mathbf{E}\mathbf{A}, \tag{4.19}$$

where V is the spatial volume,[4] and $\mathbf{E}$ is the electric field (the variable canonically conjugate to $\mathbf{A}$) which is here assumed to be approximately constant in space. Note the analogy of e^{iS} to the momentum eigenstate e^{ipx} in quantum mechanics, where p corresponds to $\mathbf{E}$ and x to $\mathbf{A}$. For the matter state χ we take a Gaussian functional, using the momentum representation for the

---

[4] Physically, this represents the volume where the electric field is non-vanishing, e.g. the volume of an appropriate cavity.

scalar fields,

$$\chi = N(t) \exp\left(-\int d^3p \, \phi^*(\mathbf{p})\Omega(\mathbf{p},t)\phi(\mathbf{p})\right). \qquad (4.20)$$

Gaussian functionals of this type can represent general vacuum states in the functional Schrödinger picture, see for example Jackiw (1988). Inserting this ansatz into (4.18) yields equations for N and Ω, which can be explicitly solved. A special application is the derivation of Schwinger's pair creation formula in this framework.

If one restricts observations to the electromagnetic field only (which thus plays the role of the "relevant" subsystem), one has to trace out the scalar field in the total state, and to discuss the reduced density operator

$$\rho_A = \text{tr}_{\phi,\phi^*} |\Psi\rangle\langle\Psi| \qquad (4.21)$$

or, alternatively, its Fourier transform $\tilde{\rho}_E$ (in its matrix representation denoted by $\tilde{\rho}(\mathbf{E},\mathbf{E}')$). For times $T \gg m/eE$ one finds, neglecting particle creation,

$$\tilde{\rho}(E,E') \approx \tilde{\rho}_0 \exp\left(-\frac{8\pi^2 V}{e^3 \bar{E}T}(E-E')^2\right), \qquad (4.22)$$

where we have taken the electric field in, say, the z-direction, $\mathbf{E} = E\mathbf{e}_z$, $\bar{E} \equiv (E+E')/2$, and ρ_0 would be the density matrix if there were no scalar field. Here, $T \equiv t_f - t_i$, where t_i (resp. t_f) denotes an asymptotic switching-on time (resp. switching-off time) for the electromagnetic field. We note that the above limit $T \gg m/eE$ is reached quite soon (for $E \approx 10^3$ V/cm in about 10^{-6} s) after the switching-on. The electric field can thus be considered as a classical quantity if the width of the "Gaussian" in (4.22) is much smaller than the mean value $\bar{E}$ of the field itself. The region where quantum effects remain important is thus the *weak field region*, a result which is consistent with the analysis of the measurement of electromagnetic field quantities in Bohr and Rosenfeld (1933). It is also interesting to note that the nondiagonal elements in (4.22) would vanish exactly in the infrared limit $V \to \infty$ ("infinite cavity"). The time dependence in (4.22) may seem strange at first glance, but one should note that (due to the coupling in the interaction part of the total Hamiltonian), the vector potential $\mathbf{A}$ (strictly speaking, the magnetic field, since $\mathbf{A}$ is not gauge invariant) is "measured" (and not the electric field, which is the conjugate variable). The coherence width with respect to E thus increases with time. However, this behaviour holds only until the kinetic term of the vector potential itself (which is neglected in the Born-Oppenheimer expansion) comes into play. If decoherence is effective, the field thus *seems to be* in one of its classical states with the corresponding probability.

One can also study the back reaction of the charged field onto the electromagnetic field by using the Wigner function. This leads to the result that the Hamilton-Jacobi equation for the electromagnetic field is corrected by an additional contribution which is just the expectation value of H_ϕ with respect to the Gaussian state (4.20), provided the fluctuation of H_ϕ with respect to this state is not too large, see Kiefer (1992).

Consider now a superposition of two such states,

$$\Psi \approx e^{-iVEA}\chi[\phi, \phi^*, A] + e^{iVEA}\chi^*[\phi, \phi^*, A]. \qquad (4.23)$$

This state describes a (possibly macroscopic) superposition of an electric field pointing "upwards" with one pointing "downwards". We now show that this superposition, should it have been generated, becomes suppressed through the correlation with the scalar field state in (4.23). Assuming that E has approximately a definite value within one component of (4.23) (this case has been discussed above), one can focus on the nondiagonal element $\rho_\pm$ of the reduced density matrix along the diagonal in **A**-space, which describes interferences between the two components in (4.23). One obtains, again neglecting particle creation,

$$
\begin{aligned}
\rho_\pm \quad = \quad & \rho_0 \, \mathrm{tr}_{\phi,\phi^\dagger}\chi^2 = \rho_0 \, \det\frac{\mathrm{Re}(\Omega)}{\Omega} \\
\approx \quad & \rho_0 \, \exp\left(-\frac{VT}{256\pi^2}\frac{(eE)^3}{m^2 + (eET)^2} - \frac{V(eE)^2}{256\pi^2 m}\arctan\frac{eET}{m}\right) \\
\xrightarrow{T\gg m/eE} \quad & \rho_0 \, \exp\left(-\frac{Ve^2E^2}{512\pi m}\right),
\end{aligned}
\qquad (4.24)
$$

where m is the mass of the scalar field. Again, this vanishes in the infrared limit $V \to \infty$ where it would give rise to an "exact superselection rule" (in the weak sense of vanishing non-diagonal terms). One can estimate from (4.24) that for, say, $E \approx 10^7$ V/cm interference effects may become noticeable only on length scales smaller than about 10^{-4} cm. This result has been confirmed by Shaisultanov (1995) with a technically different approach using the influence functional (see Chap. 5). We also note that the experimental possibility of studying a decohering influence on superpositions such as (4.23) has been discussed by Davidovich *et al.* (1995).

The above effect of the "charged matter vacuum" is time-independent, as seen from (4.24). The reason is that the effect is reversible, analogously to the effect by the Coulomb field: an initial superposition begins to disappear if the volume V of the cavity is increased, but reappears after the volume is decreased again. In this sense the situation is different from the irreversible effect of "real" decoherence. The latter situation happens if particle creation is taken into account (Kiefer 1992, Habib *et al.* 1995). The decoherence factor is then of the order of

$$\exp\left(-\frac{2VT(eE)^2}{(2\pi)^3}e^{-\pi m^2/eE}\right). \qquad (4.25)$$

Not surprisingly, it contains the typical expression $\exp(-\pi m^2/eE)$ from Schwinger's pair creation formula.

The effect of the "vacuum fluctuations" (i.e. the matter fields being in the Gaussian state (4.20)) discussed here should be considered as a lower limit. In realistic situations, real charges are present and suppress interference terms even more. In principle, however, the effect of decoherence (4.24) should be measurable. A superposition of the form (4.23) can be generated, for example, by coupling an atomic device to a capacitor. By measuring expectation values of an appropriate observable $\mathcal{O}$ according to $\langle\mathcal{O}\rangle = \mathrm{tr}(\rho\mathcal{O})$, the density matrix (4.24) is distinguishable from an ensemble of field-up and field-down configurations. This is due to the remaining coherence width for finite volume.

The above derivation breaks down if the kinetic term for the electromagnetic field (which was neglected here) becomes important. One may thus expect that the exact "pointer states" are not **A**- or **E**-eigenstates, but some generalised coherent states. In contrast to the quantum mechanical case, coherent states in field theory cannot be interpreted as localised particles, but as a localisation in the amplitudes of field modes (Anglin and Zurek 1996a) – it is the mean field (in the sense of the Born-Oppenheimer approximation) which assumes classical properties. Small excitations above this background field can, however, not be distinguished by the interaction with the environment and thus behave quantum mechanically, comprising a "quantum halo" of states. Anglin and Zurek also note that the reverse situation of matter fields measured by electromagnetic fields is different because the matter fields couple bilinearly in the Hamiltonian. One would then expect that the "pointer states" were n-particle states rather than coherent states. A similar observation had been made by Kübler and Zeh (1973) who emphasised that this is the reason why the particle aspect is dominant for fermionic fields, while the field aspect is dominant for vector fields. If fundamental charged scalar fields were found, however, they would behave – in this respect – similarly to fermionic fields.

4.2 Decoherence and the Gravitational Field

4.2.1 Emergence of Classical Spacetime

According to the Copenhagen interpretation of quantum theory, the existence of a classical world is needed from the outset in order to interpret quantum theory. Appropriate classical apparata are assumed to *define* the occurrence of quantum phenomena. The presence of such classical measurement agencies seems to be possible only if spacetime exists as a classical entity.

The discussion of the previous chapters has, however, convincingly demonstrated that quantum theory has a much wider range of applicability than the pioneers had imagined. Classical properties are not intrinsic to objects but emerge through the irreversible interaction with the environment. The experiments discussed in Chap. 3 are an impressive confirmation of this idea.

What about the structure of spacetime itself? Before the advent of the general theory of relativity, spacetime was considered to be a given, non-dynamical background structure. This is also the case in quantum field theories such as QED (Sect. 4.1). In general relativity, however, the geometry of spacetime is associated with the gravitational field and thereby becomes *dynamical*. If the gravitational field is fundamentally described by quantum theory, then spacetime cannot be a classical entity.

But has gravity to be described by quantum theory? Quite generally, it does not seem possible to find a fundamental hybrid description that couples a quantum system to a classical system in a consistent way (Kiefer 2003). This does of course not mean that there exists no *effective* theory which couples quantum to classical systems. For example, one can develop a formalism in which a decohered ("classical") system is coupled to a quantum system that does not exhibit decoherence (Halliwell 1998).

As has already been mentioned, it was important already during the early discussions between Einstein and Bohr to apply the uncertainty relations to macroscopic objects (screens, photographic plates etc.) in order to save them for microscopic systems. This is reasonable because macroscopic objects are composed of atoms. Such *consistency arguments* are at the heart of these discussions. At the Solvay conference in 1930, Bohr and Einstein had a debate concerning the time-energy uncertainty relation, $\Delta E \Delta t \geq \hbar/2$. In the discussion, Bohr had to invoke general relativity to counter Einstein's objections. But only very little structure from general relativity does in fact enter the argument; it is only the equivalence principle and therefore the curved nature of spacetime, from which the redshift of light follows as a consequence. The redshift may be derived by just applying the energy law to the expression $\hbar\omega$ for the energy of a photon. One could thus phrase Bohr's argument in the way that a violation of the uncertainty relation would entail a violation of energy conservation.

In fact, the possible violation of conservation laws often plays an important role in such consistency arguments. Eppley and Hannah (1977), for example, consider the interaction of classical gravitational waves with quantum systems. They find, as a consequence, a violation of either momentum conservation or the uncertainty relations for the quantum system, or the occurrence of signals faster than light. Since not many peculiarities of the gravitational field enter their discussion, these results hold also for other systems such as the electromagnetic field. This type of arguments is certainly enforced for the gravitational field due to its coupling to *all* other degrees of freedom. Taking then the quantum nature of the gravitational field for granted, one would expect that efficient decoherence results from this universal coupling for both the gravitational field and other variables.

In a heuristic example, where quantum theory is applied to Newtonian gravity, one finds that the gravitational field is decohered by its action with quantum matter (Joos 1986b). Suppose that a (homogeneous) gravitational field within a box of side length L is in a quantum superposition of different strengths, i.e.

$$|\psi\rangle = c_1|g\rangle + c_2|g'\rangle, \ g \neq g'. \tag{4.26}$$

A particle with mass m in a state $|\chi\rangle$, which moves through this volume, "measures" the value of g, since its trajectory depends on the metric, yielding the total state

$$|g\rangle|\chi_g(t)\rangle \ . \tag{4.27}$$

This correlation destroys the coherence between g and g', and the reduced density matrix can be estimated to assume the following form after many such interactions are taken into account:

$$\rho(g, g', t) = \rho(g, g', 0) \exp\left(-\Gamma(g - g')^2 t\right) \ , \tag{4.28}$$

where

$$\Gamma = nL^4 \left(\frac{\pi m}{2k_B T}\right)^{3/2} ,$$

for a gas with particle density n and temperature T. For example, air under ordinary conditions, and $L = 1$ cm, $t = 1$ s yields a remaining coherence width of $\Delta g/g \approx 10^{-6}$.

One can give quite general arguments that the gravitational field is fundamentally of quantum nature (Kiefer 2000, 2003):

- *Singularity theorems of general relativity:* Under very general conditions, the occurrence of a singularity, and therefore the breakdown of the unquantised theory, is unavoidable. A more fundamental theory is therefore needed to overcome this breakdown, and the natural expectation is that this fundamental theory is a quantum theory of gravity. This is similar to ordinary quantum theory preventing the singularity that classical electromagnetism would predict for atoms.
- *Initial conditions in cosmology:* This is related to the singularity theorems which predict the existence of a "big bang" where the known laws of physics break down. To fully understand the evolution of our Universe, its initial state must be amenable to a physical description.
- *Unification:* Apart from general relativity, all known fundamental theories are *quantum* theories. It would thus seem awkward if gravity, which couples to all other fields, should remain the only classical entity in a fundamental description.
- *Gravity as a regulator:* Many models indicate that the consistent inclusion of gravity in a quantum framework would automatically eliminate the divergences that plague ordinary quantum field theory.
- *Problem of time:* In ordinary quantum theory, the presence of an external time parameter t is crucial for the interpretation of the theory: "Measurements" take place at a certain time, matrix elements are evaluated at fixed times, and the norm of the wave function is conserved *in* time.

Since in general relativity, on the other hand, time as part of spacetime is a dynamical quantity (as defined by the metric), both concepts of time must be modified at a fundamental level.

But what does the "quantisation" of spacetime mean? In other words, to which classical structures does one have to apply the superposition principle, while the rest remains classical? Isham (1994a) presents the following hierarchy of structures where this decision can be made at each level:

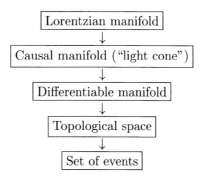

A straightforward quantisation of general relativity, for example, would dissolve spacetime as a fundamental classical entity, but would retain a fixed three-dimensional manifold in the formalism. This *canonical approach* is briefly described in the next subsection and will be the basis for the calculations presented below. Path integration, for example, would entail a superposition of different manifolds. This should also be true in a "theory of everything" (for which superstring theory is a candidate) which encompasses all interactions of Nature in a single quantum framework. In such a fundamental theory it is probably only very little structure, if any, that remains classical, although this is not yet clear, cf. Seiberg and Witten (1999).

Quantum effects of gravity are expected to become relevant at the Planck scale. This is the scale where, for an elementary particle, Schwarzschild radius and Compton wavelength coincide. The Planck mass is given by

$$m_P = \sqrt{\frac{\hbar c}{G}} \approx 10^{-5} \text{ g} , \qquad (4.29)$$

while Planck length and Planck time are given by the following expressions, respectively,

$$l_P = \sqrt{\frac{\hbar G}{c^3}} \approx 10^{-33} \text{ cm} , \quad t_P = \sqrt{\frac{\hbar G}{c^5}} \approx 10^{-44} \text{ s} . \qquad (4.30)$$

As we discuss at length in this volume, quantum effects are not a priori restricted to a particular scale. In Chap. 3 we have demonstrated that it is not the large mass by itself that provokes classical behaviour for a quantum object, but its interaction (whose strength of course depends on the mass) with the environment. Analogously, it is not the smallness of the Planck length by itself that a priori prevents quantum-gravity effects to occur at larger scales. The classical *appearance* of spacetime at larger scales should again be due to the unvoidable interaction with other degrees of freedom. It is for this reason that we can restrict ourselves in the following discussion to canonical quantum gravity, since this should be valid as an *effective* theory for scales $l \gg l_P$, independent of whether this theory is also valid at the Planck scale itself or not (in the latter case a unified theory such as string theory must be invoked).

We mention that gravity is assigned a fundamental role also in approaches which *modify* the formalism of quantum theory, see e.g. Károlyházy *et al.* (1986), Penrose (1986), as well as Chap. 8, but this will not be considered in this chapter.

4.2.2 The Formalism of Quantum Cosmology

The basic intention in the canonical approach to quantum gravity is to derive equations for wave functionals on an appropriate configuration space, analogously to the Schrödinger picture in quantum mechanics. Technically, this is achieved by foliating, in the classical theory, the classical spacetime into spatial hypersurfaces and choosing the *spatial* metric as a canonical variable (the "q"). In the spacetime which is classically constructed by dynamically developing the initial data on a particular hypersurface, the canonical momentum is linearly related to the embedding of the hypersurfaces into spacetime. (In the case of a Friedmann universe, the radius, a, is the configuration variable, while the canonical momentum corresponds to the Hubble parameter.) The postulate of nontrivial commutation relations between these quantities in quantum gravity then means that spacetime is no longer a fundamental concept, since one cannot specify both the spatial metric and the embedding. The role of spacetime is taken over by the space of all three-dimensional geometries, which is called *superspace* and which serves as the configuration space for the theory. For a detailed physical introduction into these concepts we refer to Zeh (2001); the details of the canonical formalism are presented, for example, by Wald (1984). The central kinematical quantity is thus a wave functional defined on superspace and on matter field degrees of freedom. It is often labeled $\Psi[^3\mathcal{G}, \Phi]$, where $^3\mathcal{G}$ stands for "three-dimensional geometry" (to express the fact that this wave functional is independent of particular coordinates on the three-dimensional space, as being guaranteed by the three "momentum constraints" of general relativity), and Φ symbolically denotes all non-gravitational fields. The invariance of general relativity (called invariance under coordinate transformations or under diffeomorphisms) leads to the presence of constraints: the total Hamiltonian must vanish.[5] In the

---

[5] We consider only the case of spatially closed hypersurfaces. In the asymptotically flat case, the total Hamiltonian can be written as a surface integral.

quantum theory, the constraints are implemented à la Dirac as restrictions on physically allowed wave functionals. The wave functional then obeys the Wheeler-DeWitt equation (DeWitt 1967, Wheeler 1968),

$$H\Psi = 0, \tag{4.31}$$

where H denotes the full Hamiltonian for gravity and other fields. In classical general relativity, spacetimes can be parametrised by some arbitrary time coordinates (which have lost their absolute status). Since due to the uncertainty relations no spacetimes exist anymore on the level of quantum gravity (only a wave function for spatial metrics), there is no time parameter available to parametrise them – the Wheeler-DeWitt equation is "timeless". This gives rise to the *problem of time* in quantum gravity which is extensively discussed in the literature, see e.g. Barbour (1994a,b), Isham (1992), Kuchař (1992), Zeh (1986, 2001), Kiefer and Zeh (1995), and Kiefer (2000, 2003).

We have to emphasise that this approach at present exists only on a formal level, since the explicit treatment of (4.31) is unclear.[6] In this respect the discussion in the present section is different from the rest of the book and should be considered of heuristic value only. However, from general arguments like reparametrisation invariance one would expect the fundamental equation to be of the constraint form (4.31), although the exact form of H may be different. Therefore, the main *interpretational* part of the discussion in this section would remain unaffected, and only the details of the calculations would have to be changed.

The main features of the canonical approach can already be recognised in a simple two-dimensional model – a closed Friedmann universe characterised by its scale factor a, containing a homogeneous massive scalar field φ as a representation of matter, cf. Kiefer (1988) and Halliwell (1991). Taking the units $2G = 3\pi$, the classical action for this model is the sum of the gravitational part and the matter part,

$$
\begin{aligned}
S &= \int dt\ L(a, \dot{a}, \varphi, \dot{\varphi}, N) \\
&\equiv \frac{1}{2} \int dt\ Na^3 \left(-\frac{\dot{a}^2}{N^2 a^2} + \frac{\dot{\varphi}^2}{N^2} + \frac{1}{a^2} - m^2 \varphi^2 \right).
\end{aligned} \tag{4.32}
$$

This action is invariant with respect to arbitrary reparametrisations of the time variable t, a fact which is encoded in the presence of the non-dynamical *lapse function* N which appears undifferentiated in the action. A characteristic feature of the gravitational field is the occurrence of an indefinite kinetic term in the action.

---

[6] It is known that (4.31) does not give rise to a unitary evolution in a Fock space built over three-dimensional slices.

The standard canonical formalism proceeds with the definition of the canonical momenta,

$$p_N = \frac{\partial L}{\partial \dot{N}} = 0, \; p_a = \frac{\partial L}{\partial \dot{a}} = -\frac{a\dot{a}}{N}, \; p_\varphi = \frac{\partial L}{\partial \dot{\varphi}} = \frac{a^3 \dot{\varphi}}{N} . \tag{4.33}$$

The canonical Hamiltonian is then given by

$$
\begin{aligned}
H &= p_N \dot{N} + p_a \dot{a} + p_\varphi \dot{\varphi} - L \\
&= \frac{N}{2} \left(-\frac{p_a^2}{a} + \frac{p_\varphi^2}{a^3} - a + m^2 \varphi^2 a^3 \right) \\
&\equiv \frac{N}{2} G^{AB} p_A p_B + V(a, \varphi) .
\end{aligned}
\tag{4.34}
$$

The important point is that $p_N = 0$ is a constraint that should hold at all times. Therefore, from Hamilton's equations of motion one gets $\partial H / \partial N = 0$ which gives the constraint

$$H = 0 \Leftrightarrow \dot{a}^2 = -1 + a^2(\dot{\varphi}^2 + m^2 \varphi^2) . \tag{4.35}$$

This is nothing but the classical Friedmann equation which is well known from cosmology. Variation of (4.32) with respect to a and φ give the classical equations of motion. The equation for φ, in particular, reads

$$\ddot{\varphi} + \frac{3\dot{a}}{a} \dot{\varphi} + m^2 \varphi = 0 . \tag{4.36}$$

This is the Klein-Gordon equation for a homogeneous field in an evolving universe, whose effect on φ is the second ("friction") term.

Following Dirac's procedure, the classical constraint (4.35) is then turned into the Wheeler-DeWitt equation (4.31). Using a particular factor ordering,[7] the explicit form of this equation in the present model reads

$$H\psi \equiv \left(\hbar^2 a \frac{\partial}{\partial a} \left(a \frac{\partial}{\partial a} \right) - \hbar^2 \frac{\partial^2}{\partial \varphi^2} + m^2 \varphi^2 a^6 - a^4 \right) \psi(a, \varphi) = 0 . \tag{4.37}$$

Note that the indefiniteness of the kinetic term has led to a *hyperbolic* equation for ψ – in contrast to the Schrödinger equation. In the next subsection, a more complicated model is used in which the variables a and φ play the role of the background, supplemented by additional degrees of freedom ("higher multipoles") $\{f_n\}$. (In the following we set again $\hbar = 1$.)

The Wheeler-DeWitt equation (4.31), (4.37) does not contain a classical time parameter. This is not surprising, since the classical metric is known to determine time. An *approximate* concept of time-dependence of a wave

---

[7] The chosen factor ordering is given by the Laplace-Beltrami operator in the configuration space spanned by a and φ.

function can be recovered in a Born-Oppenheimer type of approximation scheme in which part of the degrees of freedom are semiclassical (given by WKB wave functions), while the rest is fully quantum. This limit is obtained, for example, if the full wave functional in (4.31) is of the form

$$\Psi[^3\mathcal{G}, \Phi] \approx \sum_{(n)} C_{(n)}[^3\mathcal{G}] e^{iS_0^{(n)}[^3\mathcal{G}]/G\hbar} \psi^{(n)}[^3\mathcal{G}, \Phi] \equiv \sum_{(n)} \psi_0^{(n)} \psi^{(n)}, \qquad (4.38)$$

where the $S_0^{(n)}$ are solutions to the gravitational Hamilton–Jacobi equations which are fully equivalent to Einstein's field equations. The gravitational part of the total state is thus treated semiclassically. The semiclassical part may also comprise part of the matter degrees of freedom. In fact, in the discussion of decoherence in Sect. 4.2.3, the scalar field φ will belong to this part. Note the analogy to (4.17) discussed in the last section.

The sum in (4.38) runs over a whole set of indices (n) (which may also be continuous). It turns out that the matter states $\psi^{(n)}$ obey the following approximate equation in each component,

$$i\nabla S_0^{(n)} \cdot \nabla \psi^{(n)} \approx H_m^{(n)} \psi^{(n)}, \qquad (4.39)$$

where $H_m^{(n)}$ denotes the Hamiltonian for the non-gravitational fields (which of course depends on the particular solution $S_0^{(n)}$ chosen for the gravitational field). Note the analogy of $H_m^{(n)}$ to H_ϕ discussed in the last section. The expression $\nabla S_0^{(n)} \cdot \nabla \equiv \partial/\partial t^{(n)}$ is a directional derivative in the gravitational part of the full configuration space, which parametrises the family of classical spacetimes described by $S_0^{(n)}$. The parameters $t^{(n)}$ are often called *WKB times* – they control the "dynamical evolution" of the states $\chi^{(n)}$ along the WKB trajectories. Equation (4.39) is thus nothing but the Schrödinger equation, while t represents our phenomenological time. Details of the semiclassical approximation to quantum gravity are described in Kiefer (1994), see also Kiefer (1993), Giulini and Kiefer (1995).

We note that due to the central input of the Born-Oppenheimer expansion the situation here is analogous to that of Sect. 4.1.2 only ("measurement" of fields by charges), since the reverse effect (which would here correspond to "measurement" of matter by the gravitational field) is too weak to become important.

4.2.3 Decoherence in Quantum Cosmology

In quantum cosmology, all variables are fundamentally quantum and there is no classical spacetime. How does a classical spacetime emerge? It has been suggested that global degrees of freedom such as the volume of the Universe appear classical after the interaction with other degrees of freedom is taken into account (Zeh 1986). The role of such additional variables may be played by density fluctuations and gravitational waves. All these degrees of freedom are of course within the Universe, but they are "environmental" to the volume-degree of freedom in configuration space. From the viewpoint of a "local" observer who can measure the size of the Universe but has no access to small fluctuations, these other degrees of freedom have to be traced over. In this sense they are able to produce decoherence for the volume degree of freedom. We have emphasised before that the issue of classicality only arises *after* a quantum system has been chosen, for which the straightforward application of the superposition principle would lead to a macroscopically entangled state. In a sense, a classical spacetime thus arises by a "self-measurement" of the Universe.

Calculations for decoherence in quantum cosmology can be done with the help of the Wheeler-DeWitt equation (4.31), see Kiefer (1987). As a necessary prerequisite, the semiclassical approximation to quantum gravity is employed, in which an approximate Schrödinger equation is recovered for the cosmological fluctuations (see Sect. 4.2.2). The time parameter corresponding to this equation is defined by the semiclassical degrees of freedom (Halliwell and Hawking 1985). In Kiefer (1987) the relevant system was taken to be the scale factor ("radius") a of the Universe together with a homogeneous scalar field φ, cf. the model discussed in Sect. 4.2.2. The field φ plays a crucial role in modern cosmological theories where an exponential, "inflationary", expansion is assumed to have happened in an early phase of the Universe, starting about 10^{-33} s after the big bang. It is in fact the "inflaton field" φ itself that causes inflation. The inhomogeneous modes of the gravitational field and the scalar field (gravitational waves and density fluctuations) can then be shown to *decohere* the global variables a and φ.

An open problem in these early papers was the issue of regularisation; the number of fluctuations is infinite and would cause divergences, which is why a cutoff was suggested. The issue was again addressed in Barvinsky *et al.* (1999a) where a physically motivated regularisation scheme was introduced. In the following we shall briefly review this approach.

As a (semi)classical solution for a and φ one may use

$$\varphi(t) \approx \varphi, \tag{4.40}$$

$$a(t) \approx \frac{1}{H(\varphi)} \cosh H(\varphi)t , \tag{4.41}$$

where $H(\varphi) = 8\pi V(\varphi)/3m_P^2$ is the Hubble parameter generated by the inflaton potential $V(\varphi)$. It is approximately constant during the inflationary phase in which φ slowly "rolls down" the potential. We take into account fluctuations of a field $f(t, \mathbf{x})$ which can be a field of any spin. Space is assumed to be a closed three-sphere, so $f(t, \mathbf{x})$ can be expanded into a discrete series of spatial orthonormal harmonics $Q_n(\mathbf{x})$,

$$f(t, \mathbf{x}) = \sum_n f_n(t)Q_n(\mathbf{x}) . \tag{4.42}$$

One can thus represent the fluctuations by the degrees of freedom f_n.

Our intention now is to solve the Wheeler-DeWitt equation (4.31) in the semiclassical approximation. This leads to the following solution:

$$\Psi(t|\varphi, f) = \frac{1}{\sqrt{v_\varphi^*(t)}} e^{-I(\varphi)/2 + iS_{cl}(t,\varphi)} \prod_n \psi_n(t, \varphi|f_n) \ . \qquad (4.43)$$

The time t that appears here is the semiclassical ("WKB") time and is defined by the background-degrees of freedom a and φ through the "eikonal" S_{cl} which is a solution of the Hamilton-Jacobi equation; t is formally identical with the time that appears in the classical equations (4.40) and (4.41). Since φ is thus determined by a, only one variable (a or φ) occurs in the argument of Ψ. The wave functions ψ_n for the fluctuations f_n obey each an approximate Schrödinger equation (4.39) with respect to t, and their Hamiltonian H_n has the form of a ("time-dependent") harmonic-oscillator Hamiltonian. The first exponent contains the euclidean action $I(\varphi)$ from the classically forbidden region (the "De Sitter instanton") and is independent of t. Its form depends on the *boundary conditions* imposed. In the present case the so-called Hartle-Hawking condition is chosen, see e.g. Halliwell (1991), which amounts to $I(\varphi) \approx -3m_P^4/8V(\varphi)$. The detailed form is, however, not necessary for the discussion below. The function $v_\varphi(t)$ is the so-called basis function for φ and is a solution of the classical equation of motion. In the following we shall choose units such that $G = c = \hbar = 1$.

For the ψ_n we shall take – in analogy to (4.20) – Gaussian states that correspond to the so-called De Sitter-invariant vacuum state (Starobinsky 1979, Allen 1985). This is the maximally symmetric state which possesses properties very similar to the standard vacuum state in Minkowski space. (In the massless case, this state is invariant only under a subgroup of the De Sitter group.) It is given by

$$\psi_n(t, \varphi|f_n) = \frac{1}{\sqrt{v_n^*(t)}} \exp\left(-\frac{1}{2}\Omega_n(t)f_n^2\right), \qquad (4.44)$$

$$\Omega_n(t) = -ia^3(t)\frac{\dot{v}_n^*(t)}{v_n^*(t)} \ . \qquad (4.45)$$

The functions v_n are the basis functions of the De Sitter-invariant vacuum state; they satisfy the classical equation of motion

$$F_n\left(\frac{d}{dt}\right)v_n \equiv \left(\frac{d}{dt}a^3\frac{d}{dt} + a^3m^2 + a(n^2 - 1)\right)v_n = 0 \qquad (4.46)$$

with the boundary condition that they should correspond to a standard Minkowski positive-frequency function for constant a. In the simple special case of a spatially flat section of De Sitter space one would have

$$av_n = \frac{e^{-in\eta}}{\sqrt{2n}}\left(1 - \frac{i}{n\eta}\right), \qquad (4.47)$$

where η is the *conformal time* defined by $a d\eta = dt$. We note that the corresponding *negative*-frequency function enters the exponent of the Gaussian, see (4.45).

An important property of these vacuum states is that their norm is conserved *along any semiclassical solution* (4.40), (4.41),

$$\left\langle \psi_n | \psi_n \right\rangle \equiv \int df_n |\psi_n(f_n)|^2 = \sqrt{2\pi} [\Delta_n(\varphi)]^{-1/2}, \tag{4.48}$$

$$\Delta_n(\varphi) \equiv ia^3 (v_n^* \dot{v}_n - \dot{v}_n^* v_n) = \text{constant} . \tag{4.49}$$

Note that $\Delta_n(\varphi)$ is just the (constant) Wronskian corresponding to (4.46). (The corresponding Wronskian for the homogeneous mode φ is $\Delta_\varphi \equiv ia^3 (v_\varphi^* \dot{v}_\varphi - \dot{v}_\varphi^* v_\varphi)$.) We must emphasise that Δ_n is a nontrivial function of the background variable φ, since it is defined on full configuration space and not only along semiclassical trajectories (it gives the weights in the "Everett branches".) It is therefore *not* possible to normalise the ψ_n artificially to one, since this would be inconsistent with respect to the full Wheeler-DeWheeler equation (Barvinsky *et al.* 1999a).

The solution (4.43) forms the basis for our discussion of decoherence. Since the $\{f_n\}$ are interpreted as the environmental degrees of freedom, they have to be integrated out to get the reduced density matrix for φ or a (a and φ can be used interchangeably, since they are connected by t). The reduced density matrix thus reads

$$\rho(t|\varphi, \varphi') = \int df \Psi(t|\varphi, f) \Psi^*(t|\varphi', f) , \tag{4.50}$$

where Ψ is given by (4.43), and it is understood that $df = \prod_n df_n$. After the integration one finds

$$\rho(t|\varphi, \varphi') = C \frac{1}{\sqrt{v_\varphi^*(t) v_\varphi'(t)}} \exp\left[-\frac{1}{2}I - \frac{1}{2}I' + i(S - S') \right]$$
$$\times \prod_n \left[v_n^* v_n' (\Omega_n + \Omega_n'^*) \right]^{-1/2} , \tag{4.51}$$

where C is a numerical constant. The diagonal elements $\rho(t|\varphi, \varphi)$ describe the probabilities for certain values of the inflaton field to occur. In an appropriate model, one can find that these probabilities are peaked at the onset of inflation around values of φ that lead to phenomenologically satisfying results (for example, with respect to structure formation) without having to invoke the anthropic principle, see Barvinsky *et al.* (1999b) and the references therein.

It is convenient to rewrite the expression for the density matrix (4.51) in the form

$$\rho(t|\varphi, \varphi') = C \frac{\Delta_\varphi^{1/4} \Delta_\varphi'^{1/4}}{\sqrt{v_\varphi^*(t) v_\varphi'(t)}} \exp\left(-\frac{1}{2}\boldsymbol{\Gamma} - \frac{1}{2}\boldsymbol{\Gamma}' + i(S - S') \right)$$
$$\times \boldsymbol{D}(t|\varphi, \varphi'), \tag{4.52}$$

where

$$\boldsymbol{\Gamma} = I(\varphi) + \boldsymbol{\Gamma}_{1-\text{loop}}(\varphi) \tag{4.53}$$

is the full Euclidean effective action including the classical part and the one-loop part. The latter comes from the next-order WKB approximation and is important for the normalisability of the wave function with respect to φ. The last factor in (4.52) is the *decoherence factor*

$$\boldsymbol{D}(t|\varphi,\varphi') = \prod_n \left(\frac{4\text{Re}\,\Omega_n \,\text{Re}\,\Omega_n'^*}{(\Omega_n + \Omega_n'^*)^2} \right)^{1/4} \left(\frac{v_n \, v_n'^*}{v_n^* \, v_n'} \right)^{1/4} . \tag{4.54}$$

It is equal to one for coinciding arguments. While the decoherence factor is time-dependent, the one-loop contribution to (4.52) does not depend on time and may play only a role at the onset of inflation. In a particular model with non-minimal coupling (Barvinsky *et al.* 1997), the size of the non-diagonal elements is at the onset of inflation approximately equal to those of the diagonal elements. The Universe would thus be essentially quantum at this stage, i.e. in a non-classical state.

The amplitude of the decoherence factor can be rewritten in the form

$$|\boldsymbol{D}(t|\varphi,\varphi')| = \exp \frac{1}{4} \sum_n \ln \frac{4\text{Re}\,\Omega_n \,\text{Re}\,\Omega_n'^*}{|\Omega_n + \Omega_n'^*|^2} . \tag{4.55}$$

The convergence of this series is far from being guaranteed. Moreover, the divergences might not be renormalisable by local counterterms in the bare quantised action. We shall now analyse this question in more detail.

We start with a minimally coupled massive scalar field. Equation (4.46) for the basis functions reads

$$\frac{d}{dt} \left(a^3 \frac{dv_n}{dt} \right) + a^3 \left(\frac{n^2 - 1}{a^2} + m^2 \right) v_n = 0 . \tag{4.56}$$

The appropriate solution to this equation is (Barvinsky *et al.* 1992)

$$v_n(t) = (\cosh Ht)^{-1} P^{-n}_{-\frac{1}{2} + i\sqrt{m^2/H^2 - 9/4}}(i \sinh Ht) , \tag{4.57}$$

where P denotes an associated Legendre function of the first kind. The expansion of (4.57) for large masses was derived in Barvinsky *et al.* (1992). The corresponding expression for (4.45) is given by

$$\Omega_n = a^2 \left[\sqrt{n^2 + m^2 a^2} + i \sinh Ht \left(1 + \frac{1}{2} \frac{m^2 a^2}{n^2 + m^2 a^2} \right) \right] + O\left(\frac{1}{m} \right) \tag{4.58}$$

The leading contribution to the amplitude of the decoherence factor is therefore

$$\ln |\boldsymbol{D}(t|\varphi,\varphi')| \simeq \frac{1}{4} \sum_{n=0}^{\infty} n^2 \ln \frac{4a^2 a'^2 \sqrt{n^2 + m^2 a^2} \sqrt{n^2 + m^2 a'^2}}{\left(a^2 \sqrt{n^2 + m^2 a^2} + a'^2 \sqrt{n^2 + m^2 a'^2} \right)^2} \tag{4.59}$$

The first term, n^2, in the sum comes from the degeneracy of the eigenfunctions. This expression has divergences which *cannot* be represented as additive functions of a and a'. This means that no one-argument counterterm to $\boldsymbol{\Gamma}$ and $\boldsymbol{\Gamma}'$ in (4.52) can cancel these divergences of the amplitude (Paz and Sinha 1992). One might try to apply standard regularisation schemes from quantum field theory, such as dimensional regularisation. The corresponding calculations have been performed in Barvinsky *et al.* (1999a) and will not be given here. The important result is that, although they render the sum (4.59) convergent, they lead to a *positive* value of this expression. This means that the decoherence factor would diverge for $(\varphi - \varphi') \to \infty$ and thus spoil one of the crucial properties of a density matrix – the boundedness of $\operatorname{tr} \hat{\rho}^2$. The dominant term in the decoherence factor would read

$$\ln |\boldsymbol{D}| = \frac{\pi}{24}(ma)^3 + O(m^2), \quad a \gg a' \tag{4.60}$$

and would thus be unacceptable for a density matrix. Reduced density matrices are usually not considered in quantum field theory, so this problem has not been encountered before. A behaviour such as in (4.60) is even obtained in the case of massless conformally invariant fields, for which one would expect a decoherence factor equal to one, since they decouple from the gravitational background. How, then, does one have to proceed in order to obtain a sensible regularisation?

The crucial point is to perform a *redefinition* of environmental fields and to invoke a physical principle to fix this redefinition. The situation is somewhat analogous to the treatment of the S-matrix in quantum field theory: off-shell S-matrix and effective action depend on the parametrisation of the quantum fields (Vilkovisky 1984), in analogy to the non-diagonal elements of the reduced density matrix. In Laflamme and Louko (1991) and Kiefer (1992) it has been proposed within special models to rescale the environmental fields by a power of the scale factor. It was therefore suggested in Barvinsky *et al.* (1999a) to redefine the environmental fields by a power of the scale factor that corresponds to the conformal weight of the field (which is defined by the invariance of the conformally invariant wave equation). For a scalar field in four spacetime dimensions this amounts to a multiplication by a:

$$v_n(t) \to \tilde{v}_n(t) = a \, v_n(t) , \tag{4.61}$$

$$\tilde{\Omega}_n = -ia\frac{d}{dt} \ln \tilde{v}_n^* . \tag{4.62}$$

An immediate test of this proposal is to see whether the decoherence factor is equal to one for a massless conformally invariant field. In this case, the basis functions and frequency functions read, respectively,

$$\tilde{v}_n^*(t) = \left(\frac{1 + i \sinh Ht}{1 - i \sinh Ht} \right)^{\frac{n}{2}} , \tag{4.63}$$

$$\tilde{\Omega}_n = -ia\frac{d}{dt} \ln \tilde{v}_n^*(t) = n . \tag{4.64}$$

Hence, $\tilde{\boldsymbol{D}}(t|\varphi,\varphi') \equiv 1$. The same holds also for the electromagnetic field (which in four spacetime dimensions is conformally invariant). It is interesting to note that the degree of decoherence caused by a certain field depends on the spacetime dimension, since its conformal properties are dimension-dependent.

For a massive minimally coupled field the new frequency function reads

$$\tilde{\Omega}_n = \left[\sqrt{n^2 + m^2 a^2} + i \sinh Ht \left(\frac{1}{2}\frac{m^2 a^2}{n^2 + m^2 a^2}\right)\right] + O(1/m) . \qquad (4.65)$$

Note that, in contrast to (4.58), there is no factor of a in front of this expression. Since (4.65) is valid in the large-mass limit, it corresponds to modes which evolve adiabatically on the gravitational background, the imaginary part in (4.65) describing particle creation.

It turns out that the imaginary part of the decoherence factor has at most logarithmic divergences and, therefore, affects only the phase of the density matrix. Moreover, these divergences decompose into an *additive* sum of one-argument functions and can thus be cancelled by adding counterterms to the classical action S (and S') in (4.52) (Paz and Sinha 1992). The real part is simply convergent and gives a finite decoherence amplitude. This result is formally similar to the result for the decoherence factor in QED (Kiefer 1992).

For $a \gg a'$ (far off-diagonal terms) one gets the expression

$$|\tilde{\boldsymbol{D}}(t|\varphi,\varphi')| \simeq \exp\left[-\frac{(ma)^3}{24}\left(\pi - \frac{8}{3}\right) + O(m^2)\right] . \qquad (4.66)$$

Compared with the naively regularised (and inconsistent) expression (4.60), π has effectively been replaced by $8/3 - \pi$. In the vicinity of the diagonal, one obtains

$$\ln|\tilde{\boldsymbol{D}}(t|\varphi,\varphi')| = -\frac{m^3 \pi a (a - a')^2}{64} , \qquad (4.67)$$

a behaviour similar to (4.66).

An interesting case is also provided by minimally coupled massless scalar fields and by gravitons. They share the basis- and frequency functions in their respective conformal parametrisations:

$$\tilde{v}_n^*(t) = \left(\frac{1 + i \sinh Ht}{1 - i \sinh Ht}\right)^{\frac{n}{2}} \left(\frac{n - i \sinh Ht}{n + 1}\right) , \qquad (4.68)$$

$$\tilde{\Omega}_n = \frac{n(n^2 - 1)}{n^2 - 1 + H^2 a^2} - i\frac{H^2 a^2 \sqrt{H^2 a^2 - 1}}{n^2 - 1 + H^2 a^2}. \qquad (4.69)$$

They differ only by the range of the quantum number n ($2 \leq n$ for inhomogeneous scalar modes and $3 \leq n$ for gravitons) and by the degeneracies of the n-th eigenvalue of the Laplacian,

$$\dim(n)_{\text{scal}} = n^2 , \qquad (4.70)$$

$$\dim(n)_{\text{grav}} = 2(n^2 - 4). \qquad (4.71)$$

For far off-diagonal elements one obtains the decoherence factor

$$|\tilde{D}(t|\varphi,\varphi')| \sim e^{-C(Ha)^3}, \quad a \gg a', \quad C > 0, \tag{4.72}$$

while in the vicinity of the diagonal one finds

$$|\tilde{D}(t|\varphi,\varphi')| \sim \exp\left(-\frac{\pi^2}{32}(H - H')^2 t^2 e^{4Ht}\right), \tag{4.73}$$

$$\sim \exp\left(-\frac{\pi^2 H^4 a^2}{8}(a - a')^2\right), \quad Ht \gg 1. \tag{4.74}$$

These expressions exhibit a rapid disappearance of non-diagonal elements during the inflationary evolution.

It is interesting that the behaviour of fermions concerning decoherence is different from the behaviour of bosons (Barvinsky et $al.$ 1999c). Since their conformal weight is $-3/2$ in four dimensions, the environmental fermionic fields are reparametrised by a factor $a^{-3/2}$. For $m = 0$ this does, as in the bosonic case, render the decoherence factor finite and, due to conformal invariance, makes it equal to one. The situation for $m \neq 0$ is, however, different. In spite of the conformal reparametrisation, the decoherence factor is divergent. Moreover, dimensional regularisation would again spoil crucial properties of the density matrix and make it inconsistent. There remains, however, a freedom of reparametrisation in the fermionic case (Barvinsky et $al.$ 1999c): this is a Bogoliubov transformation that is analogous to a Foldy-Wouthuysen transformation in Minkowski space (the decoupling of spinor components in the nonrelativistic limit). Since it is explicitly n-dependent, it corresponds to a $nonlocal$ field redefinition. Instead of m one has now an effective n-dependent mass $\tilde{m}$ depending on the transformation. How can one fix this field redefinition? In Barvinsky et $al.$ (1999c) the principle was put forward that decoherence should be $minimal$ in the absence of particle creation. This is already implemented in the massless case. In the massive case, it means that decoherence is absent for a stationary spacetime which exhibits no particle creation. This leads to a decoherence factor

$$|\tilde{D}(t|\varphi,\varphi')| \sim \exp\left(-C'm^2 H^2 a^2 (a - a')^2\right), \quad C' > 0. \tag{4.75}$$

While decoherence is thus absent in the absence of particle creation, for bosons it is minimal in the sense that it is absent in the conformally-coupled case, but still present in the massive case – the expressions (4.66) and (4.67) do not depend on H. Formally, this is due to the fact that in the fermionic case one has m^2 instead of m^3 in the exponent; since one would expect to find factors of a in the nominator of the exponent (as is suggested by the coupling in the action), they have to be accompanied by corresponding factors of H for dimensional reason. Comparing (4.75) with (4.66) and (4.67) (which are valid for $m \gg H$), one recognises that fermions are $less$ $efficient$ in producing decoherence. In the massless case, there influence is totally absent.

The point that decoherence is linked with particle creation has been made before (Calzetta and Hu 1994, Hu and Matacz 1995). Using the influence-functional approach to decoherence, see Chap. 5, one can derive an explicit formula connecting the decoherence factor with the Bogoliubov coefficients describing particle creation (Hu and Matacz 1995).[8] Given a special initial state (a "vacuum"), this encodes the irreversible aspect of decoherence. In the massless bosonic case, (4.72) and (4.74), the effect may be interpreted as arising from a cutoff at a mode number $n \approx aH$, i.e., a cutoff of modes whose wavelength a/n is smaller than the Hubble scale H^{-1} (Halliwell 1989). As we shall see in the next subsection, these are exactly the modes that experience particle creation.

The above analysis of decoherence was based on the state (4.43). One might, however, start with a quantum state that is a superposition of many semiclassical components, i.e. many components of the form $\exp(iS_{cl}^k)$, where each S_{cl}^k is a solution of the Hamilton-Jacobi equation for a and φ. Decoherence between different such semiclassical *branches* has also been the subject of intense investigation (Halliwell 1989, Kiefer 1992). The important point is that decoherence between different branches is usually weaker than the above discussed decoherence within one branch. Moreover, it usually follows from the presence of decoherence within one branch. In the special case of a superposition of (4.43) with its complex conjugate, one can immediately recognise that decoherence between the semiclassical components is smaller than within one component: in the expression (4.54) for the decoherence factor, the term $\Omega_n + \Omega_n'^*$ in the denominator is replaced by $\Omega_n + \Omega_n'$. Therefore, the imaginary parts of the frequency functions add up instead of partially cancelling each other and (4.54) becomes smaller. One also finds that the decoherence factor is equal to one for vanishing expansion of the semiclassical universe (Kiefer 1992).

We note that the decoherence between the $\exp(iS_{cl})$ and $\exp(-iS_{cl})$ components can be interpreted as a *symmetry breaking* analogously to the case of sugar molecules, see Sect. 3.2.4 and Chap. 9. There, the Hamiltonian is invariant under space reflections, but the state of the sugar molecules exhibits chirality. Here, the Hamiltonian in (4.31) is invariant under complex conjugation,[9] while the "actual states" (i.e., one decohering WKB component in the total superposition) are of the form $\exp(iS_{cl})$ and are thus intrinsically complex. It is therefore not surprising that the recovery of the classical world follows only for complex states, in spite of the real nature of the

---

[8] The decoherence factor in the massive bosonic case, (4.66) and (4.67), comes from the adiabatic part of $\tilde{\Omega}_n$ and is not directly related to particle creation. This is not in conflict with Hu and Matacz (1995), since there the assumption is being made that the state separates between system and environment in the past, which is not the case here.

[9] We ignore here alternative approaches which use a complex Hamiltonian from the very beginning (Kiefer 1993).

Wheeler-DeWitt equation (see in this context Barbour 1993). Since this is a prerequisite for the derivation of the Schrödinger equation, one might even say that *time* (the WKB time parameter in the Schrödinger equation) arises from symmetry breaking.

The above considerations thus lead to the following picture. The Universe was essentially "quantum" at the onset of inflation. Mainly due to bosonic fields, decoherence set in and led to the emergence of many "quasi-classical branches" which are dynamically independent of each other. Strictly speaking, the very concept of time makes only sense after decoherence has occurred. In addition to the horizon problem etc., inflation also solves the "classicality problem". It remains of course unclear why inflation happened in the first place (if it really did). Looking back from our Universe (our semiclassical branch) to the past, one would notice that at the time of the onset of inflation our component would interfere with other components to form a timeless quantum-gravitational state. The Universe would thus cease to be transparent to earlier times (because there was no time). This demonstrates in an impressive way that quantum-gravitational effects are not restricted to the Planck scale.

It is interesting that a similar kind of constructive interference would occur near the turning point of a classically recollapsing universe (Kiefer and Zeh 1995). This is a direct consequence of the consistent way in which boundary conditions have to be imposed in this case. Again, this demonstrates that quantum effects are not restricted a priori to a particular scale and that it is a quantitative question referring to the dynamics when and to which extent classical properties emerge.

Our analysis has been restricted to the case where the "system" is taken to be a Friedmann universe containing a homogeneous scalar field. This is justified from phenomenological grounds, since our Universe appears isotropic and homogeneous on largest scales. Again, this may be traced back to the presumed occurrence of an inflationary phase and the validity of the cosmic no-hair conjecture. In spite of this, one can discuss decoherence in the context of anisotropic models, too (Gangui *et al.* 1991, Camacho and Camacho-Galván 1999), and find classical properties for the corresponding scale factors.

We want finally to stress the importance of decoherence for the *origin of irreversibility* in our Universe (Zeh 2001; Kiefer and Zeh 1995). Since the entropy of the present Universe (defined by its "relevant" degrees of freedom) is still extremely small compared to its maximal possible value (which would be achieved if the whole mass of the Universe were present in the form a black hole), the evolution of the Universe must have been started with a state of almost zero entropy (Penrose 1981). A possible explanation of this fact must necessarily invoke the fundamental quantum theory of gravity. It has been argued in the above references that a simple boundary condition at $a \to 0$ for the wave function of the Universe may be sufficient to explain the observed arrow of time, and may even lead to macroscopic quantum effects near the turning point of a classically recollapsing universe as well as for black holes. Such a boundary condition was proposed, for example, in Conradi and Zeh (1991). It roughly states that the wave function for small a depends only on a itself, but not on further degrees of freedom. This is consistent with the special form of the potential in the Wheeler-DeWitt equation. The wave function is thus independent, in this limit, of the "higher multipoles" introduced in this section. For increasing size of the Universe, the total state becomes entangled with these further degrees of freedom, and the decoherence for the "relevant subsystem" can be recognised after the "irrelevant" part is integrated out. The local entropy connected with the scale factor and other "relevant" variables, as calculated from the reduced density matrix in the standard way, $S = -k_B \mathrm{tr}(\rho \ln \rho)$, thus *increases* and gives rise to the observed arrow of time in the Universe. An interesting consequence is the occurrence of recoherence in the case of a classically recollapsing universe (Kiefer and Zeh 1995).

4.2.4 Classicality of Primordial Fluctuations in the Early Universe

The idea that the Universe underwent a period of exponential expansion in its early phase does not only remove the flatness and horizon problems of the standard big-bang model, but can also explain the formation of structure in a causal way. This might be considered as the most important merit of inflation. In the last subsection we have shown how different semiclassical components of the universal wave function, each describing an inflationary universe, become dynamically independent due to decoherence caused by inhomogeneous degrees of freedom. Here we shall focus attention directly on the fluctuations, now playing themselves the role of the system. They propagate on a "decohered" inflationary universe described by the scale factor $a(t)$ or, more conveniently, $a(\eta)$, where again η denotes the conformal time defined by $a d\eta = dt$, $\eta \in (-\infty, 0)$.

How can inflation explain structure formation? The basic idea is the following, see e.g. Liddle and Lyth (2000) or Börner and Gottlöber (1997) for details. The physical wavelength of a spatial fluctuation mode in an expanding Universe is given by

$$\lambda_{phys} = \frac{a}{k} \ , \tag{4.76}$$

where k denotes the (dimensionless) wave number. The scale over which physical processes can act causally is the horizon scale H^{-1}, the inverse of the Hubble parameter. Modes that enter the horizon during the radiation- or matter-dominated era have left the horizon during the inflationary era. Without inflation, they would have always been outside the horizon before entering, and their behaviour would not be amenable to a causal dynamical description, but would have been subject only to initial conditions which have to impose nonlocal relations.

It is evident that any pre-existing structure would have been damped away during the rapid inflationary expansion. The only exception are the ubiquitous *quantum fluctuations* of fundamental fields. In fact, the idea is that all observed structure in the Universe (galaxies, clusters of galaxies, etc.) have their origin in quantum fluctuations that are present during inflation. These quantum fluctuations can be observed as classical stochastic fluctuations in the anisotropy spectrum of the cosmic microwave background (CMB) radiation. Observations are made both from space through satellite missions (such as COBE or the planned mission PLANCK) as well as from balloons and earth-bound telescopes. The main issue to be addressed in this subsection is: how can the transition from quantum to classical fluctuations be understood? The analysis will consist of two parts: firstly, the dynamics of the fluctuations as an isolated system is discussed. Secondly, their interaction with a decohering environment is considered.

Since it will be sufficient to perform the discussion for small fluctuations, they will be described by time-dependent harmonic-oscillator Hamiltonians. Because of inflation, the corresponding frequency will both rapidly change and develop an imaginary part, giving rise to "particle creation". A simple example for particle creation due to rapid change of frequency is the following, see e.g. Allen (1997).

The length l of a pendulum with a mass hanging from a string is increased and its frequency thereby decreased (from ω_i to ω_f during a time T). If the length change is performed adiabatically, no particle creation occurs and the final energy is $E_f = \frac{1}{2}\hbar\omega_f$. But what happens for a non-adiabatic change, i.e. if $\dot{l}/l > \omega$? Assume that the system is initially in its ground state (mass $m = 1$),

$$\psi_i(x) = \left(\frac{\omega_i}{\pi\hbar}\right)^{1/4} \exp\left(-\frac{\omega_i}{2\hbar}x^2\right) . \tag{4.77}$$

If at $t = 0$ a sudden change of the frequency to $\omega_f < \omega_i$ is made, one can expand ψ_i with respect to the new eigenfunctions $\psi_{n,f}(x)$,

$$\psi_i(x) = \sum_{n=0}^{\infty} c_n \psi_{n,f}(x) . \tag{4.78}$$

The expectation value of the energy is then

$$E_f = \sum_{n=0}^{\infty}(n + \frac{1}{2})\hbar\omega_f|c_n|^2$$
$$= \hbar\omega_f\left(\frac{1}{2} + \frac{(\omega_f - \omega_i)^2}{4\omega_i\omega_f}\right) \stackrel{\omega_i \gg \omega_f}{\approx} \frac{1}{4}\hbar\omega_i . \tag{4.79}$$

This corresponds to a number of "created quanta" $N_f \approx \frac{\omega_i}{4\omega_f}$, which can be quite large in the case of inflation, since it is highly non-adiabatic.

In inflation, the particle-creation rate depends on both the change of ω and the emergence of an imaginary part for ω, see e.g. Mijič (1998). An illustrative example for the second feature is the case of an upside-down oscillator, see e.g. Albrecht *at al.* (1994). Its Hamiltonian reads

$$H = \frac{1}{2}(p^2 - q^2) = -\frac{\hbar}{2}(a^2 + a^{\dagger 2}) ,$$ (4.80)

where a and $a^\dagger$ are the standard annihilation and creation operators. Starting with the ground state of the normal harmonic oscillator, it is easy to show that (4.80) evolves this into

$$|\psi(t)\rangle \equiv \mathbf{S}|0\rangle \equiv \exp\left(\frac{r}{2\hbar}(a^2 e^{-2i\varphi} - \text{h.c.})\right)|0\rangle$$ (4.81)

with $r = t$ as the *squeeze factor* and $\varphi = -\pi/4$ as the *squeezing phase*. Squeezed states such as (4.81) have been discussed already in Chap. 3.3.3. The expectation value of the particle number N and its variance are given by, respectively,

$$\begin{aligned}
\langle N \rangle &= \sinh^2 r , \\
(\Delta N)^2 &= \langle N \rangle (1 + \cosh 2r) .
\end{aligned}$$ (4.82)

The system thus possesses a "growing mode" ($\propto e^r$) and a "decaying mode" ($\propto e^{-r}$), a feature that is typical for the dynamics of the modes that leave the horizon during inflation. The presence of a growing and a decaying mode is also seen in the classical equations of motion and in the quantum equations in the Heisenberg picture. In the classical case, the dynamics squeezes an initial circle in phase space into a highly elongated ellipse, while in the quantum case the same happens for contours of the Wigner function. It is therefore clear that the result of the squeezing is more or less independent of the initial state.

To be specific, we shall consider the case of a minimally-coupled scalar field in a spatially flat Friedmann universe. Its action is given by

$$S = \frac{1}{2}\int d^4x \sqrt{-g}\, \partial^\mu \phi\, \partial_\mu \phi .$$ (4.83)

This model is at the same time simple and generic. It can be readily applied to gravitational waves (tensorial contribution of the metric fluctuations): the ansatz

$$h_{ab} = e_{ab}\phi(t)e^{i\mathbf{kx}} ,$$

where h_{ab} is the perturbation part of the three-metric and e_{ab} is the polarisation tensor, leads to equations for ϕ that correspond to those of a minimally-coupled scalar field. Gravitational waves also contribute to the spectrum of the CMB background.

For matter perturbations one needs the scalar perturbations of metric and matter fields and has to combine them in a gauge-invariant fashion. This is more complicated in detail but yields similar results (Albrecht *et al.* 1994) , which is why we may restrict ourselves to the simple example (4.83).

It is convenient to work with the rescaled field variable $y \equiv a\phi$; recall that this was also the appropriate variable for the discussion in Sect. 4.2.3. Because there is no self-coupling of the fluctuations in this approximation, it is also convenient to go to Fourier space,

$$y_{\mathbf{k}} = \frac{1}{(2\pi)^{3/2}} \int d^3x \ y(\mathbf{x})e^{-i\mathbf{k}\mathbf{x}} \ . \tag{4.84}$$

Note that $y_{\mathbf{k}}$ is complex with $y_{\mathbf{k}} = y^*_{-\mathbf{k}}$. The classical equations of motion read (′ denotes a derivative with respect to η)

$$y_{\mathbf{k}}'' + \left(k^2 - \frac{a''}{a} \right) y_{\mathbf{k}} = 0 \ . \tag{4.85}$$

This "Klein-Gordon equation" corresponds to a harmonic oscillator with time-dependent frequency $\omega_k^2 = k^2 - a''/a$.[10] From

$$a = a_0 e^{Ht} = -\frac{1}{H\eta} \tag{4.86}$$

one recognises that ω_k^2 becomes negative for modes with k smaller than approximately $|\eta|^{-1}$; these are the modes that are outside the horizon, since $\lambda_{phys} > H^{-1}$. They will therefore lead to (irreversible) particle creation. Assuming for simplicity that the system can be enclosed in a volume of finite extent, the Hamiltonian can be decomposed as a sum of Hamiltonians for each mode,

$$H = \sum_{\mathbf{k}} H_{\mathbf{k}} \ . \tag{4.87}$$

Noting that the canonical momentum to y is $p = y' - (a'/a)y$, the Hamiltonians $H_{\mathbf{k}}$ are found to read

$$H_{\mathbf{k}} = \frac{1}{2} \left(p_{\mathbf{k}}p^*_{\mathbf{k}} + k^2 y_{\mathbf{k}}y^*_{\mathbf{k}} + \frac{a'}{a} [y_{\mathbf{k}}p^*_{\mathbf{k}} + p_{\mathbf{k}}y^*_{\mathbf{k}}] \right) \ . \tag{4.88}$$

One thus has effectively two degrees of freedom with momenta $\mathbf{k}$ and $-\mathbf{k}$ (or, the real and imaginary part of y). Since the modes do not couple to each other in this order of approximation, one is effectively left with a quantum-mechanical problem. Introducing annihilation and creation operators as usual, i.e.

$$a_{\mathbf{k}} = \frac{1}{\sqrt{2}} \left(\sqrt{k} y_{\mathbf{k}} + \frac{i}{\sqrt{k}} p_{\mathbf{k}} \right) \tag{4.89}$$

---

[10] The case of non-vanishing mass can easily be accomodated and leads to $\omega_k^2 = m^2 a^2 + k^2 - a''/a$.

etc., one finds for the Hamiltonians the expression

$$H_{\mathbf{k}} = \frac{k}{2}\left(a_{\mathbf{k}}a_{\mathbf{k}}^{\dagger} + a_{-\mathbf{k}}^{\dagger}a_{-\mathbf{k}}\right) + \frac{ia'}{2a}\left(a_{\mathbf{k}}^{\dagger}a_{-\mathbf{k}}^{\dagger} - a_{\mathbf{k}}a_{-\mathbf{k}}\right) . \qquad (4.90)$$

The second part of the Hamiltonian comes from the coupling of the modes to the expanding Universe. Its form is familiar from quantum optics and known to lead to a *two-mode squeezed state* (Schumaker 1986). As initial condition at the onset of inflation we shall assume again, as in the previous subsection, that the modes are all in their ground state $|0\rangle_{\text{in}}$, $a_{\mathbf{k}}|0\rangle_{\text{in}} = 0$, $\forall\mathbf{k}$. The following results are, however, rather insensitive to this choice. The important input is to assume that inflation occurred in the first place.

It is known from quantum optics that the evolution operator U can be written in the form (Schumaker 1986)

$$U = \mathbf{S}(r_k, \varphi_k)\mathbf{R}(\theta_k) , \qquad (4.91)$$

where $\mathbf{S}$ comes from the interaction part of the Hamiltonian,

$$\mathbf{S}(r_k, \varphi_k) = \exp\left(r_k[e^{-2i\varphi_k}a_{-\mathbf{k}}a_{\mathbf{k}} - \text{h.c.}]\right) \qquad (4.92)$$

and is responsible for the squeezing, while $\mathbf{R}$ comes from the free part of the Hamiltonian,

$$\mathbf{R}(\theta_k) = \exp\left(-i\theta_k[a_{\mathbf{k}}^{\dagger}a_{\mathbf{k}} + a_{-\mathbf{k}}^{\dagger}a_{-\mathbf{k}}]\right) \qquad (4.93)$$

and leads to a "rotation" of the state. One has

$$\mathbf{S}|0\rangle_{\text{in}} = \sum_{n=0}^{\infty}\frac{1}{\cosh r_k}\left(e^{-2i\varphi_k}\tanh r_k\right)^n |n, \mathbf{k}; n, -\mathbf{k}\rangle , \qquad (4.94)$$

where $|n, \mathbf{k}; n, -\mathbf{k}\rangle = \frac{1}{n!}(a_{\mathbf{k}}^{\dagger}a_{-\mathbf{k}}^{\dagger})^n|0\rangle_{\text{in}}$ describes pairs of quanta created with zero momentum. As in (4.82), one has $\langle N_k\rangle = \sinh^2 r_k$ for the expectation value of the number operator. Upon comparing the general expression for the evolution operator U with (4.91), one finds differential equations for the squeezing factor r_k, the squeezing phase φ_k, and the rotation phase θ_k (Albrecht *at al.* 1994),

$$r_k' = -\frac{a'}{a}\cos 2\varphi_k , \qquad (4.95)$$

$$\varphi_k' = -k + \frac{a'}{a}\coth 2r_k \sin 2\varphi_k , \qquad (4.96)$$

$$\theta_k' = k - \frac{a'}{a}\tanh r_k \sin 2\varphi_k . \qquad (4.97)$$

The squeezed-state formalism was first applied to inflationary cosmology in Grishchuk and Sidorov (1989). In the case of pure exponential expansion,

$a = -(H\eta)^{-1}$, these equations can be solved exactly, and the solution reads

$$r_k = \text{arsinh} \frac{1}{2k\eta} \, , \tag{4.98}$$

$$\varphi_k = -\frac{\pi}{4} - \frac{1}{2} \arctan \frac{1}{2k\eta} \, , \tag{4.99}$$

$$\theta_k = k\eta + \arctan \frac{1}{2k\eta} \, . \tag{4.100}$$

In the relevant limit of very long wavelengths, $k\eta \to 0^-$, one has $\varphi_k \to 0^-$, $r_k \to -\infty$, and $\theta_k \to k\eta - \frac{\pi}{2}$. The squeezing angle thus becomes "frozen" and, since $(\theta_k + \varphi_k)' \to 0$, "logged" to the rotation phase. The limit $r_k \to -\infty$ refers to extreme *squeezing in momentum*, i.e. broadening in position. If one wants to stick to the standard convention to have r_k positive, one should substitute $r_k \to -r_k$ in the following expressions (this is done in some of the cited references).

How large is r_k in a realistic cosmological scenario? One can estimate that $r_k \approx \ln(a_2/a_1)$, where a_2 is the scale factor when the mode leaves the horizon during the inflationary phase, and a_1 is the scale factor when the mode re-enters the horizon in the radiation- or matter-dominated era (Grishchuk and Sidorov 1989). For modes that enter now the horizon on the largest cosmological scales (corresponding to a frequency of $\nu \approx 10^{-18}$Hz), one finds $r \approx 120$. This corresponds to an enormous squeezing; for example, Meekhof *et al.* (1996) report to have generated a squeezed state in the laboratory with $r \approx 1.8$.

Acting with (4.91) on the ground state at the onset of inflation, one can write the state at any later time in the following form:

$$\psi(y,t) = \left(\frac{1}{\pi \, |f_k|^2} \right)^{1/2} \exp \left(-\frac{1 - i2F(k)}{2 \, |f_k|^2} |y_{\mathbf{k}}|^2 \right) \, . \tag{4.101}$$

The quantities f_k and $F(k)$ can be expressed in terms of the squeezing parameters,

$$|f_k|^2 = \frac{1}{2k} \left(\cosh 2r_k - \cos 2\varphi_k \sinh 2r_k \right) \, , \tag{4.102}$$

$$F(k) = \frac{1}{2} \sin 2\varphi_k \sinh 2r_k \, . \tag{4.103}$$

It should be mentioned that in the Heisenberg picture f_k plays the role of the field operator that obeys the equations of motion (Kiefer and Polarski 1998). The exponent in (4.101) can also be conveniently written in the form

$$\exp \left(-k \frac{1 - e^{2i\varphi} \tanh r}{1 + e^{2i\varphi} \tanh r} |y|^2 \right) \, . \tag{4.104}$$

We emphasise that (4.101) is the same state as (4.44); it is just more convenient here to express it in terms of squeezing parameters. The limit of large squeezing, which is obtained during inflation, yields

$$F \overset{k\eta \to 0}{\longrightarrow} \frac{1}{2k\eta} \longrightarrow -\infty \ . \tag{4.105}$$

This means that the quantum state is of an approximate *WKB form* in this limit. In this specific sense, it appears "quasiclassical", even though the Gaussian (4.101) becomes very *broad* in y, $|f|^2 \overset{k\eta \to \infty}{\longrightarrow} (2k^3\eta^2)^{-1} \to \infty$, and is in the sense of decoherence not yet classical. In fact, this state is extremely sensible to entanglement with the "environment". Because of the WKB form in the large-squeezing limit, the situation has been called "decoherence without decoherence" (Polarski and Starobinsky 1996). For $r \to \infty$ the state becomes "indistinguishable" from a classical stochastic distribution, as can be seen in gedanken experiments involving hypothetical slits through which the system is forced to pass (Kiefer and Polarski 1998, Kiefer *et al.* 1998a). For example, the quantum expectation value of the operator $pp^\dagger$ (momentum squared) is

$$\left(\frac{F}{|f|} \right)^2 + \frac{1}{4|f|^2} \ ,$$

while the corresponding classical average is

$$\left(\frac{F}{|f|} \right)^2 \ .$$

Both expressions coincide in the limit of large squeezing, where $F \to -\infty$ (Kiefer and Polarski 1998). The Wigner function becomes a stretched ellipse with negligible thickness that almost coincides with the y-axis. Because of this extreme squeezing, the result is almost independent of the initial state, since information about the initial phase-space distribution would be lost. Only for the fine-tuned case of an initial state being extremely squeezed in y would one get another result. This is of course a consequence of the special nature of inflation itself, an issue which can probably only be understood in full quantum cosmology (Zeh 2001).

After re-entering the horizon (the second Hubble-radius crossing), r starts to oscillate. This gives rise to oscillations in the power spectrum of the density perturbations, leading to "acoustic oscillations" which are observable in the CMB-spectrum (Albrecht *et al.* 1994). The high degree of squeezing is preserved for a sufficiently long time (Kiefer *et al.* 1998a), and the stretched Wigner ellipse is rotating with $\omega = 2\pi/\lambda_{phys}$.

In the Heisenberg picture, the situation can be described by the fact that the non-commutativity of $\hat{p}$ and $\hat{y}$ does not play an important role. Moreover, $\hat{y}$ commutes at different times, up to terms of the order e^{-r} (Kiefer *et al.* 1998b)

$$[\hat{y}(t_1), \hat{y}(t_2)] \approx 0 \ . \tag{4.106}$$

This condition is referred to as *quantum nondemolition measurement*, see also Chap. 3. It is responsible for the possibility to describe the situation in terms of "consistent histories" (Polarski 1999), see also Chap. 5.

As has already been mentioned, the extreme broadness of (4.101) in y makes it very sensitive to interaction with other degrees of freedom. This relates to genuine decoherence and constitutes here the second step of the quantum-to-classical transition for primordial fluctuations. Most interactions are local in field space, not in canonical-momentum space. There are exceptions (e.g. graviton-graviton scattering), but these do not play a significant role. One therefore has

$$[\hat{y}, \hat{H}_{int}] \approx 0 \ , \qquad (4.107)$$

making $\hat{y}$ a "pointer basis". Since one has in addition the condition (4.106), $\hat{y}$ commutes approximately with the *full* Hamiltonian, making it a constant of motion and a pointer basis par excellence. For this reason, it is neither the particle-number basis nor the coherent-state basis that is robust, but the field basis itself.

A quantitative calculation can be made with the help of the rate of de-separation which was introduced in Chap. 3.1.1 (and applied to primordial fluctuations by Kiefer and Polarski 1998). It describes in a general way how fast states which are initially separated become quantum entangled due to some Hamiltonian that couples the system to other fields. It reads

$$\mathcal{A} = \sum_{j \neq 0, m \neq 0} |\langle \varphi_j \phi_m | H | \varphi_0 \phi_0 \rangle|^2 \ , \qquad (4.108)$$

where φ_0 and ϕ_0 are the initial states of system and environment. The decoherence timescale is then given by

$$t_D = \frac{1}{\sqrt{\mathcal{A}}} \ . \qquad (4.109)$$

Real interactions are of course complicated and difficult to tackle. To calculate the efficiency of decoherence, it is, however, sufficient to consider a simple example. We take a linear interaction Hamiltonian (coupled oscillators), cf. also (3.267),

$$H_{int} = gk^2(y_{\mathbf{k}} z_{\mathbf{k}}^\dagger + y_{\mathbf{k}}^\dagger z_{\mathbf{k}}) = 2gk^2(y_1 z_1 + y_2 z_2) \ , \qquad (4.110)$$

where g is a dimensionless coupling constant, and $z_{\mathbf{k}} \equiv z_1 + iz_2$ is the environmental field which is assumed to possess the same "free" part as $y_{\mathbf{k}}$. This is similar to the interaction term used in the toy models of Sakagami (1988) and Brandenberger *et al.* (1990). A Hamiltonian of this type should give a lower bound on decoherence.

We assume that at some initial time the total state is a product state of the y-part and the z-part,

$$\Psi \equiv \psi_{y_1} \psi_{y_2} \psi_{z_1} \psi_{z_2} \ . \qquad (4.111)$$

For the y-part we take the squeezed state produced by inflation, while for the z-part we take for simplicity the vacuum state. The rate of de-separation then becomes (Kiefer and Polarski 1998)

$$\mathcal{A} = g^2 k^3 |f_k|^2 = \frac{g^2 k^2}{2} (\cosh 2r_k - \cos 2\varphi_k \sinh 2r_k) . \qquad (4.112)$$

We emphasise that the measure for quantum entanglement is thus essentially given by the power spectrum of the fluctuations, see (4.102). In the limit of large squeezing in p-direction ($\varphi_k \to 0$) this becomes

$$\mathcal{A} \to \frac{g^2 k^2}{2} e^{2r_k} , \qquad (4.113)$$

and the corresponding decoherence time (4.109) is then given by

$$t_D \sim \frac{a}{gke^r} \sim \frac{\lambda_{phys}}{ge^r} \sim \frac{H_I}{g} , \qquad (4.114)$$

where λ_{phys} denotes again the physical wavelength of the fluctuations and H_I is the Hubble parameter of inflation. [11] For example, for $\lambda_{phys} \sim 10^{28}$cm (present horizon scale) and $e^r \sim 10^{50}$ (squeezing factor of this mode) one has

$$t_D \sim 10^{-31} g^{-1} \text{ s} ,$$

so that decoherence would be negligible only if one fine-tuned the coupling to values $g \ll 10^{-31}$ – a totally unrealistic fine-tuning! The decoherence time scale is thus set by the Hubble parameter of inflation.

For more complicated couplings, the decoherence rate is even shorter (Kiefer and Polarski 1998). Scenarios involving nonlinear couplings have been discussed at length in Calzetta and Hu (1995). We emphasise, however, that nonlinearity is not necessary for the occurrence of decoherence, as the above discussion shows. Similar to the quantum-mechanical case of Fermi's Golden Rule and exponential decay, see Chap. 3, the occurrence of decoherence enforces the results that one gets from a treatment of the system itself. While no interference effects can be observed at the system, probabilities are not affected (this corresponds to an ideal measurement). Therefore, all classical predictions by inflationary cosmology remain unchanged.

It should be clear from the above discussion that one should not average (coarse-grain) over the squeezing angle since this would be in contradiction with the correlation between p and y. If such an averaging were performed in the squeezed state (4.101), one would obtain a reduced density matrix that is diagonal in the particle-number basis rather than the field-amplitude basis (Prokopec 1993). This would yield an entanglement entropy $S_k \approx 2|r_k|$ per mode, corresponding to a smearing out of the stretched Wigner ellipse into

---

[11] The factor a in (4.114) occurs after the physical fields $\phi = a^{-1}y$ etc. are considered.

a big circle. If this were true, there would be no acoustic oscillations in the CMB spectrum, and one of the standard predictions of inflation would fail. The presence of at least three acoustic peaks has, however, been confirmed by now (Netterfield *et al.* 2002). It was shown in Kiefer *et al.* (2000) that the entanglement entropy is in fact much smaller than its maximum value of $2|r_k|$, in accordance with observations. An interesting point is that the entropy production rate is given by the Hubble parameter, $\dot{S} \approx H$, reflecting the fact that the decoherence time is proportional to H^{-1}. During inflation, H is approximately constant, and the entropy increases linearly in t. In the post-inflationary phases, the Hubble parameter is proportional to $1/t$, and the entropy increases only logarithmically in time. This behaviour is similar to the behaviour in chaotic systems, cf. Chap. 3. The role of the Lyapunov coefficient is here played by the Hubble parameter. The Kolmogorov entropy corresponds to the entropy production rate mentioned here.

4.2.5 Black Holes, Wormholes, and String Theory

Here we shall discuss non-cosmological applications of decoherence in which the quantum nature of the gravitational field plays a role. This involves systems such as black holes, wormholes, or D-branes. We shall start with an example dealing with quantum black holes.

The quantum nature of black holes was discovered when Hawking found that black holes radiate with a temperature (setting Boltzmann's constant equal to one)

$$T_{BH} = \frac{\hbar\kappa}{2\pi} \, , \tag{4.115}$$

where κ is the surface gravity of the black hole (see e.g. Kiefer (1999) for an introduction and details). Connected with this *Hawking temperature* is the existence of a non-trivial entropy for the black hole,

$$S_{BH} = \frac{A}{4G\hbar} \, , \tag{4.116}$$

where A is the surface area of the event horizon. These expressions can only be interpreted on a fundamental level when a consistent quantum theory of gravity is available. Within the canonical approach to quantum gravity one can find semiclassical solutions to the Wheeler-DeWitt equation from which the Hawking temperature can be found. In analogy to (4.43) we consider a state of the form

$$\Psi \approx C[h_{ab}] \exp\left(iS_0[h_{ab}]\right) \psi[h_{ab}, f] \, , \tag{4.117}$$

where h_{ab} denotes the three-dimensional metric, and f denotes a scalar field (playing the role of the fluctuations $\{f_n\}$ in Sect. 4.2.3). Again, ψ is a solution of the functional Schrödinger equation in which time is defined through S_0. The analysis can be simplified for two-dimensional black holes. Such models arise in *dilaton-gravity theories*, where a scalar field φ (the "dilaton") is added to the two-dimensional metric (Callan *et al.* 1992). These models contain one dimensionful parameter λ which resembles a cosmological constant and has the dimension of an inverse length. Instead of (4.115), the black-hole temperature is now given by

$$T_{BH} = \frac{\hbar\lambda}{2\pi} \; .\tag{4.118}$$

In this case, one can give exact expressions for S_0 (i.e. an exact solution of the Hamilton-Jacobi equation) as well as for ψ (Demers and Kiefer 1996). Along a particular semiclassical trajectory, the "matter wave function" depends only on t and f. Analogously to the cosmological case treated in Sect. 4.2.4, one can start from the ground state for ψ. Solving the Schrödinger equation in the gravitational field of a collapsing star (mimicked in this model through the choice of appropriate boundary conditions), one finds again that this ground state becomes *squeezed*. The corresponding Gaussian functional reads explicitly (from now on again using $\hbar = 1$)

$$\psi(t, f] = N \, \exp\left(-2 \int_0^\infty dk \; k \coth\left(\frac{\pi k}{2\lambda} + ikt\right) |f(k)|^2\right) \; .\tag{4.119}$$

This has the form of a squeezed state, cf. in particular (4.104), with

$$\tanh r_k = e^{-\pi k/\lambda}, \quad \varphi_k = -kt - \frac{\pi}{2} \; .\tag{4.120}$$

In contrast to the case of the cosmological fluctuations, the squeezing is constant in time – this is due to the presence of a horizon and is responsible for the thermal nature of Hawking radiation. As in (4.82) we get the expectation value of the particle number from

$$\langle N_k \rangle = \sinh^2 r_k = \frac{1}{e^{2\pi|k|/\lambda} - 1} \; ,\tag{4.121}$$

which is a thermal distribution with the temperature (4.118). We emphasise that this thermal distribution has been obtained from a pure state. This is because the three-geometry has been chosen to stay outside the horizon. If, on the other hand, the hypersurface is chosen to penetrate the horizon, one obtains a mixed state by tracing out the interior of the horizon, see below.

Instead of (4.117) we now consider a state that is a superposition of (4.117) with its complex conjugate,

$$\Psi \approx e^{iS_0}\psi + e^{-iS_0}\psi^* \; .\tag{4.122}$$

This can be understood heuristically as a superposition of a black hole with a "white hole" (the time reversed white-hole state being obtained by complex conjugation). It is clear that (4.122) cannot describe classical black holes.

Again, the interference terms between the components in (4.122) become de-localised through the interaction with other degrees of freedom. The simplest example of an environment responsible for decoherence is the black hole's own Hawking radiation (Demers and Kiefer 1996). Tracing this out yields the following decoherence factor between the two components:

$$|D| \approx \exp\left(-\frac{L\lambda}{4\pi^2} F\left[\frac{2\lambda t}{\pi} \right] \right) , \qquad (4.123)$$

where F is a function which rapidly approaches a constant value (numerically about 1.65). In order to avoid infrared singularities, the black hole has been considered in a box with length L. Decoherence is thus efficient for $L\lambda > 1$, i.e. if the size of the box is bigger than the typical wavelength of the Hawking radiation (bigger than the Schwarzschild radius).

In the case of the superposition of a black hole with a no-hole state, the decoherence factor is smaller by a factor $1/4$ in the exponent. This can be understood by noting that in this case only one component in the superposition (4.122) produces Hawking radiation.

The decoherence mechanism discussed above works only for semiclassical states. It should thus not work for virtual black holes which, like virtual particles in quantum field theory, are time symmetric (i.e. contribute only to the "dressing" in the sense of Sect. 4.1.1). Hawking and Ross (1997) conclude that quantum fields may be decohered by scattering off vacuum fluctuations in which black-hole pairs appear and disappear. They speculate that particles could "fall" into virtual holes, which would then radiate other particles in a mixed state. Such a process can only happen if a direction of time is put in by hand in some hidden way. Virtual black holes by themselves should not decohere.

An interesting question is whether one can give a microscopic interpretation for the black-hole entropy (4.116). This is related to the *information-loss paradox*: where does the information about the initial state (e.g. of the collapsing star) stay after the black hole has evaporated? If only thermal Hawking radiation remained, this would be a violation of the unitary evolution for closed quantum systems. It is tempting to interpret S_{BH} as an *entanglement entropy* obtained by integrating out the degrees of freedom in the interior of the horizon. Attempts in this direction exist, see e.g. Kiefer (1998b), but there is always the problem of regularisation, having to choose an appropriate cutoff to obtain the desired result (except for an approach within "induced gravity" that gives the correct result directly). As was, however, argued in Kiefer (2001), the mixed appearance of Hawking radiation is already obtained for hypersurfaces that stay outside the horizon – precisely due to decoherence. There might thus be no information-loss paradox in the first place, and Hawking radiation would always be in a pure state.

In string theory, the expression (4.116) was obtained by counting states of so-called D-branes. These are extended quantum objects that automatically arise in string theory. One has $S_{BH} = \ln \mathcal{D}$, where $\mathcal{D}$ is the degeneracy of D-branes, see e.g. Horowitz (1997) for a review. In order for such a formula to be applicable, the D-brane states must be assumed to have already decohered, although this is usually not explicitly discussed. Therefore, black holes that are "formed" by such D-branes are in a mixed state (Myers 1997, Amati and Russo 1999). In Ellis *et al.* (1998) one finds a discussion of how an entanglement entropy arises between D-branes and particles by scattering processes.

Let us again consider the state (4.119). As in the cosmological case, it may be generated from the standard (Gaussian) vacuum through the application of the squeezing operator (4.92). Apart from a phase, the state can thus be written in the form

$$|\psi\rangle = \prod_{\mathbf{k}} \sqrt{1 - e^{-2\pi k/\lambda}} \sum_{n=0}^{\infty} e^{-n\pi k/\lambda} |n, \mathbf{k}; n, -\mathbf{k}\rangle \ . \qquad (4.124)$$

Such states are also found in the BCS-theory of superconductivity. The two modes are there the electron ("Cooper") pairs with zero total momentum near the Fermi surface. For a state that is quantum entangled between the interior and the exterior of a black hole, one can take one of the n to correspond to the exterior, the other n to correspond to the interior region. If one integrates over all modes that are *inside* the horizon, one finds from (4.124) a mixed state in the exterior region,

$$\rho_{\text{outside}} = \prod_{\mathbf{k}} \left(1 - e^{-2\pi k/\lambda}\right) \sum_{n=0}^{\infty} e^{-2\pi nk/\lambda} |n, \mathbf{k}\rangle\langle n, \mathbf{k}| \ . \qquad (4.125)$$

This density matrix describes a canonical ensemble with temperature (4.118). The same state emerges for a uniformly accelerated observer in Minkowski space. Because of the presence of horizons, such an observer is restricted to one quarter of flat spacetime, which is why he observes the global vacuum state as a mixed state with temperature $a/2\pi$, where a is the acceleration. The decoherence for an accelerated detector, equivalent to a "fixed" detector in a gravitational field, was discussed in Audretsch *et al.* (1995).

In Chap. 3 it was shown how classical properties (localisation) for quantum systems can emerge due to the interaction with irrelevant, environmental, degrees of freedom. An important example is a macroscopic object in interaction with air molecules. Instead of using such "ordinary" environments, it was also suggested that the origin of decoherence may itself be caused by some fundamental gravitational influence. Some proposals invoke collapse models which explicitly modify the Schrödinger equation, see Chap. 8. However, here we only want to discuss models that do not abolish the unitary evolution for closed quantum systems.

Explicit proposals for quantum-gravitational environments are virtual black holes, wormholes, or excited modes in non-critical string theory. The details are of course very speculative and much less clear than the processes described in earlier sections. Still, such models have been discussed at length in the literature and have even been related to possible observations. Since they nonetheless exhibit interesting general features, we shall now give a brief overview of some of these discussions.

Many approaches deal with *wormholes*. They are related to quantum-gravitational tunneling effects that might lead to "baby universes" branching off our big "parent" universe. They can thereby carry quantum correlations with the parent universe. This entanglement then leads to a mixed state and to decoherence in our universe. Wormholes are euclidean configurations that describe such tunneling events and that contribute to the corresponding path integral (semiclassically or not). They should not be confused with another type of wormholes that are three-dimensional configurations of non-trivial topological character. An explicit example for a wormhole consisting of axions and gravitons can be found in Giddings and Strominger (1988a).

We shall describe the influence of wormholes by some effective action, assuming that the wormhole scale is between the Planck scale and the scale of laboratory physics ("dilute-gas approximation"), see Coleman (1988a). This effective action reads

$$S_{\text{eff}} = S_0 + \sum_i S_i(a_i + a_i^\dagger) \,, \tag{4.126}$$

where S_0 is the action of quantum fields without wormholes, and S_i are expressions that may depend on ordinary field operators. The important part are the creation and annihilation operators a_i and $a_i^\dagger$ which describe the creation and annihilation of baby universes of type i. They enter symmetrically, since creation and annihilation are indistinguishable. They do not depend on x, since baby universes are not at specific points in spacetime.

We introduce the operators $A_i = a_i + a_i^\dagger$ (which are proportional to the position operators). Since $[A_i, A_j] = 0$, they can be simultaneously diagonalised,

$$A_i|\alpha_a, \alpha_2, \ldots\rangle = \alpha_i|\alpha_1, \alpha_2, \ldots\rangle \equiv \alpha_i|\{\alpha\}\rangle \,. \tag{4.127}$$

We consider a model with a scalar field and wormholes. The euclidean path integral ("partition sum") then reads

$$Z = \langle 0_\phi|\langle 0_{WH}| \int D\phi \, e^{-S_0[\phi] - \sum_i S_i A_i}|0_{WH}\rangle|0_\phi\rangle \,, \tag{4.128}$$

where $|0_\phi\rangle$ and $|0_{WH}\rangle$ denote the ground states for scalar field and wormholes, respectively. We expand the wormhole groundstate in terms of the states (4.127),

$$|0_{WH}\rangle = \int \prod_i d\alpha_i |\{\alpha\}\rangle\langle\{\alpha\}|0_{WH}\rangle \,. \tag{4.129}$$

Inserting this into (4.128), one obtains

$$Z = \langle 0_\phi | \int \prod_j d\alpha'_j \langle 0_{WH} | \{\alpha'\} \rangle \langle \{\alpha'\} | \times$$

$$\int D\phi \, e^{-S_0[\phi] - \sum_i S_i \alpha_i} \int \prod_i d\alpha_i \langle \{\alpha\} | 0_{WH} \rangle | \{\alpha\} \rangle | 0_\phi \rangle$$

$$= \langle 0_\phi | \int \prod_i d\alpha_i |\langle 0_{WH} | \{\alpha\} \rangle|^2 \int D\phi \, e^{-S_0[\phi] - \sum_i S_i \alpha_i} | 0_\phi \rangle$$

$$\equiv \int d\alpha \, P(\alpha) Z(\alpha) \,, \tag{4.130}$$

where $P(\alpha) \equiv |\langle 0_{WH} | \{\alpha\} \rangle|^2$. The expression $Z(\alpha)$ in (4.130) is the partition sum for the scalar field with *shifted* coupling constants, $g \to g + \alpha$. For example, there might be a term ϕ^2 in one of the S_i, say S_1. Because there is a mass term $m^2 \phi^2 / 2$ in S_0, the presence of wormholes yields a shift of the mass according to $m^2/2 + \alpha_1$. This lies at the origin of the claim that wormholes can drive the cosmological constant to zero (Coleman 1988b).

Let us consider two examples (for simplicity, only one wormhole species is considered). In the first example, the wormhole ground state is just the ordinary vacuum (no-wormhole state), i.e. $|0_{WH}\rangle = |0\rangle$. Then,

$$P(\alpha) = |\langle \alpha | 0 \rangle|^2 = \frac{1}{\sqrt{\pi}} e^{-\alpha^2/2}$$

and therefore

$$Z \propto \langle 0_\phi | \int D\phi \, e^{-S_0 + S^2/2} | 0_\phi \rangle \,. \tag{4.131}$$

The term S^2 corresponds to a nonlocal contribution to the action, similarly to the Feynman-Vernon influence functional, see Chap. 5. In the second example we take $|0_{WH}\rangle = |\alpha_0\rangle$, i.e. an eigenstate to the wormhole operator A_0 (the index 0 denoting the single wormhole species being present). Then, $P(\alpha) = \delta(\alpha - \alpha_0)$ and

$$Z = \langle 0_\phi | \int D\phi \, e^{-S_0 - S\alpha_0} | 0_\phi \rangle \,. \tag{4.132}$$

No nonlocal contribution to the action is present, and the only consequence is a shift in the coupling constants. Such a shift corresponds to the usual back reaction caused by an environment, compare for example the change in frequency of the oscillator in quantum Brownian motion (Caldeira and Leggett 1983a), cf. Chap. 3.

Is there a loss of coherence due to the presence of wormholes, i.e. the branching off of baby universes? Some authors answer this question in the affirmative and derive master equations for the mixed state in the parent universe, see e.g. Hawking and Ross (1997) and Ellis *et al.* (1990). Others, such as Coleman (1988a) and Giddings and Strominger (1988b), reject this possibility. Their argument goes as follows. Since one has no control over the wormhole wave function (being determined by the boundary conditions on the wave function of the universe), one cannot observe decoherence because the state might have always been mixed. Still, it would mean that certain interference terms would *not* be observable locally if entanglement with baby universes were present. (It is often claimed that one is effectively in an eigenspace of A, so that there would be only a shift in coupling constants, no loss of coherence; another problem is that the creation and annihilation of baby universes is not irreversible and would only contribute to a "dressing" of the local state, so that no decoherence would result, similarly to dressed charges in QED.)

In spite of the speculative nature of such interactions, a phenomenological approach has been formulated in order that a detailed comparison with observations can be performed. One can formulate an effective master equation arising from wormholes (Ellis *et al.* 1984) or higher-order modes in non-critical string theory (Ellis *et al.* 1996). This is different from our earlier examples of decoherence, in which a master equation is *derived* from the exact quantum state of system *plus* environment. One can make an ansatz

$$\dot{\rho} = i[\rho, H] + \delta H \,\rho \,. \tag{4.133}$$

The simplest case is a two-state system with energy levels $E \pm \Delta E/2$, described by the Hamiltonian $H = E + \Delta E \sigma_z$. The normal quantum-mechanical evolution (without the last term in (4.133)) is unitary and given by

$$\rho(t) = \frac{1}{2} \begin{pmatrix} 1 & e^{-i\Delta Et} \\ e^{i\Delta Et} & 1 \end{pmatrix} \,. \tag{4.134}$$

The additional term δH in (4.133) can be described by a 4×4-matrix $f_{\alpha\beta}$ that can be parametrised in the following form:

$$f_{\alpha\beta} = \begin{pmatrix} 0 & 0 & 0 & 0 \\ 0 & -\alpha & -\beta & 0 \\ 0 & -\beta & -\gamma & 0 \\ 0 & 0 & 0 & 0 \end{pmatrix} \tag{4.135}$$

with $\alpha, \gamma > 0$ and $\alpha\gamma > \beta^2$ following from probability and energy conservation. Instead of (4.134) one has now

$$\rho(t) = \frac{1}{2} \begin{pmatrix} 1 & e^{-(\alpha+\gamma)t/2}e^{-i\Delta Et} \\ e^{-(\alpha+\gamma)t/2}e^{i\Delta Et} & 1 \end{pmatrix} \xrightarrow{t\to\infty} \frac{1}{2}\,\mathbb{1} \,. \tag{4.136}$$

The following possible tests of this non-unitary evolution have been suggested:

-   Influence of wormholes on SQUIDs (Ellis *et al.* 1990). This enables one to make a comparison with the usual dissipative effect described in Caldeira and Leggett (1983b). While the latter leads to a linear time dependence in the exponent of the Gaussian contribution to the density matrix, wormholes produce a t^2-dependence. This is similar to the effect by interaction with long-range forces, see Chap. 3, in contrast to localisation by scattering which produces a dependence linear in t.

– Neutron interferometry (Ellis *et al.* 1984).

– Oscillations of K mesons (Ellis *et al.* 1984, 1996).

We note that the emphasis is here not on the emergence of classical properties for macroscopic objects, but on the decohering influence of microscopic systems (or mesoscopic systems, as for SQUIDs) from exotic sources such as wormholes. We shall briefly describe the results for the investigation of the K mesons. The particles K^0 and $\bar{K}^0$ are produced by the (strangeness conserving) strong interaction, while their decay is ruled by the weak interaction, proceeding from CP-eigenstates, not strangeness eigenstates. The latter are the superpositions

$$|K_S\rangle = \frac{1}{\sqrt{2}} \left(|K^0\rangle + |\bar{K}^0\rangle \right) , \tag{4.137}$$

$$|K_L\rangle = \frac{1}{\sqrt{2}} \left(|K^0\rangle - |\bar{K}^0\rangle \right) . \tag{4.138}$$

These states have different masses, which is why the intensities for K^0 and $\bar{K}^0$ oscillate with frequency $\Delta m \sim 10^{-6}$ eV (from which one finds the strongest constraints on a possible CPT-violation).

The additional term in (4.133) has an effect on the coherence of these oscillations (and possibly leads to CPT-violation). One can find limits on the parameters α, β and γ in (4.135) from measurements of different decay asymmetries of K-mesons. In Ellis *et al.* (1996), the following limits are given:

$$\alpha < 4.0 \times 10^{-17} \text{ GeV} , \tag{4.139}$$

$$\beta < 2.3 \times 10^{-19} \text{ GeV} , \tag{4.140}$$

$$\gamma < 3.7 \times 10^{-21} \text{ GeV} . \tag{4.141}$$

Since one might expect a possible effect to be of order $m_K^2/m_P \approx 10^{-20}$GeV, these limits could be very close to an actual observation. It is, however, also imaginable that possible effects are proportional to a power of this ratio, which would prevent any observation in the foreseeable future. For some suggestions to use atom interferometry for observations see Benatti and Floreanini (2002).

There exist other interesting examples for decoherence in this context. Scattering of two light particles in a D-brane background, for example, leads to decoherence for both the particles and the D-branes due to the emerging entanglement between the corresponding quantum states (Ellis *et al.* 1998). They all show the universal occurrence of decoherence, which is inextricably linked to the nature of quantum theory.

5 Consistent Histories and Decoherence

C. Kiefer

In our presentation of decoherence in the previous chapters, an essential ingredient was the separation of a quantum system into a subsystem (called the "relevant" or "distinguished" part) and its environment (called the "irrelevant" or "ignored" part). Provided certain initial conditions hold, "coarse-graining" with respect to the irrelevant degrees of freedom may lead to the emergence of classical behaviour in the relevant part, since (almost) all information about quantum phases has migrated into correlations with the environment and is thus no longer accessible in observations of the subsystem alone.

It is the purpose of the present chapter to introduce and discuss the notion of *consistent histories* and to clarify their relation to decoherence. Consistent histories are often mentioned in this context and also sometimes called "decoherent histories." It is *not* the purpose of this chapter to present a detailed review of the use of consistent histories beyond decoherence, although we shall make some remarks on their general interpretation.

What are the main motivations for introducing consistent histories? One common aspect is the desire to find a *language* which is also suitable for closed quantum systems, including all observers and measurement apparata, but where no explict reference to measurements or observers must be made. Ultimately, of course, this refers to the whole Universe as the only system which is strictly closed. It is for this reason why this notion is sometimes encountered in the framework of quantum cosmology.[1]

The central aim is to find sets of "consistent histories" for a given system, i.e. sets of "histories" which exhibit vanishing (or negligible) interference with respect to each other (the precise definition of a consistent history is given in Sect. 5.2). One can then formally assign, at least approximately, probabilities to such histories, which obey the usual rules of probability theory. *If*, in addition, a division into some "system" and its "environment" is possible and *if*, furthermore, the latter has a decohering influence in the sense discussed in our previous chapters, consistent histories can be used as a tool to *describe* the occurrence of decoherence and are then appropriately called *decoherent histories*. The main emphasis here is thus the description of a *method* for discussing decoherence. It should also be mentioned that there is a connection of this notion with the concept of quantum state diffusion (Brun 2002).

---

[1] "Quantum mechanics is best and most fundamentally understood in the framework of quantum cosmology" (Gell-Mann and Hartle 1990).

There are other motivations given in the literature. One is the interest to implement appropriate "coarse-grainings" with respect to *time*. This is frequently employed, for example, if one performs spacetime averages over field quantities (like in the traditional Bohr–Rosenfeld analysis). Tracing out part of the configuration variables is, however, performed at each instant of time and does therefore not describe any coarse-graining *of* time, i.e. dividing the real time axis into intervals, Δt, of finite length. Therefore a more general formalism is sought for.

Another motivation is provided by the hope that the framework of consistent histories is general enough to cope with situations where no distinguished external time parameter is available, such as is the case in quantum gravity (see Sect. 4.2) or other reparametrisation-invariant theories like string theories where no privileged set of spacelike hypersurfaces exists. One thus aims at constructing decoherent histories for spacetime domains (Halliwell 2002). This includes the calculations of probabilities for crossing a surface during a finite interval of time.

The notion of consistent histories does not necessarily pre-assume the traditional Hilbert space structure of ordinary quantum theory. Instead, in the particular case of the non-relativistic theory one can recover this Hilbert space structure, and the notion of consistent histories becomes *fully equivalent* to the standard language, although its interpretation may differ in some important respects. We shall deal with this case in most of the present chapter and refer the discussion of the "timeless" case to the last section.

We shall begin our discussion with an introduction into the notion of *influence functionals* in quantum mechanics. This serves two purposes. First, it provides a special example for a decoherence functional, the general definition of which will be given in the second section. The central concept in the first section is the Feynman path integral in configuration space. Second, it is as close as possible to our discussion of decoherence in the previous chapters. In fact, the use of Feynman integrals is only *technically* different from our former discussion, but is an implementation of the same physical ideas. All what will be said in the first section can thus be rephrased in the language of wave functions, but is in our view useful as a heuristic introduction into the concept of histories. After all, Feyman's influence functional was historically the first case of a decoherence functional, although it was not interpreted in this way. As an explicit application we shall discuss in detail the example of quantum Brownian motion. In the second section we shall give the general definition of the *decoherence functional* and discuss some of its properties. Feyman's influence functional is recovered as a special case at the end of that section. The connection between the reduced density matrix and the histories approach, as well as its connection to the Zwanzig formalism, is discussed in Sect. 5.3. The last section is then devoted to the question of whether the consistent histories formulation contains an arrow of time. We also include there a discussion of the "timeless" situation of quantum gravity. We shall argue that this concept cannot be extrapolated to quantum gravity in a straightforward manner. We then conclude with some remarks on the measurement problem in the light of consistent histories and on the general interpretation of the formalism, which appears somewhat problematic.

5.1 Influence Functional and Its Application to Quantum Brownian Motion

We shall reformulate in the following the sum over environmental degrees of freedom and the evolution of the reduced density matrix in the language of path integrals. As mentioned above, this is primarily a technical tool and could be done alternatively with the methods of Chap. 3. However, this language possesses technical advantages and, moreover, can serve to motivate the definition of consistent histories in Sect. 5.2. This method was introduced by Feynman and Vernon (1963) and elaborated in particular by Caldeira and Leggett (1983a). We shall mainly follow these papers, with elaborations, as well as the exposition by Gell-Mann and Hartle (1993). A formal treatment of continuous measurements by path integral methods was made by Mensky (1979, 1993). We emphasise that our discussion in this section is – independent from its motivation for the decoherence functional formalism – interesting in its own right, since path integral methods are frequently, and efficiently, used in studies of decoherence.

We consider two interacting systems, A and B, and consider A as the relevant part and B as the irrelevant part, i.e. B contains the environmental degrees of freedom (heat bath, reservoir, ...). For simplicity we shall call in the following the relevant part "the system". The physical situation is thus precisely the same as in the last two chapters and only the technical discussion is different. We assume that the total system can be described by a Hamiltonian which can be written as a sum of two parts referring respectively to A and B and a third part describing the interaction:

$$H = H_A + H_B + H_I. \tag{5.1}$$

The total density matrix for system and environment evolves unitarily in time according to

$$\rho(t_f) = \exp(-iHt_f)\rho(t_i)\exp(iHt_f), \tag{5.2}$$

where $\rho(t_i) \equiv \rho_i$ denotes the density matrix at some initial time t_i. The density matrix can be written in the coordinate representation as follows (we write capital letters for the environment and small letters for the system and suppress additional indices which may be attached to these variables),

$$\langle x_f R_f \mid \rho(t_f) \mid y_f Q_f \rangle = \int dx_i \, dy_i \, dR_i \, dQ_i \, \langle x_f R_f \mid \exp(-iHt_f) \mid x_i R_i \rangle$$
$$\times \langle x_i R_i \mid \rho(t_i) \mid y_i Q_i \rangle \langle y_i Q_i \mid \exp(iHt_f) \mid y_f Q_f \rangle. \tag{5.3}$$

The transition amplitudes which appear on the right-hand side of (5.3) can then be expressed as path integrals, i.e.,

$$\langle x_f R_f \mid \exp(-iHt_f) \mid x_i R_i \rangle = \int \mathcal{D}x(t)\,\mathcal{D}R(t)\,\exp(iS[x(t), R(t)]),$$

$$\langle y_i Q_i \mid \exp(iHt_f) \mid y_f Q_f \rangle \;\; = \int \mathcal{D}y(t)\,\mathcal{D}Q(t)\,\exp(-iS[y(t), Q(t)]).\text{(5.4)}$$

Here S is the total action,

$$S = S_A + S_B + S_I, \tag{5.5}$$

and the sum is over all paths which satisfy the boundary conditions,

$$x(t_i) = x_i, \;\; x(t_f) = x_f, \;\; \text{etc.} \tag{5.6}$$

We are, as in our previous chapters, interested in the evolution of the reduced density matrix, ρ_A, for the system, which is defined as usual by tracing out the environmental variables,

$$\rho_A(x_f, y_f, t_f) = \int dR_f\, dQ_f\, \delta(R_f - Q_f)\langle x_f R_f \mid \rho(t_f) \mid y_f Q_f \rangle. \tag{5.7}$$

Using (5.3) and (5.4), this can be expressed as follows,

$$\rho_A(x_f, y_f, t_f) = \int dx_i\, dy_i\, dR_i\, dQ_i\, dR_f\, \langle x_i R_i \mid \rho(t_i) \mid y_i Q_i \rangle$$

$$\times \int \mathcal{D}x\,\mathcal{D}y\,\mathcal{D}R\,\mathcal{D}Q\, \exp\left(iS_A[x(t)] - iS_A[y(t)] + iS_B[R(t)] - iS_B[Q(t)]\right.$$

$$\left. +iS_I[x(t), R(t)] - iS_I[y(t), Q(t)]\right). \tag{5.8}$$

This expression still contains the total initial density matrix $\rho_i(x_i, R_i; y_i, Q_i)$. It is very convenient, and has thus been assumed by many authors, that ρ_i factorises into a system part and an environmental part (just as in the last two chapters),

$$\rho_i(x_i, R_i; y_i, Q_i) = \rho_A(x_i, y_i)\rho_B(R_i, Q_i), \tag{5.9}$$

and ρ_B is often chosen to be the density matrix of a canonical ensemble (corresponding to the description of a heat bath). The factorisation condition (5.9) is an expression of the physical assumption that quantum correlations between the system and its environment are absent (or, at least, can be neglected) at the initial time. It corresponds to a Sommerfeld condition if the degrees of freedom of the environment are scattering, see Chap. 3. While this assumption may be unrealistic in many circumstances, it does not alter the main conclusions to be drawn in the course of this section, as far as the occurrence of an "inverse branching" is negligible. Moreover, it is straightforward, although less transparent, to generalise the discussion to initial conditions which do not factorise, see for example Grabert *et al.* (1988) who show that more general initial conditions may be encoded by means of an additional "euclidean part" of the path integral. It has also been argued by Hu *et al.* (1992) that the factorising initial condition (5.9) may produce artefacts like an initial "jolt" in the diffusive coefficient describing some rapid decoherence which may be absent for more general initial conditions. We note again that a special initial condition like (5.9) may play a crucial role in quantum cosmology, see Zeh (2001).

Under the assumption (5.9) we arrive at our final expression for the evolution of the reduced density matrix, which reads

$$\rho_A(x_f, y_f, t_f) = \int dx_i \, dy_i \, J(x_f, y_f, t_f; x_i, y_i, t_i) \rho_A(x_i, y_i; t_i), \qquad (5.10)$$

where the "propagator" J is given by

$$J(x_f, y_f, t_f; x_i, y_i, t_i)$$
$$= \int \mathcal{D}x \, \mathcal{D}y \, \exp\left(iS_A[x(t)] - iS_A[y(t)]\right) \mathcal{F}[x(t), y(t)]. \qquad (5.11)$$

The reduced dynamics thus corresponds to a *pre-master equation* for ρ_A with the special initial condition (5.9). In the expression (5.11) we have introduced the *influence functional*

$$\mathcal{F}[x(t), y(t)] = \int dR_i \, dQ_i \, dR_f \, \rho_B(R_i, Q_i)$$
$$\times \int \mathcal{D}R \, \mathcal{D}Q \, \exp\left(iS_B[R(t)] - iS_B[Q(t)]\right.$$
$$\left. + iS_I[x(t), R(t)] - iS_I[y(t), Q(t)]\right) \equiv \exp(iW[x(t), y(t)]) \qquad (5.12)$$

as well as the *influence "phase"* W (but note that W may be complex!) which is often more convenient to use than the influence functional itself. The boundary conditions on the paths $R(t)$ and $Q(t)$ to be summed over in (5.12) are such that they start at the respective points R_i and Q_i, but meet at the common end point $R_f = Q_f$ at $t = t_f$. Since we have imposed the factorising condition (5.9) the influence functional does not depend explicitly on the initial system coordinates x_i and y_i, but only implicitly through the dependence of the action on the paths. This is no longer true in the general case. The influence functional does, however, depend on the initial state of the environment. $\mathcal{F}$ is in general highly nonlocal (the behaviour is non-Markovian) – it depends on the whole "histories" $x(t)$ and $y(t)$ which are inextricably mixed in the propagator J (5.11). The influence functional comprises the whole influence of the environment onto the system. If it turned out that different environments yield the same influence functional for the system, the reduced dynamics of the system would be physically indistinguishable.

We note the following properties for $\mathcal{F}$:

(1) If the interaction is switched off, $\mathcal{F}$ is equal to one. In this case, J (see (5.11)) is just the Green function for the von Neumann equation of system A. That $\mathcal{F} = 1$ holds in this case can be seen by expanding the environmental density matrix into its normalised eigenstates ψ_n, i.e.

$$\rho_B(R_i, Q_i) = \sum_n p_n \psi_n^*(Q_i)\psi_n(R_i) \tag{5.13}$$

and inserting this into the expression (5.12) for the influence functional with $S_I \equiv 0$,

$$\mathcal{F} = \sum_n p_n \int dR_f\, \psi_n^*(R_f, t_f)\psi_n(R_f, t_f) = 1. \tag{5.14}$$

(2) In the general case, the absolute square of $\mathcal{F}$ is bounded from above by one,

$$|\mathcal{F}| \le 1, \tag{5.15}$$

since there is no initial correlation between system and environment (this implements an arrow of time for the system, see Chap. 2). This property is important for its interpretation as a decoherence functional, as we shall see in the following. Condition (5.15) means that the imaginary part of the influence phase W is always non-negative. It vanishes if the interaction is switched off. Note the analogy to a refraction index or an optical potential.

The proof of the property (5.15) is straightforward, see e.g. Brun (1993): By inserting again the expansion (5.13) for ρ_B into (5.12), one obtains

$$\mathcal{F} = \sum_n p_n \int dR_f\, \psi_n^{[x]}(R_f)\psi_n^{*\,[y]}(R_f)$$
$$= \sum_n p_n \langle \psi_n^{[x]} \mid \psi_n^{[y]} \rangle, \tag{5.16}$$

where $\psi_n^{[x]}(R_f)$ denote the wave functions which are obtained from $\psi_n(R_i)$ under the unitary time evolution generated by S_B and S_I according to

$$\psi_n^{[x]}(R_f) = \int dR_i \int \mathcal{D}R \exp(iS_B[R] + iS_I[x, R])\psi_n(R_i) . \tag{5.17}$$

They depend on the path $x(t)$ of the distinguished system, as we have indicated by our notation. We thus have

$$|\mathcal{F}| \le \sum_n p_n |\langle \psi_n^{[x]} \mid \psi_n^{[y]} \rangle| \le \sum_n p_n \le 1. \tag{5.18}$$

For coinciding paths $x(t)$ and $y(t)$ we immediately find that $\mathcal{F} = 1$ (vanishing imaginary part of W), i.e. the absolute value of the influence functional in configuration space has a local maximum and the off-diagonal elements in its vicinity are suppressed. There can, of course, exist paths $x(t) \neq y(t)$ for which there is again a maximum and, thus, no suppression of interferences. It has, however, been argued that this does not happen if the environment has a huge number of degrees of freedom as occurs in realistic cases, see Brun (1993).

(3) From its definition one immediately finds

$$\mathcal{F}[x, y] = \mathcal{F}^*[y, x], \tag{5.19}$$

or, if expressed in terms of the influence phase $W = W_R + iW_I$,

$$W_R[x, y] = -W_R[y, x] \tag{5.20}$$
$$W_I[x, y] = W_I[y, x]. \tag{5.21}$$

In most cases the influence functional (5.12) cannot be evaluated exactly, since the interaction between system and reservoir is too complicated. One must therefore rely on simple models if one is interested in exact expressions, or apply numerical methods in realistic cases. It is thus not surprising that most authors have considered actions like the following, which are at most quadratic in its variables,

$$S = \int_{t_i}^{t_f} dt \left(\frac{M}{2}(\dot{x}^2 - \Omega_0^2 x^2) + \sum_n \frac{m}{2}(\dot{Q}_n^2 - \omega_n^2 Q_n^2) - x \sum_n C_n Q_n \right). \tag{5.22}$$

Here the distinguished variable, x, corresponds to a harmonic oscillator with mass M which is linearly coupled to a bath of harmonic oscillators described by variables Q_n. (For simplicity we attribute an equal mass m to all environmental oscillators.) We assume that the environmental oscillators are initially in thermal equilibrium at temperature T. This is certainly a realistic assumption for the quantum Brownian motion case, but may be totally unrealistic, for example, in cosmology. Eventually, however, the very existence of thermal states has to be justified by decoherence. The initial density matrix of the heat bath thus describes a canonical ensemble, i.e.

$$\rho_B(\mathbf{R}, \mathbf{Q}) = \prod_n \frac{m\omega_n}{2\pi \sinh(\omega_n/kT)}$$

$$\times \exp\left(-\frac{m\omega_n}{2\sinh(\omega_n/kT)}[(R_n^2 + Q_n^2)\cosh(\omega_n/kT) - 2R_n Q_n] \right). \tag{5.23}$$

The influence functional (5.12) can then be exactly evaluated (Feynman and Vernon 1963), since only integrations over quadratic forms are involved, with the result (choosing the initial time to be zero)

$$
\mathcal{F}[x, y] = \exp \left(-\frac{i}{2} \int_0^{t_f} dt \int_0^t ds \, [x(t) - y(t)] k_R(t - s)[x(s) + y(s)] \right.
$$
$$
\left. -\frac{1}{2} \int_0^{t_f} dt \int_0^t ds \, [x(t) - y(t)] k_I(t - s)[x(s) - y(s)] \right), (5.24)
$$

where k_R and k_I are, respectively, the real and imaginary part of a complex kernel. They are explicitly given by the expressions

$$
k_R(t - s) = -\sum_n \frac{C_n^2}{2m\omega_n} \sin \omega_n(t - s)
$$
$$
\equiv -\int_0^\infty d\omega \, I(\omega) \sin \omega(t - s) \tag{5.25}
$$

and

$$
k_I(t - s) = \sum_n \frac{C_n^2}{2m\omega_n} \coth \frac{\omega_n}{2kT} \cos \omega_n(t - s)
$$
$$
\equiv \int_0^\infty d\omega \, I(\omega) \coth \frac{\omega}{2kT} \cos \omega(t - s). \tag{5.26}
$$

We have introduced in these expressions the *spectral density* $I(\omega)$ of the environmental oscillators,

$$
I(\omega) \equiv \sum_n \delta(\omega - \omega_n) \frac{C_n^2}{2m\omega_n}. \tag{5.27}
$$

The effect of the environment can be described by such a spectral density *if* the reduced density matrix obeys a master equation, i.e., if there are no recurrences (or too long to be of interest) and no Zeno effect, see Chap. 3. We note that only k_I depends on the temperature of the bath. More generally, we write

$$
I(\omega) \approx \rho_D(\omega) \frac{C^2(\omega)}{2m\omega} \tag{5.28}
$$

with a smooth density ρ_D.

It turns out to be appropriate to introduce the following classification for the spectral density. If in the physical range (i.e., for frequencies ω smaller than some cutoff frequency Ω) $I(\omega)$ is proportional to ω, the environment is classified as *ohmic*. This is the case most frequently studied in the literature, since it leads to a dissipative force linear in the velocity of the system. We shall also restrict ourselves to this case below. If $I(\omega)$ is proportional to ω^n with some power $n > 1$ ($n < 1$), the environment is classified as *supraohmic* (*subohmic*). We already note here that this classification is frequently used in the algebraic approach to superselection rules. Amann (1991b), for example, found in his studies of the spin-boson model (see the remarks in Sect. 4.1) that there is a "superselection rule" (a total suppression of interference terms for the two-state system) only in the subohmic case. If integrals such as (5.25) and (5.26) formally diverge, it is necessary to employ some appropriate regularisation. This is conveniently done, for example, by including a factor $\exp(-\omega^2/\Omega^2)$ in the expression for the spectral density, see Hu *et al.* (1992).

What is the interpretation of k_R and k_I? If the environment is a heat bath, k_R is related to *dissipation*, while k_I describes *fluctuations* ("noise")[2] and, most important for the present discussion, *decoherence*. They are connected by the *fluctuation-dissipation theorem*. We shall encounter in the next section an explicit example of these relations. Note that k_R leads to a change of the classical action S_A in (5.11). This corresponds to a *back reaction* of the environment onto the system, which can sometimes be absorbed in a renormalisation of system parameters, e.g. the frequency. Its effect is thus identical to the effect of the phase of the reduced density matrix, as it was discussed in the preceding chapters.

As a specific model we shall treat the one by Caldeira and Leggett (1983a), who choose a continuum of oscillators which are distributed according to the following density,

$$\rho_D(\omega)C^2(\omega) = \begin{cases} 4Mm\gamma\omega^2/\pi & \text{if } \omega < \Omega \\ 0 & \text{if } \omega > \Omega \end{cases}, \tag{5.29}$$

where Ω is, again, a high-frequency cutoff, and γ is an effective coupling between the Brownian particle and the reservoir. It will appear below in the damping term for the effective classical equation of motion for the system coordinate (γ^{-1} is the relaxation time). The value of the cutoff Ω is given by the number of environmental degrees of freedom. If its value is converted into a temperature, one obtains the Debye temperature.

The resulting expression for the reduced propagator (5.11) is especially simple in the limit of high temperature, $kT \gg \Omega \gg \Omega_R$ (sometimes referred to as the Fokker–Planck limit), where Ω_R is the renormalised oscillator frequency of the distinguished oscillator. The explicit expressions for k_R and k_I read in this limit

$$k_R = -4M\gamma\delta'(t-s), \tag{5.30}$$

$$k_I = 8M\gamma kT\delta(t-s). \tag{5.31}$$

The integrations in (5.24) are then reduced to single time integrations.

---

[2] Although this term is frequently used, we note that it is somewhat misleading, since it is a classical notion, cf. Sect. 9.2.

One thus obtains

$$J(x_f, y_f, t_f; x_i, y_i, 0) = \int \mathcal{D}x\,\mathcal{D}y\ \exp(iS_R[x(t)] - iS_R[y(t)])$$

$$\exp\left(-iM\gamma \int_0^{t_f} dt\,(x(t) - y(t))(\dot{x}(t) + \dot{y}(t))\right.$$

$$\left.-2M\gamma kT \int_0^{t_f} dt\,[x(t) - y(t)]^2\right). \tag{5.32}$$

Here we have denoted the action of the distinguished oscillator with respect to the renormalised parameters by S_R. One recognises in (5.32) a Gaussian factor which exponentially suppresses the contribution of different paths $x(t)$ and $y(t)$ – the "paths decohere". One can get from this Gaussian a rough estimate of the efficiency of decoherence. The exponent in this factor becomes approximately equal to one after a "decoherence time" of

$$t_{\text{dec}} \approx \frac{1}{\gamma}\left(\frac{1}{d\sqrt{2MkT}}\right)^2 \equiv \frac{1}{\gamma}\left(\frac{\lambda_{DB}}{d}\right)^2, \tag{5.33}$$

where d is a typical separation scale of both paths, and λ_{DB} is the de Broglie-wavelength. If one inserts the values, say, $d = 1$ cm, $T = 300$ K, $M = 1$ g, one finds (Zurek 1986) that the ratio of t_{dec} to the relaxation time γ^{-1} is approximately 10^{-40}! This demonstrates, again, the efficiency of decoherence on macroscopic scales, see also Sect. 3.2.2.3.

With the above explicit expression for the propagator J one can derive an effective equation ("master equation") for the reduced density matrix of the distinguished oscillator. The physical relevance of such equations has already been discussed in Chap. 3. Here we emphasise that they can conveniently be derived from the influence functional. In the high-temperature limit $kT \gg \Omega \gg \Omega_R$ the result was given by Caldeira and Leggett (1983a). The master equation reads

$$i\hbar\frac{\partial \rho_A}{\partial t}(x, y, t) = \left\{-\frac{\hbar^2}{2M}\left(\frac{\partial^2}{\partial x^2} - \frac{\partial^2}{\partial y^2}\right) + \frac{\Omega_R^2}{2}(x^2 - y^2)\right.$$

$$\left.-i\hbar\gamma(x - y)\left(\frac{\partial}{\partial x} - \frac{\partial}{\partial y}\right) - i\frac{2M\gamma kT}{\hbar}(x - y)^2\right\}\rho_A(x, y, t), \tag{5.34}$$

where we have re-inserted $\hbar$ for illustration. The various terms in (5.34) can be easily interpreted. The first two terms on the right-hand side just make up the free Hamiltonian for the distinguished oscillator ("von Neumann part"), the only imprint of the environment being the renormalised frequency in the potential term. If there were no interaction, these would be the only terms. The third term describes *dissipation*. It is a direct consequence of the imaginary part of the influence functional and contains the parameter γ which measures the strength of the interaction with the heat bath, see (5.29). The last term of the master equation is responsible for *decoherence*, since it only contributes for off-diagonal elements, $x \neq y$, of the density matrix. As can be immediately seen from (5.32), this term has its roots in the real part of the decoherence functional. Upon comparison with a similar master equation which was derived by Joos and Zeh (1985) and discussed in Chap. 3 we note that the dissipative term is absent in their equation. The reason for this is that they concentrated on decoherence and did not take into account the back reaction of the environmental degrees of freedom onto the massive body, which formally corresponds to the limit $M \to \infty$ or $\gamma \to 0$.

An exact master equation for arbitrary temperature has been derived by Hu *et al.* (1992) using the same methods as Caldeira and Leggett (1983a). They have discussed a general form of the environment (ohmic, supraohmic, and subohmic). We only quote their result:

$$
i\hbar \frac{\partial \rho_A}{\partial t}(x, y, t) = \left\{ -\frac{\hbar^2}{2M} \left(\frac{\partial^2}{\partial x^2} - \frac{\partial^2}{\partial y^2} \right) + \frac{\Omega_R^2}{2}(x^2 - y^2) \right.
$$
$$
-i\hbar\Gamma(t)(x - y) \left(\frac{\partial}{\partial x} - \frac{\partial}{\partial y} \right) - i\Gamma(t)h(t)(x - y)^2
$$
$$
\left. +\hbar\Gamma(t)f(t)(x - y) \left(\frac{\partial}{\partial x} + \frac{\partial}{\partial y} \right) \right\} \rho_A(x, y, t). \tag{5.35}
$$

Here, $\Gamma(t), h(t)$, and $f(t)$ are nonlocal functions of time which are given by certain integral expressions involving k_R and k_I. They express the non-Markovian behaviour of the system. They have been calculated for several instructive examples in Hu *et al.* (1992) to which we refer the reader for more details. The interpretation of the first four terms on the right-hand side of (5.35) is analogous to (5.34). We note the presence of a new term (the last term on the right-hand side) which also leads to decoherence. A similar term was found by Unruh and Zurek (1989). Again, the main contribution to decoherence comes from the $(x - y)^2$ term in the above master equation. An alternative derivation of the master equation for a general environment using the reduced Wigner function is presented in Calzetta *et al.* (2001).

We conclude this section with some remarks on the physical meaning of the paths which are summed over in the above functional integrals. Since the path integral is only an expression for a quantum mechanical amplitude, albeit a very convenient one, the paths which are summed over are fictitious and thus have no existence in any sense. This becomes especially clear in an illustrative example provided by Teitelboim (1991) who considers the free particle in a parametrised form. He shows that it is perfectly consistent to choose a gauge for this parametrised system such that "time does not evolve at all" and the whole amplitude emerges through a surface term in the action at final time t_f. The fictitious paths in the path integral can be related to real spacetime trajectories only if the system is under continuous observation

(Aharonov and Vardi 1980), i.e. only if the histories decohere from each other in the sense that nondiagonal elements of the influence functional are small. Classical trajectories may only emerge as a derived and approximate concept, through decoherence, and play no role at a fundamental level.

5.2 Definition and Properties of Consistent Histories

We have introduced in the last section the influence functional and used it as an efficient tool to study quantum Brownian motion. Here we shall present a general definition of *consistent histories* which can, in some sense, be thought as an appropriate generalisation of the influence functional. A *history* is defined to be a time-ordered sequence of quantum-mechanical "events". But what does "event" mean in this context? As some authors emphasise (e.g. Griffiths 1994), "event" is here introduced only as a formal concept (a "possible event"), in contrast to "dynamical events" such as "quantum jumps", for example. Therefore the word "proposition" is occasionally used instead. The important question is, however, whether such formal events can also occur *dynamically*. We return to this question at the end of this chaper, where also comments about the relevance of these concepts for the measurement problem are made, but already note here that this can happen only via some assumed "collapse of the wave function" (in which case one comes back to the Copenhagen interpretation of quantum mechanics), or by adopting some new, bizarre, interpretation, see d'Espagnat (1989, 1995) and Sect. 2.4.

A "proposition" or "event" is here defined by a projection operator, i.e. a hermitean operator P with $P^2 = P$ acting on the quantum mechanical Hilbert space for the *total* system to produce an appropriate subspace of this Hilbert space. Such subspaces may be states describing "the position x of an electron to lie within the range Δx", "its momentum p to lie within the range Δp", or "its spin to point in z-direction". It is clear that, due to the possibility of interference terms, the very formulation of quantum theory does *not* allow, in general, to assign probabilities to such "events" at different times. The extent to which this can still be formally achieved is the subject of this section. The corresponding "classical" histories are called *consistent* or *decoherent*. We emphasise that these projectors (which act on the Hilbert space of states) must not be confused with the Zwanzig projection $\hat{P}$ (which acts on the density operator), although they define special Zwanzig projections (cf. also the remarks in Chap. 2). We shall return to this issue in the next sections.

Consistent histories were first discussed by Griffiths (1984) who intended to develop an appropriate language (as close as possible to a classical stochastic theory) where some aspects of quantum theory seem to lose their "paradoxical" appearance and can be discussed without referring to concepts such as "measurement". An example is the famous EPR gedankenexperiment, which was described in this language by Griffiths (1987), see also Griffiths (1994) for a review of his ideas. Omnès (1988, 1992, 1999) has stressed the "logical" aspects of quantum theory and used the histories formulation, together with some rigorous results in the semiclassical theory, to develop his "logical interpretation" of quantum theory.

The notion of consistent histories was connected with decoherence by Gell-Mann and Hartle (1990) who showed how the emergence of quasi-classical behaviour can be formulated in the language of consistent histories. According to them, a central issue is to determine *all* sets of alternative histories to which one can meaningfully attach probabilities. This serves to explore the possible "quasiclassical domains" of the Universe. These authors also emphasise that this notion gives a definite meaning to Everett's branches. They therefore call this a "post-Everett interpretation". Since we are mainly concerned with decoherence, the following discussion basically follows, with elaborations, their paper. As already emphasised, one common aim of all authors is to develop a formulation of quantum theory applicable to closed systems, with no reference to external measurement agencies.

The formal definition of a history is given, in the Heisenberg picture, by a chain of projectors,

$$C_\alpha = P^n_{\alpha_n}(t_n) P^{n-1}_{\alpha_{n-1}}(t_{n-1}) \cdots P^1_{\alpha_1}(t_1) \tag{5.36}$$

with a time ordering $t_n > t_{n-1} > \cdots > t_1$. We see already here that the presence of an external time parameter is an essential ingredient (as already indicated by the word "history"). The superscript k in $P^k_{\alpha_k}(t_k)$ denotes the set of projectors at time t_k, and α_k denotes the particular alternative which has been chosen. For example, k may stand for "spin projection", and α_k for "spin up". The letter α is a shorthand notation for the set of all alternatives α_1 to α_n. One thus allows for the possibility of using different kinds of projectors at different times, for example a position projector at one time and a momentum projector at another time. We also allow the histories to be explicitly *branch-dependent*, that is we allow the possibility that the set of alternatives at time t_k depends on the realisation of *earlier* alternatives $\alpha_1 \ldots \alpha_{k-1}$. This of course introduces an *arrow of time* into the formulation unless we extend the branch dependence onto future alternatives as well. (Histories can also be formulated in a time-symmetric way, see Griffiths (1993) and Sect. 5.4.) The chain of projections can thus be written as

$$C_\alpha = P^n_{\alpha_n}(t_n; \alpha_{n-1}, \ldots, \alpha_1) P^{n-1}_{\alpha_{n-1}}(t_{n-1}; \alpha_{n-2}, \ldots, \alpha_1) \cdots P^1_{\alpha_1}(t_1). \tag{5.37}$$

This is important in particular for the discussion of Schmidt histories (see Sect. 5.3) and the arrow of time (see Sect. 5.4). In most of the following expressions we will suppress this extra dependence of the projectors on earlier alternatives.

We demand that the alternatives at a given time be *exhaustive*, i.e.

$$\sum_{\alpha_n} P^n_{\alpha_n} = I, \tag{5.38}$$

and *mutually exclusive*, i.e.

$$P_{\alpha_n}^n P_{\beta_n}^n = \delta_{\alpha\beta} P_{\alpha_n}^n.\qquad(5.39)$$

Sometimes use is made of projectors which obey some of these properties only approximately like the "Gaussian slits" used in Dowker and Halliwell (1992). This is very convenient for practical calculations without changing the physical conclusions. From (5.38) it follows immediately that

$$\sum_\alpha C_\alpha = I.\qquad(5.40)$$

It may be helpful to think of the C_α as describing branches of the initial density matrix. In the special case of a pure state one can, for example, write

$$|\psi\rangle = \sum_\alpha C_\alpha |\psi\rangle.$$

One can then define partitions of the histories $\{\alpha\}$ into classes $\{\bar\alpha\}$ which correspond to a *coarse-graining* of the original histories, i.e. a member of $\{\bar\alpha\}$ is a whole set of histories contained in $\{\alpha\}$. This coarse-grained set of histories is represented by operators $C_{\bar\alpha}$, where

$$C_{\bar\alpha} = \sum_{\alpha \in \bar\alpha} C_\alpha = \sum_{(\alpha_1,\ldots,\alpha_n)\in\bar\alpha} P_{\alpha_n}^n(t_n)\cdots P_{\alpha_1}^1(t_1).\qquad(5.41)$$

Such operators, which are sums of chains of projections, are also called *class operators*. Note that it is in general *not* possible to represent the $C_{\bar\alpha}$ themselves as a chain of projectors. Under which circumstances, then, is it possible to assign probabilities to histories? Since the probability for a single "event" (in the sense of the "collapse interpretation" of quantum mechanics) is given by the usual formula

$$p_k = \langle P_k \rangle = \mathrm{tr}(P_k^\dagger \rho_0 P_k),\qquad(5.42)$$

where ρ_0 denotes the density matrix, its obvious generalisation to a history C_α would be

$$p(\alpha) = \mathrm{tr}(C_\alpha^\dagger \rho_0 C_\alpha).\qquad(5.43)$$

Consider now a special example of (5.41) in which the coarse-grained history is just given by a sum of two histories:

$$C_{\bar\alpha} = C_\alpha + C_\beta.\qquad(5.44)$$

Then,

$$\begin{aligned}p(C_{\bar\alpha}) &= p(C_\alpha) + p(C_\beta) + 2\mathrm{Re}\ \mathrm{tr}(C_\alpha^\dagger \rho_0 C_\beta)\\ &\neq p(C_\alpha) + p(C_\beta).\end{aligned}\qquad(5.45)$$

The coarse-grained history (5.44) may thus only be assigned a probability if the last term in (5.45) (the quantum mechanical interference term) is zero or, at least, very small.

The well-known double-slit experiment may also in this context serve to illustrate the essential features (Hartle 1991): If measurements of the electron are made only at the screen behind the slits, it will definitely *not* be possible to assign probabilities to the alternatives that the electron went through the upper or lower slit, respectively. If, instead, a gas is present between the slits and the screen, the phase relations are delocalised by decoherence, and it is consistent to assign probabilities to the two alternative histories of the electron passing through the upper or the lower slit. This formal consistency, however, does *not* mean that these histories can be interpreted in a direct, realistic, sense of a classical electron passing through one of the slits. In fact, if "fine-grained histories" of the whole system comprising the electron and the gas molecules were considered, no consistency would be reached.

This example motivates the introduction of the *decoherence functional* $\mathcal{D}(\alpha, \alpha')$ which is a functional of the two histories α and α':

$$\begin{aligned}
\mathcal{D}(\alpha, \alpha') &= \operatorname{tr}(C_\alpha^\dagger \rho_i C_{\alpha'}) \\
&= \operatorname{tr}(P_{\alpha_n}^n(t_n) \cdots P_{\alpha_1}^1(t_1) \rho_i P_{\alpha_1'}^1(t_1) \cdots P_{\alpha_n'}^n(t_n)).
\end{aligned} \tag{5.46}$$

Here ρ_i denotes the initial density matrix. We emphasise that there appears *no* final density matrix in (5.46) (in other words, there appears only the trivial density matrix $\rho_f = 1$). This implements an arrow of time into the formalism, whose aspects will be discussed in more detail in Sect. 5.4.

In the "collapse interpretation", the diagonal element, $\mathcal{D}(\alpha, \alpha)$, would just give the total probability for n subsequent collapses into the states described by the corresponding projectors. In this context it is sometimes called "Wigner's formula" (Wigner 1963, d'Espagnat 1995). Since such collapses are *not* assumed to happen by most proponents of consistent histories, $\mathcal{D}(\alpha, \alpha)$ can describe probabilities only under very restrictive conditions.

From (5.45) we conclude that this can be done, i.e. that probabilities can be assigned to histories, if

$$\operatorname{Re}\mathcal{D}(\alpha, \alpha') = 0, \ \forall \alpha \neq \alpha'. \tag{5.47}$$

This condition is sometimes referred to as "weak decoherence." Griffiths (1984) had originally employed an even weaker condition, since he demanded that (5.47) be valid *only* for sums of chains which can again be written as a projector. The present condition (5.47) must also hold for arbitrary *sums* of chains of projections. Its validity is sufficient to assign probabilities to histories, but it is by no means sufficient to explain classical behaviour. The reason is that there may be highly nonclassical histories which are nevertheless consistent (an example will be given in the next section). Gell-Mann and Hartle (1993) impose a stronger condition which demands that the non-diagonal elements of the *whole* decoherence functional be zero:

$$\mathcal{D}(\alpha, \alpha') = 0, \ \forall \alpha \neq \alpha'. \tag{5.48}$$

They refer to this condition as "medium decoherence", now also preferred by Griffiths (1994). Since this condition can also be fulfilled for very non-classical situations (see Sect. 5.3), we prefer to call this a *consistency condition* to avoid confusion with the physical concept of decoherence discussed in our previous chapters. If (5.48) holds, we can assign to a history C_α the probability

$$p(\alpha) = \mathcal{D}(\alpha, \alpha). \tag{5.49}$$

We note again that two histories are already different if they differ in at least one alternative, i.e. if $\alpha_k \neq \alpha'_k$ for at least one value of k. The validity of the condition (5.49) thus depends very sensitively on the choice of alternatives. It is clear from its definition (5.46) that in the presence of only one alternative the consistency condition is automatically fulfilled – one only has to exploit the cyclic property of the trace in the definition (5.46). This is nothing but an expression of the fact that quantum mechanics always allows to assign probabilities to a single "event". We note that there is a subjective element in the whole framework, in the sense that the choice of alternatives is made by hand to arrive at the description one wants to make, see d'Espagnat (1995).

In the case of a pure initial state, $\rho_i = |\psi\rangle\langle\psi|$, medium decoherence is equivalent to the existence of "generalised records" for each history, see Gell-Mann and Hartle (1993). This happens since, if (5.48) holds, there exist orthogonal projection operators $\{R_\alpha\}$ that project on the various branches of the pure initial state,

$$R_\alpha|\psi\rangle = C_\alpha|\psi\rangle. \tag{5.50}$$

The C_α are, of course, not necessarily projectors themselves. If one then assigns these projection operators R_α to the string of projections (5.36) at any time $t > t_n$, the diagonal elements of the resulting decoherence functional contain a joint probability correlating the events in the original history with the R_α which are therefore referred to as generalised records (information about the histories $\{\alpha\}$ is "stored there"). The converse is also true: The existence of generalised records leads to medium decoherence of the corresponding set of histories. An example is provided by the α-particle in the Wilson chamber: The successive positions of the particle lead to generalised records in the atoms of the chamber.

We briefly note that the analogous condition to (5.50), if fulfilled for an impure state, is called "strong decoherence condition" by Gell-Mann and Hartle (1993).[3] Since it turns out that this imposes a too restrictive condition on the allowed set of histories, it will not be considered here. While the physical situation of decoherence can be adequately described in terms of consistent

---

[3] In Gell-Mann and Hartle (1995) they use this concept for a more restrictive condition.

histories (see below), there exist many more consistent histories without decoherence. A trivial example is provided by a sequence of projectors which project on an eigenstate of the density matrix.

We note the following properties of the decoherence functional, which immediately follow from its definition:

$$\mathcal{D}(\alpha, \alpha') \qquad = \mathcal{D}^*(\alpha', \alpha), \tag{5.51}$$

$$\sum_{\alpha, \alpha'} \mathcal{D}(\alpha, \alpha') = \mathrm{tr}\rho_i = 1, \tag{5.52}$$

$$\mathcal{D}(\alpha, \alpha) \qquad \geq 0, \tag{5.53}$$

$$\sum_{\alpha} \mathcal{D}(\alpha, \alpha) = 1, \tag{5.54}$$

$$|\mathcal{D}(\alpha, \alpha')|^2 \leq \mathcal{D}(\alpha, \alpha)\mathcal{D}(\alpha', \alpha'). \tag{5.55}$$

The second-to-last of these properties guarantees the correct normalisation of the probability (5.49). The last property was shown by Dowker and Halliwell (1992) and means, in particular, that there can be no interference with a history whose decoherence functional has a vanishing diagonal element.

We note that there are even weaker consistency conditions in the literature. Goldstein and Page (1995), for example, consider "linearly positive histories" which are a broad extension of the histories discussed above. Since they allow one to assign formal probabilities even to histories which quantum mechanically interfere, their use in the context of quantum mechanics remains unclear.

While a history is composed of projection operators, it is itself usually not a projector. An alternative scheme, in which the history *is* a projector, has been discussed in Isham *et al.* (1998) and the references therein. This method may be directly applicable to quantum field theory, but we shall not consider it here.

There is a close connection of the decoherence functional with the influence functional defined in Sect. 5.1, if projections are made onto alternative ranges of the configuration variables. Note that the notion of influence functional is very restrictive, since it usually refers to configuration space only. In the simplest case these projections just correspond to tracing out part of the configuration variables. The coarse-graining then consists in ignoring the environmental degrees of freedom and focusing attention on the distinguished variable only. The only difference between the decoherence functional and the influence functional is that the definition of the former, see (5.46), comprises in addition a multiplication with the initial density matrix of the distinguished variable as well as a sum over end points of the sytem variables, see e.g. (5.57) below. It also contains the total action of system plus environment. It is instructive to discuss the example of quantum Brownian motion (Sect. 5.1) in the framework of consistent histories, see Gell-Mann and Hartle (1993). Instead of the variables x and y we use for the distinguished oscillator the following variables:

$$X = \frac{1}{2}(x + y),$$

$$\xi = x - y. \tag{5.56}$$

The decoherence functional then reads (compare (5.11))

$$
\begin{aligned}
\mathcal{D}[X(t), \xi(t); X_i, \xi_i) &= \int dX_f \, d\xi_f \, \delta(\xi_f) \exp\left(iS_A \left[X(t) + \frac{\xi(t)}{2}\right]\right. \\
&\quad \left. -iS_A \left[X(t) - \frac{\xi(t)}{2}\right] + iW[X(t), \xi(t)]\right) \\
&\quad \times \rho_A \left(X_i + \frac{\xi_i}{2}, X_i - \frac{\xi_i}{2}\right). \tag{5.57}
\end{aligned}
$$

Note that this functional is still completely fine-grained in the variables of the distinguished oscillator (only the heat bath variables have been summed over). Note also that due to the trace in (5.46) one has to sum over the common end point $x_f = y_f$ of the paths. It will be convenient for the following to introduce the functional $\mathcal{A}$,

$$A[X(t), \xi(t)] = S_A \left[X(t) + \frac{\xi(t)}{2}\right] - S_A \left[X(t) - \frac{\xi(t)}{2}\right] + W[X(t), \xi(t)]. \tag{5.58}$$

If one then, *in addition* to the sum over environmental variables, performs a coarse-graining in the distinguished variable, the decoherence functional (5.57) reads

$$\mathcal{D}(\alpha, \alpha') = \int_{(\alpha, \alpha')} \mathcal{D}X \, \mathcal{D}\xi \, \delta(\xi_f) \exp(iA)\rho_A. \tag{5.59}$$

Since we expect decoherence between the paths $x(t)$ and $y(t)$, the ξ- integration in (5.59) only makes a contribution in the neighbourhood of $\xi = 0$. We thus consider this integration in the following as being effectively unrestricted by the coarse-graining in the distinguished variable.

Using the explicit form of the action (5.22), one can after some integrations by part write the functional $\mathcal{A}$ as follows (recall that $\xi_f = 0$),

$$
\begin{aligned}
A[X(t), \xi(t)] &= -M\dot{X}_i\xi_i + \int_0^{t_f} dt \, \xi(t)e(t) \\
&\quad + \frac{i}{4} \int_0^{t_f} dt \int_0^{t_f} ds \, \xi(t)k_I(t - s)\xi(s), \tag{5.60}
\end{aligned}
$$

where

$$e(t) = -M\ddot{X}(t) - M\Omega_0^2 X(t) - \int_0^t ds \, k_R(t - s)X(s), \tag{5.61}$$

and the upper limit of the s-integration in (5.60) has been extended to the final time t_f, taking k_I to be symmetric. As assumed above, the ξ- integration in (5.59) can be performed without restrictions. The integral over the initial value ξ_i in (5.59) yields the Wigner distribution for the initial values of coordinates and momenta, i.e.,

$$\int d\xi_i \, \exp(-M\dot{X}_i\xi_i)\rho_A\left(X_i + \frac{\xi_i}{2}, X_i - \frac{\xi_i}{2}\right) \equiv w(X_i, M\dot{X}_i). \qquad (5.62)$$

One thus obtains from (5.49) for the probability $p(\alpha)$ of a history in the coarse-grained set of histories,

$$p(\alpha) = \mathcal{D}(\alpha, \alpha) \approx \int_\alpha \mathcal{D}X \left(\det\frac{k_I}{4\pi}\right)^{-1/2}$$

$$\times \exp\left(-\int_0^{t_f} dt \int_0^{t_f} ds \, e(t)k_I^{-1}(t-s)e(s)\right) w(X_i, M\dot{X}_i). \qquad (5.63)$$

The interpretation of this expression is that, given initial values X_i and $\dot{X}_i$, the histories with the largest probabilities are those for which the Gaussian in (5.63) is approximately equal to one, i.e. for which

$$e(t) = -M\ddot{X}(t) - M\Omega_0^2 X(t) - \int_0^t ds \, k_R(t-s)X(s) = 0. \qquad (5.64)$$

This is the effective equation of motion for the distinguished oscillator. Its interaction with the heat bath is encoded in a nonlocal force term which is the back reaction of the heat bath on the system and which directly arises from the phase of the decoherence functional in (5.59). It describes dissipation, since the energy for the distinguished oscillator alone is not conserved. For the very special case of the Fokker–Planck limit, using the expression (5.30) for k_R, this additional term is equal to the well known damping term $2M\gamma\dot{X}$.

Although the Wigner function can become negative, the probability (5.63) is manifestly positive, since it is the diagonal element of the decoherence functional, see (5.53). It may, however, happen that small negative values in this expression occur due to the implementation of some approximation.

We note that the real and imaginary parts of the kernel appearing in the influence functional (5.24) can always be related by the *fluctuation-dissipation theorem*. In the simple case of the above model, see (5.25) and (5.26), one has

$$k_I(t) = \int_{-\infty}^{\infty} ds \, f(t-s)\gamma(s), \qquad (5.65)$$

where

$$k_R(t) = \frac{d}{dt}\gamma(t),$$

$$f(t) = \frac{1}{\pi}\int\limits_0^\infty d\omega\,\omega\coth\frac{\omega}{2kT}\cos\omega t. \qquad (5.66)$$

The energy loss of the distinguished oscillator due to dissipation can then be conveniently written as

$$\Delta E = -\frac{1}{2}\int\limits_0^t ds'\int\limits_0^t ds\,\gamma(s-s')\dot{X}(s)\dot{X}(s'),$$

where the integration is over the whole history of the system. One recognises that only the even part of γ is responsible for dissipation (the odd part contributes to renormalisation). The fluctuation-dissipation theorem demonstrates that, in fact, decoherence, noise ("fluctuations"), and dissipation have the same physical origin (they result from the same amount of coarse-graining and the same interaction with the "bath") and are thus intimately connected to each other. Just because of this strong linkage, however, classical behaviour requires a certain balance between decoherence and noise. This was already discussed in Chap. 3 with the help of the Wigner function corresponding to the reduced density matrix. If the influence of the environment is very strong, decoherence is very efficient but the noise may also be so strong that the distinguished system is "kicked off its classical trajectory." One thus has to demand a sufficient *inertia* for the system. In the above quantum Brownian motion example, this corresponds to the demand for a high mass, M, of the distinguished oscillator.[4] As can be recognised from (5.33), the decoherence time decreases with increasing temperature of the heat bath. The "degree of classicality" can then be characterised in this model by the parameter M/T.

The above formalism may applied in a straightforward way to models with nonlinear coupling, see Gell-Mann and Hartle (1993), and Brun (1993). The basic input is the expansion of the influence phase $W[X(t),\xi(t)]$ in powers of $\xi(t)$,

$$W[X(t),\xi(t)] = W[X(t),0] + \int\limits_0^{t_f} dt\,\xi(t)\,\frac{\delta W}{\delta\xi}\bigg|_{\xi(t)=0}$$

$$+\frac{1}{2}\int\limits_0^{t_f} dt\int\limits_0^{t_f} ds\,\xi(t)\,\frac{\delta^2 W}{\delta\xi(t)\delta\xi(s)}\bigg|_{\xi(t)=0}\xi(s)+\dots. \quad (5.67)$$

---

[4] In more general models, higher inertia can be achieved by increasing the amount of coarse-graining, for example smearing out over a larger volume.

The coefficients in (5.67) can be expressed with respect to the expectation values of an appropriate "force operator" and its fluctuations (This force operator is found from the interaction Hamiltonian in (5.1)). The relations (5.63) and (5.64) can then easily be generalised to the nonlinear case. Explicit results for various choices of interaction potentials have been derived by Hu *et al.* (1993) from the corresponding influence functionals. They also derived the respective master equations for the reduced density matrix.

5.3 Reduced Density Matrix and Decoherence

In the previous section we have discussed how the physical mechanism of decoherence can be formulated within a framework whose basic entities are temporal sequences of projection operators. The aim of this section is to build a bridge between this language and the language of the reduced density matrix which played the central role in Chaps. 3 and 4. We thus assume that the total, closed, system for which the decoherence functional is formulated can be conveniently separated into a subsystem (simply called "system" from now on) and an environment. Our treatment will roughly follow the discussion of Paz and Zurek (1993), see also Twamley (1993) for a somewhat different way of comparing decoherence with consistent histories. We shall, in particular, address the question of which class of consistent histories exhibits classical behaviour. As we have argued above, consistency alone (vanishing nondiagonal elements of the decoherence functional) does not yet guarantee the approximate validity of classical equations.

Since decoherence was discussed in Chaps. 3 and 4 in the Schrödinger picture, it will be convenient to rewrite the expression (5.46) for the decoherence functional in the same way. We write

$$P^1_{\alpha_1}(t_1) = U^\dagger(t_i, t_1) P^1_{\alpha_1}(t_i) U(t_i, t_1) \tag{5.68}$$

and assign the time dependence to the density matrix,

$$\rho(t_1) = U(t_i, t_1) \rho(t_i) U^\dagger(t_i, t_1) \equiv K^{t_1}_{t_i}[\rho(t_i)]. \tag{5.69}$$

The decoherence functional can then be written as

$$\mathcal{D}(\alpha, \alpha') = \mathrm{tr}\left(P^n_{\alpha_n} U(t_{n-1}, t_n) P^{n-1}_{\alpha_{n-1}} \cdots U(t_1, t_2) P^1_{\alpha_1} \rho(t_1) \right. \tag{5.70}$$

$$\left. P^1_{\alpha'_1} U^\dagger(t_1, t_2) \cdots U^\dagger(t_{n-1}, t_n) P^n_{\alpha'_n} \right).$$

$$= \mathrm{tr}\left(P^n_{\alpha_n} K^{t_n}_{t_{n-1}}(\cdots P^1_{\alpha_1} K^{t_1}_{t_i}[\rho(t_i)] P^1_{\alpha'_1} \cdots) P^n_{\alpha'_n} \right), \tag{5.71}$$

where K is the evolution operator for the "path projected" density matrix, i.e. the density matrix which is obtained from (5.69) by consecutively applying a projector and a unitary time evolution operator. If the symbol $P^k_{\alpha_k}$ without argument is used, it refers to the initial time. The path projected density

matrix must not be confused with the time evolved reduced density matrix
(5.10), since one deals here with concepts referring to the total system.

How, then, does the description of decoherence in terms of the functional
(5.71) correspond to the description in terms of the *reduced* density matrix,
i.e., the density matrix which is obtained by tracing out part of the degrees
of freedom? As mentioned above, one cannot, in general, interpret the action
of the above projection operators on the density operator, which is of the
form $P\rho P$, as a Zwanzig projection. The former map a pure state into a pure
state while a coarse-graining of the Zwanzig type in general leads to a mixed
state. One can, however, consider a Zwanzig projection $\hat{P}$ of the special form

$$\hat{P}\rho = \sum_\alpha P_\alpha \rho P_\alpha, \qquad (5.72)$$

where the $\{P_\alpha\}$ project on a complete set of states, cf. (2.17). We shall show
in the next section that the validity of a master equation for $\hat{P}\rho$ corresponds
to the consistency of the various histories built from the projectors P_α. Here
we want to address the question under which circumstances one can find
projection operators which act on the system *alone*, i.e. which are of the form
$P_{\alpha_k}^k(t_k) = P_{\alpha_k}^{k(S)} \otimes \mathbf{I}$ (with the unit operator $\mathbf{I}$ for the "environment"), and
which describe the behaviour of the reduced density matrix. For simplicity
we shall omit the superscript (S) in the following. Of course, (5.72) cannot be
exactly valid, since initially the reduced density matrix can have nondiagonal
elements (which, however, rapidly vanish because of decoherence).

The physical picture behind this idea is the following. Consider the example of the dust particle in interaction with light, which was extensively
discussed in Chap. 3. Tracing out the photonic degrees of freedom leads to a
reduced density matrix for the particle, which is approximately diagonal in
the position representation with the lower bound on the localisability given
roughly by the de Broglie wavelength of the particle. The position basis is
thus an approximate pointer basis. One would then be interested to find a set
of projection operators which "project out" a consistent history (an approximate classical trajectory described by an effective equation such as (5.64)) for
the dust particle. These projectors would describe a spatial coarse-graining
over a region approximately given by the de Broglie wavelength. It is clear
that this is a direct consequence of the fact that the pointer states are approximate position states. A particular consistent history would then correspond
to one particular branch out of the many branches of the total state, which
decohere with respect to each other (although, of course, all branches are still
present). In the example of the quantum Brownian motion model a spatial
coarse-graining of this type was invoked in the form of "Gaussian slits" by
Dowker and Halliwell (1992). We emphasise again that the justification for
such a procedure comes from environment-induced decoherence.

One would, however, not expect that one can find in the general case
such projection operators which would enable one to write the decoherence

functional solely in terms of reduced quantities. The operators P_α in (5.72) act in general on the whole system. It can, however, be done if the correlations which are set up between the system and its environment affect the dynamics of the system only instantaneously, i.e. if the behaviour of the system is, at least approximately, Markovian. The demonstration is straightforward (Paz and Zurek 1993). One attempts to write down the decoherence functional in a form similar to (5.71), but with respect to reduced quantities only,

$$\mathcal{D}(\alpha, \alpha') = \text{tr}_S \left(P^n_{\alpha_n} \tilde{K}^{t_n}_{t_{n-1}} (\cdots P^1_{\alpha_1} \tilde{K}^{t_1}_{t_i} [\rho_{red}(t_i)] P^1_{\alpha'_1} \cdots) P^n_{\alpha'_n} \right), \qquad (5.73)$$

where $\tilde{K}$ is an operator acting in the system alone. Since the trace in (5.71) is a trace in the whole Hilbert space, $\text{tr} \equiv \text{tr}_S \text{tr}_\mathcal{E}$, it is clear that this reformulation can only be achieved if it is possible to *move* the trace over the environmental degrees of freedom through all the intermediate terms up to the initial density matrix. This is the case if

$$\text{tr}_\mathcal{E} \left(U(t_{n-1}, t_n) P^{n-1}_{\alpha_{n-1}} \cdots \rho(t_1) \cdots P^{n-1}_{\alpha'_{n-1}} U^\dagger(t_{n-1}, t_n) \right)$$
$$= \tilde{K}^{t_n}_{t_{n-1}} P^{n-1}_{\alpha_{n-1}} \text{tr}_\mathcal{E} \left(U(t_{n-2}, t_{n-1}) \cdots \rho(t_1) \right.$$
$$\left. \cdots U^\dagger(t_{n-2}, t_{n-1}) \right) P^{n-1}_{\alpha'_{n-1}}, \qquad (5.74)$$

etc. The condition (5.74) is fulfilled if the correlations between system and environment evolve in a Markovian way. This happens, for example, in the high temperature limit of the quantum Brownian motion model, see the master equation (5.34). Of course, this is also possible for certain non-Markovian systems, but restriction to Markovian systems is made for simplicity.

One recognises from (5.74) how one can generate consistent histories. If the projectors $P^k_{\alpha_k}$ project on the instantaneous eigenstates of the path projected reduced density matrix

$$\rho_S \equiv \tilde{K}^{t_k}_{t_{k-1}} [P^{k-1}_{\alpha_{k-1}} \tilde{K}^{t_{k-1}}_{t_{k-2}} \cdots \rho_{red}(t_1) \cdots P^{k-1}_{\alpha'_{k-1}}] \qquad (5.75)$$

for all $k = 1, \ldots, n$, they commute with this density matrix and thus act on the $P^k_{\alpha'_k}$ on the right-hand side of $\rho_{red}(t_i)$ in (5.73) to yield $\delta_{kk'}$. One gets a diagonal decoherence functional and thus an exactly consistent history. The instantaneous eigenbasis of the density matrix is, of course, the Schmidt basis which we have encountered earlier. The corresponding histories are therefore labelled "Schmidt histories" by Paz and Zurek (1993). These Schmidt histories are in general highly branch dependent, see (5.37), since the eigenstates of the path projected density matrix at time t_k depend on the choice of earlier alternatives, i.e. the choice of earlier projectors. One can thus get a drastically different Schmidt history by choosing a different time sequence (this may already happen if one adds or omits one projection operator at a single instant of time). Schmidt histories are thus in general highly unstable and do *not* automatically describe classical behaviour.

Under which conditions are Schmidt histories stable? One recognises from the above discussion that they are stable if the eigenbasis of the path-projected density matrix is time independent, since in that case the commutation properties remain unaltered under the addition or omission of projectors. This, in turn, is satisfied if the environment distinguishes a robust preferred basis for the system (the pointer basis) with respect to which the reduced density matrix is approximately diagonal. We emphasise again that this is a dynamical, time-dependent process which must not be confused with the instantaneous diagonalisation of the density matrix. In the Markovian case, then, there exist projectors acting on the system state space only and projecting on members of the pointer basis such that the decoherence functional can be written in the form (5.73). However, as we recall from our discussion in Chap. 3, this can only be achieved if the time differences $\Delta t = t_{k+1} - t_k$ are chosen to be bigger than the decoherence time scale, see also Albrecht (1993). Usually this scale is very tiny, see (5.33), but it may be increased in testable situations to demonstrate the instability of Schmidt histories, see the example of an atom coupled to the electromagnetic field, which is discussed in Paz and Zurek (1993).

In the general case the projectors $\{P_{\alpha_k}^k\}$ which mimic the influence of the environment on the system can still be assumed to exist, but they necessarily have to act on the total system which includes the environment. Here again the Schmidt states are of use. As in Chap. 3 we can invoke the Zwanzig projection (assuming for simplicity that it acts on a pure state)

$$\hat{P}_{cl}(|\psi\rangle\langle\psi|) = \sum_k |c_k|^2 |\phi_k\rangle\langle\phi_k| \otimes |\Phi_k\rangle\langle\Phi_k|, \qquad (5.76)$$

where ϕ_k (Φ_k) is an orthonormal basis of the system (environment). The projectors used in the decoherence functional may then be chosen such that

$$P_\alpha(|\psi\rangle\langle\psi|)P_\alpha = |\phi_\alpha\rangle\langle\phi_\alpha| \otimes |\Phi_\alpha\rangle\langle\Phi_\alpha|, \qquad (5.77)$$

where $|\phi_\alpha\rangle \otimes |\Phi_\alpha\rangle \equiv c_k|\phi_k\rangle \otimes |\Phi_k\rangle$, c.f. (5.72) and (2.17).

Of course, one has still to assume that $\Delta t > t_{dec}$. In this case, the reduced density matrix defined by the Zwanzig projection (5.76) satisfies a master equation with a special initial condition, and the pointer basis approximately agrees with the Schmidt basis.

It should be clear from the above discussion that the primary use of consistent histories in the context of decoherence is the *description* of the physical process of decoherence in terms of a temporal sequence of quantum events. The task is to find appropriate projection operators (a choice which is made by hand in the theoretical description of a given process) which project on members of the (fictitious) ensemble as it emerges in the coarse-grained sense through decoherence. As stated above, one can choose infinitely many other projectors which lead to consistent histories in the formal sense, but which do not describe classical behaviour. If one is interested in describing quasi-classical behaviour this would, however, correspond to an unnatural choice which does not reflect the dynamics of the system.

5.4 Consistent Histories, Arrow of Time, and Quantum Gravity

In the definition of the decoherence functional (5.46) the initial instant t_i is distinguished from the final instant t_f at which the trace over the whole Hilbert space is performed – while the choice of an initial density matrix, ρ_i, sets the "preparation" of the system, there appears no final density matrix ρ_f in this expression (in other words, there appears only the "final state of complete ignorance" $\rho_f = 1$).

We shall discuss in this section the extent to which this asymmetry in the expression for the decoherence functional actually reflects the presence of an *arrow of time* in quantum theory or thermodynamics, i.e. a fundamental irreversibility in the formulation of the theory itself. We shall find that the very concept of consistent histories and decoherence functional does not imply any intrinsic time asymmetry, i.e. before any particular choice of boundary conditions has been made. Irreversibility comes into play only through a special initial condition in our Universe.

The central arguments go back to the work of Aharonov *et al.* (1964) who showed that a probability distribution in quantum mechanics (using the collapse interpretation) can be made time-symmetric if the corresponding ensembles are constructed time-symmetrically, i.e., if they are conditioned on an initial as well as on a final condition. This can, of course, only be done for microscopic systems, but not for the macroscopic world. The authors consider the special example of a sequence of one-dimensional projectors (we use our notation) $A(t_i), P^1_{\alpha_1}(t_1) \ldots P^n_{\alpha_n}(t_n), B(t_f)$ which project on non-degenerate states (and are furthermore assumed to be time-independent for simplicity). We denote the initial and final projectors by $A = |a\rangle\langle a|$ and $B = |b\rangle\langle b|$, respectively. In this example, the initial and final states are assumed to be pure. The probability that the corresponding eigenvalues, $\alpha_1 \ldots \alpha_n$, of the intermediate observables are observed, under the *condition* that the initial eigenvalue is a *and* the final eigenvalue is b, is then given by the expression

$$p(\alpha_1, \ldots, \alpha_n \mid a, b) = \frac{p(\alpha_1, \ldots, \alpha_n, b \mid a)}{p(b \mid a)}$$
$$= N \text{tr}\left(B P^n_{\alpha_n} \cdots P^1_{\alpha_1} A P^1_{\alpha_1} \cdots P^n_{\alpha_n}\right), \qquad (5.78)$$

where

$$N^{-1} = \sum_{\alpha_1, \ldots, \alpha_n} \text{tr}\left(B P^n_{\alpha_n} \cdots P^1_{\alpha_1} A P^1_{\alpha_1} \cdots P^n_{\alpha_n}\right), \qquad (5.79)$$

and $p(x \mid y)$ denotes the conditional probability for the occurrence of x given y. This expression is manifestly time symmetric since it can also be written in the form

$$p(\alpha_1, \ldots, \alpha_n \mid a, b) = N\mathrm{tr}\left(AP^1_{\alpha_1} \cdots P^n_{\alpha_n} BP^n_{\alpha_n} \cdots P^1_{\alpha_1}\right). \qquad (5.80)$$

If one asks, for example, for the conditional probability that the outcome of a measurement is α_k, given the outcome α_{k-1}, one finds from these expressions

$$p(\alpha_k \mid \alpha_{k-1}; a, b) = \frac{p(\alpha_k, \alpha_{k-1} \mid a, b)}{p(\alpha_{k-1} \mid a, b)} = \frac{p(\alpha_{k-1} \mid \alpha_k) p(\alpha_k \mid a, b)}{p(\alpha_{k-1} \mid a, b)} =$$

$$\frac{\sum_{\alpha_{k+1}, \ldots, \alpha_n} \mathrm{tr}\left(P^{k-1}_{\alpha_{k-1}} P^k_{\alpha_k} P^{k+1}_{\alpha_{k+1}} \cdots P^n_{\alpha_n} BP^n_{\alpha_n} \cdots P^{k+1}_{\alpha_{k+1}} P^k_{\alpha_k}\right)}{\sum_{\alpha_k, \ldots, \alpha_n} \mathrm{tr}\left(P^{k-1}_{\alpha_{k-1}} P^k_{\alpha_k} P^{k+1}_{\alpha_{k+1}} \cdots P^n_{\alpha_n} BP^n_{\alpha_n} \cdots P^{k+1}_{\alpha_{k+1}} P^k_{\alpha_k}\right)} = |\langle \alpha_{k-1} \mid \alpha_k \rangle|^2$$

$$\times \frac{\sum_{\alpha_{k+1}, \ldots, \alpha_n} \mathrm{tr}\left(P^k_{\alpha_k} P^{k+1}_{\alpha_{k+1}} \cdots P^n_{\alpha_n} BP^n_{\alpha_n} \cdots P^{k+1}_{\alpha_{k+1}}\right)}{\sum_{\alpha_k, \ldots, \alpha_n} \mathrm{tr}\left(P^{k-1}_{\alpha_{k-1}} P^k_{\alpha_k} P^{k+1}_{\alpha_{k+1}} \cdots P^n_{\alpha_n} BP^n_{\alpha_n} \cdots P^{k+1}_{\alpha_{k+1}} P^k_{\alpha_k}\right)}. \qquad (5.81)$$

Note that the result would have been just $|\langle \alpha_{k-1} \mid \alpha_k \rangle|^2$ if no postselection had been made – in accordance with the standard quantum mechanical formula for the transition element (this can immediately be recovered from (5.81) by putting $B = 1$). Note also that the conditional probability (5.81) is independent of the history of measurements *prior* to $P^{k-1}_{\alpha_{k-1}}$. Conversely, if we had asked for the conditional probability to find α_{k-1} *given* α_k (the outcome of the measurement immediately following), we would have obtained the time-reversed expression of (5.81), i.e., an expression depending *only* on the history of measurements prior to $P^{k-1}_{\alpha_{k-1}}$ – the initial selection A as well as the outcomes $P^1_{\alpha_1} \cdots P^{k-2}_{\alpha_{k-2}}$. Of course, in all of the above expressions the complete disappearance of phase relations after "measurements" has been assumed (otherwise the corresponding histories would not be consistent and there would be no justification to assign probabilities to diagonal elements of the decoherence functional). Here, of course, the diagonal elements of the decoherence functional are interpreted in the sense of "Wigner's formula".

Aharonov *et al.* (1964) also make the interesting observation that the unitary evolution according to the Schrödinger equation between two measurements can be interpreted consistently in a time-reversed, "acausal" sense. This is based on the equality of the matrix elements $\langle \phi'_k \mid \phi_n(\Delta t) \rangle = \langle \phi'_k(-\Delta t) \mid \phi_n \rangle$, see also Penrose (1979), and Zeh (2001). If one sticks to the collapse interpretation one usually says that a state unitarily evolves from ϕ_n at $t = 0$ to a state which collapses, due to a measurement, into a state ϕ'_k at time Δt. With the same justification one could, however, say that a state ϕ_n collapses at $t = 0$ into a state which then unitarily evolves into ϕ'_k just before the measurement at time Δt is being made. Such an interpretation can only be excluded if one wishes to extrapolate an intuitive kind of causality to a region where by definition no observation is being made. This is, of course, pure convention. It is *not* a matter of convention if the system is being continuously measured (which requires an arrow of time), since then the corresponding pointer basis must always be regarded as *given*, see Zeh (2001).

In analogy to the time-symmetric expressions (5.78) – (5.80) it has been suggested to define a time-symmetric version of the decoherence functional (5.46), see Gell-Mann and Hartle (1994), Page (1993), and – in a slightly different context – Griffiths (1993):

$$\mathcal{D}(\alpha, \alpha') = \text{tr}\left(\rho_f C_\alpha \rho_i C_{\alpha'}^\dagger\right) / \text{tr}\left(\rho_f \rho_i\right), \qquad (5.82)$$

where $\rho_f = \rho(t_f)$, etc. As above, the insertion of a nontrivial final density matrix drastically changes the probabilities $p(\alpha) = \mathcal{D}(\alpha, \alpha)$ (in the case where they are defined, i.e., if the consistency conditions are fulfilled). Gell-Mann and Hartle (1994) find that the possibility of "medium decoherence" (i.e., the consistency condition (5.48)) is rather limited if such a final density matrix is included. The possibility of effective decoherence for subsystems, however, remains essentially unchanged as long as one is not too close to the final time where conspirative phenomena may seem to arise due to the final condition. As demonstrated by Page (1993), the probabilities

$$p(\alpha) = \mathcal{D}(\alpha, \alpha) = \text{tr}\left(\rho_f C_\alpha \rho_i C_\alpha^\dagger\right) / \text{tr}\left(\rho_f \rho_i\right), \qquad (5.83)$$

if they are defined, are equal to the probabilities for the time-reversed history $\tilde{\alpha}$ (more generally, to the CPT-reversed history), if both the initial and final density matrices are invariant under CPT and commute with each other. It is not necessary to demand that ρ_f is the CPT-reversed density matrix of ρ_i. One would, however, not expect these probabilities to be equal in general.

In the above expression (5.80) it was not necessary to invoke extra conditions on the initial and final projections A and B, since they also have been interchanged by going from (5.78) to the time-reversed expression (5.80). In the present case, however, one deals with distinguished initial and final density matrices (no interchange).

As this discussion demonstrates, there is no built-in time asymmetry in the formulation of quantum theory itself (before any boundary conditions are imposed), either in the standard formulation or in the consistent histories formulation. An arrow of time comes into play only through the choice of a special initial condition, i.e., the choice of a non-trivial initial density matrix ρ_i, but a trivial final density matrix $\rho_f = 1$ like in (5.46). The justification for such a choice must ultimately come from a *cosmological* boundary condition satisfied by our particular Universe (more precisely, by a boundary condition on the wave function of the Universe), see Sect. 4.2, and Zeh (2001).

In the case where such a special initial condition is imposed and a master equation holds for a coarse-grained density matrix, the asymmetric expression

$$p_{\alpha_1 \ldots \alpha_n} = \text{tr}\left(C_\alpha \rho_i C_\alpha^\dagger\right) \qquad (5.84)$$

can be cast into a symmetric *form* (Kiefer and Zeh 1995), even though the histories are asymmetric. Consider a coarse-graining of the specific form (see

(5.76))

$$\hat{P}\rho = \sum_\alpha P_\alpha \rho P_\alpha, \tag{5.85}$$

and *assume* the approximate validity of a master equation (describing a monotonic increase of the corresponding entropy $S = -k_B \mathrm{tr}(\hat{P}\rho \ln \hat{P}\rho)$)

$$\left(\frac{d(\hat{P}\rho)}{dt}\right) \approx \left(\frac{d(\hat{P}\rho)}{dt}\right)_{\mathrm{master}} := \frac{\hat{P}e^{-i\hat{L}\Delta t}\hat{P}\rho - \hat{P}\rho}{\Delta t} \approx -i\hat{P}\hat{L}\hat{P}\rho - \hat{G}_{\mathrm{ret}}\hat{P}\rho, \tag{5.86}$$

with the Liouville operator $\hat{L} := [H, \ldots]$, a positive time scale Δt which is larger than some "relaxation time", and a positive operator $\hat{G}_{\mathrm{ret}}$.

The expression (5.84) for the probability of a consistent history can then be written in the alternative *form*

$$p_{\alpha_1 \ldots \alpha_n} = \left(\mathrm{tr}\left[\rho_f C_\alpha \rho_i (C_\alpha)^\dagger\right]\right)^{1/2}. \tag{5.87}$$

This can be seen as follows. Using (5.84), (5.85), the cyclic property of the trace, and $P_{\alpha_k} P_{\alpha_l} = \delta_{kl} P_{\alpha_k}$, one has

$$\mathrm{tr}\left(\rho_f C_\alpha \rho_i C_\alpha^\dagger\right)$$
$$= \mathrm{tr}\left(e^{-iH(t_n - t_i)}\rho_i e^{iH(t_n - t_i)} P_{\alpha_n} \cdots \rho_i \cdots P_{\alpha_n}\right)$$
$$= \mathrm{tr}\left(P_{\alpha_n}\rho(t_n) P_{\alpha_n} P_{\alpha_n} \cdots \rho_i \cdots P_{\alpha_n}\right)$$
$$= \mathrm{tr}\left(\hat{P}\rho(t_n) P_{\alpha_n} \cdots \rho_i \cdots P_{\alpha_n}\right).$$

In the present case we can write $P_{\alpha_n} = |n\rangle\langle n|$, where $\{|n\rangle\}$ denotes an orthonormal basis in the full Hilbert space (this projection may be thought as projecting on the pointer basis and its relative state, see (5.77)). We find upon performing the trace that the above expression is equal to

$$\langle n \mid \hat{P}\rho(t_n) \mid n\rangle\langle n \mid P_{\alpha_n} \cdots \rho_i \cdots P_{\alpha_n} \mid n\rangle = (p_{\alpha_1 \ldots \alpha_n})(p_{\alpha_1 \ldots \alpha_n}).$$

The symmetric form (5.87) is thus based on the factual asymmetry which is represented by the master equation and caused by the special initial condition. The probabilities would in general be changed drastically, however, by inserting $\rho_f \neq 1$ in this or a similar symmetric expression if the master equation did *not* hold as an approximation through all times from t_i to t_f, see (5.82) and thereafter. It would, in particular, not hold if the entropy with respect to a fixed relevance concept $\hat{P}$ were low both at t_i and t_f, thus forming a thermodynamically time-symmetric (though possibly unitarily evolving) universe.

In the Heisenberg picture (and without any collapse) ρ_i and ρ_f are identical, while the introduction of two *independent* density matrices ρ_i and ρ_f in order to define a symmetric "transition probability" $\mathrm{tr}(\rho_f \rho_i)$ (with or without considering "histories"), would interpret the whole Universe as *one* probabilistic "scattering event." This is, of course, dubious in the absence of external observers.

The above discussion refers to the situation of standard quantum theory where an external time parameter is available. In fact, the very concept of consistent histories seems to break down if no external time exists. And yet, as we have discussed in Sect. 4.2 such a time parameter can only be assumed to exist in an approximate sense — in the case where the gravitational field is in a semiclassical state. Especially in cosmology, where the issue of boundary conditions plays a central role, such an assumption is illusory. In particular, the extent to which an "initial" density matrix can be distinguished from a "final" density matrix in a timeless context remains unclear. Expressions such as (5.82) can thus not be immediately extrapolated to quantum gravity.

How, then, should one interpret consistent histories in quantum gravity? We already mentioned at the beginning of this chapter that one of the main motivations for introducing this concept is, for some authors, its possible application in a "timeless" context. If we first focus on the influence functional of Sect. 5.1, it is obvious that expressions like (5.12) cannot be used in quantum gravity. The reason is the non-validity of a composition law for quantum cosmological path integrals, which in turn is a consequence of the absence of an external time parameter. Quantum cosmological path integrals behave more like energy Green functions than propagators, see Kiefer (1991a). The validity of the composition law, however, lies at the heart of the derivations presented in Sect. 5.1. Only in situations where the gravitational field is semiclassical can one, through a steepest descent approximation in the path integral, arrive at expressions which resemble the standard ones. This was explicitly discussed in Kiefer (1991b) to which we refer the reader for more details. The influence functional $\mathcal{F}$ can then be written only as a function of the endpoint values of the semiclassical gravitational variables, since only the classical paths for these variables enter into the action in the steepest descent approximation. This may seem to be a drawback of this approach, but one must keep in mind that the occurrence of classical properties remains, to the very least, doubtful in situations where gravity is fully quantum and where thus no external spacetime is available.

Nevertheless, there exist attempts to introduce the notion of consistent histories on the level of full quantum gravity, see Hartle (1991, 1995), and Isham (1994b). The only ingredients of such a "quantum mechanics of spacetime" are supposed to be the concepts of fine-grained histories, coarse-graining, and decoherence functional. Starting point is a path integral formulation of quantum gravity, where a matrix element between two three-geometries described by metrics h' and h'' with matter fields χ' and χ'' on them is obtained by integrating over all four-metrics and matter fields which can be matched between these boundary values. Of course, such a formulation is not yet known to exist and can thus be understood only in a very heuristic sense. Each four-metric (plus matter fields) in this sum-over-histories is a special example of a fine-grained history. As explained in Sect. 5.2, coarse-grained histories are obtained by assembling sets of fine-grained histories into classes whose members can no longer be distinguished in the coarse-grained sense. Formally,

$$\langle h'', \chi'' \mid C_\alpha \mid h', \chi' \rangle \equiv \int_\alpha \mathcal{D}g\, \mathcal{D}\phi\, \mu[g, \phi] \exp\left(iS[g, \phi]\right), \qquad (5.88)$$

where g and ϕ denote the four-dimensional metric and a matter field, respectively. The functional $\mu[g, \phi]$ is a complicated expression built from gauge fixing conditions and the Faddeev–Popov determinant. The sum in (5.88) is over all histories that begin at h', χ', end at h'', χ'', and *belong to a class specified by the coarse-graining* (as indicated by the class operator C_α on the left-hand side).

Hartle (1995) then attaches an "initial" wave functional $\psi_{in}[h', \chi']$ and a "final" wave functional $\psi_{fin}[h'', \chi'']$ to the above matrix element, thereby defining

$$\langle \psi_{fin} \mid C_\alpha \mid \psi_{in} \rangle = \psi_{fin}[h'', \chi''] \circ \langle h'', \chi'' \mid C_\alpha \mid h', \chi' \rangle \circ \psi_{in}[h', \chi'], \quad (5.89)$$

where $\circ$ denotes an appropriate inner product on superspace. In the more general case of initial and final density matrices $\sum_i p_{in}^i \psi_{in}^{i*} \psi_{in}^i$, etc., the decoherence functional is proposed to read

$$\mathcal{D}(\alpha, \alpha') = N \sum_{i,j} p_{fin}^i \langle \psi_{fin}^i \mid C_{\alpha'} \mid \psi_{in}^j \rangle \langle \psi_{fin}^i \mid C_\alpha \mid \psi_{in}^j \rangle^* p_{in}^j . \qquad (5.90)$$

For the actual calculation, lattice versions of general relativity such as Regge calculus (where the geometries are discretised) have been suggested.

What is the interpretation of this decoherence functional? In particular, what is the meaning of the initial and final wave functions in this expression? As was explained at length in Sect. 4.2, quantum gravity does not contain an external time parameter which would enable one to distinguish between an "initial" and a "final" state. The quantum cosmological path integral can thus not propagate wave functions with respect to such a time parameter. One might think that it instead propagates the wave function with respect to some intrinsic degree of freedom like the scale factor of the Universe. As shown by Halliwell and Ortiz (1993), this cannot be the case, since the sum-over-histories formulation of quantum gravity does not satisfy a composition law with respect to such an "intrinsic time" either.

One may alternatively try to interpret the path integral in a first step as representing a propagation with respect to a *formal* parameter t according to $\Psi(t) = e^{-iHt}\Psi(0)$, where H is the Wheeler–DeWitt Hamiltonian, followed by a projection onto the corresponding zero-frequency mode, that is, onto a solution of the Wheeler–DeWitt equation $H\Psi = 0$ by integrating over t from

$-\infty$ to $+\infty$. This would amount to writing

$$\psi_{fin}[h'', \chi''] = \int_{-\infty}^{+\infty} dt \int \mathcal{D}h\,\mathcal{D}\chi \, \langle h'', \chi'' \mid h', \chi'; t \rangle \psi_{in}[h', \chi'], \qquad (5.91)$$

with the formal propagator $\langle h'', \chi'' \mid h', \chi'; t \rangle$. It would propagate a formal "initial" wave function ψ_{in} that has to be given on the full configuration space (including *all* three-metrics). Although this function may be *chosen artificially* to contain a factor such as $\delta[h']$, perhaps multiplied by a constant in χ', such an assumption would not represent a natural choice of a boundary for this formal "dynamics." Any information about a direction of propagation in formal time t would be lost by the integration. The meaning of the expression (5.90) thus remains unclear. The analogy with the relativistic particle is misleading since, due to the simple structure of the Klein-Gordon equation, a composition law with respect to the Klein-Gordon inner product *is* valid. For these reasons, the decoherence functional (5.90) is interpreted directly in full superspace, and no "propagation" from spacelike surface to spacelike surface is employed (Hartle, personal communication). Thus, there would be no need for the validity of the composition law.

One can of course apply a semiclassical approximation to the formal path integral (5.88). This leads to the standard result that non-gravitational fields propagate on a spacetime given by the steepest-descent approximation, in accordance with the discussion presented in Kiefer (1991b).

In summary, it still remains unclear how to write down a sensible expression for a decoherence functional in quantum gravity. Only if gravity is in a semiclassical state does it seem possible to define histories in a straightforward way through the eikonal functional $S_0[h]$ which emerges in the steepest descent approximation. With respect to these histories one can then meaningfully investigate a sensible notion of decoherence functional. In our opinion, this indicates that semiclassical gravitational states are a prerequisite for the emergence of classical properties in quantum theory.

We want to conclude this chapter with some brief comments on the relevance of consistent histories beyond decoherence. It is important to stress again the fact that the class of consistent histories is much larger than the class which exhibits quasi-classical behaviour as a consequence of decoherence. We have argued that in the presence of decoherence the language of consistent histories may be conveniently employed to *describe* this situation and, moreover, that it can provide technical advantages. The latter point was demonstrated in detail by our discussion of quantum Brownian motion.

What, then, is the use of consistent histories *beyond* their description of decoherence? Can they, in particular, provide a solution to the measurement problem as it is sometimes claimed? One of the motivations to develop this framework was the attempt to use a classical, realistic, language as much as possible (Griffiths 1984). Since one knows that there are no local hidden variable theories which are consistent with quantum mechanics (and with experiments), the concept of reality of histories must be quite unusual (d'Espagnat

1989, 1995), see also Kent (1997) and the response by Griffiths and Hartle (1998) and, again, Kent (1998), as well as Bassi and Ghirardi (1999). In fact, a realistic interpretation in the usual sense would immediately lead to paradoxes: There exist consistent histories which cannot simultaneously be real. A simple example of a single spin can allow the formulation of consistent histories, where in one history the spin points in a direction $\hat{n}_1$, and in another one in a direction $\hat{n}_2 \neq \hat{n}_1$ at the *same* time. "Reality" therefore only has meaning with respect to some *given* consistent history. In this sense, the "events" in (5.36) are only constructs of thought, not dynamical entities. Omnès (1999) considers the consistent-histories formulation as a "more elaborate version" of the Copenhagen interpretation.

Because it provides a mere description of the behaviour of quantum mechanical systems (e.g., the times in the decoherence functional are "chosen" by the physicist, and the behaviour of histories may change drastically by choosing different times), the notion of consistent histories *cannot*, by itself, explain the dynamical origin of "superselection rules", i.e. the fact that certain superpositions are never observed in Nature – in spite of their natural occurrence in the formalism of quantum mechanics. But this *can* be dynamically understood by the irreversible mechanism of decoherence. By its very nature, decoherence is only approximate, since phase relations are never destroyed (as long as one sticks to ordinary quantum mechanics – alternatives are discussed in Chap. 8). An insistence on *exact* consistency conditions as a fundamental description of the world would be doomed to fail from the very beginning (see in this connection Dowker and Kent 1995, 1996), i.e. it would *not* be sufficient to recover the quasiclassical nature of the observed world. Moreover, the wish to recover a single consistent history as a description of *the* world as it is can only be achieved by going *beyond* quantum mechanics.

6 Superselection Rules and Symmetries

D. Giulini

The task set by the decoherence program is to derive all classical proper-ties of a quantum system as consequences of its (practically) irreversible interaction with the environment. This program gave rise to the notion of *environment-induced superselection rules*, – an expression coined by Zurek (1982) –, which goes back to the very early papers on the subject of deco-herence by Zeh (1970) and Kübler and Zeh (1973, in particular Sect. 4). In our book this concept was already introduced in Chap. 2 and explained in more detail in Chap. 3.1.3. There we saw that the local absence of certain superposition states can be accounted for by the decoherence mechanism. In effect this puts the notion of environment-induced superselection rules in strong analogy to the so-called "exact" (or "kinematical") superselection rules (hence the name chosen for the former), which were postulated much earlier in a kinematical framework and on a more axiomatic basis in the con-text of elementary-particle phenomenology by Wick et al. (1952). (See e.g. Wightman's (1995) review article for a lucid historical account and the rele-vant references.) It seems natural to probe for a deeper theoretical connection between these two notions. In particular, one may ask whether at least some of the kinematical superselection rules may find a natural dynamical explana-tion once the dynamical correlations to the environment are properly taken into account (Zeh 1970), thereby essentially relaxing some of the strict (and in that respect possibly overidealized) kinematical assumptions made in the ef-fective description.[1] According to the more ambitious program formulated in Chap. 2 and Sect. 3.1.3, environmentally induced superselection rules should be the *only* cause for apparent inhibitions of the superposition principle. In particular, according to this program, exact kinematical superselection rules should not exist at all.

On the other hand, there are mathematically rigorous arguments, based on certain hypotheses, for the existence of *exact* superselection rules. These arguments appear also to be physically strong, since the hypotheses on which they rest only involve basic symmetry principles. Such arguments were al-ready given in the first paper by Wick et al. (1952). Here the first symmetry principle invoked was time-reversal invariance, which was shown to lead to the superselection rule between integer and half-integer spin states: the so-called "univalence" superselection rule. It was also mentioned (but not shown

---

[1] For example, symmetries may only be approximately valid. This will be further discussed below.

explicitly) in their paper that the same conclusion could be drawn by using $SO(3)$-rotational invariance. This was later proven by Hegerfeldt et al. (1968). We shall also offer a self-contained proof in Sect. 6.2.3. Wick et al. (1952) also discussed the superselection rule of electric charge, this time using the symmetry under global phase transformations of charged (i.e. complex-valued) fields.

Is it possible to fully reduce this kinematical concept of superselection rules to the aforementioned dynamical one by carefully taking into account all interactions with the environment? We do not know a straight and generally valid answer to this important question. But any attempt in this direction must first start with scrutinizing the formal hypotheses on which the deductions of the exact rules rest. For example, one might argue that there is often no real justification to require the implementability of classically established symmetry groups in quantum theory; compare e.g. pp. 90-91 in (Weinberg 1996). This possibility was already mentioned in footnote 13 of Sect. 3.1.3, where it was pointed out that the replacement of $SO(3)$ by $SU(2)$ would already avoid the implication of a univalence superselection rule. Another example, to be discussed in the present chapter and App. 6, concerns the superselection rule for the overall mass in non-relativistic quantum mechanics. We will argue that it does not even make sense to propose such a superselection rule without first formulating a theory in which mass is a dynamical variable.[2] But doing this, at least in the most straightforward manner, shows that one should not ask for an implementation of the 10-parameter Galilei group, but rather for an implementation of the 11-dimensional so-called Bargmann group which, in contrast, will *not* give rise to a mass superselection rule. In this case, the larger group emerges naturally already on a classical level once dynamical consistency is required.

One might wonder whether these examples stand for general principles. With respect to the first we can ask: can we always replace the symmetry group that gave rise to superselection rules by a "slightly" larger one so as to avoid the unwanted kinematical inhibition of the superposition principle? The well known answer to this question is "yes" and connected with some technical constructions, known under the name of "central extensions", which we will discuss in Sect. 6.2. Given this, we can further ask at what price this is possible, i.e. how many classically unknown symmetries do we have to add? And then: what is the physical meaning of the extra "quantum-symmetries" so acquired? These questions, too, will be addressed.

Somewhat similar to the example of the superselection rule for mass is that for electric charge in QED. Clearly, this field-theoretic case contains extra technical features due to the infinity of degrees of freedom. But there is also a clear analogy visible if one approaches the charge superselection rule

---

[2] This example is given to exclusively illustrate this *theoretical* point and its consequences. Since we know that the Galilei group is not a fundamental symmetry group, no deeper phenomenological significance is attributed to this example.

from a similar viewpoint, i.e. to first seek a formulation in which charge, or rather their conjugate variable (the "global phase"), correspond to a physical degree of freedom. We will argue that such a formulation is not only a matter of taste: it is mandatory in order to allow the classical variational principle to have long-ranged (i.e. charged) field configurations as stationary points. The charge then generates a physically meaningful symmetry by translating the system in the global-phase degree of freedom. This is to be distinguished from a mere gauge transformation to which – by definition – no physical degree is associated. Erroneously conflating these two notions is often responsible for thinking that the charge superselection rule is essentially the implication of gauge invariance (and hence strict), whereas it is rather connected to a physical symmetry plus a locality requirement, of which the latter is the essential ingredient that renders invisible the global degree of freedom. Since the distinction between these different notions of "symmetry" is a point of general importance, which becomes crucial for the argument just outlined, we shall include a detailed discussion of it in Sect. 6.3.

Outside the field of decoherence, the notion of superselection rules (in the exact sense) plays a dominant role in axiomatic and constructive approaches to the theory of quantum fields. See, for example, the book by Bogolubov et al. (1990) and the more recent status report by Buchholz (2000). These are formulated in quite a different language than the environment-induced superselection rules. Since, as outlined above, we intend to study the question of their connection, we also need to clarify the theoretical origin of the kinematical superselection rules and how their existence is expressed in the theory. This necessitates a fairly detailed structural understanding of spaces of observables next to spaces of states, which we shall give in an attempted pedagogical fashion in Sect. 6.1.2. Readers who first wish to learn more about the heuristics of the ideas mentioned above can skip most of the more formal material and directly jump to Sects. 6.3 and 6.4.

6.1 States, Observables, and Superselection Rules

Exact superselection rules are typically derived from kinematical constraints, like symmetry requirements. Such requirements could not even be implemented if we had no means to kinematically move the system in a physically well defined way in its space of states. Here "physically well defined" means that we have an interpretation a priori of at least a preferred class of motions, like e.g. spatial rotations. Formally, motions in the quantum mechanical state space are generated by self-adjoint operators.[3] Those operators which we deem physically well defined in the sense above are called observables.

---

[3] For example, assume we know that the self adjoint operator R generates rotations with respect to a physically specified frame and axis. Then a well defined rotation of a state ψ_0 is obtained by solving $i\hbar\partial_s\psi(s) = R\psi(s)$ with initial condition $\psi(s = 0) = \psi_0$, where s is an angular parameter. The formal solution is $\psi(s) =$

Hence we assume that we know which observable to pick in order to, for example, "rotate the system". We stress that we assume this knowledge to be given independently of any knowledge of how to actually *dynamically* realize this motion by means of interactions with other systems. For example, in order to carry out a real rotation (relative to something) in space over a finite time one needs to exchange angular momentum between the system to be rotated and some other agent which does the rotation. This relational and ultimately dynamical character of concepts which here are introduced only on a kinematical level will be a central theme in later discussions. In this subsection we first wish to understand how the arguments for the existence of exact superselection rules work and therefore restrict attention to kinematical structures and arguments.

6.1.1 Spaces of States

Throughout this chapter pure quantum states of a system S will be identified with rays (one-dimensional subspaces) in a Hilbert space $\mathcal{H}_S$. We assume this Hilbert space to be unambiguously associated with S. For example, $\mathcal{H}_S$ is often taken to be the space of square-integrable functions over the classical configuration space. Two non-zero vectors in $\mathcal{H}_S$ which are proportional represent the same state of S, i.e. are indistinguishable by observations on S. There is a family of two real parameters of different superpositions of two states (they form a two-dimensional sphere), corresponding to the different rays in the plane spanned by them. In Hilbert space this simply corresponds to first taking all linear combinations (which forms a family of two complex parameters) and then canceling the information that resides in the *overall* complex factor, thereby leaving a one-complex- or two-real-parameter family of locally distinguishably different superpositions. Here we understand "locally" as "on the system which is fully described by the Hilbert space in which the superposition is formed".

The identification of rays rather than vectors as descriptors of (local) states will make a difference when one implements (local) symmetries. In the former case, it is sufficient to implement symmetries by ray representations, which is a strictly weaker requirement than implementability by proper linear representations on the vector space. For example, in non-relativistic quantum mechanics, the Galilei group can be implemented by ray- but not by proper representations. The same holds for the rotation group $SO(3)$ on half-integer angular-momentum states. This will be discussed in detail below.

Let us now turn to statistical states. As is well known, a major difference between statistical states in classical and quantum physics is that the latter generically do not allow for an ensemble interpretation (cf. Chap. 2). Formally this is due to the fact that a density matrix does generally not de-

---

$\exp(R/i\hbar)\psi_0$. R is then interpreted as angular momentum with respect to the specified frame and axis.

termine a unique collection of rays, which could then be interpreted as the underlying statistical ensemble. This fact is precisely encoded in the mathematical structure of the space of statistical states. Since this directly relates to superselection rules, we now give a brief review of this.

Let us assume we are content with predictions of probabilities of outcomes of measurements. To this end we accept von Neumann's projection postulate[4] and do not care – at this moment – how it might be derived from fundamental dynamical laws. Granted this, the mathematical task for a statistical state in quantum mechanics is then to provide a probability measure on the set $\mathcal{P}(\mathcal{H})$ of projection operators onto subspaces of the Hilbert space $\mathcal{H}$ (Beltrametti and Casinelli 1981). Here $\mathcal{P}(\mathcal{H})$ is taken to include $P_{\mathcal{H}}$, the projection onto $\mathcal{H}$ (i.e. $P_{\mathcal{H}} = 1$), and P_0, the projection onto the null space (i.e. $P_0 = $ constant map onto $0 \in \mathcal{H}$). By a measure on $\mathcal{P}(\mathcal{H})$ we simply mean a function $E : \mathcal{P}(\mathcal{H}) \to [0,1]$ which (i) maps $P_{\mathcal{H}}$ to 1, (ii) maps P_0 to 0, and (iii) is additive on orthogonal projectors, i.e., if $\{P_m,\ m \in M\}$ is a (countable) set of projectors satisfying $P_m P_n = \delta_{mn} P_m$ (no summation), then $E(\sum_m P_m) = \sum_m E(P_m)$. Any set $\{P_m,\ m \in M\}$ of projectors which is exclusive ($P_m P_n = \delta_{mn} P_m$) and exhaustive ($\sum_m P_m = 1$) kinematically characterize a "measuring device", in the sense of Sect. 2.2, i.e. an operator up to arbitrary reparametrizations of the pointer scale[5]. A probability measure E is therefore sufficient to calculate the probability for any of the possible measurement outcomes. In fact, due to a famous theorem of Gleason (Gleason 1957, Varadarajan 1985, Parthasarathy 1992), *any* probability measure on $\mathcal{P}(\mathcal{H})$, for $\dim(\mathcal{H}) \geq 3$, is of the form

$$E(P) = \mathrm{tr}(\rho P) \,, \tag{6.1}$$

for some self-adjoint, positive[6], trace-class operator ρ satisfying $\mathrm{tr}(\rho) = 1$. Such operators are called "density matrices" or "statistical operators" and will be denoted by $\mathcal{D}(\mathcal{H})$. The subset of density matrices of rank one are projectors onto one-dimensional subspaces and hence correspond to pure states; they will be denoted by $\mathcal{S}(\mathcal{H}) \subset \mathcal{D}(\mathcal{H})$. Density matrices represent a statistical state precisely insofar as they define the probability measure E. Their functional analytic properties are discussed in App. 7.7. Expectation values of self-adjoint operators can be calculated by applying (6.1) to their spectral resolution. This results in the standard formula

$$\langle A \rangle_\rho = \mathrm{tr}(\rho A) =: \omega_\rho(A) \,, \tag{6.2}$$

which, for given ρ, defines a continuous, positive, and normalized linear functional on the space $\mathcal{B}(\mathcal{H})$ of all bounded operators on the Hilbert space $\mathcal{H}$

---

[4] Whenever we speak of von Neumann's projection postulate we shall mean its specification given by Lüders (1951).

[5] Let $p(A, \Delta; \rho)$ be the probability that a measurement of the observable A on the system in state ρ results in a value within $\Delta \subset \mathbb{R}$, then we can express such probabilities for the function $f(A)$ by $p(f(A), \Delta; \rho) = p(A, f^{-1}(\Delta); \rho)$.

[6] An operator A is said to be positive, $A \geq 0$, if $\langle \psi \mid A \mid \psi \rangle \geq 0\ \forall \psi$.

which we labeled ω_ρ. Here positivity refers to the condition $\omega_\rho(A) \geq 0$ whenever A is positive $(A \geq 1)$ and normality to $\omega_\rho(1) = 1$. It is sufficient to restrict ourselves to the formally more convenient *bounded* operators, since unbounded operators A may, without loss of information, always be replaced by $f(A)$, where f is a continuous invertible function with bounded range. Mathematically the pairing expressed in (6.2) leads to a certain duality between states and observables whose precise meaning will also be explained in more detail in App. 7.7. It suggests a generalization of the word "state" by abstraction from its definition (6.2) as follows: Given a subalgebra $\mathcal{A} \subseteq \mathcal{B}(\mathcal{H})$, we call *any* continuous, positive, and normalized functional on $\mathcal{A}$ a "state of $\mathcal{A}$". There is a slight technical subtlety here in that this definition actually comprises more such functionals than could be obtained via (6.2); the reason being that the functionals defined in (6.2) also satisfy $\omega_\rho(\sum_m P_m) = \sum_m \omega_\rho(P_m)$ for any countable family of orthogonal projectors. If we add this condition[7] in the abstract definition we can, however, remove this abundance.

As regards the physical act of the (local) measurement, von Neumann's projection postulate declares that *after* the measurement the (local) state of the system is collapsed to a mixture of states defined by the measurement device, i.e. by the commuting projectors from its spectral resolution and the given probability measure. This mixture may then be taken as an ensemble of states, in just the same way as one interprets the ensemble of gases in Gibbs' formulation of statistical mechanics[8]. But before the measurement, and without knowledge how the initial state was prepared, the quantum mechanical state represented by the density matrix ρ alone could not be identified with such an ensemble. Rather, it defines such an ensemble *for each* measuring device, which is only realized *after* having applied von Neumann's projection postulate. On the other hand, without having applied the projection postulate, we maintain the whole family of "virtual ensembles" which in their totality do not allow an interpretation as a single ensemble. The reason is that two different measuring devices are generally incompatible. Mathematically this is simply expressed by the failure of the projectors defined by the first device to commute with the projectors defined by the second device. This has the drastic consequence that quantum mechanics is not a classical stochastic theory (Koopman 1957, Nelson 1967). (This issue will be taken up again in App. 5.) So, in quantum mechanics, without the information on

---

[7] In the mathematical literature "states" are defined without this condition. If this condition is added the states are called *normal* (not to be confused with *normalized*).

[8] Recall that this ensemble is imaginary insofar as it contains many copies of the gas-system under investigation with only one copy being eventually identified with the real system in the laboratory. The usage of such an ensemble is necessitated by our ignorance of the state which may be due to incomplete knowledge of the initial state and/or incomplete knowledge of the (possibly fundamentally indeterministic) dynamical evolution.

what measurement (or preparation) has been performed, we cannot speak of statistical states and ensembles in the usual sense.

Let us briefly introduce a more precise and intuitive terminology that helps to capture and illustrate this important point in a geometric fashion. Given a set of n points (the "outcomes of measurements" or simply "events") and a probability measure $\{p_m, \ m = 1, .., n \mid p_m \geq 0, \ \sum_{m=1}^{n} p_m = 1\}$, we can identify the space of all such measures with the subset in $\mathbb{R}^n$ whose coordinates are just the p_m's. This is just the definition of an $(n-1)$-simplex. For example, a 1-simplex is a closed interval, a 2-simplex a triangle including boundary edges and vertices, and a 3-simplex a solid tetrahedron with boundary faces, edges, and vertices. A simplex is a special example of a convex set (thought of as subset in a vector space), i.e. a space that for any two of its points also contains the line joining them. In fact, if $\boldsymbol{p}_1$ and $\boldsymbol{p}_2$ are two points of a simplex, then their convex combination $\lambda \boldsymbol{p}_1 + (1 - \lambda)\boldsymbol{p}_2$, where $\lambda \in [0, 1]$, describes their connecting line segment which clearly lies entirely within the simplex. Those points of a convex set which do not lie on a line connecting two other points, i.e. which cannot be written non-trivially as a convex combination, are called extremal. The only extremal points of a simplex are its vertices. It is clearly true for a general convex set that any of its points can be written as convex combination of extremal points. But the crucial and in fact defining property of simplices is that this combination is *unique*. It is precisely this uniqueness that allows an ensemble interpretation.

Now, quite generally, statistical states (classical and quantum) must form a convex set in order to allow for mixtures. The extremal points, which themselves cannot be written as a non-trivial mixture, are the pure states. In classical physics, spaces of statistical states are always simplices so that non pure states can at least correspond to unique mixtures of pure states which are then interpreted as members of an ensemble. At this point we should mention that simplices can also be defined for infinite dimensional convex spaces and refer to (Alfsen 1971) for the technical details.[9]

In contrast, in quantum mechanics the space of density matrices is convex but *not* a simplex. For example, the most general density matrix for a spin-$\frac{1}{2}$ system is given by

$$\rho = \tfrac{1}{2}(\mathbb{1} + \boldsymbol{a} \cdot \boldsymbol{\sigma}), \qquad \|\boldsymbol{a}\| \leq 1 , \tag{6.3}$$

which is uniquely labeled by $\boldsymbol{a}$. This identifies the space of density matrices with the solid ball $B^3 = \{\boldsymbol{a} \in \mathbb{R}^3 \mid \|\boldsymbol{a}\| \leq 1\}$, which is clearly a convex set. Its extremal points are given by the boundary two-sphere whose points

---

[9] We also mention that statistical states in classical statistical mechanics can also be considered as positive linear functionals over a certain algebra of observables, which is abelian. In fact, under certain technical assumptions (namely that the algebra of observables is C^*-algebra with unit) one may prove the general result that the space of states over an algebra of observables is a simplex if and only if the algebra is abelian (Takesaki 1979, Sect. 4.6, Exercise 2.b)

represent pure states. This is obvious from the geometric picture and also easy to derive algebraically: ρ is pure iff[10] it is a matrix of rank one, and this is true iff its determinant is zero (no ρ has rank zero), which in turn is equivalent to $\|a\| = 1$. The geometry also makes it obvious that any non-pure state can be written in uncountably many different ways as the convex combination of pure states. Note from (6.3) that antipodal points on the boundary two-sphere correspond to (projectors onto) orthogonal rays in the two-dimensional Hilbert space. If we pick two orthogonal rays defined by the unit eigenvectors $|e, +\rangle$ and $|e, -\rangle$ of angular momentum in e direction, we obtain the probabilities

$$p(e, \pm) = \langle e, \pm \mid \rho \mid e, \pm \rangle = \tfrac{1}{2}(1 \pm e \cdot a) . \qquad (6.4)$$

This shows that once e is fixed the relevant information contained in ρ consists only in $e \cdot a \in [-1, 1]$. This interval, which is a 1-simplex, represents the space of probability measures on the two orthogonal rays defined by $|e, \pm\rangle$. But ρ clearly contains more information than this, which is only fixed if $p(e, \pm)$ is given for three linearly independent choices of e.[11]

This geometric picture nicely illustrates the point made above, namely why the space of density matrices fails to be a simplex: it is a union of 1-simplices given by the radial intervals in $\mathbb{R}^3$ which make up the ball B^3. The reason why we have to consider more than two orthogonal rays may be seen in the superposition principle, which requires any other ray in their span to also represent a state. The failure of the space of density matrices to be a simplex can hence be traced back to the superposition principle. It is therefore also responsible for the failure of ρ to define a statistical state with ensemble interpretation. This is the point where superselection rules come in: their partial suppression of the superposition principle partially restores the uniqueness of the extremal decomposition of statistical states and hence also partially restores the possibility of an ensemble interpretation. This will be further discussed in Sect. 6.1.3.

In the general case, i.e. in higher dimensional Hilbert spaces, the picture remains the same. This means that a general density matrix ρ can be

---

[10] For the rest of this chapter we use "iff" as abbreviation for "if and only if".

[11] This, by the way, shows explicitly that Gleason's theorem is not valid in two dimensions. Indeed, if we identify the space of projectors onto one-dimensional subspaces of $\mathcal{H} = \mathbb{C}^2$ with S^2, where antipodal points correspond to projectors onto mutually perpendicular subspaces, then any function $E : S^2 \to [0, 1]$ which satisfies $E \circ \alpha = 1 - E$, where α is the antipodal map of S^2, uniquely defines a probability measure on $\mathcal{P}(\mathcal{H})$ after the obvious extension $E(P_{\mathcal{H}}) = 1$ and $E(P_0) = 0$. Note that no other properties of E are required like, for example, continuity. But there are obviously many more such functions E than density matrices (6.3). For our discussion this exception from Gleason's theorem is just an inessential peculiarity.

decomposed into pure states ρ_n forming extremal points:

$$\rho = \sum_n \lambda_n \rho_n, \quad \lambda_n > 0, \quad \sum_n \lambda_n = 1 , \tag{6.5}$$

but this decomposition is generally far from unique; see Hughston et al. (1993) for a general classification of this non-uniqueness. Pure states are density matrices ρ which are rank-one projection operators, i.e. satisfy $\rho^2 = \rho$. Alternatively, in terms of (6.2) we may say that ω_ρ is pure iff there exists a rank one projector P such that $\omega_\rho(P) = 1$. That is, there exists an elementary[12] property which, upon measurement, will be found with certainty.

We have stressed the relational character of the statistical interpretation of ρ, namely that only with reference to a exhaustive set of compatible projectors does ρ define a probability distribution. We also mentioned the physical status of this prescription: whereas the projection operators kinematically specify a measuring device, von Neumann's postulate tells us that *after* performing the measurement with this device the (local) system actually lies in one of the projector eigenspaces, although it does not specify which one. Instead it gives a probability distribution on the projectors according to (6.1). At this level von Neumann's postulate cannot be understood as anything more than a phenomenological prescription for a complicated dynamical process. But granted its effective validity, it allows us – after the measurement – to interpret the probabilities as being distributed over actual properties, in just the same way as we interpret the ensemble of gases in statistical thermodynamics according to Gibbs.

6.1.2 Spaces of Observables

In quantum mechanics, where one starts with a Hilbert space $\mathcal{H}$, it is usually tacitly assumed that any self-adjoint operator corresponds to a physical observable. Self-adjointness is first of all imposed in order to have diagonalizability, or its generalization (in case of continuous spectra) in form of the spectral theorem. The latter is also required in order to define more general than just polynomial functions of operators. Clearly, complex linear combinations of such operators are also admissible and the resulting space may be given the structure of an appropriate algebra (see below and Chap. 7). On the other hand, it is certainly an over-idealization to assume that *all*

---

[12] Here the term "elementary" alludes to the simple fact that the set $\mathcal{P}(\mathcal{H})$ is partially ordered: $P_1 \leq P_2 \Leftrightarrow P_1 P_2 = P_2 P_1 = P_1$. In words: P_1 implies P_2 iff P_1 projects into a subspace of P_2. Clearly, P_0, the projector onto the null space, implies everything and is implied only by itself whereas $P_\mathcal{H}$, the projector onto $\mathcal{H}$, implies only itself and is implied by everything. We call P "elementary" if P is only implied by P_0 and itself. Since we did not mention the lattice structure of $\mathcal{P}(\mathcal{H})$ we also avoided the proper technical term *atomic* used for this property in lattice theory.

self-adjoint operators are realizable by actual physical processes on a given system. Hence one is led to study smaller spaces of operators, whose size and structure encode on a purely kinematical level much of the physical properties of the system under study and its possible interactions with the environment. In this subsection we review and motivate some of the general properties of spaces of observables that directly relate to superselection rules. As usual, we denote by $\mathcal{B}(\mathcal{H})$ the space of bounded operators on $\mathcal{H}$. The antilinear operation of taking the adjoint is denoted by $*$ (rather than $\dagger$), which makes $\mathcal{B}(\mathcal{H})$ a $*$-algebra. Given a subset of bounded operators, $\{A_\lambda \mid \lambda \in \Lambda\} \subset \mathcal{B}(\mathcal{H})$, where Λ is some index set, $\{A_\lambda\}' \subset \mathcal{B}(\mathcal{H})$ denotes the set of all operators in $\mathcal{B}(\mathcal{H})$ which commute with each element A_λ; it is called the *commutant* of the set $\{A_\lambda\}$ in $\mathcal{B}(\mathcal{H})$:

$$\{A_\lambda\}' := \{B \in \mathcal{B}(\mathcal{H}) \mid BA_\lambda = A_\lambda B, \forall \lambda \in \Lambda\}. \tag{6.6}$$

$\{A_\lambda\}'$ will be a $*$-subalgebra of $\mathcal{B}(\mathcal{H})$ if the set $\{A_\lambda\}$ is invariant under $*$. As direct consequence of (6.6) we have for any two subsets $\mathcal{A}, \mathcal{B}$ of $\mathcal{B}(\mathcal{H})$

$$\mathcal{A} \subseteq \mathcal{B} \Rightarrow \mathcal{B}' \subseteq \mathcal{A}' . \tag{6.7}$$

Dirac was the first who spelled out certain rules concerning the spaces of states and observables (Dirac 1930). He introduced the notion of *compatible* (i.e. simultaneously performable) observations, which mathematically are represented by a set of commuting observables, and the notion of a *complete set* of such observables, which is meant to say that there is precisely one state for each set of simultaneous "eigenvalues". Starting from the hypothesis that states are faithfully represented by rays, Dirac argued that a complete set of such mutually compatible observables existed. But, technically speaking, this only makes sense if all the observables in question have purely discrete spectra. In the general case one has to proceed differently: we heuristically define Dirac's requirement as the statement, that "there exists at least one complete set of mutually compatible observables" and show how it can be rephrased mathematically so as to apply to all cases. In doing this we essentially follow the exposition by Jauch (1960). To develop some intuition we will first describe some of the consequences of Dirac's requirement in the most simplest case: a finite dimensional Hilbert space.

Gaining Intuition in Finite Dimensions

Let $\mathcal{H}$ be an n-dimensional complex Hilbert-space, then $\mathcal{B}(\mathcal{H})$ is the algebra of complex $n \times n$ matrices. Physical observables are represented by hermitean matrices in $\mathcal{B}(\mathcal{H})$, but we will explicitly *not* assume the converse, namely that *all* hermitean matrices correspond to physical observables. Rather we assume that the physical observables are somehow given to us by some set $\mathcal{S}$ of hermitean matrices. This set does not form an algebra, since taking products and complex linear combinations does not preserve hermiticity. But for mathematical reasons it would be convenient to have such an algebraic structure, and just work with the algebra $\mathcal{O}$ generated by this set, called the *algebra of observables*. (Note the usual abuse of language, since only the hermitean elements in $\mathcal{O}$ are observables.) But for this replacement of $\mathcal{S}$ by $\mathcal{O}$ to be allowed, $\mathcal{S}$ must be a set of hermitean matrices which is uniquely determined by $\mathcal{O}$, for otherwise we cannot reconstruct the set $\mathcal{S}$ from its artificial enlargement $\mathcal{O}$. To grant us this mathematical convenience we assume that $\mathcal{S}$ was already maximal, i.e. that $\mathcal{S}$ already contains all the hermitean matrices that it generates. But we stress that there seems to be no obvious reason why in a particular practical situation the set of physically realizable observables should be maximal in this sense.

We may choose a set $\{O_1, \ldots O_m\}$ of hermitean generators of $\mathcal{O}$. Then $\mathcal{O}$ may be thought of as the set of all complex polynomials in these (generally non-commuting) matrices. But note that we need not consider higher powers than $(n-1)$ of each O_i, since each complex $n \times n$ matrix O is a zero of its own characteristic polynomial p_O, i.e. satisfies $p_O(O) = 0$, by the theorem of Cayley-Hamilton. Since this polynomial is of order n, O^n can be re-expressed by a polynomial in O of order at most $(n-1)$. For example, the $*$-algebra generated by a single hermitean matrix O can be identified with the set of all polynomials of degree at most $(n-1)$ and whose multiplication law is as usual, followed by the procedure of reducing all powers n and higher of O via $p_O(O) = 0$.

Now let $\{A_1, \cdots, A_m\} =: \{A_i\}$ be a complete set of mutually commuting observables. It is not difficult to show that commutativity implies the existence of an observable A and of polynomials p_i, $i = 1, \cdots, m$ such that $A_i = p_i(A)$; see e.g. Isham (1995) for a simple proof. Hence the algebra generated by $\{A_i\}$, which we call $\mathcal{A}$, must be a subalgebra of the n-dimensional algebra of polynomials of degree at most $n-1$ in A. We will see below that the requirement of completeness implies that it must, in fact, be identical to the latter. This algebra is abelian, which is equivalently expressed by saying that $\mathcal{A}$ is contained in its commutant (compare (6.6)):

$$\mathcal{A} \subseteq \mathcal{A}' \qquad \boxed{\text{``}\mathcal{A}\text{ is abelian''}} \,. \tag{6.8}$$

Now comes the requirement of "completeness", meaning that simultaneous eigenspaces for the observables in $\mathcal{A}$ are non-degenerate, i.e. one-dimensional, so that a set of "quantum numbers" uniquely fixes a ray in Hilbert space. In terms of A it is easy to see that it is equivalent to the condition that A has a simple spectrum (i.e. the eigenvalues are pairwise distinct). This has the following consequence: let B be an observable that commutes with A, then B is also a function of A, i.e. $p_B(B) = A$ for some polynomial p_B. The proof is simple: we simultaneously diagonalize A and B with eigenvalues α_a and β_a, $a = 1, \cdots, n$. We wish to find a polynomial of degree $n-1$ such that $p_B(\alpha_a) = \beta_a$. Writing $p_B(x) = c_{n-1}x^{n-1} + \cdots + c_0$, this leads to a system of n linear equations ($\alpha_a^b := b^{\text{th}}$ power of α_a)

$$\sum_{b=0}^{n-1} \alpha_a^b c_b = \beta_a, \qquad \text{for } a = 1, \cdots, n \tag{6.9}$$

for the n unknowns ($c_0, \cdots, c_{n-1}$). Its determinant is the so-called "Vandermonde determinant" for the n tuple ($\alpha_1, \cdots, \alpha_n$):

$$\det\{\alpha_a^b\} = \prod_{a<b} (\alpha_a - \alpha_b), \tag{6.10}$$

which is non-zero iff the spectrum of A is simple. This implies that every observable that commutes with $\mathcal{A}$ is already contained in $\mathcal{A}$. (It follows from this that the algebra generated by $\{A_i\}$ is equal to, and not just a subalgebra of, the algebra generated by $\{A\}$, as stated above.) Since a $*$-algebra is generated by its self-adjoint elements (observables), $\mathcal{A}$ cannot be properly enlarged as abelian $*$-algebra by adding more commuting generators. In other words, $\mathcal{A}$ is *maximal*. Since $\mathcal{A}'$ is a $*$-algebra, this can be equivalently expressed by

$$\mathcal{A}' \subseteq \mathcal{A} \quad \boxed{\text{"$\mathcal{A}$ is maximal"}} . \tag{6.11}$$

By derivation, equations (6.8) and (6.11) together are equivalent to Dirac's condition in finite dimensions. But the point of this reformulation is that it will remain applicable in the general case. Hence Dirac's condition can now be stated in the following form, first given by Jauch (1960): the algebra of observables $\mathcal{O}$ contains a maximal abelian $*$-subalgebra $\mathcal{A} \subseteq \mathcal{O}$, i.e.

$$\boxed{\text{Dirac's requirement, } 1^{\text{st}} \text{ version: } \exists\, \mathcal{A} \subseteq \mathcal{O} \text{ satisfying } \mathcal{A} = \mathcal{A}'} . \tag{6.12}$$

This may seem as if Dirac's requirement could be expressed in purely algebraic terms. But this is deceptive, since the very notion of "commutant" (compare (6.6)) makes direct reference to the Hilbert space $\mathcal{H}$ through $\mathcal{B}(\mathcal{H})$. Without further qualification the term "maximal" always means maximal *in* $\mathcal{B}(\mathcal{H})$.[13]

This reference to $\mathcal{H}$ can be further clarified by yet another equivalent statement of Dirac's requirement. Since $\mathcal{A}$ consists of polynomials in the observable A, which has a simple spectrum, the following is true: there exists a vector $|g\rangle \in \mathcal{H}$, such that for *any* vector $\phi \in \mathcal{H}$ there exists a polynomial p_ϕ such that

$$p_\phi(A)|g\rangle = |\phi\rangle . \tag{6.13}$$

---

[13] The condition for an abelian $\mathcal{A} \subseteq \mathcal{O}$ to be maximal in $\mathcal{O}$ would be $\mathcal{A} = \mathcal{A}' \cap \mathcal{O}$. Such abelian subalgebras *always* exist (by Zorn's Lemma), in contrast to those $\mathcal{A} \subseteq \mathcal{O}$ which satisfy the stronger condition to be maximal in the ambient algebra $\mathcal{B}(\mathcal{H})$, which need not exist for a given $\mathcal{O} \subset \mathcal{B}(\mathcal{H})$.

Such a vector $|g\rangle$ is called a generating or *cyclic vector* for $\mathcal{A}$ in $\mathcal{H}$. The proof is again very simple: let $\{\phi_1, \cdots, \phi_n\}$ be the pairwise distinct, non-zero eigenvectors of A (with any normalization); then choose

$$|g\rangle = \sum_{i=1}^{n} |\phi_i\rangle. \tag{6.14}$$

Equation (6.13) now defines again a system of n linear equations for the n coefficients $c_{n-1}, \cdots, c_0$ of the polynomial p_ϕ, whose determinant is again the "Vandermonde determinant" (6.10) for the n eigenvalues $\alpha_1, \cdots, \alpha_n$ of A. Conversely, if A had an eigenvalue, say α_1, with eigenspace $\mathcal{H}_1$ of two or higher dimensions, then such a cyclic $|g\rangle$ cannot exist. To see this, suppose to the contrary that it did exist, and let $|\phi_1^\perp\rangle \in \mathcal{H}_1$ be orthogonal to the projection of $|g\rangle$ into $\mathcal{H}_1$. Then $\langle\phi_1^\perp|p(A)g\rangle = 0$ for all polynomials p. Thus $|\phi_1^\perp\rangle$ is unreachable, contradicting our initial assumption. Hence a simple spectrum of A is equivalent to the existence of a cyclic vector.

The General Case

In infinite dimensions we have to care more about technical issues, mainly concerning the topology on the space of observables, since there are many inequivalent ways to generalize the finite dimensional case. The natural choice for the topology is the so-called "weak topology", which is characterized by declaring that a sequence $\{A_i\}$ of observables converges to the observable A iff the sequence $\langle\phi|A_i|\psi\rangle$ of complex numbers converges to $\langle\phi|A|\psi\rangle$ for all $|\phi\rangle, |\psi\rangle \in \mathcal{H}$. The justification for this choice is obvious from a physicists point of view, for whom all the relevant information about an observable is exclusively contained in its matrix elements. Accordingly one requires that the algebra of observables is weakly closed (i.e. closed in the weak topology). Such a weakly closed $*$-subalgebra of $\mathcal{B}(\mathcal{H})$ is called a W^*- or von-Neumann-algebra (we shall use the first name for brevity).

A crucial and extremely convenient fact is, that the weak topology is fully encoded in the operation of taking the commutant (see (6.6)) in the following sense: let $\{A_\lambda\}$, $\lambda \in \Lambda$ (some index set) be any $*$-invariant subset of $\mathcal{B}(\mathcal{H})$, then $\{A_\lambda\}'$ is automatically weakly closed (see Jauch (1960) p. 716 for a simple proof) and hence a W^*-algebra. Moreover, the weak closure of a $*$-algebra $\mathcal{A} \subseteq \mathcal{B}(\mathcal{H})$ is just given by $\mathcal{A}''$ (the commutant of the commutant). Hence we can characterize a W^*-algebra purely in terms of commutants: $\mathcal{A}$ is W^* iff $\mathcal{A} = \mathcal{A}''$.

This allows to generalize the notion of "algebra generated by observables": Let $\{O_\lambda\}$ be a set of self-adjoint elements in $\mathcal{B}(\mathcal{H})$, then $\mathcal{O} := \{O_\lambda\}''$ is called the (W^*-) algebra generated by this set. This definition is natural since $\{O_\lambda\}''$ is easily seen to be the smallest W^*-algebra containing $\{O_\lambda\}$, for if $\{O_\lambda\} \subseteq \mathcal{B} \subseteq \mathcal{O}$ for some W^*-algebra $\mathcal{B}$, then taking the commutant twice yields $\mathcal{B} = \mathcal{O}$.[14]

Now we see that Dirac's requirement in the form (6.12) directly translates to the general case if all algebras involved (i.e. $\mathcal{A}$ and $\mathcal{O}$) are understood as W^*-algebras. Now we also know what a "complete set of (bounded) commuting observables" is, namely a set $\{A_\lambda\} \subseteq \mathcal{B}(\mathcal{H})$ whose generated W^*-algebra $\mathcal{A} := \{A_\lambda\}''$ is maximal abelian: $\mathcal{A} = \mathcal{A}'$. This latter condition is again equivalent to the existence of a cyclic vector $|g\rangle \in \mathcal{H}$ for $\mathcal{A}$, where in infinite dimensions the definition of cyclic is that $\{A|g\rangle \mid A \in \mathcal{A}\}$ is *dense* in (rather than equal to) $\mathcal{H}$. It is also still true that there is an observable A such that all A_λ are functions (in an appropriate sense, not just polynomials of course) of A (von Neumann 1931). But since the spectrum of A may be (partially) continuous, there is no direct interpretation of a "simple" spectrum as in finite dimensions. Rather, one now *defines* simplicity of the spectrum of A by the existence of a cyclic vector for $\mathcal{A} := \{A\}''$.

Now we come to our final reformulation of Dirac's condition. Namely, looking at (6.12), we may ask whether we can reformulate the existence of such a maximal abelian $\mathcal{A}$ purely intrinsically in terms of the algebra of observables $\mathcal{O}$ (we already noted that (6.12) makes reference to $\mathcal{H}$ via $\mathcal{B}(\mathcal{H})$). This is indeed possible. First recall the definition of the *center*, $\mathcal{O}^c$, of an algebra $\mathcal{O}$:

$$\mathcal{O}^c := \{A \in \mathcal{O} \mid AB = BA, \quad \forall B \in \mathcal{O}\} . \tag{6.15}$$

Now, we have $\mathcal{A} \subseteq \mathcal{O} \Rightarrow \mathcal{O}' \subseteq \mathcal{A}' = \mathcal{A} \subseteq \mathcal{O}$, hence $\mathcal{O}' \subseteq \mathcal{O}$. Since $\mathcal{O} = \mathcal{O}''$, the last condition is equivalent to saying that $\mathcal{O}'$ is abelian ($\mathcal{O}' \subseteq \mathcal{O}''$), or to saying that $\mathcal{O}'$ is equal to the center $\mathcal{O}^c$ of $\mathcal{O}$, since by (6.6) and (6.15) the center can also be written as $\mathcal{O}^c = \mathcal{O} \cap \mathcal{O}'$. Now, conversely, it is shown in Jauch and Misra (1961) that an abelian $\mathcal{O}'$ implies the existence of a maximal abelian $\mathcal{A} \subseteq \mathcal{O}$. Hence we have the following alternative formulation of Dirac's requirement, first spelled out, independently of (6.12), by Wightman (1959) who called it the "hypothesis of commutative superselection rules", for reasons to become clear soon:

$$\boxed{\text{Dirac's requirement, 2}^{\text{nd}} \text{ version: } \mathcal{O}' \text{ is abelian}} . \tag{6.16}$$

There are several interesting ways to interpret this condition. From its derivation we know that it is equivalent to the existence of a maximal abelian $\mathcal{A} \subseteq \mathcal{O}$. But we can in fact make an apparently stronger statement, which also relates to the earlier footnote 13, namely: (6.16) is equivalent to the condition, that *any* abelian $\mathcal{A} \subseteq \mathcal{O}$ that is maximal in $\mathcal{O}$, i.e. satisfies $\mathcal{A} = \mathcal{A}' \cap \mathcal{O}$, is also maximal in $\mathcal{B}(\mathcal{H})$.[15]

---

[14] Note: for any $M \subseteq \mathcal{B}(\mathcal{H})$, definition (6.6) immediately yields $M \subseteq M''$ and hence $M' \supseteq M'''$ (by (6.7)). But also $M' \subseteq M'''$ (by replacing $M \to M'$); therefore $M' = M'''$ for any $M \subseteq \mathcal{B}(\mathcal{H})$.

[15] Proof: We need to show that ($\mathcal{O}'$ abelian) $\Leftrightarrow$ ($\mathcal{A} = \mathcal{A}' \cap \mathcal{O} \Rightarrow \mathcal{A} = \mathcal{A}'$). "$\Rightarrow$": $\mathcal{O}'$ abelian implies $\mathcal{O}' \subseteq \mathcal{O}'' = \mathcal{O}$ and $\mathcal{A} \subseteq \mathcal{O}$ implies $\mathcal{O}' \subseteq \mathcal{A}'$, so that $\mathcal{O}' \subseteq \mathcal{A}' \cap \mathcal{O}$.

How General Is Dirac's Requirement?

Another way to understand (6.16) is via its limitations on *gauge-symmetries*. To see this, we mention that any W^*-algebra is generated by its unitary elements. Hence $\mathcal{O}'$ is generated by a set $\{U_\lambda\}$ of unitary operators. Each U_λ commutes with *all* observables and therefore generates a one-parameter group of gauge-transformations. Condition (6.16) is then equivalent to saying that the total gauge group, which is generated by all U_λ, is *abelian*. Note also that an abelian $\mathcal{O}'$ implies that the gauge-algebra, $\{U_\lambda\}'' = \mathcal{O}'$, is contained in the observables, $\mathcal{O}' \subseteq \mathcal{O}'' = \mathcal{O}$, so that $\mathcal{O}' = \mathcal{O}^c$. From this one infers the following general statement:

> Dirac's requirement implies that gauge- and sectorial structures are fully determined by the center $\mathcal{O}^c$ of the algebra of observables $\mathcal{O}$.   (6.17)

One might think that the restriction to abelian gauge groups is too restrictive. To take a proper quantum-mechanical example, consider the system of $n > 2$ identical spinless particles with n-particle Hilbert space $\mathcal{H} = L^2(\mathbb{R}^{3n})$ on which the permutation group $G = S_n$ of n objects acts in the obvious way (permuting the coordinate triples associated to different particles) by unitary operators $U(g)$. That these particles are identical means that observables must commute with each $U(g)$. Without further restrictions on observables one would thus define $\mathcal{O} := \{U(g), \ g \in G\}'$. Hence $\mathcal{O}'$ is the W^*-algebra generated by all $U(g)$, which is clearly non-abelian, thus violating (6.16). But does this generally imply that general particle statistics cannot be described in a quantum-mechanical setting which fulfills Dirac's requirement? The answer to this question is "no". Let us explain why.

If we decompose $\mathcal{H}$ according to the unitary, irreducible representations of G we obtain (Galindo et al. 1962)

$$\mathcal{H} = \bigoplus_{i=1}^{p(n)} \mathcal{H}_i \ , \tag{6.18}$$

where i labels the $p(n)$ inequivalent, unitary, irreducible representations D_i of G of dimension d_i. Each $\mathcal{H}_i$ has the structure $\mathcal{H}_i \cong \mathbb{C}^{d_i} \otimes \tilde{\mathcal{H}}_i$, where S_n acts irreducibly via D_i on $\mathbb{C}^{d_i}$ and trivially on $\tilde{\mathcal{H}}_i$, whereas $\mathcal{O}$ acts irreducibly via some *-representation π_i on $\tilde{\mathcal{H}}_i$ and trivially on $\mathbb{C}^{d_i}$. π_i and π_j are inequivalent if $i \neq j$. Hence we see that $\mathcal{H}_i$ furnishes an irreducible representation for $\mathcal{O}$, iff $d_i = 1$, i.e. for the Bose and Fermi sectors only. Pure states from these sectors are just the rays in the corresponding $\mathcal{H}_i$. In contrast, for $d_i > 1$, given a non-zero vector $\phi \in \tilde{\mathcal{H}}_i$, all non-zero vectors in the d_i-dimensional

---

Hence $\mathcal{A} = \mathcal{A}' \cap \mathcal{O}$ implies $\mathcal{O}' \subseteq \mathcal{A}$, which implies $\mathcal{A}' \subseteq \mathcal{O}'' = \mathcal{O}$, and hence $\mathcal{A} = \mathcal{A}'$. "$\Leftarrow$": $(\mathcal{A} = \mathcal{A}' \cap \mathcal{O} \Rightarrow \mathcal{A} = \mathcal{A}')$ is equivalent to $\mathcal{A}' \subseteq \mathcal{O}$, which implies $\mathcal{O}' \subseteq \mathcal{A}'' = \mathcal{A}$ and hence that $\mathcal{O}'$ is abelian.

subspace $\mathbb{C}^{d_i} \otimes \phi \subset \mathcal{H}_i$ define the *same* pure state, i.e. the same expectation-value-functional on $\mathcal{O}$. Furthermore, a vector in $\mathcal{H}_i \cong \mathbb{C}^{d_i} \otimes \tilde{\mathcal{H}}_i$ which is not a pure tensor product defines a non-pure state, since the restriction of $O \in \mathcal{O}$ to $\mathcal{H}_i$ is of the form $\mathbb{1} \otimes \tilde{O}$, which means that a vector in $\mathcal{H}_i$ defines a state given by the reduced density matrix obtained by tracing over the left (i.e. $\mathbb{C}^{d_i}$) state space. But from elementary quantum mechanics we know that the resulting state is pure iff the vector in $\mathcal{H}_i$ was a pure tensor product (i.e. of rank one). Hence in those $\mathcal{H}_i$ where $d_i > 1$ not all vectors correspond to pure states, and those which do, represent pure states in a redundant fashion by higher dimensional subspaces, sometimes called "generalized rays" in the older literature on parastatistics (Messiah and Greenberg 1964).

However, the factors $\mathbb{C}^{d_i}$ are completely redundant as far as physical information is concerned, which is already fully encoded in the irreducible representations π_i of $\mathcal{O}$ on $\tilde{\mathcal{H}}_i$; no further physical information is contained in d_i-fold repetitions of π_i. Hence, without loss of information, we may regard the truncated Hilbert space

$$\tilde{\mathcal{H}} := \bigoplus_{i=1}^{p(n)} \tilde{\mathcal{H}}_i . \tag{6.19}$$

This procedure has also been called "elimination of the generalized ray" in the older literature on parastatistics (Hartle and Taylor 1969 ; see also Giulini 1995b for a more recent discussion). Since every pure state in $\mathcal{H}$ is also contained in $\tilde{\mathcal{H}}$, just without repetition, these two sets are called "phenomenological equivalent" in the literature on QFT; see e.g. Chap. 6.1.C of (Bogolubov et al. 1990). The point is that pure states are now *faithfully* labeled by rays in the $\tilde{\mathcal{H}}_i$ and that $\mathcal{O}'$ – where the commutant is now taken in $\mathcal{B}(\tilde{\mathcal{H}})$ rather than $\mathcal{B}(\mathcal{H})$ – is generated by $\mathbb{1}$ and the $p(n)$ (commuting!) projectors into the $\tilde{\mathcal{H}}_i$'s. Hence Dirac's requirement is satisfied. But clearly the original gauge group has now no action on $\tilde{\mathcal{H}}$ anymore; however, there is also no physical reason why one should keep it.[16] It served to define $\mathcal{O}$, but then only its irreducible representations π_i are of interest. Only a residual action of the center of G still exists, but the gauge group generated by the projectors into the $\tilde{\mathcal{H}}_i$ consists in fact of the continuous group of $p(n)$ copies of U(1), one global phase change for each sector. Its meaning is simply to induce the separation into the different sectors $(\tilde{\mathcal{H}}_i, \pi_i)$, and that in accordance with Dirac's requirement.

To sum up: we have seen that even if a theory is initially formulated via non-abelian gauge groups, we can give it a physically equivalent formulation

---

[16] In (Galindo et al. 1962) Dirac's requirement together with the *additional* requirement that the physical Hilbert space must carry an action of the gauge group has been used to "prove" the absence of parastatistics. The second requirement is, in fact, hidden in many such "proofs' to the same effect. In our opinion there seems to be no physical reason to accept this additional requirement and hence any such "proof"; this is further argued for in (Hartle and Taylor 1969 and Giulini 1995b).

that has at most a residual abelian gauge group left and hence obeys Dirac's requirement. Hence the "obvious" counterexamples to Dirac's requirement turn out to be harmless. This is generally true in quantum mechanics, but in quantum field theory there are genuine possibilities to violate Dirac's condition which we will ignore here.[17]

6.1.3 Superselection Rules

Mathematically the simplest way to state the existence of kinematical superselection rules is via the algebra of observables $\mathcal{O}$: one says that superselection rules exist iff $\mathcal{O}$ has a nontrivial center, that is, iff $\mathcal{O}^c$ is strictly larger than the identity.[18] The self-adjoint elements in $\mathcal{O}^c$ are then called the superselection observables. If $\mathcal{O} = \mathcal{B}(\mathcal{H})$, i.e. if any self-adjoint operator is an observable (recall that the self-adjoint elements in $\mathcal{B}(\mathcal{H})$ generate $\mathcal{B}(\mathcal{H})$), no superselection rules exist. But if the space of observables is smaller than that, superselection rules generically appear. Projection operators in $\mathcal{O}^c$ then define subspaces in $\mathcal{H}$ which are invariant under the action of $\mathcal{O}$. More precisely, for W^*-algebras we can simultaneously diagonalize all observables in $\mathcal{O}^c$ and accordingly write $\mathcal{H}$ as direct integral over the real line with respect to some (Lebesgue-Stieltjes) measure $d\mu(\lambda)$:

$$\mathcal{H} = \int_{\mathbb{R}}^{\oplus} d\mu(\lambda)\, \mathcal{H}(\lambda). \qquad (6.20)$$

This is to be understood in the following sense: each $\mathcal{H}(\lambda)$ has an inner product $(\cdot,\cdot)_\lambda$ and elements $f \in \mathcal{H}$ are given by maps $f : \lambda \mapsto f(\lambda) \in \mathcal{H}_\lambda$

---

[17] An abelian $\mathcal{O}'$ implies that $\mathcal{O}$ is a W^*-algebra of type I (see Chap. 8 of Dixmier 1981), whereas truly infinite systems in QFT are often described by type III algebras.

[18] In "Algebraic Quantum Mechanics" one associates to each quantum system an abstract C^*-algebra, $\mathcal{C}$, which is thought of as being the mathematical object that fully characterizes the system in isolation, i.e. its intrinsic or "ontic" properties. But this is not yet what we call the algebra of observables. This latter algebra is not uniquely determined by the former. It is obtained by studying faithful representations of $\mathcal{C}$ in some Hilbert space $\mathcal{H}$, such that $\mathcal{C}$ can be identified with some subalgebra of $\mathcal{B}(\mathcal{H})$. This is usually done by choosing a reference state (positive linear functional) on $\mathcal{C}$ and performing the GNS construction. Closing $\mathcal{C}$ in the weaker topology of $\mathcal{B}(\mathcal{H})$ makes it a W^*-algebra which corresponds to our $\mathcal{O}$. Technically speaking, $\mathcal{O}$ is a von Neumann algebra which properly contains an embedded copy of $\mathcal{C}$. The added observables, i.e. those in $\mathcal{O} - \mathcal{C}$, do not describe intrinsic but *contextual* properties, dependent on the particular properties of the chosen representation (i.e. reference state in the GNS construction). For example, it may happen that $\mathcal{O}$ has a non-trivial center whereas $\mathcal{C}$ has a trivial center. In this case, the superselection rules described by $\mathcal{O}$ are called "contextual". See Primas (2000) and literature therein for a more extended exposition of this approach.

with finite inner product in $\mathcal{H}$; the latter is given by

$$(f, g) = \int_{\mathbb{R}} (f(\lambda), g(\lambda))_\lambda \, d\mu(\lambda) . \tag{6.21}$$

Operators in $\mathcal{O}$ respect this decomposition in the sense that each $A \in \mathcal{O}$ defines a map $A : \lambda \mapsto A(\lambda) \in \mathcal{B}(\mathcal{H}_\lambda)$ such that

$$(f, Ag) = \int_{\mathbb{R}} (f(\lambda), A(\lambda)g(\lambda))_\lambda \, d\mu(\lambda) . \tag{6.22}$$

For $A \in \mathcal{O}^c$ we have $A(\lambda) = \phi(\lambda) \, \mathbb{1}_\lambda$, where ϕ is a complex-valued function and $\mathbb{1}_\lambda$ is the unit operator in $\mathcal{H}_\lambda$. Moreover, given that Dirac's requirement holds, the set $\{A(\lambda) \mid A \in \mathcal{O}\}$ (λ fixed) acts irreducibly on $\mathcal{H}(\lambda)$.[19] Hence, provided that Dirac's requirement is satisfied, (6.20) is the generally valid statement of how $\mathcal{H}$ decomposes into irreducible components with respect to $\mathcal{O}$. But note that $\mathcal{H}_\lambda$ will not be a linear subspace of $\mathcal{H}$ if $d\mu$ has continuous support around λ. Rather, any non-null, closed, $\mathcal{O}$-invariant subspace of $\mathcal{H}$ is of the form

$$H_\Delta = \int_\Delta \mathcal{H}_\lambda \, d\mu(\lambda) , \tag{6.23}$$

where $\Delta \subseteq \mathbb{R}$ is a $d\mu$-measurable subset of non-zero measure. The corresponding projection operators P_Δ are just given by $P_\Delta f(\lambda) = \chi_\Delta(\lambda) f(\lambda)$, where $\chi_\Delta(\lambda) = 1$ for $\lambda \in \Delta$ and $= 0$ for $\lambda \notin \Delta$ (the characteristic function of Δ). Accordingly, two vectors f and g in $\mathcal{H}$ are said to be *disjoint*, or *separated by a superselection rule*, if and only if their component functions $f(\lambda)$ and $g(\lambda)$ have disjoint support on $\mathbb{R}$ (up to measure-zero sets). Note how much stronger disjointness is than mere orthogonality: matrix elements (6.22) between disjoint vectors vanish *for all $A \in \mathcal{O}$*.

Let us now simplify the discussion by restricting attention to the special case of purely discrete spectra for all superselection observables. The integral (6.20) then turns into a countable sum

$$\mathcal{H} = \bigoplus_{m \in M} \mathcal{H}_m , \tag{6.24}$$

where each direct summand $\mathcal{H}_m$ is now a proper subspace of $\mathcal{H}$, representing $\mathcal{O}$ irreducibly. The countable family of projection operators $P_m : \mathcal{H} \to \mathcal{H}_m$ generates the center of $\mathcal{O}$. It is now easy to see that the following statements concerning observables and states are equivalent (Bogolubov et al. 1990):

(i) Any density matrix ρ satisfies the identity

$$\rho = \sum_m P_m \rho P_m . \tag{6.25}$$

---

[19] This irreducibility statement indeed depends crucially on the fulfillment of Dirac's requirement. In general, the $O(\lambda)$'s will act irreducibly on $\mathcal{H}(\lambda)$ for each λ, iff $\mathcal{O}^c$ is maximal abelian *in* $\mathcal{O}'$, i.e. iff $\mathcal{O}^c = (\mathcal{O}^c)' \cap \mathcal{O}'$. But we already saw that (6.16) also implies $\mathcal{O}^c = \mathcal{O}'$ and hence the latter condition.

(ii) Any $A \in \mathcal{O}$ satisfies the identity

$$A = \sum_m P_m A P_m .$$ (6.26)

(iii) Any density matrix commutes with all P_m, $m \in M$.
(iv) The algebra of observables is given by the commutant of $\{P_m \mid m \in M\}$:

$$\mathcal{O} = \{A \in \mathcal{B}(\mathcal{H}) \mid P_m A = A P_m, \ \forall m \in M\} .$$ (6.27)

For given ρ, let $M' \subseteq M$ be the subset of all indices for which $\mathrm{tr}(P_m \rho) \neq 0$. Define $\lambda_m := \mathrm{tr}(P_m \rho)$ and $\rho_m := P_m \rho P_m / \lambda_m$ for $m \in M'$. Then (6.25) reads

$$\rho = \sum_{m \in M'} \lambda_m \rho_m ,$$ (6.28)

which is a non-trivial convex combination if M' contains more than one element. Hence the only pure ρ are projectors onto rays which lie entirely within one superselection sector $\mathcal{H}_m$, i.e. onto rays in the union

$$\bigcup_{m \in M} \mathcal{H}_m .$$ (6.29)

Rays in $\mathcal{H}$ not contained in the union (6.29) define a special kind of non-pure states: their decomposition (6.28) represents them *uniquely* as convex combination of extremal points (pure states). Hence rays in $\mathcal{H}$ can be properly thought of as mixtures, in the sense of an ensemble whose components are uniquely determined. In the general case the decomposition (6.28) according to the central projectors P_m is still unique, but further decompositions of the ρ_m now again suffer from the typical non-uniqueness discussed at the beginning. Hence one can say that in the presence of superselection rules the system acquires some classical properties, namely those represented by the central observables.

As regards the physical meaning of superselection rules the obvious question is how to *physically* motivate the size of the algebra of observables. As already emphasized in Chap. 2, if we start the theoretical description with a given Hilbert space $\mathcal{H}$ we are naturally led to consider all self-adjoint operators as observables, which identifies $\mathcal{O}$ with $\mathcal{B}(\mathcal{H})$. Hence $\mathcal{O}^c = \{\mathbb{1}\}$ and there are no superselection rules in the exact sense just described.

But this does not hinder us to model phenomenologically existing superselection rules by *choosing* $\mathcal{O}$ much smaller, so as to only contain physically realizable observables. With respect to this choice superselection rules acquire the mathematical status of theorems, though the ultimate measure of their physical status must be based on the physical hypotheses that (implicitly) underlie the "choice" of $\mathcal{O}$. In general, this choice should be understood as partial answer to a dynamical question, since answering what part $\mathcal{O}$ of

$\mathcal{B}(\mathcal{H})$ is appropriate for the phenomenological description of a specific situation must eventually involve considerations on the measurement theory of the system involving all its environmental couplings. Realistic couplings then often lead to "environment-induced superselection rules". Mathematically this means that in the course of time evolution, certain states develop *approximately* into disjoint states. Exact disjointness will only be possible in the $t \mapsto \infty$ limit (Primas 2000). The precise meanings of convergence that are involved here are further discussed in Chap. 7. In this respect we recall Wightman's and Glance's (1989) summary, which originally was meant to apply to molecular physics only, but which also seems appropriate in a much wider context: "The theoretical results [concerning superselection rules] currently available fall into two categories: rigorous results on approximate models and approximate results on realistic models".

But what about superselection rules whose origin is of kinematical rather than dynamical nature, which can be rigorously derived from symmetry requirements? At first glance there thus seems to be no chance for a physical explanation along the lines of a dynamical, and hence emerging, character of such superselection rules. This will be analyzed in more detail in the next section.

6.2 Symmetries and Superselection Rules

The main motivation for postulating a group G as symmetry group in quantum theory is often drawn from the fact that G is already a symmetry group of the underlying classical theory. Since, in our formulation, quantum mechanical states are given by rays, G need only act by ray representations on the Hilbert space. This simple mathematical fact now offers the possibility to "derive" superselection rules. The argument, which we will present in greater detail below, can be summarized as follows: given two inequivalent ray representations on two Hilbert spaces $\mathcal{H}_1$ and $\mathcal{H}_2$, one shows that there is no ray representation on $\mathcal{H} = \mathcal{H}_1 \oplus \mathcal{H}_2$ which restricts to the given ray-subrepresentations on $\mathcal{H}_{1,2}$. If, by hypothesis, one requires that the physical state space carries an action of G, one must conclude that non-trivial linear combinations of vectors in $\mathcal{H}_1$ with vectors in $\mathcal{H}_2$ cannot represent physical states – end of the argument. Below we shall explicitly give such a derivation for the case of the univalence superselection rule, where $G = SO(3)$, and also discuss the case where G is the Galilei group in App. 6. We end with the remark that the conclusions just drawn delicately depend on the choice of G. In fact, as will be explained below, it turns out that instead of G we could always employ a "slightly" larger group $\bar{G}$, which acts on the Hilbert space by proper representations rather than ray-representations. Replacing G by $\bar{G}$ would then invalidate the argument given above so that no superselection rules result in connection with $\bar{G}$. Without further input one then seems to face the following two alternatives: either to work with quantum theories which exhibit just the classical symmetries and which, as a consequence of these symmetries, have superselection sectors in good agreement with present phenomenological knowledge, or to adopt theories with larger symmetry groups, which do not induce superselection sectors. These two alternatives in fact exist in the case of spatial rotations[20] (see next section). On the other hand, the case of Galilei invariance provides an example where the extension $\hat{G}$ (the Bargmann group) of the Galilei group G naturally appears as symmetry group already on the classical level. Therefore derivations based on G seem not convincing to us. This is further discussed in App. 6 and in more detail in Giulini (1996).

6.2.1 Symmetries

In this section we investigate in more detail the restriction of the superposition principle due to the (postulate of the) existence of a symmetry group G. We shall use the notion of "symmetry" in a formal way, as structure preserving maps (automorphisms), like e.g. canonical transformations in classical mechanics. We do not imply special compatibility with dynamical evolution (implying constants of motion) and shall also not consider gauge symmetries. These issues will be addressed later.

Projective Representations

Let us denote our space of statistical states by $\mathcal{D}$, which is a convex subset of the space of density matrices $\mathcal{D}(\mathcal{H})$. We explicitly allow $\mathcal{D}$ to be strictly smaller than $\mathcal{D}(\mathcal{H})$, since $\mathcal{D}$ may be restricted by condition (6.25) if superselection rules exist (assuming their discreteness for simplicity). Pure states in $\mathcal{D}$ are denoted by $\mathcal{S}$ so that $\mathcal{S} = \mathcal{D} \cap \mathcal{S}(\mathcal{H})$.

A symmetry of $\mathcal{D}$ is now simply defined by a bijective mapping $\alpha : \mathcal{D} \to \mathcal{D}$ which preserves the convex structure:

$$\alpha(\lambda_1 \rho_1 + \lambda_2 \rho_2) = \lambda_1 \alpha(\rho_1) + \lambda_2 \alpha(\rho_2), \quad \lambda_{1,2} \geq 0, \quad \lambda_1 + \lambda_2 = 1 . \quad (6.30)$$

No additional constraint is necessary – see Kadison (1965), Hunziker (1972) and Bogolubov et al. (1990). We call such maps automorphisms of $\mathcal{D}$. They preserves the metric that derives from the trace norm (7.87):

$$\|\alpha(\rho_1) - \alpha(\rho_2)\|_1 = \|\rho_1 - \rho_2\|_1 . \quad (6.31)$$

Automorphisms map pure states onto pure states and after restriction to $\mathcal{S}$ (6.31) reduces to the conservation of the transition probability; see App. 7.7.

---

[20] This is a point where, because we take states to be rays rather than vectors, our conclusions differ from those of Chaps. 2 and 3. If states are truly represented by vectors we cannot allow for ray representation so that the spin-$\frac{1}{2}$ states cannot be said to transform under $SO(3)$ but only under $SU(2)$.

Hence the definition of symmetry given above is just a straightforward generalization of Wigner's definition to general statistical states. Wigner's theorem (Wigner 1931 pp 251-254, Bargmann 1964) now implies that any symmetry transformation can be written as

$$\alpha(\rho) = U\rho U^\dagger, \qquad (6.32)$$

with a unitary or anti-unitary operator U on $\mathcal{H}$. The symmetry transformation (6.32) of the state space is equivalent to a linear transformation β of the algebra of observables $\mathcal{O}$

$$\beta(A) := U^\dagger A U , \qquad (6.33)$$

such that

$$\mathrm{tr}(\alpha(\rho)A) = \mathrm{tr}(\rho\beta(A)) . \qquad (6.34)$$

If the symmetries of a system form a group G then we have a family α_g, $g \in G$, of automorphisms of $\mathcal{D}$, which satisfy

$$\alpha_{g_1} \circ \alpha_{g_2} = \alpha_{g_1 g_2} \quad \text{for all } g_1, g_2 \in G , \qquad (6.35)$$

and the identity $e \in G$ is represented by the trivial automorphism

$$\alpha_e(\rho) = \rho \quad \text{for all } \rho \in \mathcal{D} . \qquad (6.36)$$

We emphasize that symmetries are already realized on the set of pure states since each automorphism α_g maps $\mathcal{S}$ onto itself. But in an algebraic context, which uses the duality (6.2), it is more natural to use $\mathcal{D}$. From Wigner's theorem we know that for each α_g there is a unitary operator $U(g)$ satisfying (6.32). Applying (6.35) twice then shows that $U(g_1)U(g_2)U^\dagger(g_1 g_2)$ commutes with each $\rho \in \mathcal{D}$. This implies that

$$U(g_1)U(g_2)U^\dagger(g_1 g_2) = \sum_m \omega_m(g_1, g_2)\, P_m , \qquad (6.37)$$

where P_m is the projector onto the m-th superselection sector and where $\omega_m : G \times G \to \mathrm{U}(1)$ (here $\mathrm{U}(1)$ denote the complex numbers of unit modulus) is an arbitrary function for each m (in particular, the ω_m need a priori not be continuous). Let us restrict attention to one sector (dropping the index m). Then we have

$$U(g_1)U(g_2) = \omega(g_1, g_2)\, U(g_1 g_2) . \qquad (6.38)$$

Evaluating the law of associativity, $\big(U(g_1)U(g_2)\big)U(g_3) = U(g_1)\big(U(g_2)U(g_3)\big)$, yields the so-called "cocycle condition" for ω:

$$\omega(g_1, g_2)\, \omega(g_1 g_2, g_3) = \omega(g_1, g_2 g_3)\, \omega(g_2, g_3) , \qquad (6.39)$$

for all $g_1, g_2, g_3 \in G$. One says that ω is a $\mathrm{U}(1)$-valued 2-cocycle on G. The cocycle condition together with (6.38) defines what is called a *projective unitary representation* and the function ω is called its *multiplier*.

Note that $U(g)$ is far from being uniquely determined. First, without loss of generality, we may require $U(e) = \mathbb{1}$ so that $\omega(1, g) = \omega(g, 1) = 1$. Second, given any function $\phi : G \to U(1)$ with $\phi(e) = 1$ we may redefine U by

$$U(g) \mapsto U'(g) := \phi(g)U(g) . \tag{6.40}$$

This results in the following redefinition of ω:

$$\omega(g_1, g_2) \mapsto \omega'(g_1, g_2) := \omega(g_1, g_2)\phi(g_1 g_2)(\phi(g_1))^{-1}(\phi(g_2))^{-1} . \tag{6.41}$$

Two multipliers are said to be *similar* iff (6.41) holds for some function $\phi(g)$. Note that simple multiplication in $U(1)$ makes the set of multipliers an abelian group in which the multipliers similar to the identity form a (necessarily invariant) subgroup. The quotient is called $H^2(G, U(1))$, the second cohomology group of G with coefficients in $U(1)$, and faithfully labels the similarity classes of multipliers (similarity is clearly an equivalence relation).

Central Extensions

It turns out that any projective unitary representations of a group G can always be turned into a proper unitary representation of some other, larger group $\bar{G}$, which is a central extension of G. This will become an important point in our discussion of the derivations of superselection rules from symmetry requirements.

Let G be a group and A an abelian group. A group $\bar{G}$ is called an *extension* G by A, iff $\bar{G}$ contains A as normal subgroup such that the quotient $\bar{G}/A$ is isomorphic to G. It is called a *central extension* iff A lies in the center of $\bar{G}$. We refer to A as the "extending" and to G as the "extended" group. We summarize this relation between the three groups $A, G, \bar{G}$ by the following short sequence of groups and maps:

$$A \xrightarrow{\ \ i\ \ } \bar{G} \xrightarrow{\ \ p\ \ } G \tag{6.42}$$

which is to be read in the following way: i is an (injective) homomorphism that embeds A in $\bar{G}$ and p is the canonical (surjective) projection homomorphism that maps $\bar{G}$ to its quotient $\bar{G}/A = G$. In the sequel we shall often simply speak of A when we actually mean its isomorphic image $i(A)$ in $\bar{G}$.

A concrete and generic way to construct such extensions is as follows: let $\omega : G \times G \to A$ be a function that satisfies $\omega(e, g) = \omega(g, e) = 1$ (here 1 denotes the identity in A) and also the cocycle condition (6.39). This just slightly generalizes the notion of multipliers mentioned above, by allowing them to take values in any abelian group A (not just $U(1)$). We call them A-multipliers unless $A = U(1)$. Next we consider the set $A \times G$ and make it into a group by defining the following multiplication law:

$$(a_1, g_1)(a_2, g_2) := (a_1 a_2 \omega(g_1, g_2) , g_1 g_2) . \tag{6.43}$$

The unique unit element is $(1, e)$ and the inverse of a general element (a, g) is uniquely given by[21]

$$(a, g)^{-1} = \left(1/(a \; \omega(g^{-1}, g)) \; , \; g^{-1}\right) \; . \tag{6.44}$$

Finally, associativity of the multiplication (6.43) directly follows from (6.39), which completes showing that $\bar{G}$ is indeed a group. Moreover, the set of elements $\{(a, e) \mid a \in A\}$ forms a subgroup in the center of $\bar{G}$ which is isomorphic to A. In contrast, the set $\{(1, g) \mid g \in G\}$ is generally *not* a subgroup, since multiplying two such elements results in a non-trivial first entry according to (6.43). But the quotient $\bar{G}/A$ is clearly isomorphic to G.

Two central A-extensions $\bar{G}$ and $\bar{G}'$ are said to be equivalent, iff there is a group isomorphism $\bar{G} \to \bar{G}'$ whose restriction to the extending group A is the identity and which also induced the identity on the quotient group G. Note that these latter conditions make this notion of equivalence much stronger than just mere isomorphicity of groups. We will encounter a relevant example for this in our discussion of the Galilei group. In fact, extensions are equivalent iff they correspond to similar multipliers. To see this, consider two A-multipliers ω and ω' which are similar, meaning that (6.41) holds for some function $\phi : G \to A$ with $\phi(e) = 1$. Then we have the equivalence $I : \bar{G} \to \bar{G}'$, $(a, g) \mapsto (\phi(g)a \, , \, g)$. This map restricts to the identity on $A \subset \bar{G}$ and obviously induces the identity on the quotient G. Conversely, it can be shown that *any* central A-extension arises via (6.43) for some multiplier ω and that equivalent extensions correspond to similar multipliers; see (Raghunathan 1994). Hence the equivalence classes of central A-extensions of G are in bijective correspondence to the similarity classes of A-valued multipliers on G. Like for the special case $A = U(1)$, the latter have a natural abelian group structure denoted by $H^2(G, A)$, called the second cohomology class of G with coefficients in A.

Now, given a unitary projective representation U of G on $\mathcal{H}$ with multiplier ω, we can easily define a proper unitary representation $\bar{U}$ of the corresponding central $U(1)$-extension $\bar{G}$ on $\mathcal{H}$ whose restriction to G is just U. This is achieved by

$$\bar{U}(z, g) := z \cdot U(g) \; , \tag{6.45}$$

where $z \in U(1)$.

This works for a single multiplier, like for irreducible projective representations, where the sum (6.37) necessarily contains only one term. In more general cases with many multipliers, i.e. more than one term in (6.37), it is still true that for any projective representation U of G there exits a pair $\bar{G}, \bar{U}$ with the above mentioned properties, but $\bar{G}$ will now be some central A-extension of G with A generically different form $U(1)$. In fact, there are even so-called *universal* central extensions where one and the same group $\bar{G}$

---

[21] To see that the left and right inverses are identical use $\omega(g^{-1}, g) = \omega(g, g^{-1})$, which follows from (6.39) for $g_1 = g_2^{-1} = g_3 = g$.

has the property that for *any* projective unitary representation U of G there is a proper unitary representation $\bar{U}$ of $\bar{G}$ which restricts to U on $G \subset \bar{G}$. Let us explain this important concept in more detail.

Universal Central Extensions

A central extension $\hat{G}$ of G by A is called a *universal central extension* of G, summarized by

$$\hat{A} \xrightarrow{\ \hat{i}\ } \hat{G} \xrightarrow{\ \hat{p}\ } G \tag{6.46}$$

iff for any central U(1)-extension, summarized by (6.42) with $A = \mathrm{U}(1)$, there are homomorphisms $h : \hat{A} \to A$ and $H : \hat{G} \to \bar{G}$ such that $i \circ h = H \circ \hat{i}$ and $\hat{p} = p \circ H$. This can be summarized by the following diagram of groups and maps in which compositions of maps that form a square commute:

$$\begin{array}{ccccc}
\hat{A} & \xrightarrow{\ \hat{i}\ } & \hat{G} & \xrightarrow{\ \hat{p}\ } & G \\
\downarrow{\scriptstyle h} & & \downarrow{\scriptstyle H} & & \downarrow{\scriptstyle \mathrm{id}} \\
\mathrm{U}(1) & \xrightarrow{\ i\ } & \bar{G} & \xrightarrow{\ p\ } & G
\end{array} \tag{6.47}$$

The crucial fact is now the following

Theorem 1. *Any group G possesses a universal central extension $\hat{G}$.*

For a proof see e.g. Raghunathan (1994). Now, given a unitary projective representation U of G and a proper unitary representation $\bar{U}$ of some central extension $\bar{G}$ which induces U on $G = \bar{G}/A$, we also get a proper unitary representation $\hat{U}$ of $\hat{G}$ which induces U on $G = \mathcal{U}(G)/\hat{A}$: just set $\hat{U} := U \circ H$. That this is a proper unitary representation is obvious and one also infers immediately from the commutativity of right square in (6.47) that it induces the same projective representation on G as $\bar{U}$, which is U.

Theorem 1 makes no assertion concerning uniqueness of $\hat{G}$, and, in fact, it is generally highly non unique. Abstract constructions without detailed knowledge of G usually end up with very large groups $\hat{G}$, i.e. very large extending groups $\hat{A}$. An obvious question is whether there is, in some sense, a minimal choice $\hat{G}_{\mathrm{min}}$. Physically this question is of interest since it asks for the fewest additional symmetries to be added to G in order to always guarantee implementability by unitary representations in quantum mechanics. Assuming $\hat{G}_{\mathrm{min}}$ was unique, one could, with some justification, call it the *quantum symmetry group* associated to G (no relation to quantum groups!) and address the central elements in the extending $\hat{A}_{\mathrm{min}} \subset \hat{G}_{\mathrm{min}}$ as *pure quantum symmetries* which do not have classical counterparts. Note again that the group of *classical symmetries*, G, then forms a quotient but not a subgroup of the quantum symmetry group.

In general, it is not easy to determine $\hat{G}_{\mathrm{min}}$, if at all possible. However, results concerning existence, uniqueness, as well as explicit constructions of

$\hat{G}_{\min}$ are known for many cases of physical interest. They show that $\hat{G}_{\min}$ is often surprisingly small. An example of such a result is the following

Theorem 2. *Let G be a connected semisimple Lie group. Then its minimal universal extension is given by its universal cover.*

See again Raghunathan (1994) for a proof. Note that the universal cover of G is an extension of G by a *discrete* group, namely the fundamental group of G. A well known example is $G = SO(3)$, whose universal cover group is $SU(2)$, which is an extension of $SO(3)$ by its fundamental group $\mathbb{Z}_2$ of just two elements. The latter is generated by a rotation through $360°$ which in this example makes up the only "pure quantum symmetry" (there is only one rotation by $360°$ in $SU(2)$). A related example covered by Theorem 2 is the proper orthochronous Lorentz group, whose universal cover is the group $SL(2, \mathbb{C})$. There exist generalizations of Theorem 2 which include more general groups, notably the group of euclidean motions or the inhomogeneous Lorentz group, which are not semisimple but still perfect (generated by commutators).

In general the complexity of $\hat{G}$ (which is an algebraic concept) is not at all controlled by the fundamental group of G (which is a topological concept). This is only true if the Lie algebra of G has only trivial central extensions (like in the semisimple case) so that all possibilities to extend G reside in the global topological features of G. An example where no topology is involved is the planar translation group, $\mathbb{R}^2$, which plays a rôle in solid state physics. It is clearly its own universal cover, but its minimal universal central extension is the 3-dimensional Heisenberg group. Hence, in this case, the "quantum symmetry group" gains a full continuum worth of symmetries over and above the "classical" ones. See Divakaran (1994) for a more comprehensive discussion.

6.2.2 Superselection Rules from Symmetries

The central mathematical input into the derivation of superselection rules from symmetry requirements is the following

Theorem 3. *Two projective unitary representations can be subrepresentations of a (reducible) projective representation iff they have similar multipliers.*

To see this, let U' and U'' be the two projective representations on $\mathcal{H}'$ and $\mathcal{H}''$ with multipliers ω' and ω'' respectively. Now set $U = U' \oplus U''$ on $\mathcal{H}' \oplus \mathcal{H}''$ and consider

$$U(g_1)U(g_2) = \omega'(g_1, g_2)\, U'(g_1g_2) \oplus \omega''(g_1, g_2)\, U''(g_1g_2) \,. \tag{6.48}$$

U is again a projective representation iff for all $g_1, g_2 \in G$ the right hand side of (6.48) can be written in the form

$$\omega(g_1, g_2)\, (U'(g_1g_2) \oplus U''(g_1g_2)) \tag{6.49}$$

with some multiplier ω, for some choice of multipliers ω' and ω'' within their similarity class. But this is clearly possible iff the multipliers ω' and ω'' are similar, which proves the claim.

The existence of superselection rules is now a direct consequence of Theorem 3, *given the hypothesis of a definite symmetry group G*. The argument is just this: if the physical state space $\mathcal{D}$ admits G as symmetry group, then each G-invariant subspace of states also carries a projective representation since it satisfies the conditions for Wigner's theorem. Hence, by Theorem 3, $\mathcal{D}$ cannot contain pure states corresponding to superpositions of pure states from G-invariant subspaces whose projective subrepresentations have non-similar multipliers. In this sense G-symmetry inhibits their superposition.

A classification of non-similar multipliers of a group G therefore yields superselection rules for a theory. We already mentioned that this classification problem is mathematically equivalent to the computation of the second cohomology group $H^2(G, \mathrm{U}(1))$ of G with coefficients in $\mathrm{U}(1)$ or, equivalently, to the construction of the universal central extensions of G, which for $G = SO(3)$ is simply its universal-cover group $SU(2)$. For the Galilei group, which is not semisimple, the universal central extension is not its universal-cover group and follows from a more complicated construction, see App. 6. The investigations of these problems started with Bargmann's paper (1954). A systematic treatment with respect to superselection rules has been given by Divakaran (1994) and Raghunathan (1994).

The best known example of a superselection rule derived from a symmetry requirement is the univalence superselection rule between integer and half-integer angular momentum spaces. It emerges from the *postulate* that the rotation group $SO(3)$ is a physical symmetry. This superselection rule persists if $SO(3)$ is embedded into any larger group, such as the Galilei group or the Poincaré group. The second cohomology group $H^2(SO(3), \mathrm{U}(1))$ can be shown to be isomorphic to $\mathbb{Z}_2$ so that we obtain two superselection sectors, which can be identified with the subspaces of integer and of half-integer angular-momentum states. To illustrate this result, we give an explicit derivation of the univalence superselection rule.

6.2.3 An Example: The Univalence Superselection Rule

Let $\mathcal{H}_{\frac{1}{2}} = \mathbb{C}^2$ be the spin-$\frac{1}{2}$ representation space and $\mathcal{H}_1 = \mathbb{C}^3$ the spin-1 representation space of $SO(3)$. We will show that it is impossible to define a representation α_g, $g \in SO(3)$, on the state space $\mathcal{D}(\mathcal{H}_{\frac{1}{2}} \oplus \mathcal{H}_1)$ which restricts to the given subrepresentations on the spin-$\frac{1}{2}$ and spin-1 subspaces. Following the arguments given above we have to prove that $SO(3)$ is realized in both subspaces with projective representations whose multipliers are non-similar.

Generically speaking, to prove that two multipliers are non-similar is a very difficult task, since we have to show that no function $\phi : G \to \mathrm{U}(1)$ exists which satisfies (6.41). Note that except for the condition $\phi(e) = 1$

the function ϕ is completely arbitrary; this generality renders proofs of its nonexistence even more difficult. In this respect the following observation is very helpful[22]: Let G' be a subgroup of G; then multipliers on G become multipliers on G' after restriction to G'. Moreover, if two multipliers ω and ω' on G are similar they are also similar as multipliers on G'; just restrict the function ϕ in (6.41) to G'. Hence, conversely, in order to prove non-similarity of multipliers on G it is sufficient to prove that they are non-similar after restriction to some convenient subgroup $G' \subset G$.

In case of $G = SO(3)$ we make use of this observation by choosing as subgroup G' the abelian, four-element group $D_2 \subset SO(3)$, the so-called dihedral group of index 2, whose non-trivial elements are the three $180°$-rotations about three perpendicular axes, which we call C_a, $a = 1, 2, 3$; the identity is called E. The group is characterized by

$$C_a^2 = E , \quad C_a C_b = C_b C_a = C_c \quad a, b, c \text{ cyclic} , \tag{6.50}$$

which shows that, as abstract group, it is isomorphic to $\mathbb{Z}_2 \times \mathbb{Z}_2$. The defining representation via 3×3 matrices will be called the spin-1 representation of D_2. It is a faithful representation with trivial multiplier (i.e. $\omega \equiv 1$). The spin-$\frac{1}{2}$ representation of this group is defined as usual by

$$C_a \rightarrow I_a := \exp\left(-i\pi \frac{\sigma_a}{2}\right) = -i\sigma_a , \quad E \rightarrow \mathbb{1} , \tag{6.51}$$

where the σ_a are the Pauli matrices and $\mathbb{1}$ is the unit 2×2-matrix. A trivial calculation gives:

$$I_a^2 = -\mathbb{1} , \quad I_a I_b = -I_b I_a = I_c , \quad a, b, c \text{ cyclic} , \tag{6.52}$$

showing that (6.51) is a projective representation of D_2 with multiplier

$$\omega(C_a, C_a) = -1 , \quad \omega(C_a, C_b) = -\omega(C_b, C_a) = 1 , \quad a, b, c \text{ cyclic} . \tag{6.53}$$

Since we think of D_2 as being embedded in $SO(3)$ we do not consider anti-unitary representations. Now let us change the phases of this representation by some function ϕ, trying to achieve trivial multipliers. From (6.41) we have

$$\omega'(C_a, C_b) = \omega(C_a, C_b)\phi(C_a C_b)(\phi(C_a))^{-1}(\phi(C_b))^{-1} . \tag{6.54}$$

Now comes the crucial observation: Since D_2 is abelian we have $\phi(C_a C_b) = \phi(C_b C_a)$, so that the second part of (6.53) implies that the redefined multiplier is again antisymmetric: $\omega'(C_a, C_b) = -\omega'(C_b, C_a)$, for $a \neq b$. Hence no choice of ϕ can achieve $\omega' = 1$. This shows that the multipliers are not similar on D_2 and therefore also not similar on $SO(3)$. This completes the proof of the univalence superselection rule from the hypothesis of $SO(3)$-symmetry.

---

[22] I learned this and the following proof for the non-similarity of the multipliers for the spin-$\frac{1}{2}$ and spin-1 projective representations of $SO(3)$ from J. Kupsch

Let us illustrate Theorem 3 by showing explicitly that already the $D_2 \subset SO(3)$ symmetry forbids a definite phase relation between spin-$\frac{1}{2}$ and spin-1 states. For this we take a spin-$\frac{1}{2}$ vector $\Psi \in \mathcal{H}_{\frac{1}{2}} = \mathbb{C}^2$ and a spin-1 vector $F \in \mathcal{H}_1 = \mathbb{C}^3$. The direct-sum representation of D_2 is

$$U(C_a)(\Psi + F) = \phi_{\frac{1}{2}}(C_a)\, I_a\Psi + \phi_1(C_a)\, C_a F\ , \tag{6.55}$$

where we allowed arbitrary phase functions ϕ_j, $j = \frac{1}{2}, 1$. If D_2 were a symmetry, then (for a, b, c cyclic)

$$U(C_a)U(C_b)(\Psi + F) = \phi_{\frac{1}{2}}(C_a)\phi_{\frac{1}{2}}(C_b)\, I_c\Psi + \phi_1(C_a)\phi_1(C_b)\, C_c F \tag{6.56}$$

$$U(C_b)U(C_a)(\Psi + F) = -\phi_{\frac{1}{2}}(C_a)\phi_{\frac{1}{2}}(C_b)\, I_c\Psi + \phi_1(C_a)\phi_1(C_b)\, C_c F \tag{6.57}$$

should represent the same state, i.e. differ at most by an overall phase factor. But this is true iff $\Psi = 0$ or $F = 0$. Hence the spaces $\mathcal{H}_{\frac{1}{2}}$ and $\mathcal{H}_1$ form superselected sectors.

6.2.4 Discussion and Caveats

To derive the univalence superselection rule it was essential that $SO(3)$, and consequently its subgroup D_2, were symmetries. If instead of $SO(3)$ we assumed $SU(2)$ as symmetry group– which is, in fact, the above mentioned universal central extension of $SO(3)$ –, no superselection rule would result since $SU(2)$ acts by some proper unitary representations $SU(2) \ni g \mapsto U_g$ on $\mathcal{H}_{\frac{1}{2}} \oplus \mathcal{H}_1$. This follows, for example, from a trivial application of Theorem 3, since $SU(2)$ already acts on $\mathcal{H}_{\frac{1}{2}}$ and $\mathcal{H}_1$ by proper unitary representations, that is, by projective unitary representations with trivial – and hence similar – multipliers. Hence there are symmetry automorphisms $\alpha_g(\rho) = U_g \rho U_g^\dagger$ of the full state space which satisfy (6.35) and (6.36) for all $\rho \in \mathcal{D}(\mathcal{H}_{\frac{1}{2}} \oplus \mathcal{H}_1)$. All this means that for $G = SU(2)$ no superselection rules exist, which nicely illustrates that the existence of superselection rules from symmetries can be very unstable against minor variations of the size of the symmetry group. Since the usual hamiltonians which describe interactions between bosons and fermions do not discriminate between an $SO(3)$- or an $SU(2)$-symmetry, it would be dynamically consistent to postulate a fundamental $SU(2)$ rather than $SO(3)$ invariance. Accordingly, the physical significance of the formal derivation of the univalence superselection rule has been questioned (e.g. Mirman 1970). These arguments receive apparent support from experimental results about the observability of the sign change of spinors under 2π-rotations (Rauch et al. 1975, Werner et al. 1975, Rauch et al. 1978, and Greenberger 1983). But one should also point out that these experiments dynamically rotate only certain components (subspaces) in the Hilbert space of the fermion and leave others fixed. This is *not* the motion generated by the kinematically defined rotations, which act globally on that Hilbert space (cf. Hegerfeldt and Kraus 1968). Hence these experiments are still compatible with the invariance of the fermionic system under spatial $SO(3)$-rotations. Moreover, a phase between a half-integer and an integer angular momentum state is not measured, such that these experiments do not contradict the univalence superselection. At this point we should also mention that current ideas in *local* quantum field theory also clearly favor the existence of this superselection rule. More precisely, in order to violate the univalence superselection rule, one would need a spinorial observable to measure the relative phase between bosonic and fermionic degrees of freedom. But the Wightman formalism of quantized fields does not admit local observables with half-integer spin, which commute for spacelike separations, see e.g. Lüders and Zumino (1958) and the discussion in Chap. 9. of Bogolubov et al. (1990). Hence such spinorial observables would lead to serious problems with causality.

For two-dimensional systems, which are of interest in solid state physics, superselection rules have been derived from the classification of inequivalent central extensions of symmetry groups by Divakaran (1990) and Divakaran and Rajagopal (1991). For example, in the case of an infinite plane with homogeneous magnetic field, the symmetry group is $E(2) = \mathbb{R}^2 \rtimes SO(2)$. There is a one real parameter family of inequivalent central extensions which are just those which extend the translation subgroup $\mathbb{R}^2$ to the Heisenberg group.[23] We note that the different non-trivial extensions all lead to isomorphic groups which nevertheless classify as inequivalent extensions. This gives a simple example of our remark above (p. 282), that the classification of extensions is much finer than that according to mere group isomorphicity. The classifying parameter, which mathematically is an element of $H^2(\mathbb{R}^2, U(1))$, has the physical interpretation of charge times magnetic flux through a surface element of unit area.

Instead of looking for the representation of a symmetry in the state space, one may also look for the representation of a symmetry group as automorphism group of the algebra of observables. Such a program has been presented by Landsman (1990) for the quantum mechanics of particles on a homogeneous configuration space $Q = G/H$, where G is the symmetry group of the system.

The basic assumption of this section is the existence of an exact symmetry group on the physical state space. But most symmetries in nature are not absolutely exact. For example, consider a finite crystal: its symmetries are broken by defects, boundary effects, the external gravitational field, etc. But we can describe the physics of this system approximately by a theory which exhibits the exact symmetry. Then this theory must have the corresponding superselection sectors because of mathematical consistency. Within a "bet-

---

[23] This is a particular property of *two* dimensions. In $n > 2$ dimensions, there is a $\frac{1}{2}n(n-1)$ - parameter family of inequivalent central extensions of the translation subgroup $\mathbb{R}^n$, none of which survives the embedding into $E(n)$. This is also true for the inhomogeneous pseudo-orthogonal groups, like the Poincaré group. See Chap. 12 of (Hamermesh 1989).

ter" theory which takes into account small deviations from the symmetry the superselection sectors will no longer exist. But if the symmetry-respecting theory was already a good approximation, the off-diagonal elements of the density matrices would have to be very small. On the other hand, some symmetries are thought to be exact, like those associated to fundamental gauge freedoms. Do they give rise to exact superselection rules? This will be discussed in the following sections.

6.3 Physical Symmetries Versus Gauge Transformations

So far we used the notion of *symmetry* only in the formal sense of "automorphisms of the state space". But there are other, more refined notions attached to this word which we need to clarify. Roughly speaking, and in some anticipation of what is to follow, we wish to achieve a clear and unambiguous distinction between proper physical symmetries on one hand, and gauge symmetries or mere automorphisms of the mathematical scheme on the other. To do this, we shall recall some elementary concepts concerning the notion of redundant state spaces and then explain and delimit our notion of symmetry in the following section. Besides for obvious reasons of clarification, this separation seems necessary to properly interpret certain derivations of superselection rules in quantum mechanics and QED, which we discuss at the end of this and the next section, respectively.

6.3.1 Configuration Spaces and Spaces of State

A standard theoretical construction in physics is to associate to each physical system, S, a state space, $\mathcal{S}$, whose points label uniquely its different states. The symbol S is understood to incorporate all the kinematical and dynamical data of the system in question. $\mathcal{S}$ is endowed with the necessary structures (e.g. C^∞-manifold, with (pseudo) Riemannian and/or symplectic structures) needed to discuss dynamics on it. In case the dynamics allows for a well posed initial value formulation it is common practice to identify $\mathcal{S}$ with the space of initial data or, equivalently, with the space of solutions to the equations of motion. Here the notion of "state" clearly comprises all the dynamically relevant information. Related to but different from the state space is the configuration space, $\mathcal{Q}$, whose points label the (generalized) positions. For classical hamiltonian systems $\mathcal{S}$ is the phase space whose coordinates are given by the generalized positions, q^i, together with the corresponding momenta, p_i. In a more mathematical terminology, $\mathcal{S}$ is given by the cotangent bundle, $T^*\mathcal{Q}$, over $\mathcal{Q}$. There also exist examples of physical state spaces which are not cotangent bundles over any configuration spaces, i.e. in which no global split into position coordinates and momentum coordinates exists. Many such examples are provided by the reduced configuration spaces of classical constrained systems. Moreover, the application of these concepts is not restricted to classical physics. The quantum mechanical dynamics (i.e. the Schrödinger equation) for a strictly isolated system can also be naturally cast into the form of a hamiltonian dynamical system, but now on a different, quantum mechanical, space of states. It is given by the space of rays in some Hilbert space, i.e. in mathematical terminology, by some complex projective space of generally infinite dimension. In particular, this state space is not a cotangent bundle. For finite dimensions this is briefly outlined in App. 5. In any case, the important point here is that we associate a state space to each physical system, and that this association depends on the dynamical laws. As an illustrative example we take n point-particles moving in $\mathbb{R}^3$ under the rule of some classical dynamical law. The state space is then the familiar space of $6n$ linear position and momentum coordinates, $\mathbb{R}^{6n}$. However, if we think of the quantum mechanical description of n point particles, endowed with the corresponding Schrödinger dynamics, the state space is given by the space of rays in the Hilbert space $L^2(\mathbb{R}^{3n})$ of square integrable functions on the classical configuration space.

In the sequel we shall denote any state space by $\mathcal{S}$, irrespectively of whether it has a configuration space or not. If it does, the latter will be called $\mathcal{Q}$.[24]

Clearly, $\mathcal{S}$ should be chosen big enough to accommodate all the states of S. On the other hand, one would also like it to be small enough in order to avoid redundant labelings. That is, two different points in $\mathcal{S}$ should label two different states of S. Operationally speaking, to say that S is in a state $s \in \mathcal{S}$ is taken to mean that there exists a set of functions (observables) on $\mathcal{S}$ whose values uniquely characterize a point on $\mathcal{S}$. This is, for example, the case if there exist sufficiently many (compatible) observables that form a coordinate system on $\mathcal{S}$, like e.g. the position and momentum functions (q^i, p_i) for the phase spaces in classical mechanics. So far it is not necessary to imply that any algebraic combination of these functions also correspond to observables. This step is usually implied in classical mechanics, where at least all polynomials in the (q^i, p_i) are called observables. Moreover, by taking the completion

---

[24] For the abstract machinery of analytical canonical mechanics to work there is clearly no need to ever talk about a configuration space. However, any separation into "coordinates" and "momenta" reintroduces that split, which is necessary to relate the canonical variables to familiar space-time concepts; for example, momentum measurements are often done by first relating momentum to velocity, and then using a spatio-temporal measurement to determine the latter. Quite generally, the configuration space is endowed with a special interpretational status which is hidden by general canonical invariance that treats positions and momenta on equal footing. A split is also required in the canonical quantization procedure, where the quantum state depends on only "half" the phase-space variables. In this context the splitting is referred to as "choice of polarization" and amounts to the choice of that half of the phase space directions on which the state function does not depend.

of this function space, the space of observables is usually identified with a much bigger function space which, for example, also contains all functions which assume the constant value 1 on some arbitrary domain $U \subset \mathcal{S}$ and zero on its complement (i.e. the characteristic function of U). These are then identified (in a standard, idealizing fashion) with "yes-no" measurements, responding to the question: "does the system occupy a state within U?" Clearly, these "projection-observables" allow to separate any two points in $\mathcal{S}$ in the obvious way and are therefore sometimes taken as corresponding to elementary measurements ("yes-no" measurements, or "filters") out of which every physical statement is to be built. But, as stressed above, they are not necessary to fully justify the notion of state space.

Let us turn to the quantum theory of a strictly isolated system. Its state space $\mathcal{S}$ is given by the space of rays in the Hilbert space of L^2-functions on the classical configuration space. The Schrödinger equation then defines a hamiltonian system on this space with states and observables corresponding to certain functions thereon. This is is discussed in more detail in App. 5. All this sounds as if quantum mechanics is just a special hamiltonian system. But we know that there clearly must be strong limitations to this analogy and it is instructive to see what they are in this formally similar setting. First of all, the quantum mechanical state space is infinite dimensional even if the state space of the underlying classical system has finite dimensions. Second, and more important, the functions on the quantum mechanical state space that actually correspond to observables provided by quantum mechanics are indeed very few out of the space of all smooth functions (unlike a classical situation). In particular, they do not include functions whose support is a proper subset of the state space. Clearly, the set of functions corresponding to observables is still rich enough such that their values uniquely characterize any point in $\mathcal{S}$ (otherwise we were not even justified to speak of $\mathcal{S}$), but there is no single set of mutually *compatible* observables which does that. A given complete subset of mutually compatible observables characterizes *some* (in fact very few) states out of $\mathcal{S}$ (the set of simultaneous eigenstates), but not any in the complement. Conversely, given a state, there is *some* complete subset of mutually compatible observables which can characterize the given state (this is just Dirac's requirement discussed in Section 6.1.2).

Finally we wish to draw attention to the following major difference: in classical physics a state space of a composite system is given by the cartesian product of the individual state spaces, whereas the quantum mechanical state space of a composite system is the space of rays in the tensor product of the individual Hilbert spaces. This space is much larger than just the cartesian product of the individual state spaces which it contains as the subset (not a linear subspace) of pure (i.e. rank one) tensor product vectors. This is an expression of the fact that a (possibly mixed) state of a composite system in quantum mechanics is generally not determined by the individual reduced states. More details may be found in App. 3.

6.3.2 Symmetries and Redundant State Spaces

Let G be a group with smooth action[25] on $\mathcal{S}$:

$$G \times \mathcal{S} \to \mathcal{S} \ . \tag{6.58}$$

The dynamical laws of $\mathcal{S}$ select possible curves (in general parameterized, unless one works with a reparametrization-invariant theory) on $\mathcal{S}$ from among all curves. We call G a *symmetry* group of S, if every solution curve on $\mathcal{S}$ is mapped to a solution curve under the action of G. For classical systems, a situation often encountered in practice is that the symmetry group G already acts on the configuration space $\mathcal{Q}$, so that the action on $\mathcal{S}$ can be understood as a lifted[26] action. Note that any group action on $\mathcal{Q}$ has a natural lift to the cotangent bundle, just by pulling back the covector with the inverse map. The characteristic feature of lifted actions is that the phase-space function that generates the corresponding infinitesimal group motions (via Poisson bracket) are linear-homogeneous in the momenta, like the generators for spatial translations and rotations, which are given by the expressions of total linear- and angular momentum respectively. However, there are also important exceptions. A familiar example is the $SO(4)$-symmetry of the Kepler problem, which not only contains the spatial $SO(3)$ symmetry generated by the angular-momentum functions but also the Lenz-Runge vector which is non-homogeneous and quadratic in the momenta. Only the spatial $SO(3)$ acts already on the configuration space $\mathcal{Q}$. Other examples are boost transformations for the Galilei or Lorentz group as well as time translations. None of the corresponding generating functions is linear homogeneous in the momenta and, accordingly, the symmetries do not separately act on $\mathcal{Q}$. Finally, time-reparametrization-invariant theories display constraints which are typically quadratic polynomials in the momenta whose coefficients generally depend on configuration space.

As outlined above, the state space $\mathcal{S}$ was assumed to be faithful in the sense that any two different points on $\mathcal{S}$ correspond to two different states of S. However, a situation often encountered in practice is that instead of $\mathcal{S}$ a larger space, $\bar{\mathcal{S}}$, is employed, which labels physical states of S in a *redundant* fashion. That is, there is a surjective map

$$\tau : \bar{\mathcal{S}} \to \mathcal{S} \tag{6.59}$$

so that points in the set $\tau^{-1}(p) \subset \bar{\mathcal{S}}$, $p \in \mathcal{S}$, label the *same* physical state. Here also, the redundant state space $\bar{\mathcal{S}}$ is usually given as cotangent bundle

---

[25] We use the term action in the mathematical sense, i.e. there is a map $\phi : G \times \mathcal{S} \to \mathcal{S}$, $(g, s) \mapsto \phi(g, s)$, so that for all $s \in \mathcal{S}$ (i) $\phi(e, s) = s$, where $e \in G$ is the identity, and for all $g, h \in G$ (ii) $\phi(g \cdot h, s) = \phi(g, \phi(h, s))$. The action is called smooth if ϕ is smooth.

[26] An action $\phi : G \times \mathcal{S} \to \mathcal{S}$ is called *lifted*, if there is an action $\varphi : G \times \mathcal{Q} \to \mathcal{Q}$ so that $\pi(\phi(g, s)) = \varphi(g, \pi(s))$ for all s, g, where $\pi : \mathcal{S} \to \mathcal{Q}$ denotes the projection map.

over some redundant configuration space $\bar{\mathcal{Q}}$. This is the case for gauge theories or general relativity, where the configuration spaces are given by the space of gauge potentials and the space of Riemannian metrics respectively, both defined on a 3-dimensional hypersurface $(t = \text{const.})$ that serves as carrier space for the initial data.

It turns out that in many cases of interest the sets $\tau^{-1}(p)$ can be identified with the orbits of some group K acting freely[27] on $\bar{\mathcal{S}}$, $\bar{\mathcal{S}} \times K \to \bar{\mathcal{S}}$. Then, technically, $\bar{\mathcal{S}}$ can be given the structure of a principal fiber bundle with base $\mathcal{S}$ and group K such that the map τ is just the bundle projection:

$$(\bar{s}, k) \to \bar{s} \cdot k, \quad \tau(\bar{s} \cdot k) = \tau(\bar{s}) \quad \forall k \in K. \tag{6.60}$$

Two subsets U and $U \cdot k$ of $\bar{\mathcal{S}}$ then contain *identical* physical states. It is also true that K transforms solution curves of the dynamical equations on $\bar{\mathcal{S}}$ into solution curves. But these would be physically indistinguishable, in contrast to the action of a symmetry, which relates physically *different* solutions. To make this distinction we call K the *redundancy* group. For gauge theories or general relativity, K is generally *not* given by all gauge transformations or diffeomorphisms respectively but rather by some appropriate (infinite dimensional) subgroup (Giulini 1995a). These are the subgroups generated by the so-called Gauß constraints in the case of gauge theories and the so-called diffeomorphism constraint in the case of general relativity. This will be explained in more detail below. In these two cases we can in fact consider the group K as acting on $\bar{\mathcal{Q}}$ where the corresponding relations (6.60) equally hold.

Now, the problem with theories like those of Yang–Mills or general relativity is precisely that their dynamical equations are formulated on such redundant state spaces and that it is technically highly non-trivial to find a parametrization of the proper state space $\mathcal{S}$ (and the proper configuration space $\mathcal{Q}$). Suppose that for this (or other) reasons we worked with the redundant label spaces $\bar{\mathcal{S}}$ (and $\bar{\mathcal{Q}}$). This now poses the problem of how we should define proper symmetries in this setting. Suppose for a moment that we could construct $\mathcal{S}$ and we had been given an action of a symmetry group G on it. How would this fact be recognized on $\bar{\mathcal{S}}$? We cannot generally expect G to also act on $\bar{\mathcal{S}}$. Rather, it will merge with K into some larger group $\bar{G}$ which must contain the redundancy group K as a normal (i.e. invariant) subgroup. The last condition is necessary to ensure that the quotient $\bar{G}/K = G$ has a well defined action on $\mathcal{S}$, as assumed. Hence $\bar{G}$ must be a K extension of G (cf. Sect 6.2.1):

---

[27] For example, the action of asymptotically trivial gauge transformations (or diffeomorphisms) is free. This is a natural restriction for open manifolds. For closed manifolds one either needs to restrict the gauge transformations to those which are the identity at a distinguished point, or restrict oneself to the subset of irreducible connections (Booss and Bleecker 1985 pp 358, Isham 1981). Here we primarily think of open manifolds.

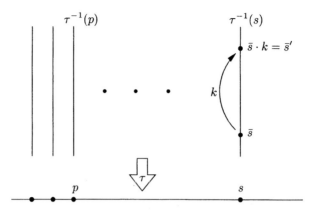

Fig. 6.1. The relation between the redundant phase space $\bar{S}$ (collection of vertical lines) and the true phase space S (lower horizontal line) is illustrated. The vertical lines correspond to the sets $\tau^{-1}(p)$ and contain physically indistinguishable points. A vertical line is generated by applying the group K to any one point in it. In that fashion, two points $\bar{s}$ and $\bar{s}'$ in the same line (the fiber) are connected by a unique group element k so that $\bar{s}' = \bar{s} \cdot k$. The figure is topologically deceptive in the sense that $\bar{S}$ is generally not a topological product space $K \times S$.

$$K \xrightarrow{\ i\ } \bar{G} \xrightarrow{\ p\ } G \qquad (6.61)$$

where i denotes the injective inclusion homomorphism and p the surjective projection homomorphism.

Note that as in the definition of a (true) symmetry, $\bar{G}$ also transforms solution curves on $\bar{S}$ into other solution curves on that space. However, due to the redundancy of state labelings by $\bar{S}$, it would be misleading to also address $\bar{G}$ as symmetry group in the same sense, since it combines a redundancy group K and a proper symmetry group G, as depicted in Fig. 6.2. To account for this hybrid structure we will simply call it the *invariance* group. The knowledge that S and not $\bar{S}$ is the true state space is then equivalent to splitting the invariance group $\bar{G}$ into a redundancy part K (a normal subgroup) and a symmetry part G (the quotient $\bar{G}/K$). Note how crucially this distinction in interpretation influences the whole kinematical and dynamical setup. For example, we would only regard K-invariant functions on $\bar{S}$ (i.e. functions which are constant along the orbits of K) as observables, since only those project onto the true state space S. This means that the observables contain only functions whose Poisson brackets with the generators of all redundancy transformations vanish. This is precisely what one does in gauge theories, when one requires all observables to (Poisson-) commute with the Gauß constraint.

For the two examples cited above, gauge theories and general relativity, we are given large groups (group of gauge transformations, diffeomorphism

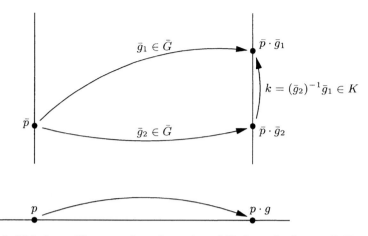

Fig. 6.2. This figure illustrates how the action of K along the (vertical) fibers and the action of G on the (horizontal) base space combine into an action of the group $\bar{G}$ on the total space. Whereas K separately acts on $\bar{S}$ (Fig. 6.1) this is generally not true for G.

group) primarily as invariance groups of the dynamical system. The formalism then tells us (via the Gauß constraint and the momentum constraint respectively), that large parts of these groups (the asymptotically trivial gauge transformations or diffeomorphisms respectively) must be considered as redundancy. Depending on the boundary conditions at "infinity", the redundancy part does not exhaust all of the invariance group. The remaining transformations must then be interpreted as symmetries;[28] that is, motions generated by the corresponding symmetry generators are proper physical motions which, in principle, implies the existence of observables that detect this motion. More technically speaking, the set of observables must contain phase space functions whose Poisson brackets with the symmetry generators are non-zero. In general relativity, for example, a rotation of an isolated system by an angle different from an integer multiple of 2π defines a diffeomorphism which is not asymptotically trivial and is therefore a symmetry rather than a redundancy. In fact, were it required to be a redundancy, i.e. were the generator of this transformation (angular momentum) required to vanish, the state space would not contain solutions with non-vanishing angular momentum (at "spatial infinity"), like e.g. the Kerr black-hole. To speak of non-zero angular momentum thus also implies that, in principle, we have observables

---

[28] There is an additional slight subtlety here due to the existence of gauge transformations which are asymptotically trivial but not connected to the identity (so-called "large" gauge transformations). They too are not generated by the Gauß constraint and may be interpreted as genuine (i.e. observable) symmetries (see e.g. Giulini 1995a). In the sequel we shall ignore this point which is not relevant for our main discussion.

at our disposal that are sensitive to the spatial orientation of the isolated system with respect to the reference frame at "infinity". The same holds for any global charge, like the total electric charge, measured "at infinity". Here it is more difficult to interpret the variables that do not commute with the total charge. This will be seen in more detail below.

In the corresponding quantum (field)theories – where we think in terms of the Schrödinger representation – the difference of redundancies and symmetries manifest itself in a similar fashion. Starting the quantization from the classically redundant formulation must lead to redundancies in the corresponding quantum theory. This results in an auxiliary state space which carries an action of $\bar{G}$. However, any physical state function is required to be strictly invariant under the action of K, whereas it might well change under a symmetry transformation. The invariance under K and the action of $\bar{G}$ then imply an action of the symmetry group G on the physical state space. Moreover, observables must commute with the generators of K but not necessarily with those of G.

We already mentioned that quantum mechanics itself can be considered as a hamiltonian dynamical system on some state space. As just outlined, the quantum mechanical state space is redundant if the underlying classical theory has been formulated in a redundant fashion. Apart from this kind of inherited redundancy, the traditional Hilbert space formulation of strictly isolated quantum mechanical systems also contains an intrinsic redundancy which is completely independent of the inherited one just discussed. The idea is that vectors in Hilbert space label states in a redundant fashion due to the physically undetermined overall complex factor of the state vector.[29] Put into our general framework we say that the Hilbert space is a redundant state space that carries an action of the redundancy group which is given by the multiplicative group of non-zero complex numbers, denoted by $\mathbb{C}^\times$. This group is isomorphic to $\mathbb{R}_+ \times \mathrm{U}(1)$, where the $\mathbb{R}_+$ represents multiplication with positive real numbers and $\mathrm{U}(1)$ the multiplication with complex numbers of unit modulus. The true state space is then given by taking the quotient with respect to this action which results in the space of rays. To relate to standard terminology, we mention that we can first reduce the $\mathbb{R}_+$ action by restriction to normalized vectors. This results in the still redundant and non-linear intermediate state space of unit vectors (forming a sphere). The remaining redundancy is then labeled by all $\mathrm{U}(1)$ transformations (phases). For the classification of the different implementations of the symmetry group on Hilbert space it is irrelevant whether we work with normalized or unnor-

---

[29] The general preference for working with the redundant Hilbert space is due to the evidently much greater formal simplicity that the superposition principle assumes in conjunction with the linearity of the Schrödinger equation. Compare e.g. Roberts and Roepstorff (1969) for a formulation of the superposition principle not using the Hilbert space formalism.

malized vectors. To conform with standard terminology we shall henceforth understand state vectors to be normalized and thus stick to U(1).

In Sect. 6.2.1 we used this redundancy interpretation of the overall phase of the quantum state that allows for the more general ray representations in addition to ordinary representations on Hilbert space. This is because a ray representation of a group G still assures an action of G on the space of rays, which is sufficient if the overall phase is considered a redundant labeling. Since this just fits into the scheme developed above, we can state this equivalently using diagram (6.61): if G is considered as a symmetry group of a quantum mechanical system, we can on the Hilbert space only expect an action of a group $\bar{G}$, given by a U(1)-extension of G, but not G itself. These two notions are in fact equivalent as was already discussed in Sect. 6.2.1. In App. 6 we discuss the implementation of the Galilei group according to this scheme. There we demonstrate that different mass parameters require inequivalent U(1)-extensions, which in turn leads to the famous Bargmann superselection rule for total mass by an argument very similar to the one given for the rotation group in Sect. 6.2. (Bargmann 1954, Levy-Leblond 1963, Kaempffer 1965, Galindo and Pascual 1990). We stress that the precise form of the implementation depends on the laws of motion relative to which it forms a symmetry. Consequences of the so-established implementation should therefore not be considered as independent of the given dynamical laws.[30] In this respect it is therefore illuminating to go through the derivations in App. 6 in detail to see how the dynamical law is invoked.

6.3.3 Symmetries, Redundancies, and Superselection Rules

Given the refinement of the symmetry concept as developed above, we wish to revisit the question of how symmetries are related to superselection rules. We explained how in a classical theory with invariance group the kinematical setup separates between the physically non-existent degrees of freedom, corresponding to the action of the redundancy group, and those connected with the symmetry group as physically existent ones. A quantum theory based on such a classically redundant description necessarily inherits these redundancies and must seek its own prescriptions to deal with the spurious degrees of freedoms.[31] For example, in the quantum theory of gauge theories, this is

---

[30] We consider this an important point of general validity. The "right" kinematical implementation of a symmetry group eventually relies on fundamental dynamical laws with respect to which it becomes a *dynamical* symmetry group. Hence, fundamentally speaking, there is no kinematical symmetry (in the sense of being independent of dynamical laws).

[31] This "extrinsic quantization" becomes a necessity if the "intrinsic quantization", which consists of first eliminating these degrees of freedom, cannot be performed. At first sight a natural requirement on an extrinsic quantization scheme is that in cases where the "intrinsic" quantization is available it leads to a unitarily equivalent quantum theory. But there are known cases where there are many

usually attempted by imposing the Gauß constraint in some suitable operator form, which then says that the formal conjugate momentum operator (generating the redundancy) annihilates the state functional and thus demands it to be independent of the physically non-existent degrees of freedom. In contrast, the momentum generating a symmetry is not a priori required to annihilate permissible state functionals. These might therefore depend on the corresponding conjugate coordinate in arbitrary ways. Now, what typically happens in gauge theories is that the invariance group is implemented to act locally on the redundant set of fields whereas the symmetry group, generated by the global charge, generates motions in highly non-local degrees of freedom, sometimes called "gauge transformations at infinity", though they are symmetries and not gauge redundancies as we shall argue. This will be discussed explicitly in the next section. Formally this means that strictly local observables commute with the symmetry generator and that any such local observable has vanishing matrix elements between eigenstates for different eigenvalues of the symmetry generator. In other words, we encounter a superselection rule *relative* to local observables; see Haag (1992) for a general discussion of the locality principle. A heuristic reason for this may be seen in the fact that the conserved charges can be written as surface integrals over two-spheres in the limit of infinite radius (compare next section). But let us stress again at this point that superselection operators obtained in this fashion always correspond to generators of symmetries which, according to the general theory, always generate motions in *physically existing* degrees of freedom.

Different from the typically field-theoretical mechanisms for superselection rules, which involve concepts of locality, is that explained in detail in Sect. 6.2.2, which occurs if a symmetry group allows for inequivalent central U(1)-extensions. There we already presented the example of the univalence superselection rule. More suited for the actual point we wish to make is another example from Galilei-invariant quantum mechanics, namely that of the mass superselection rule. Let us briefly discuss its essential features and defer the formal details to App. 6.

We consider N non-relativistic particles interacting through some unspecified Galilei-invariant potential. The Galilei group turns out to be implemented via some non-trivial extension labeled by the total mass of the system. Ac-

---

"extrinsic" quantizations of which only one coincides with the "intrinsic" one. An interesting relaxation of the equivalence requirement is then to maintain the possible relevance of the "extrinsic" schemes for the quantization problem. This seems a viable strategy if the redundancy group is discrete, in which case it does not automatically give rise to a constraint to be imposed on the quantum mechanical state, and which can therefore carry non-trivial representations. This, for example, gives rise to the possibility for quantizing a system of identical particles in different, non-equivalent ways: as bosons, fermions, or paraparticles. Here only the bosonic sector (trivial representation) corresponds to the intrinsic quantization. For a general class of examples, see Giulini (1995b).

cording to the general argument, superpositions of N-particle states with different overall mass cannot therefore represent physical states. The common universal cover of the different central U(1)-extensions (resulting in an $\mathbb{R}$-extension) gives our universal central extension which acts as common invariance group on all sectors. We call this group $\hat{G}$; it is an 11-dimensional Lie group (sometimes referred to as the Bargmann group). Its infinitesimal generators on Hilbert space are operators $\{\mathbf{D}, \mathbf{V}, \mathbf{A}, \mathbf{B}, \mathbf{Z}\}$, where the first four symbols comprise the generators of rotations, boosts, space-, and time-translations respectively. $\mathbf{Z}$ is the generator of the extending group which is just $\mathbb{R}$ (the universal cover of U(1)). The phase changes generated by $\mathbf{Z}$ are considered as unobservable. Accordingly, $\mathbf{Z}$ must commute with all observables and therefore be represented as a multiple of the unit operator. Indeed, for an overall mass M it is represented by multiplication with $\frac{i}{\hbar}M$. In this language, the superselection rule states that only eigenvectors of the mass-operator $-i\hbar\mathbf{Z}$ correspond to pure physical states. Note that since the orbit of $\mathbf{Z}$ is non-compact, the mass spectrum is continuous.[32] Alternatively, this is just saying that we cannot observe the relative phase between eigenvectors of different eigenvalues since the only observables that would do so cannot commute with the mass operator. We thus have a continuous superselection rule with respect to mass (cf. Sect. 6.1.3).

Such simple arguments can however be criticized from several perspectives. Let us first look at the example just mentioned. We apparently have an unambiguous mathematical derivation, according to the principles outlined in Sect. 6.2.2, of a superselection rule for the overall mass in Galilei-invariant quantum mechanics. Taken literally, it claims that the superposition of mass eigenstates, $\psi_+ = \psi_M + \psi_{M'}$ for $M \neq M'$, represents a mixture.[33] But what is the system of which ψ_M and $\psi_{M'}$ represent possible states? In non-relativistic quantum mechanics the mass is not considered as a dynamical variable, but as a parameter that is part of the characterization of the system. Two N-particle systems with different (overall) masses are usually considered to be *different* systems. What does it mean to superpose their states? In order to make sense of a superposition $\psi_M + \psi_{M'}$ we must be able to regard it as state of a *single* system. In other words, M must correspond to a *dynamical* variable. Now, it is indeed possible to join the mass parameters of the N-particle system to the dynamical variables (Giulini 1996). What happens then is that the classical- and quantum theory have the *same* symmetry group, which is implementable on Hilbert space by ordinary linear unitary representations. It is just given by the $\mathbb{R}$-extension $\hat{G}$ of the Galilei group, generated by $\{\mathbf{D}, \mathbf{V}, \mathbf{A}, \mathbf{B}, \mathbf{Z}\}$. $\mathbf{Z}$ generates translations in a new, $\mathbb{R}$-valued coordinate, λ, which is canonically conjugate to the total mass. The hypoth-

---

[32] The term *eigenvector* therefore has to be understood in the usual, generalized, sense.

[33] Let $\rho_+ = |\psi_+\rangle\langle\psi_+|$ and $\rho_{\mathrm{mix}} = |\psi_M\rangle\langle\psi_M| + |\psi_{M'}\rangle\langle\psi_{M'}|$, then $\mathrm{tr}(A\rho_+) = \mathrm{tr}(A\rho_{\mathrm{mix}})$, for all observables A.

esis of the derivation, namely that the Galilei group should act on the set of physical states, is now not even classically fulfilled within this dynamical setup since $\mathbf{Z}$ acts non-trivially even classically. This shows how the classical symmetry group is changed when mass is treated dynamically, as it should be. Since $\hat{G}$ is now represented on Hilbert space by proper unitary representations there are now no superselection rules deriving from the new postulate that $\hat{G}$ (and not G) acts on the set of physical states.[34] In view of this model, it seems unmotivated to a priori request an action of G rather than $\hat{G}$.

The fact that there are now no a priori reasons for a superselection rule deriving from purely kinematical and group-theoretical requirements does not imply that we could not account for an apparent mass superselection rule, should it exist. It only means that in the present setting its origin must be a *dynamical* one. To see this, we recall that such a superselection rule for overall mass would say that pure states must be eigenstates to the mass operator (the momentum for λ) and therefore cannot localize the system in the λ-coordinate. On the other hand, λ represents an existing degree of freedom within the dynamical setup which cannot just be declared to be unobservable in principle (i.e. redundant). For example, translating a state with non-zero mass in real time along the λ direction costs non-vanishing action.[35]

---

[34] One may be inclined to attempt the same argument in the case of the univalence superselection rule which, as already discussed, would simply disappear if $SU(2)$ instead of $SO(3)$ were required to act on physical states (cf. Sect. 6.2.3 and Sect. 3.1.3). Leaving alone the question of whether the mere existence of spinor fields already rules out $SO(3)$ as group of (physically meaningful) rotations (Aharonov and Susskind 1967b, Hegerfeldt and Kraus 1968, Mirman 1970) as advocated in Sect. 3.1.3 of this book, we remark that the situation with univalence is not entirely analogous to the one discussed here, where the extension of the Galilei group was connected to an additional degree of freedom in the classical theory. A more analogous situation would consist in inventing models whose classical configuration space carried an $SU(2)$ but no $SO(3)$ action. Such classical models do, in fact, exist in general relativity (Friedman and Sorkin 1980, Giulini 1995a).

[35] The motto throughout should be clear: The theory decides on the existing degrees of freedom! If, for example, in classical mechanics we chose only relational quantities to be physical, we would need to cancel the degrees of freedom that refer to absolute space. Mechanical model-systems based on purely relational space concepts have been studied many times throughout the 20th century (Reißner 1915, Schrödinger 1925, Barbour and Bertotti 1977, 1982). The cancellation of degrees of freedom referring to absolute space can be achieved either by using only relational quantities for its formulation (intrinsic method) or by using absolute space coordinates together with a gauge principle that implicitly eliminates all but the relational degrees of freedom via constraints (extrinsic method).

General Critique

We were led to criticize the standard "derivations" of the mass superselection rule for not fulfilling the necessary requirement of being placed within a context where mass is a dynamical variable and, consequently, for hiding the real physical assumptions that must underlie such a derivation. In contrast, a derivation within a dynamical context must be due to the measurement theory of this system. As far as our model system is concerned, there is no a priori reason to give special status to the Galilei group, which is only a quotient of the actual symmetry group, and there is absolutely nothing inconsistent with localized wave packets in the λ coordinate. However, to form such superpositions we must coherently superpose mass eigenstates. On the other hand, different masses produce different long-ranged gravitational fields which would presumably strongly decohere such superpositions in realistic situations, just as it happens for different charges in electromagnetism (cf. Chap. 4 and Sect. 6.4).

A related general criticism is the following: suppose we couple a system S and a measuring device A to form a strictly closed system $S + A$. Suppose further that these systems share a strictly conserved additive physical quantity $Q = Q^{(S)} \otimes \mathbb{1} + \mathbb{1} \otimes Q^{(A)}$. It is easy to see that if the density matrix for $A + S$ commutes with Q then the reduced density matrices for S and A commute with $Q^{(S)}$ and $Q^{(A)}$ respectively. Also, a well known theorem in quantum mechanics says that A cannot measure any quantity P of S that does not commute with $Q^{(S)}$ in an ideal (i.e. von Neumann) measurement[36] (Araki and Yanase 1960). Restricting to ideally measurable observables for the moment, this says that all such observables of S must commute with $Q^{(S)}$ whose eigenspaces thus define superselection sectors. Since these arguments apply

---

[36] Assuming discrete spectra, a simple proof can be given as follows: let $\{|s_n\rangle\}$ be a complete set of eigenstates for P with eigenvalues $\{p_n\}$ (to save notation we ignore degeneracies, which can be easily accounted for), and let $|a_0\rangle$ denote the neutral state of the apparatus. The ideal von Neumann measurement is driven by the unitary evolution operator U. By hypothesis we have $[U, Q] = 0$ and $U|s_n\rangle|a_0\rangle = |s_n\rangle|a_n\rangle$, where the apparatus states $\{|a_n\rangle\}$ are normalized and hence satisfy $\langle a_n|a_m\rangle \neq 1$ whenever $p_n \neq p_m$, for otherwise there would not be a measurement at all. Now, for any index pair (n, m) such that $p_n \neq p_m$ we have:

$$
\begin{aligned}
(p_n - p_m)\langle s_n|Q^{(S)}|s_m\rangle &= \langle s_n|[P, Q^{(S)}]|s_m\rangle \\
&= \langle s_n|\langle a_0|[P \otimes \mathbb{1}, Q]|s_m\rangle|a_0\rangle \\
&= (p_n - p_m)\langle s_n|\langle a_0|U^\dagger Q U|s_m\rangle|a_0\rangle \\
&= (p_n - p_m)\langle s_n|\langle a_n|(Q^{(S)} + Q^{(A)})|s_m\rangle|a_m\rangle \\
&= (p_n - p_m)\langle s_n|Q^{(S)}|s_m\rangle\langle a_n|a_m\rangle \\
\Leftrightarrow \langle s_n|Q^{(S)}|s_m\rangle &= 0 \; \forall \, n, m \quad \Leftrightarrow \quad [P, Q^{(S)}] = 0
\end{aligned}
$$

to any conserved additive quantity we should expect an abundance of such superselection rules, not only for electric charge but, for example, also for momentum and angular momentum, which seems paradoxical. However, besides for the somewhat unrealistic restriction to ideal measurements, a characteristic flaw in this line of argumentation was pointed out long ago (Aharonov and Susskind 1967a, Lubkin 1970, Mirman 1969 and 1970): whereas the argument as given *is* formally correct, it is also irrelevant for the establishment of superselection rules. To give a simple but representative example we ask: what does it mean to determine the "z-component" of angular momentum using a Stern-Gerlach magnet? The interaction hamiltonian is rotationally invariant, so how could it be that in view of the theorem just mentioned we can measure an operator that does not commute with angular momentum in, say, the x-direction? The simple answer, of course, is that we do *not* measure such an operator. What we measure is a quantity that does commute with angular momentum, namely the projection of the angular momentum vector onto the direction of the magnetic field. It is this purely relational quantity that one means when, with some abuse of language, one refers to "the z axis". Finally, it is sometimes argued that although one cannot make ideal measurements for quantities not commuting with the conserved quantity, they can at least be measured individually up to any desired accuracy by so-called approximate measurements. Taking our example again, if we wanted to infer any information from the relational direction to an absolute angular position, say with accuracy of order ϵ, we had to prepare the magnet in a coherent superposition of angular momentum eigenstates with spread larger than $\hbar/\epsilon$ (Araki and Yanase 1960, Yanase 1961), which simply says that we have to localize the magnet's angular orientation better than ϵ, which of course just delays the problem.

Quite generally, such simple formal arguments for superselection rules refer to the overall quantities as well as the partial quantities of S and A only relative to an absolute frame of reference with respect to which the combined system $S + A$ is described. But it does not say anything about the, say, x-component of angular momentum of S *relative* to a reference system defined by the apparatus A. And it is only the latter quantity that eventually becomes important in realistic measurements. It has thus been argued that it is this characteristic misidentification of the relevant quantities that renders standard kinematical "proofs" for superselection rules irrelevant (Aharonov and Susskind 1967a, Mirman 1969 and 1970). Note that the relative quantities cannot be strictly conserved if the reference system is to define a reference frame.[37] See also Giulini et al. (1995) for an alternative discussion of the charge superselection rule.

---

[37] We distinguish between a reference *frame* and a reference *system*. By a frame, like coordinates, we understand a purely abstract notion of reference. By a system we mean a real physical dynamical system. The system defines the frame but is not identical to it; just as material bodies and light rays can define coordinates

We agree that it is a priori insufficient to derive superselection rules for quantities which are defined relative to an absolute reference frame, that is, a frame that cannot be dynamically acted upon by the system under study, without checking quantitatively that the latter assumption is dynamically justified. We stress that we think of reference frames as being defined by physical systems, material or otherwise, that make no exception to generally valid dynamical principles, such as the principle of actio and reactio. In particular this means that the object system and the frame-defining reference system must be dynamically enabled to share correlations, over and above the pure reference relation that they share in any case. Hence the reference system acts more like a reservoir, just like a material spatial reference frame acts as a reservoir for (angular) momentum or a charged capacitor as a reference system for the conjugate phase (Aharonov and Susskind 1967a). That the total system $S + A$ is not localizable in absolute space due to being in a momentum eigenstate does not mean that we cannot localize S with respect to A. For this to be possible we only require superpositions of *relative* momentum eigenstates. This viewpoint is essential in any attempt to understand existing superselection rules as dynamically emerging phenomena and part of the decoherence program.

In field theory, a similar discussion arises where the rôle of the reference frame is played by certain asymptotic structures which appear as one imposes boundary conditions for any asymptotically isolated configuration. In gauge theories such configurations are locally ambiguous due to gauge freedoms, but at spatial infinity these gauge freedoms are restricted by asymptotic boundary conditions. This introduces absolute elements into the theory which can be exhibited in form of collective coordinates parameterizing the gauge configuration at "infinity". Asymptotically non-trivial gauge transformations (modulo local ones) thus fall under the category of symmetries which are generated by charges that one should in principle regard as observable. This will be discussed in the next section.

6.4 Superselection Rules in Field Theory

In this section we wish to illustrate the remarks made in the last subsection within the context of electromagnetism. We also make a brief comparison with general relativity. In both cases we are interested in the description of isolated systems, which we want to obtain as stationary points of a variational principle. Due to the inclusion of charged configurations we must deal with long-ranged fields. In such a situation one has to be aware of the following: whenever one wishes to obtain long-ranging configurations as stationary

---

but are not identical to them. It should also be remembered that Newton's laws only assert the *existence* of an absolute frame. Its determination within Newtonian mechanics relies on the use of material bodies and the dynamical equations which together define the reference system (Lange 1885).

points of a variational principle, one must choose the action functional so as to include the sought for configurations in its domain of differentiability. It is inconsistent to just check that certain functions satisfy differential equations which were obtained by varying the action functional within a function class *not* containing these functions. In practical terms this means that the non-vanishing surface terms that one might encounter in the variational procedure have to be canceled by some extra surface contributions which one needs to add in order to restore differentiability. In addition, it was suggested long ago by Dirac (1964) (see also Wadia 1977) that boundary conditions be treated dynamically through such surface terms. In this way one naturally displays field degrees of freedom which are attached to the boundary and whose existence might have been forgotten otherwise (for example, by posing additional locality requirements). We stress that this is *not* a matter of presentation or interpretation. A consistent formulation of a theory should leave no ambiguity as to what the degrees of freedom are.

6.4.1 Charge and Asymptotic Flux-Distribution in QED

Let us now consider electromagnetism. It is formulated in Minkowski space, M^4, with coordinates $\{x^\mu\} = \{t, x, y, z\}$, Lorentz metric $\eta_{\mu\nu} = \text{diag}(-1, 1, 1, 1)$, gauge potential $A^\mu = (\phi, \boldsymbol{A})$, field strength $F_{\mu\nu} = \partial_\mu A_\nu - \partial_\nu A_\mu$, and current $j^\mu = (\rho, \boldsymbol{j})$. We use rationalized units so that no factors of 4π and c appear. Greek indices range from 0 to 3 and latin indices from 1 to 3. We define a cylinder[38] in Minkowski space, $Z = \{x^\mu \in M^4 \mid x^2 + y^2 + z^2 \leq R^2\}$, which defines the region where we want to solve Maxwell's equations and which we interpret as the history of a ball-like spatial region containing the isolated charge.[39] For this to be well defined we need to pose boundary conditions on ∂Z. We call the spatial region $\{x^\mu \mid t = \text{const.}\} \cap Z = \Sigma$ and its boundary $\partial \Sigma = S_R$. The frame $\{x^\mu\}$ defines a split into magnetic and electric components, $B_i = \frac{1}{2}\epsilon_{ijk}F^{jk}$ and $E_i = F_{i0}$. We require that no charge shall leave the boundary, i.e. the normal component of $\boldsymbol{j}$ vanishes on ∂Z, and also that the tangential components of the magnetic field vanish thereon. The standard lagrangean density without boundary terms is given by $\mathcal{L} = -\frac{1}{4}F_{\mu\nu}F^{\mu\nu} + j^\mu A_\mu$. It is easy to check that the corresponding action functional does not have any stationary points if the normal components of the electric field do not vanish

---

[38] The cylinder Z is not invariant under Lorentz transformations so that the whole constructions explicitly breaks Lorentz covariance. A Lorentz-invariant boundary would be given by the hyperboloid $\{x^\mu \mid x^\mu x_\mu \leq R^2\}$ (Gervais and Zwanziger 1980). However, the points we wish to make are more conveniently presented in the hamiltonian formulation which relies on a space-time-split anyhow.

[39] Eventually one might take the boundary to spatial infinity, $R \to \infty$. But as we will explain below, this entails a certain conceptual danger of regarding it as being outside the physical universe. For the moment we therefore prefer it to be at a finite distance. Essentially, the point is that for long-ranged field configurations one cannot push the boundary outside the physical universe.

on the spatial boundaries. But this cannot be required if one wants to obtain long-ranged field configurations in the set of stationary points. However, a suitable action principle can be obtained by adding the following boundary term (Gervais and Zwanziger 1980):

$$\int_{\partial Z} dt \, d\omega \, (\dot{\lambda} + \phi) f \; , \tag{6.62}$$

where ω abbreviates the polar angles θ and φ on the spatial two-sphere-boundary S_R and $d\omega = \sin\theta \, d\theta \, d\varphi$. λ and f are two new real-valued fields on ∂Z whose interpretation will become clear soon. In terms of three-dimensional quantities the Lagrange function, $L = \int_{\Sigma} d^3x \, \mathcal{L}$, can now be written in the form

$$\begin{aligned}
L = & \int_{\Sigma} d^3x \, \dot{\boldsymbol{A}} \cdot (-\boldsymbol{E}) + \int_{S_R} d\omega \, \dot{\lambda} f \\
& - \int_{\Sigma} d^3x \, \left[\tfrac{1}{2} \left(\boldsymbol{E}^2 + \boldsymbol{B}^2 \right) + \phi(\rho - \boldsymbol{\nabla} \cdot \boldsymbol{E}) - \boldsymbol{A} \cdot \boldsymbol{j} \right] \\
& - \int_{S_R} d\omega \, \left[\phi \left(R^2 \boldsymbol{n} \cdot \boldsymbol{E} - f \right) \right] \; .
\end{aligned} \tag{6.63}$$

Functions on the two-sphere S_R are conveniently represented in terms of their components with respect to the spherical harmonics Y_{lm}:

$$E_{lm} = \int_{S_R} d\omega \, Y_{lm}(\omega) \, (R^2 \boldsymbol{n} \cdot \boldsymbol{E}(\omega)) \; , \tag{6.64}$$

$$\lambda_{lm} = \int_{S_R} d\omega \, Y_{lm}(\omega) \, \lambda(\omega) \; , \tag{6.65}$$

and likewise for f and ϕ. The Lagrange function then reads:

$$\begin{aligned}
L = & \int_{\Sigma} d^3x \, \dot{\boldsymbol{A}} \cdot (-\boldsymbol{E}) + \sum_{l=0}^{\infty} \sum_{m=-l}^{l} \dot{\lambda}_{lm} f_{lm} \\
& - \int_{\Sigma} d^3x \, \left[\tfrac{1}{2} \left(\boldsymbol{E}^2 + \boldsymbol{B}^2 \right) + \phi(\rho - \boldsymbol{\nabla} \cdot \boldsymbol{E}) - \boldsymbol{A} \cdot \boldsymbol{j} \right] \\
& - \sum_{l=0}^{\infty} \sum_{m=-l}^{l} \phi_{lm} \left(E_{lm} - f_{lm} \right) \; .
\end{aligned} \tag{6.66}$$

From this we infer the canonical coordinate pairs $\bigl(\boldsymbol{A}(\boldsymbol{x}), -\boldsymbol{E}(\boldsymbol{x})\bigr)$, (λ_{lm}, f_{lm}), and the hamiltonian:

$$\begin{aligned}
H = & \int_{\Sigma} d^3x \, \left[\tfrac{1}{2} \left(\boldsymbol{E}^2 + \boldsymbol{B}^2 \right) + \phi(\rho - \boldsymbol{\nabla} \cdot \boldsymbol{E}) - \boldsymbol{A} \cdot \boldsymbol{j} \right] \\
& + \sum_{l,m} \phi_{lm} \left(E_{lm} - f_{lm} \right) \; .
\end{aligned} \tag{6.67}$$

The hamiltonian equations of motion now follow:

$$\dot{A} = \frac{\delta H}{\delta(-E)} = -E - \nabla\phi , \tag{6.68}$$

$$\dot{E} = -\frac{\delta H}{\delta A} = j - \nabla \times B , \tag{6.69}$$

$$\dot{\lambda}_{lm} = \frac{\partial H}{\partial f_{lm}} = -\phi_{lm} , \tag{6.70}$$

$$\dot{f}_{lm} = -\frac{\partial H}{\partial \lambda_{lm}} = 0 . \tag{6.71}$$

An important point to note here is that, *because* of the boundary term, H is (functionally) differentiable with respect to E. The boundary term in the variation of the space integral is just canceled by the variation of the boundary term that we added. It is automatically (functionally) differentiable in A since here the boundary term is zero due to the condition of vanishing tangential B-field. As in the full action, we still have to vary with respect to the scalar potential, that is, with respect to $\phi(x)$ and ϕ_{lm}. This leads to the Gauß constraints:

$$G(x) := \nabla \cdot E(x) - \rho(x) = 0 , \tag{6.72}$$

$$G_{lm} := E_{lm} - f_{lm} = 0 . \tag{6.73}$$

Equation (6.69) together with charge conservation, $\dot{\rho} + \nabla \cdot j = 0$, shows that (6.72) is preserved in time. The conditions that $n \times B$ and $n \cdot j$ vanish on the boundary imply that $\dot{E}_{lm} = 0$ and, together with (6.71), that (6.73) is also conserved in time.

Next we wish to discuss the quantum theory of this system on a heuristic level in the Schrödinger representation. Thus we consider state functions on configuration space, that is, functionals of $A(x)$ and functions on λ_{lm}: $\Psi[A(x), \lambda_{lm}]$. The conjugate momenta are then represented by

$$\hat{E}(x) = -\mathrm{i}\,\frac{\delta}{\delta A(x)} , \tag{6.74}$$

$$\hat{f}_{lm} = -\mathrm{i}\,\frac{\partial}{\partial \lambda_{lm}} . \tag{6.75}$$

The states have to satisfy the constraints (6.72-6.73) in operator form. Since in (6.72) there is one equation for each value of the continuous parameter x, it is appropriate to smear it with a test-function $\eta(x)$ of support inside and away from the boundary. This leads to ($\rho(\eta) = \int d^3x\, \rho\eta$ etc.)

$$\hat{G}(\eta)\Psi = 0 \quad \Leftrightarrow \quad -\mathrm{i}\,X^\eta\Psi = \rho(\eta)\Psi , \tag{6.76}$$

$$\hat{G}_{lm}\Psi = 0 \quad \Leftrightarrow \quad \hat{E}_{lm}\Psi = -\mathrm{i}\,\frac{\partial}{\partial\lambda_{lm}}\Psi , \tag{6.77}$$

where

$$X^\eta = \int_\Sigma d^3x \, \boldsymbol{\nabla}\eta(\boldsymbol{x}) \cdot \frac{\delta}{\delta \boldsymbol{A}(\boldsymbol{x})} \,, \tag{6.78}$$

and

$$\hat{E}_{lm} = \int_{S_R} d\omega \, R^2 Y_{lm}(\omega)\boldsymbol{n} \cdot \hat{\boldsymbol{E}}(\omega). \tag{6.79}$$

Note that the vector fields X^η just generate gauge transformations $\boldsymbol{A} \to \boldsymbol{A} + \boldsymbol{\nabla}\eta$. If the charge density were expressed in terms of the field variables of a charged field, e.g. $\rho = \psi^\dagger\psi$, the constraint would express the invariance of Ψ under simultaneous gauge transformations of $\boldsymbol{A}$ and ψ. The operators $\hat{E}_{lm}$ essentially measure the multipole moments of $\boldsymbol{E}$'s normal component on the boundary 2-sphere. In particular, the charge operator is given by $\hat{Q} = \sqrt{4\pi}\hat{E}_{00}$, so that

$$\hat{Q}\Psi = \sqrt{4\pi}\hat{f}_{00}\Psi = -\mathrm{i}\sqrt{4\pi}\frac{\partial}{\partial\lambda_{00}}\Psi. \tag{6.80}$$

A charge superselection rule, stating that all accessible observables commute with the charge operator, is equivalent to the impossibility of localizing the system in the λ_{00}-coordinate. For example, the restriction to local observables, that is observables of the form $\boldsymbol{A}(\eta)$, where η is zero in a neighborhood of the spatial boundary, enforces all $\hat{f}_{lm}$ to lie in the center of the algebra of observables (the center is thus infinite dimensional). This rich superselection structure, labeled by all multipole components of $\boldsymbol{n} \cdot \boldsymbol{E}$, can also be derived on a more rigorous basis (Buchholz 1982, Fröhlich et al. 1979, Zwanziger 1976). The restriction to local observables is usually motivated by taking the limit $R \to \infty$. But for long-ranged field configurations the need to introduce the variables λ and f persists! The associated degrees of freedom exist, even if one pushes them to "infinity". So even in this mathematically idealizing case the boundary should be considered as being *in* and not *of* the physical world. Also, since amongst the multipoles only charge is Lorentz-invariant, it is clear that the individual sectors with non-vanishing multipoles higher than the zeroth order cannot reduce the Lorentz group (Fröhlich et al. 1979). Let us first concentrate on the monopole sectors.

We have seen that the charge operator generates translations in the coordinate λ_{00}. We stressed the necessity to include the corresponding degree of freedom in the dynamical description. This implies that the motion generated by $\hat{Q}$ must in principle change the physical state. In other words, it is a symmetry and not a gauge redundancy. This can in fact be seen directly using the constraint (6.72). Integrating it over a test field η and taking $R \to \infty$ leads to

$$G(\eta) = -\rho(\eta) - \int_{\mathbb{R}^3} d^3x \, \boldsymbol{\nabla}\eta \cdot \boldsymbol{E} + \lim_{R\to\infty} \int_{S_R} d\omega \, R^2\eta \, \boldsymbol{n} \cdot \boldsymbol{E}. \tag{6.81}$$

This phase space functional generates gauge transformations iff the surface integral vanishes, which for long-ranging fields ($\|\boldsymbol{E}(r \to \infty)\| \sim 1/r^2$) implies

$\eta(r \to \infty) = 0$. The redundancy group which, by definition, is generated by the constraints thus consists of all asymptotically trivial gauge transformations whose group we call K, using the notation introduced in Sect. 6.3.2. Since the whole theory is clearly covariant under all gauge transformations in $\bar{G}$, the quotient $\mathcal{S} = \bar{G}/K$ of "global" U(1) phase transformations corresponds to proper physical symmetries, as discussed before. This is the basic and crucial difference between local and global gauge transformations.

We have so far focused on the charge superselection operator $\hat{f}_{00}$, although the foregoing considerations make it clear that by the same argument any two different asymptotic flux distributions also define different superselection sectors of the theory. Do we expect these additional superselection rules to be physically real? First note that for $l > 0$ the f_{lm} are not directly related to the multipole moments of the charge distributions, as the latter fall-off faster than $\frac{1}{r^2}$ and are hence not detectable on the sphere at infinity. Conversely, the higher multipole moments f_{lm} are not measurable (in terms of electromagnetic fields) within any finite region of space-time, unlike the charge, which is tight to massive particles; any finite sphere enclosing all sources has the same total flux. But the f_{lm} can be related to the kinematical state of a particle through the retarded Coulomb field. In fact, given a particle with constant momentum $\boldsymbol{p}$, charge e and mass m, one obtains for the electric flux distribution, $\varphi_{\boldsymbol{p}}$, at time t on a sphere centered at the instantaneous (i.e. at time t) particle position:[40]

$$\varphi_{\boldsymbol{p}}(\boldsymbol{n}) = \frac{em^2}{4\pi} \frac{[\boldsymbol{p}^2 + m^2]^{\frac{1}{2}}}{[(\boldsymbol{p} \cdot \boldsymbol{n})^2 + m^2]^{\frac{3}{2}}} \ . \tag{6.83}$$

---

[40] Formula (6.83) requires a little more explanation: for a particle with general trajectory $\boldsymbol{z}(t)$ let t' be the retarded time for the space-time point $(\boldsymbol{x}, t)$, i.e. $t' = t \quad \|\boldsymbol{x} \quad \boldsymbol{z}(t')\|$ ($c = 1$ in our units). Now we can use the well known formula for the retarded electric field (e.g. eq. (14.14) in the book by Jackson (1975)) and compute the flux distribution on a sphere which lies in the space of constant time t, where it is centered at the retarded position $\boldsymbol{z}(t')$ of the particle. This flux distribution can be expressed as function of the retarded momentum $\boldsymbol{p}' := \boldsymbol{p}(t')$ and the retarded direction $\boldsymbol{n}' := [\boldsymbol{x} - \boldsymbol{z}(t')]/\|\boldsymbol{x} - \boldsymbol{z}(t')\|$ as follows ($E' := \sqrt{\boldsymbol{p}'^2 + m^2}$):

$$\varphi'_{\boldsymbol{p}'}(\boldsymbol{n}') = \frac{em^2}{4\pi} \frac{1}{[E' - \boldsymbol{p}' \cdot \boldsymbol{n}']^2}. \tag{6.82}$$

If the particle moves with *constant* velocity $\boldsymbol{v} := \dot{\boldsymbol{z}}$, the expression for the retarded Coulomb field can be rewritten in terms of the instantaneous position $\boldsymbol{z}(t)$ by using $\boldsymbol{z}(t) = \boldsymbol{z}(t') + \boldsymbol{v}\|\boldsymbol{x} - \boldsymbol{z}(t')\|$. With respect to this center it is purely radial. Then one calculates the flux distribution on a sphere which again lies in the space of constant time t, but now centered at $\boldsymbol{z}(t)$ rather than $\boldsymbol{z}(t')$. This function can be expressed in terms of the instantaneous direction $\boldsymbol{n} := [\boldsymbol{x} - \boldsymbol{z}(t)]/\|\boldsymbol{x} - \boldsymbol{z}(t)\|$ and the instantaneous momentum $\boldsymbol{p} := \boldsymbol{p}(t)$. One obtains (6.83).

Hence different incoming momenta would induce different flux distributions and therefore lie in different sectors. Given that these sectors exist, this means that different incoming momenta cannot be coherently superposed and no incoming localized states be formed, unless one also adds the appropriate incoming infrared photons to just cancel the difference of the asymptotic flux distributions. This is achieved by imposing the "infrared coherence condition" of Zwanziger[41] (1976), the effect of which is to "dress" the charged particles with infrared photons which just subtract their retarded Coulomb fields at large spatial distances. Hence coherent superpositions of particles with different momenta can only be formed if they are dressed by the right amount of incoming infrared photons.

As a technical aside we remark that this can be done without violating the Gupta-Bleuler transversality condition $k^\mu a_\mu(k)|in\rangle = 0$ in the zero-frequency limit, precisely because of the surface term (6.62), as was first pointed out by Gervais and Zwanziger (1980). This resolved an old issue concerning the compatibility of the infrared coherence condition on one hand, and the Gupta-Bleuler transversality condition on the other (Haller 1978, Zwanziger 1978). From what we said earlier concerning the consistency of the variational principle in the presence of charged states, such an apparent clash of these two conditions had to be expected: without the surface variables one cannot maintain gauge invariance at spatial infinity (i.e. in the infrared limit) and at the same time include charged states. In the charged sectors the longitudinal infrared photons correspond to real physical degrees of freedom and it will naturally lead to inconsistencies if one tries to eliminate them by imposing the Gupta-Bleuler transversality condition also in the infrared limit. However, a gauge symmetry in the infrared limit can be maintained if one adds the asymptotic degrees of freedom in the form of surface terms.

These remarks illustrate how the rich superselection structure associated with different asymptotic flux distributions f_{lm} renders the problem of characterizing state spaces in QED for charged sectors fairly complicated. This problem has been studied within various formalisms including algebraic QFT (Buchholz 1982) and lattice approximations, where the algebra of observables can be explicitly presented (Kijowski et al. 1997). However, all this takes for granted the existence of the superselection rules, whereas we would like to understand dynamically whether localizations of the system in the degrees of freedom labeled by λ_{lm} are unstable against decoherence. In this case decoherence will prevent us from forming incoming localized wave packets of charged, *undressed* (in the sense above) particles, i.e. coherent superpositions of asymptotic flux distributions from the sectors with $l \geq 1$.

---

[41] Basically it says that the incoming scattering states should be eigenstates to the photon annihilation operators $a_\mu^{in}(k)$ in the zero-frequency limit.

6.4.2 Poincaré Charges in General Relativity

Let us digress slightly and perform a comparison with general relativity. Without going into too much detail we note that in the hamiltonian formulation the canonical variables for the gravitational field are given by a riemannian metric, $g_{ik}(x)$, on a 3-manifold Σ (the instant $t = \text{const.}$) and its conjugate momentum $\pi^{ik}(x)$. The long-ranged field situation corresponds here to asymptotically flat metrics and an $\sim 1/r^2$ fall-off for $\pi^{ik}(x)$. That is, in a suitable coordinate system we have for large r

$$g_{ik}(x) = \delta_{ik} + \frac{a_{ik}(x/r)}{r} + O(1/r^2) , \tag{6.84}$$

$$\pi^{ik}(x) = \frac{b^{ik}(x/r)}{r^2} + O(1/r^3) . \tag{6.85}$$

Analogous to the Gauß constraint in Maxwell theory is the so-called diffeomorphism constraint in general relativity. It reads

$$D^k(x) := 2\nabla_i \pi^{ik}(x) - T^{0k}(x) = 0 , \tag{6.86}$$

where $T^{0k} = t_\mu T^{\mu k}$, $t^\mu = \text{unit normal to } \Sigma$, is the time-space component of the energy momentum tensor of the matter fields. ∇_k denotes the covariant derivative on Σ. It should be understood as a local (but not ultralocal, since first derivatives occur) functional of $g_{ik}(x)$. Smearing it with a test vector field $\eta_k(x)$ we obtain a functional that is differentiable with respect to the canonical variables, provided the obvious surface term vanishes:

$$I^\eta = \lim_{R \to \infty} 2 \int_{S_R} d\omega \ R^2 n_i \eta_k \pi^{ik} = 0 . \tag{6.87}$$

Consistency with the fall-off conditions (6.84-6.85) implies $\eta^k(r \to \infty) = 0$. If this is satisfied one can compute the Poisson bracket of $C(\eta) = \int d^3x \ C^k \eta_k$ with the canonical variables and show that it just generates asymptotically trivial diffeomorphisms, just as the Gauß constraint was shown to generate asymptotically trivial gauge transformations. These form the redundancy group K in the sense of Sect. 6.3.2. Again, the invariance group is larger and given by all diffeomorphisms on Σ that preserve the asymptotic form (6.84-6.85). This is the case for the asymptotically euclidean transformations

$$\eta_i = t_i + O(1/r), \ t_i = \text{const.} , \tag{6.88}$$

$$\eta_i = \varepsilon_{ijk}\omega^j x^k + O(1/r), \ \omega^j = \text{const.} , \tag{6.89}$$

representing translations and rotations respectively. If the constraints are satisfied, the canonical generators for asymptotic translations and rotations are thus given by

$$t^i P_i = \lim_{R \to \infty} t^i 2 \int_{S_R} d\omega \ R^2 n^k \pi^{ki} , \tag{6.90}$$

$$\omega^i J_i = \lim_{R \to \infty} \omega^i 2 \int_{S_R} d\omega \ R^2 \varepsilon_{ijk} x^j n^l \pi^{lk} , \tag{6.91}$$

whose interpretation is that of total linear and angular momentum.

Thus, in general relativity, total momentum and angular momentum are expressible as asymptotic charges which are determined by surface integrals over arbitrarily large spheres, just as the electric charge in the previous example.[42] There are many well known solutions with non-vanishing such "asymptotic" charges, like for example a rotating black hole. A hamiltonian formulation quite analogous to the one given above can also be performed here. The extra configuration variables are given by six translational $\{X^i\}$ and rotational $\{\Theta^i\}$ coordinates canonically conjugate to P_i and J_i respectively. Their interpretation is to spatially locate the system relative to the asymptotic coordinate system with respect to which the conditions (6.84-6.85) of asymptotic flatness were posed. Here the variables $\{X^i, \Theta^i\}$ are analogous to λ_{00} in the previous case. Here, as there, a shift in these coordinates costs action and must be interpreted as a change in physically existent degrees of freedom, thus representing a symmetry. In a quantum theory of this system the state function Ψ then also depends on the asymptotic "embedding variables" $\{X^i, \Theta^i\}$ besides being a functional of $g_{ik}(x)$ (and other matter fields, which we neglect here): $\Psi = \Psi[g_{ik}(x), X^i, \Theta^i)$. In full analogy to (6.77), additional constraints will relate their canonically conjugate momenta to the well known ADM-expressions for the overall momentum (Regge and Teitelboim 1974)

$$-\mathrm{i}\,\frac{\partial}{\partial X^i}\Psi = \hat{P}_i\psi, \qquad (6.92)$$

where $\hat{P}_i$ is some suitable operator construct of expression (6.90) built from $\hat{g}_{ik}$ and $\hat{\pi}^{ik}$. The analog of (6.76) is given by imposing the diffeomorphism constraint (6.86) on Ψ, which implies that Ψ takes the same value on metrics g_{ij} which are related by an asymptotically trivial diffeomorphism. This is sometimes expressed by saying that Ψ depends on the metric g_{ik} only through the geometry $^3\mathcal{G}$ defined by it (cf. Sect. 4.2).

At this point is should be clear that we can attempt the same argument for superselection rules as in the electromagnetic case. Since total linear and angular momentum are given by surface integrals over arbitrarily large two-spheres, they should commute with all local observables. These observables cannot localize the system in the $\{X^i, \Theta^i\}$-frame, just as they could not localize the electromagnetic system in the λ_{00} coordinate. In comparison, for gravity the question whether these charges are observables is connected to the question about the physical realizability of the asymptotic coordinate system. Here we take the view that a positioning with respect to some "absolute space" is meaningless unless it is really meant with respect to some *physically* defined asymptotic space. A statement of non-localizability in $\{X^i, \Theta^i\}$-space is then always understood in an approximate sense ("looking at fixed stars in

---

[42] A more careful analysis shows that in fact one can implement the whole Poincaré group at infinity (Regge and Teitelboim 1974, Beig and Ó Murchadha 1987).

a cloudy night"). Superpositions of different momenta are then, in principle, possible and no fundamental superselection rule holds for them.

Although the interpretation of, say, λ_{00} is less obvious, the situation there seems in principle entirely analogous.[43] Since, after all, we do experimentally encounter a charge superselection rule, we must now ask for its dynamical origin.

6.4.3 Decoherence and Charge Superselection Rules

An obvious candidate for a dynamical origin of the (apparent) charge super-selection rule lies in the ubiquitous mechanism of decoherence (Giulini et al. 1995). As already explained in Sect. 4.1, there is strong evidence for the effectiveness of the decoherence mechanism when applied to a system of charges and fields. Here we are interested in the case where the field constitutes the irrelevant variables, which was already discussed in Sect. 4.1.1.

How could this apply to the case of isolated charged configurations as discussed above? We have seen that superpositions of different charged states could be distinguished from a corresponding mixture if localizations in the variable λ_{00} could be performed, which deprives the charge from lying in the center of the algebra of observables. But we have seen that this would necessarily involve the vector potential at spatial "infinity". An example is the Mandelstam variable (Mandelstam 1962)

$$\Phi(\boldsymbol{x}) = \phi(\boldsymbol{x}) \exp\left(-\mathrm{i} \int_{\infty}^{\boldsymbol{x}} \boldsymbol{A} \cdot d\boldsymbol{s}\right), \qquad (6.93)$$

where the integration goes along an arbitrary path coming from "infinity" to the end point $\boldsymbol{x}$. For more general variables of this type, see Bogolubov et al. (1990), eq. (10.90a). It is easily seen to be invariant under asymptotically trivial gauge transformations of the form $\phi \to \phi \exp(\mathrm{i}\lambda)$, $\boldsymbol{A} \to \boldsymbol{A} + \boldsymbol{\nabla}\lambda$, which implies that it (Poisson-) commutes with the Gauß constraint. We therefore call it an observable[44] in the sense that it maps physical states – those annihilated by the Gauß constraint – to physical states. However, it is also easily

---

[43] Outside field theory, a formal analogy between charge and linear or angular momentum has been used in the context of superselection rules to derive the apparently paradoxical result that the same arguments that lead to charge superselection rules also imply to such rules for the momenta (Aharonov and Susskind 1967a, Lubkin 1970).

[44] The term *observable* here should not be understood as implying a knowledge of how to realistically construct the necessary measurement device. In contexts where one intends to restrict oneself by hand to local observables the following terminology is sometimes employed: (Bogolubov et al. 1990, Chap. 10): Quantities commuting with the full gauge group are called *observables* whereas those that commute only with the Gauß constraint (i.e. asymptotically trivial gauge transformations) are called *physical quantities*. This makes the Mandelstam variable a "physical quantity" which is "non-observable". We shall not follow this terminology.

seen *not* to be invariant under long-ranged gauge transformations, like those generated by $\hat{Q}$ (i.e. constant λ). It has therefore non-vanishing matrix elements between different charge sectors. In particular, it may generate charged states from the vacuum, which is impossible for local observables (Ferrari et al. 1974 and 1977). That charge commutes with local observables on physical states may indeed be interpreted as a direct consequence of Gauß' law, as already anticipated in (Haag 1963), although its application to operator-valued quantities has to be defined carefully (Orzalesi 1970). Granted this, the argument may in fact be summarized in the following most simple form (cf. Strocchi and Wightman 1974), where we shall drop the hats on operators: The charge

$$Q = \lim_{R \to \infty} \int_{\|\boldsymbol{x}\| \leq R} d^3x \; \rho(\boldsymbol{x}, t) \tag{6.94}$$

can be rewritten in terms of the electric field by using the operator equation $\boldsymbol{\nabla} \cdot \boldsymbol{E} = \rho$ in the appropriate way. Given two physical states, $| \, \Phi \rangle$ and $| \, \Psi \rangle$, we then have for any local observable A

$$\langle \Psi \, | \, [Q, A] \, | \, \Phi \rangle = \lim_{R \to \infty} \left\langle \Psi \left| \int_{\|\boldsymbol{x}\| \leq R} d^3x \; [\rho(\boldsymbol{x}, t), A] \right| \Phi \right\rangle$$
$$= \lim_{R \to \infty} \left\langle \Psi \left| \int_{S_R} d\omega \; R^2 \; \boldsymbol{n} \cdot [\boldsymbol{E}(\boldsymbol{x}, t), A] \right| \Phi \right\rangle = 0 \; . \tag{6.95}$$

The last equality ($=0$) follows because eventually S_R lies entirely in the causal complement of the support of A. It is of course a physical question whether non-local operators are practically available and whether such phases at "infinity" can actually be measured in realistic experiments. (For comparison, in the Kaluza-Klein approach to electric charge, λ_{00} would essentially be the intrinsic scale of the fifth dimension.) But we emphasize that the variables at infinity are not only relevant for the classical variational principle. As we already discussed above, they play an important rôle for the infrared structure of QED.

One can, however, envisage an effective mechanism that prevents us from seeing such superpositions of charge states. Because of Gauß' law, Coulomb fields carry information about the charge at any distance, thereby decohering a superposition of different charges in any bounded subsystem of an infinite universe. This "instantaneous" action of decoherence at an arbitrary distance by means of the Coulomb field gives it the appearance of a kinematical effect, while it is in fact based on the dynamical law of charge conservation. Such a time-independent charge must then always have been "measured" by the asymptotic field and thus has always been "decohered". Here, as in the argument in (6.95), we emphasize the importance that charge is measurable on spheres S_R of arbitrarily large radius. If the Gauß constraint did not hold (and, consequently, Coulomb fields did not exist), this would not be possible since then one would have only had radiation fields which propagate along the future light cone.

Whereas the arguments so far apply to global superpositions of charged states, the more relevant case for laboratory physics is given by local superpositions. Consider the state Ψ of an isolated system – decohered by its Coulomb field – of fixed charge Q. With respect to two subsystems the state may be decomposed into individual charge eigenstates:

$$\Psi = \int dq \; f(q)|q\rangle \otimes |Q - q\rangle. \tag{6.96}$$

Of course, a global gauge transformation no longer leads to different phases for the various components of such a state, but only to one global phase factor. Consequently, one does not have to invoke a non-commuting variable at "infinity" to verify this superposition. An example of such a "local superposition" can be found in the BCS state of superconductivity,

$$\Psi_{\mathrm{BCS}} = \prod_k \left(1 + v(k)a_k^\dagger a_{-k}^\dagger\right) | 0\rangle. \tag{6.97}$$

The relative phases of two superconductors (contained in $v(k)$) can even be measured in Josephson junctions.

If it were possible to spatially separate a charge from its anti-charge in order to recombine them later, it would only be due to the irreversible "measurements" by the field that "information" about their temporary separation (that is, about their dipole moment and higher moments) remained present in the Universe. In particular, the components of a globally neutral superposition such as

$$\phi_+(1)\phi_-(2) + e^{i\alpha}\phi_-(1)\phi_+(2) , \tag{6.98}$$

with spatially separated states 1 and 2, would *remain* decohered if they were brought together. Thus, although charge is always decohered in the subsystems of a state like (6.96), (as for any conserved additive quantity, see Lubkin (1970)), the important question is whether these subsystems can be coherently brought together again after having been spatially separated. Decoherence due to Coulomb fields is reversible, while the influence of radiation fields ("escaping photons") is irreversible and leads to an upper bound for the coherent separation of charge pairs. See Chap. 4.1.1 for more details and quantitative estimations, and also Chap. 12 of (Breuer and Petruccione 2002a).

If one succeeded in reversibly preparing the above α-dependent superposition, one could also "measure" the charge of the state $\phi(1)$ by means of an interaction with a spinor field χ:

$$\phi_\pm(1)\chi_0 \rightarrow \phi_\pm(1)\chi_\pm , \tag{6.99}$$

where $\chi_\pm$ indicates charge dependence of the final state, not charge itself. After recombination of the charges one would be left with an observable superposition

$$\phi_{\text{neutral}}(\chi_+ + e^{i\alpha}\chi_-) \,, \tag{6.100}$$

and would thus have verified the existence of the above superposition.

We also note that, interestingly, decoherence is able to distinguish between action-at-a-distance theories (like the absorber theory by Wheeler and Feynman) and local field theories: in the former case decoherence would only be achieved after absorption has taken place, i.e. usually much later than in the latter case.

Finally we wish to comment on mass superselection in the context of general relativity (cf. Dominguez et al. 1997). We said above that in fact all ten Poincaré charges can be written as surface integrals over S_R for $R \to \infty$ (Beig and Ó Murchadha 1987). The expression for total mass is given by (restoring the gravitational constant G and velocity of light c)

$$M = \lim_{R \to \infty} \left\{ \frac{1}{16\pi G c^2} \int_{S_R} d\omega \ R^2 n^i (\partial_k g_{ik} - \partial_i g_{kk}) \right\} \,, \tag{6.101}$$

which in the canonical formalism appears as conjugate momentum to a global "clock" variable at "infinity", which we call λ, since it corresponds to the variable λ in the previous section. As discussed there, in order to observe states localized in λ one needs to observe coherent superpositions of mass eigenstates. But here the same kind of arguments that apply to the case of charge superpositions can be applied, at least qualitatively. On the quantitative side one has of course to take into account the different interaction strength. For example, as far as the interaction with retarded fields is concerned, one would expect the typical electromagnetic radiation factor, $2e^2/3c^3$, to be replaced by the typical gravitational radiation factor, $GM^2/5c^5$.

7 Open Quantum Systems

J. Kupsch

To understand quantum dynamics it is essential to realize that observable systems are always subsystems of a larger one – ultimately of the whole universe. The notion of an isolated system makes sense only as an approximation depending on the concrete situation. This point of view has been emphasized in the preceding chapters as the basis of the whole book. This chapter presents a self-contained discussion of the exact dynamics of a subsystem and of the Markovian approximation including the stochastic Schrödinger equations. These equations are then formally identical with stochastic modifications of the Schrödinger equation which have been proposed by some authors as a priori modifications of the Hamiltonian dynamics and which will be discussed in Chap. 8.

The mathematical notations of this chapter are shortly reviewed in Sect. 7.7.1. For a Hilbert space $\mathcal{H}$ the set of all trace class or nuclear operators is denoted by $\mathcal{T}(\mathcal{H})$. The subset of statistical operators or density matrices is $\mathcal{D}(\mathcal{H})$; these are positive operators with trace 1. The set $\mathcal{T}(\mathcal{H})$ is a Banach space: a closed linear space with the trace norm (7.87). Observables are represented by self-adjoint operators on $\mathcal{H}$. The $\mathbb{R}$-linear space $\mathcal{B}_{\mathbb{R}}(\mathcal{H})$ of all bounded self-adjoint operators is a Banach space with the operator norm (7.84).

In this chapter we use the notion "open system" for a system S which interacts with an "environment" E, such that the total system $S + E$ satisfies the usual unitary dynamics generated by a Hamiltonian. The system S is singled out by the fact that all observations refer only to this subsystem. The Hilbert space $\mathcal{H}_{S+E}$ of the total system $S + E$ is taken as the tensor space $\mathcal{H}_S \otimes \mathcal{H}_E$ of the Hilbert spaces for S and for E. The projection methods presented in Sect. 7.2 allow more general types of subsystems; but the main results of this chapter refer to the tensor construction. If the state of the total system is $\rho \in \mathcal{D}(\mathcal{H}_{S+E})$, then the state of the system S is given by the reduced statistical operator $\rho_S = \mathrm{tr}_E \rho$. The partial trace

$$\rho \in \mathcal{D}(\mathcal{H}_{S+E}) \to \mathrm{tr}_E \rho \in \mathcal{D}(\mathcal{H}_S) \tag{7.1}$$

is defined by the identity $\langle \Psi \mid \mathrm{tr}_E \rho \mid \Phi \rangle = \sum_n \langle \Psi \otimes f_n \mid \rho \mid \Phi \otimes f_n \rangle$, where f_n, $n = 1, 2, \ldots$ is a complete orthonormal system of $\mathcal{H}_E$ and Ψ and Φ are arbitrary elements of $\mathcal{H}_S$. The expectation values of observables $A \otimes I_E \in \mathcal{B}_{\mathbb{R}}(\mathcal{H}_{S+E})$ with $A \in \mathcal{B}_{\mathbb{R}}(\mathcal{H})$ then satisfy

$$\mathrm{tr}_S \rho_S A = \mathrm{tr}_{S+E} \rho(A \otimes I_E). \tag{7.2}$$

The reduced statistical operator is uniquely determined by this relation. The reduced statistical operator and the partial trace have already been extensively used in Chaps. 2 and 3. Since all information about a physical subsystem is given by a statistical operator, we shall here refer to the statistical operator ρ_S as the "state" of the subsystem.[1]

The space of observables $\mathcal{B}_{\mathbb{R}}(\mathcal{H})$ can be extended to the complex space $\mathcal{B}(\mathcal{H}) = \mathcal{B}_{\mathbb{R}}(\mathcal{H}) \oplus i\, \mathcal{B}_{\mathbb{R}}(\mathcal{H})$ of all bounded linear operators on $\mathcal{H}$. The relation (7.2) has first been established for self-adjoint $A \in \mathcal{B}_{\mathbb{R}}(\mathcal{H}_S)$, but it is valid for all $A \in \mathcal{B}(\mathcal{H}_S)$. This possibility to use the larger space $\mathcal{B}(\mathcal{H}_S)$ instead of $\mathcal{B}_{\mathbb{R}}(\mathcal{H}_S)$ is a general property of quantum mechanics. This extension simplifies some mathematical arguments not only because $\mathcal{B}(\mathcal{H}_S)$ is a complex linear space, but also because $\mathcal{B}(\mathcal{H}_S)$ is a $\mathbf{C}^*$-algebra. Thereby the usual operator product is the product of the algebra and the transition to the adjoint operator $A \to A^\dagger$ is the antilinear $*$-operation of this algebra.

As mentioned above we assume the usual unitary dynamics for the total system, i.e. $\rho(t) = U(t)\rho(0)U^\dagger(t)$ with the unitary group $U(t)$, generated by the total Hamiltonian. Except for the trivial case that S and E do not interact, the dynamics of the reduced statistical operator

$$\rho_S(t) = \mathrm{tr}_E U(t)\rho(0)U^\dagger(t) \tag{7.3}$$

is no longer unitary, and it is the purpose of this chapter to evaluate (7.3) in some detail. Many of these investigations go back to the study of irreversibility and thermodynamic behaviour of physical systems. But during the last twenty years this subject has entered with increasing importance the study of the measurement process and, moreover, the interpretation of quantum mechanics in general. The dynamics of open systems will be investigated in Sects. 7.1–7.6 with an emphasis on fundamental aspects in the following steps:

1. The mathematical structure of the reduced dynamics (7.3) is studied in some detail. Then more general projection techniques from a large system $S + E$ to a subsystem S are introduced. The resulting dynamical equation for the reduced statistical operator is usually called a generalized or premaster equation (Zwanzig 1960a, b, Favre and Martin 1968, Kübler and Zeh 1973, Haake 1973).

2. The Markovian approximation of the dynamics of the subsystem and the resulting quantum dynamical semigroups for the time evolution of the states of the system S are discussed. The Markovian approximation simplifies the equations considerably. It is valid only under very specific initial conditions, which seem to be satisfied not only for some phenomenological models of nonequilibrium thermodynamics, but also for the dynamics of subsystems on a more fundamental level.

---

[1] See also Sect. 6.1.; it should be kept in mind that the state of the total system cannot be recovered from these "states" of the subsystem.

3. An investigation of quantum stochastic processes – usually called stochastic Schrödinger equations – for effective wave functions of the system S is presented. The expectation values calculated from these stochastic functions agree with the Markovian semigroup dynamics of the statistical operator. Stochastic Schrödinger equations have also been introduced as a priori modifications of the unitary dynamics in order to represent the "collapse" of the wave functions. This approach will be discussed in detail in Chap. 8.
4. Environmentally induced superselection sectors are investigated on the basis of exactly soluble Hamiltonian models. Thereby the Hamiltonian determines whether such sectors emerge, but the time scale of this process is mainly determined by the initial state of the environment.

7.1 Reduced Dynamics

In this section we derive some structural properties of the reduced dynamics (7.3). These properties are considered as essential for any dynamics in quantum theory, and also approximations as discussed in Sect. 7.4 should obey them.

We assume that the Hilbert space of the total system $S + E$ is the tensor space $\mathcal{H}_S \otimes \mathcal{H}_E$ of the Hilbert spaces for S and for E. The dynamics (7.3) of the reduced statistical operator $\rho_S(t)$ is uniquely determined, if the initial state $\rho(0)$ of the total system is given. But for the usual applications one would like to calculate $\rho_S(t)$ as a functional of the initial state $\rho_S(0) = \mathrm{tr}_E \rho$ of the system S. For that purpose we need an injective mapping $\mathbf{F} : \rho_S \in \mathcal{D}(\mathcal{H}_S) \to \rho = \mathbf{F}(\rho_S) \in \mathcal{D}(\mathcal{H}_S \otimes \mathcal{H}_E)$. Then

$$\rho_S(0) \to \rho_S(t) = \Phi_t(\rho_S) := \mathrm{tr}_E U(t)\, \mathbf{F}\left(\rho_S(0)\right) U^\dagger(t) \tag{7.4}$$

is the required time evolution of the system S. An additional restriction for $\mathbf{F}$ follows from (7.4) at time zero: $\mathrm{tr}_E \mathbf{F}\left(\rho_S(0)\right) = \rho_S(0)$. In the literature the usual solution for this problem is the product ansatz

$$\mathbf{F}(\rho_S) = \rho_S \otimes \rho_E \tag{7.5}$$

with a fixed reference state $\rho_E \in \mathcal{D}(\mathcal{H}_E)$ of the environment, see e. g. (Favre and Martin 1968, Davies 1976, Martin 1979). In this case $\mathbf{F}$ is a linear mapping from $\mathcal{T}(\mathcal{H}_S)$ into $\mathcal{T}(\mathcal{H}_S \otimes \mathcal{H}_E)$, and consequently the time evolution (7.4) is linear. Assuming linearity then it has been shown in (Kupsch, Smolyanov, and Sidorova 2001) that there is no other possibility for $\mathbf{F}$ than the product ansatz (7.5); all mappings $\mathbf{F} : \mathcal{D}(\mathcal{H}_S) \to \mathcal{D}(\mathcal{H}_S \otimes \mathcal{H}_E)$ which commute with mixing, i. e. $\mathbf{F}(\alpha_1 \rho_S^1 + \alpha_2 \rho_S^2) = \alpha_1 \mathbf{F}(\rho_S^1) + \alpha_2 \mathbf{F}(\rho_S^2)$ with $\alpha_{1,2} \geq 0$, $\alpha_1 + \alpha_2 = 1$, and which satisfy $\mathrm{tr}_E \mathbf{F}(\rho_S) = \rho_S$ have the structure (7.5). Hence, if we require that mixing commutes with the dynamics, the most general type of a reduced dynamics is

$$\rho_S \in \mathcal{D}(\mathcal{H}_S) \to \rho_S(t) = \Phi_t(\rho_S) := \mathrm{tr}_E U(t) \left(\rho_S \otimes \rho_E\right) U^\dagger(t) \in \mathcal{D}(\mathcal{H}_S), \tag{7.6}$$

where $\rho_E \in \mathcal{D}(\mathcal{H}_E)$ is a fixed reference state of the environment.[2]

It should be pointed out that the essential restriction in (7.6) is not the linearity, but the factorization with the lack of all correlations between system and environment. Also if ρ_E would be determined as a functional of ρ_S, e. g. $\rho_E = V\rho_S V^\dagger$ with a suitable isometric mapping $V : \mathcal{H}_S \to \mathcal{H}_E$, or as a more complicated non-linear function of ρ_S, the initial state factorizes. Starting from that factorizing initial state the dynamics (7.6) is exact and takes into account all correlations and (back-) reactions which are generated by the dynamics. The product ansatz for the initial state and the role of the reference state will further be discussed in Sects. 7.3 and 7.6.

We continue with some additional mathematical considerations. The reduced dynamics (7.6) $\Phi_t(\rho_S)$ is well defined for arbitrary $\rho_S \in \mathcal{T}(\mathcal{H}_S)$ and has the obvious properties [3]

$$\begin{aligned} &\Phi_t(\rho_S) \geq 0 \text{ if } \rho_S \geq 0, \\ &\|\Phi_t(\rho_S)\|_1 \leq \|\rho_S\|_1, \\ &\mathrm{tr}_S\Phi_t(\rho_S) = \mathrm{tr}_S\rho_S. \end{aligned} \tag{7.7}$$

Here $\|\cdot\|_1$ is the trace norm (7.87). Inserting complete orthogonal systems in (7.6) one obtains the following representation for Φ_t

$$\Phi_t(\rho_S) = \sum_n V_n(t)\rho_S V_n^\dagger(t), \tag{7.8}$$

where the number of terms can be infinite. The time dependent linear operators $V_n(t) \in \mathcal{B}(\mathcal{H}_S)$ are bounded and satisfy[4]

$$\sum_n V_n^\dagger(t)V_n(t) = 1_S. \tag{7.9}$$

From (7.6) and the identity $\mathrm{tr}_S\rho_S(t)A(0) = \mathrm{tr}_S\rho_S(0)A(t)$ we obtain the dynamics in the Heisenberg picture $A \in \mathcal{B}(\mathcal{H}_S) \to A(t) = \Phi_t^+(A) \in \mathcal{B}(\mathcal{H}_S)$ with[5]

$$\Phi_t^+(A) := \mathrm{tr}_E U^\dagger(t)\,(A \otimes I_E)\,U(t)\rho_E. \tag{7.10}$$

This linear mapping has the obvious properties

$$\begin{aligned} &\Phi_t^+(A) \geq 0 \text{ if } A \geq 0, \\ &\|\Phi_t^+(A)\| \leq \|A\|, \\ &\Phi_t^+(I_S) = I_S, \end{aligned} \tag{7.11}$$

---

[2] The linearity of the dynamics $\rho \to \Phi_t(\rho) = \rho(t)$ is often considered as fundamental property, see Axiom IX in (Mackey 1963) or (Primas 2000).

[3] A self-adjoint operator A on a Hilbert space $\mathcal{H}$ is denoted as *positive* if $\langle f \mid Af \rangle \geq 0$ for all $f \in \mathcal{H}$ in the domain of A. The usual notation for this property is $A \geq 0$.

[4] The convergence of this sum is meant as strong operator convergence.

[5] The notation $\Phi_t^+(A)$ emphasizes the duality relation $\mathrm{tr}_S\rho_S\Phi_t^+(A) = \mathrm{tr}_S\Phi_t(\rho_S)A$ between the Heisenberg and the Schrödinger picture.

where $\| \, . \, \|$ is the operator norm. From (7.8) we obtain the equivalent representation for $\Phi_t^+(A)$

$$\Phi_t^+(A) = \sum_n V_n^\dagger(t) A V_n(t). \tag{7.12}$$

The general properties (7.7) and (7.11) are a consequence of this representation and the normalization (7.9). The representations (7.8) and (7.12) can be considered as an essential structural attribute of quantum dynamics that has to be preserved also if approximations are made. In the mathematical literature linear transformations on the space of bounded operators, which have the representation (7.12) with bounded linear operators V_n for which the series $\sum_n V_n^\dagger(t) V_n(t)$ is (strongly) convergent, are called *completely positive*. They map positive operators into positive operators, but they are not fully characterized by this property. A simple example of a linear transformation $\Phi^+ : \mathcal{B}(\mathbb{C}^2) \to \mathcal{B}(\mathbb{C}^2)$ which maps positive 2×2 matrices into positive matrices, but which is not completely positive, is given in Sect. 7.7.2. The importance of complete positivity in quantum dynamics was first emphasized by Kraus (1971, 1983).

The mapping (7.8) for the statistical operator is just the dual version of a completely positive map. In finite dimensional Hilbert spaces the sum $\sum_n V_n V_n^\dagger$ converges, and the mapping (7.8) for the statistical operator is also completely positive, since in the definition the operators V_n and $V_n^\dagger$ can be interchanged. In infinite dimensional spaces it is possible that $\sum_n V_n V_n^\dagger$ diverges. Then (7.8) cannot be extended to arbitrary bounded operators, and Φ is not completely positive. In general, linear transformations $\Phi(\rho) : \mathcal{D}(\mathcal{H}) \to \mathcal{D}(\mathcal{H})$, which have a completely positive dual Φ^+, are called *quantum dynamical maps*, see (Alicki and Lendi 1987).

In the next section we present a more general formulation for a subsystem than just taking the partial trace, and in Sect. 7.3 we derive an integro-differential equation for the dynamics of a subsystem.

7.2 Projection Methods

The problem of the dynamics of a subsystem can be formulated in a rather general context. Let $\mathcal{H}$ be the Hilbert space of a system and $\mathcal{D}(\mathcal{H}) \subset \mathcal{T}(\mathcal{H})$ be the corresponding space of the states of the system, represented by statistical operators. Then Nakajima (1958) and Zwanzig (1960a, b) have introduced a projection technique to split the information contained in a statistical operator $\rho \in \mathcal{D}(\mathcal{H})$ into a "relevant" and an "irrelevant" part. The reduction to the relevant part which we are interested in for further considerations is obtained by a linear projection operator $\hat{P}$ defined on the Banach space $\mathcal{T}(\mathcal{H})$. There are also nonlinear projection techniques we shall not consider here; but see the comment at the end of this section. In the following we assume
  (i) $\hat{P}$ is a linear operator on $\mathcal{T}(\mathcal{H})$,
  (ii) $\hat{P}^2 = \hat{P}$,

(iii) $\hat{P}\rho \geq 0$ if $\rho \geq 0$,

(iv) $\operatorname{tr} \hat{P}\rho = \operatorname{tr}\rho$.

The first two properties define a projection operator on the linear space $T(\mathcal{H})$. The properties (iii) and (iv) imply that $\hat{P}$ is a continuous operator on $T(\mathcal{H})$ with $\left\|\hat{P}\rho\right\|_1 \leq \|\rho\|_1$ at least for self-adjoint $\rho \in T(\mathcal{H})$. Moreover these assumptions guarantee that $\hat{P}\rho$ is a state if ρ is one.

The operator $\hat{I} - \hat{P}$, where $\hat{I}$ is the identity mapping on $T(\mathcal{H})$, is again a projector satisfying $\hat{P} + (\hat{I} - \hat{P}) = \hat{I}$, $\hat{P}(\hat{I} - \hat{P}) = (\hat{I} - \hat{P})\hat{P} = 0$. As a consequence the Banach space $T(\mathcal{H})$ is split into two linearly independent, closed subspaces $T(\mathcal{H}) = \hat{P}T(H) \oplus (\hat{I} - \hat{P})T(H)$. For any statistical operator $\rho \in \mathcal{D}(\mathcal{H})$ we have

$$\rho = \hat{P}\rho + (\hat{I} - \hat{P})\rho \tag{7.13}$$

with the relevant part $\rho_{rel} = \hat{P}\rho \in \mathcal{D}(\mathcal{H})$ and the irrelevant part $\rho_{irrel} = (\hat{I} - \hat{P})\rho \in T(\mathcal{H})$. The statistical operator $\hat{P}\rho$ contains only part of the information which we may obtain from ρ, since it is restricted to a proper subspace of $T(\mathcal{H})$. On the other hand the specific choice of the projection introduces also information, therefore the von Neumann entropy of $\hat{P}\rho$ may eventually be larger than the entropy of ρ as we shall see below.

Given a Zwanzig projection $\hat{P}$ on $T(\mathcal{H})$ we can define a projection operator $\hat{Q} : B(\mathcal{H}) \to B(\mathcal{H})$ for the observables, which satisfies the duality relation

$$\operatorname{tr}(\hat{P}\rho)A = \operatorname{tr}\rho(\hat{Q}A) \tag{7.14}$$

for all $\rho \in T(\mathcal{H})$ and $A \in B(\mathcal{H})$. It is easy to verify that such an operator is a linear projection operator on $B(\mathcal{H})$ with the properties $\hat{Q}^2 = \hat{Q}$, $\hat{Q}A \geq 0$ if $A \geq 0$ and $\hat{Q}I = I$.

Since the projections $\rho \to \hat{P}\rho$ or $A \to \hat{Q}A$ are the starting point for a dynamical equation, we assume that $\hat{Q}$ is completely positive, or equivalently

(v) $\hat{P}$ is a quantum dynamical map, i. e. it has the representation

$$\hat{P}\rho = \sum_n V_n \rho V_n^\dagger \tag{7.15}$$

with bounded operators $V_n \in B(\mathcal{H})$, which satisfy $\sum_n V_n^\dagger V_n = I_\mathcal{H}$. This assumption is an additional demand, which does not follow from (i) - (iv).[6] The condition (v) is usually not stated explicitly, but it seems that all examples discussed in the literature satisfy this constraint.

The list of constraints (i) - (v) imposed on $\hat{P}$, or the equivalent list of constraints imposed on $\hat{Q}$, do not include hermiticity, because the spaces $T(\mathcal{H})$ and $B(\mathcal{H})$ do not have an inner product. But there is a subclass of

---

[6] In Sect. 7.7.2 we give an example of a projection $\hat{P}$ on $\mathcal{D}(\mathbb{C}^2)$ which satisfies (i) - (iv) and the hermiticity constraint (7.16), but (7.15) is not true.

Zwanzig projections, which are hermitean with respect to the inner product (7.86) of Hilbert–Schmidt operators

$$\mathrm{tr}(\hat{P}\rho_1)\rho_2 = \mathrm{tr}\,\rho_1(\hat{P}\rho_2). \tag{7.16}$$

In this case, the operator $\hat{P}$ is the restriction of $\widehat{Q}$ to the set of all trace class operators.

As the first type of projection operators we consider the restriction to the (block-) diagonal terms. Let P_n, $n = 1, 2, \ldots$ be a finite (or countable) exhaustive family of mutually exclusive self-adjoint projection operators on the Hilbert space $\mathcal{H}$ with $\sum_n P_n = I$ and $P_m P_n = 0$ if $m \neq n$. For instance, P_n can be chosen as projections on an orthonormal basis of $\mathcal{H}$, but we might also take projections on larger, even infinite dimensional subspaces. Then

$$\hat{P}\rho = \sum_n P_n \rho P_n \tag{7.17}$$

defines a projection on the space $\mathcal{T}(\mathcal{H})$ satisfying conditions (i)–(v): It selects the (block-) diagonal part of the statistical operator ρ as the relevant contribution.

The information contained in $P_n \rho P_n$ can be further restricted to $\mathrm{tr}\,(P_n \rho)$ which leads to the modified projection

$$\hat{P}\rho = \sum_n (\mathrm{tr}\,P_n \rho)\omega_n. \tag{7.18}$$

Here ω_n, $n = 1, 2, \ldots$ is a family of states $\omega_n \in \mathcal{D}(\mathcal{H})$ such that $P_n \omega_n = \omega_n$, i. e. the statistical operators ω_n are arbitrary states concentrated on the subspaces $P_n \mathcal{H}$. Only if the subspaces $P_n \mathcal{H}$ are finite dimensional, one can choose the uniform distribution $\omega_n = P_n(\dim P_n \mathcal{H})^{-1}$, and (7.18) is a *coarse graining* projection.

There is another type of projection operators which uses the tensor construction of Sect. 7.1. In order to reduce a state $\rho \in \mathcal{D}(\mathcal{H}_S \otimes \mathcal{H}_E)$ to $\rho_S \in \mathcal{D}(\mathcal{H}_S)$ there is the standard technique of taking the partial trace (7.1) which is a linear mapping $\rho \in \mathcal{T}(\mathcal{H}_S \otimes \mathcal{H}_E) \to \rho_S = \mathrm{tr}_E \rho \in \mathcal{T}(\mathcal{H}_S)$. But to obtain a mapping with image in $\mathcal{T}(\mathcal{H}_S \otimes \mathcal{H}_E)$ an additional "reference state" $\rho_E \in \mathcal{D}(\mathcal{H}_E)$ is used, and

$$\hat{P}\rho = (\mathrm{tr}_E \rho) \otimes \rho_E \tag{7.19}$$

is the definition for a linear projection operator on $\mathcal{T}(\mathcal{H}_S \otimes \mathcal{H}_E)$, which satisfies (i)-(v), see (Favre and Martin 1968, Davies 1976, Martin 1979). The operator (7.19) depends on the "reference state" ρ_E, but nevertheless the projection selects exactly the information of the reduced state $\mathrm{tr}_E \hat{P}\rho = \mathrm{tr}_E \rho = \rho_S$. The reference state $\rho_E \in \mathcal{D}(\mathcal{H}_E)$ is so far an arbitrary element of $\mathcal{D}(\mathcal{H}_E)$. But within a given physical context there are natural choices for this state, see Sect. 7.3. The uniform distribution $\rho_E = I_E(\dim \mathcal{H}_E)^{-1}$ is admitted only if $\mathcal{H}_E$ is a finite dimensional space. The dual projection $\widehat{Q}$ of (7.19) is

$$\widehat{Q}A = (\mathrm{tr}_E A\rho_E) \otimes I_E. \tag{7.20}$$

It should be noticed that the additional condition of hermiticity with respect to the Hilbert–Schmidt form (7.16) is satisfied by the following projections: the projection (7.17), the coarse graining projection (7.18) if $\omega_n = P_n(\dim P_n\mathcal{H})^{-1}$ for all n, and the projection (7.19) if $\rho_E = I_E(\dim \mathcal{H}_E)^{-1}$. The choice of the uniform (microcanonical) ensemble in the last two cases is of course only possible, if the dimensions of the spaces $P_n\mathcal{H}$ or $\mathcal{H}_E$ are finite.

A commonly accepted measure for the information content of a statistical operator is the *von Neumann entropy* $S(\rho) = -\mathrm{tr}(\rho \ln \rho)$. Since the projection $\rho \to \hat{P}\rho$ annihilates information, one expects an increase of this information entropy

$$S(\hat{P}\rho) \geq S(\rho). \tag{7.21}$$

But this inequality is not always satisfied by a (linear) Zwanzig projection, and we have to elucidate measures of information in more detail. A counter example to (7.21) can be easily constructed for projections on a finite dimensional Hilbert space $\mathcal{H}$, $\dim \mathcal{H} = N$. Then $S(\rho)$ has a unique absolute maximum at the uniform (microcanonical) distribution $\rho_{mc} = \frac{1}{N}I_{\mathcal{H}}$ with $S(\rho_{mc}) = \ln N$. Hence any projection on finite dimensional spaces with $\hat{P}\rho_{mc} \neq \rho_{mc}$ will decrease the entropy at least in the neighbourhood of ρ_{mc}.[7] One can derive the increase (7.21) of the von Neumann entropy for all Zwanzig projectors, which satisfy the assumptions (i) – (iv) and are hermitean with respect to the Hilbert–Schmidt form (7.16), see Sect. 7.7.3.[8] Hence the projector (7.17), the coarse graining operator (7.18) with finite rank projections P_n and the choice $\omega_n = P_n(\dim P_n\mathcal{H})^{-1}$, and the projection (7.16) with a finite dimensional reference space $\mathcal{H}_E$ and $\rho_E = I_E(\dim \mathcal{H}_E)^{-1}$ satisfy the inequality (7.21).

For all dynamical maps Λ and consequently for all projections (7.15) one can formulate a general statement about the loss of information due to the transition $\rho \to \Lambda\rho$. For that purpose one defines a more general entropy functional: the *relative entropy*

$$H(\rho \mid \rho') = \mathrm{tr}(\rho \ln \rho - \rho \ln \rho'). \tag{7.22}$$

---

[7] It is therefore easy to see that the projectors (7.18) and (7.19) violate (7.21), if $P_n\mathcal{H}$ and $\mathcal{H}_E$ are finite dimensional spaces and the reference states do not coincide with the uniform distributions, i. e. $\rho_{En} \neq P_n(\dim P_n\mathcal{H})^{-1}$ or $\rho_E \neq I_E(\dim \mathcal{H}_E)^{-1}$.

[8] The hermiticity (7.16) implies that $\hat{P}$ is contractive in the operator norm, $\left\|\hat{P}\rho\right\| \leq \|\rho\|$. This property and complete positivity can also be used to derive the inequality (7.21), see Proposition 2.1 in (Olkiewicz 2000).

This functional is positive for all states ρ, $\rho' \in \mathcal{D}(\mathcal{H})$: $H(\rho \mid \rho') \geq 0$, including $H = +\infty$, and the equality $H(\rho \mid \rho') = 0$ holds if and only if $\rho = \rho'$ (Lindblad 1973), see also as general references about entropy (Wehrl 1978) and (Thirring 1983).

The definition (7.22) is the quantum version (operator version) of the gain of information which measures the "distance" between two probability distributions in information theory, see (Kullback and Leibler 1951, Rényi 1961) and also the article by Cover in (Halliwell, Pérez-Mercader and Zurek 1994). For the use of relative entropy in classical statistical mechanics see (Jauch and Baron 1972) and (Lasota and Mackey 1994)[9]. If the statistical operators in (7.22) are related by mixing, there is a simple interpretation of the relative entropy. Let $\rho_i \in \mathcal{D}(\mathcal{H})$ be a family of states and $\lambda_i \geq 0$, $\sum_i \lambda_i = 1$. Then the von Neumann entropy $S(\rho)$ of the mixed state $\rho = \sum_i \lambda_i \rho_i$ is calculated as

$$S(\rho) = \sum_i \lambda_i S(\rho_i) + \sum_i \lambda_i H(\rho_i \mid \rho) \qquad (7.23)$$

and $H(\rho_i \mid \rho)$ is the additional amount of entropy obtained by mixing. If ρ is decomposed into pure states $\rho = \sum_i \lambda_i P_i$ with rank one projectors P_i then (7.23) allows to calculate the von Neumann entropy from the relative entropies as $S(\rho) = \sum_i \lambda_i H(\rho_i \mid \rho)$.

For statistical operators defined on a tensor space $\rho \in \mathcal{D}(\mathcal{H}_S \otimes \mathcal{H}_E)$ the relative entropy $H(\rho_S \otimes \rho_E \mid \rho)$ is the entropy of the correlations which appear in ρ, but are absent in the product $\rho_S \otimes \rho_E$ of the reduced statistical operators $\rho_S = \mathrm{tr}_E \rho$ and $\rho_E = \mathrm{tr}_S \rho$

$$\begin{aligned} S(\rho) &= S(\rho_S \otimes \rho_E) + H(\rho_S \otimes \rho_E \mid \rho) \\ &= S(\rho_S) + S(\rho_E) + H(\rho_S \otimes \rho_E \mid \rho), \end{aligned} \qquad (7.24)$$

see (Wehrl 1978). Given the statistical operators $\rho_S \in \mathcal{D}(\mathcal{H}_S)$ and $\rho_E \in \mathcal{D}(\mathcal{H}_E)$ the factorizing state $\rho_S \otimes \rho_E$ is therefore the state of lowest entropy among all states $\rho \in \mathcal{D}(\mathcal{H})$ which have ρ_S and ρ_E as their reduced states.

In dynamical theories of open systems the relative entropy measures the production of entropy within the system apart from the entropy flow coming from the environment (Schlögl 1966, Spohn 1978). Only for a system with a finite dimensional Hilbert space $\mathcal{H}$, $\dim \mathcal{H} = N$, are the von Neumann entropy and the relative entropy directly related. In that case the uniform or microcanonical distribution $\rho_{mc} = \frac{1}{N} I_\mathcal{H}$ is an admitted state with $S(\rho_{mc}) = \ln N$ and we obtain from (7.22) for arbitrary $\rho \in \mathcal{D}(\mathcal{H})$

$$S(\rho) = \ln N - H(\rho \mid \rho_{mc}). \qquad (7.25)$$

---

[9] This reference denotes the relative entropy as conditional entropy, and it uses a different sign convention.

Lindblad (1975) has derived the following general rule for the loss of information caused by a quantum dynamical map[10]

$$H(\Lambda\rho \mid \Lambda\rho') \leq H(\rho \mid \rho'). \qquad (7.26)$$

This rule applies consequently to all Zwanzig projections with a representation (7.15). The increase of the von Neumann entropy (7.21) follows from (7.26) with the help of (7.25) for systems with finite dimensional Hilbert spaces, if the uniform distribution ρ_{mc} is preserved, $\hat{P}\rho_{mc} = \rho_{mc}$.

In the next section we investigate the dynamics of a subsystem with the help of a projection operator $\hat{P}$. In this context we have to discuss time–reversal invariance. Let T be the usual antiunitary time reflection operator on $\mathcal{H}$ – for $\mathcal{L}^2$-spaces without spin it is just complex conjugation of all wave functions – then we shall call a projection operator $\hat{P}$ time–reversal invariant if $T\hat{P}T^\dagger=\hat{P}$. For the projection (7.17) this means $TP_nT^\dagger = P_n$, i. e. the subspaces $P_n\mathcal{H}$ have to be invariant under T. For (7.18) we need in addition $T\omega_nT^\dagger = \omega_n$, and for (7.19) the reference state has to be invariant, $T\rho_ET^\dagger = \rho_E$.

In the following we shall always assume that the definition of relevance, i. e. the definition of $\hat{P}$, does not distinguish between the directions of time. Then the measures of information for the relevant part, $S_{\mathrm{rel}}(\rho) = S(\hat{P}\rho)$ and $H_{\mathrm{rel}}(\rho \mid \rho') = H(\hat{P}\rho|\hat{P}\rho')$ have to be invariant under time reversal. But from either of the identities $S_{\mathrm{rel}}(\rho) = S_{\mathrm{rel}}(T\rho T^\dagger)$ or $H_{\mathrm{rel}}(\rho \mid \rho') = H_{\mathrm{rel}}(T\rho T^\dagger \mid T\rho'T^\dagger)$ we obtain $T\hat{P}T^\dagger=\hat{P}$. We therefore assume in the following that this symmetry is satisfied.

As already mentioned, one can generalize the Zwanzig projection method to nonlinear projections, e. g. to a separating ansatz

$$\hat{P}\rho=(\mathrm{tr}_E\rho) \otimes (\mathrm{tr}_S\rho), \qquad (7.27)$$

which satisfies the entropy inequality (7.21). Then the exact equation for the dynamics of $\rho_{\mathrm{rel}} = \hat{P}\rho$ becomes nonlinear, like the Boltzmann equation. Moreover, the Markovian approximation is also a nonlinear equation for ρ_{rel}, see (Grabert and Weidlich 1974). To some extent semigroup techniques are also available to treat such master equations (Alicki and Lendi 1987). But we shall not follow these approaches for two reasons. First, the usual dynamics in quantum mechanics commutes with the mixing property $\rho_{1,2} \in \mathcal{D}(\mathcal{H}) \to \alpha_1\rho_1 + \alpha_2\rho_2 \in \mathcal{D}(\mathcal{H})$ where $\alpha_{1,2} \geq 0$, $\alpha_1 + \alpha_2 = 1$. To obtain linear evolution equations for the statistical operator of a subsystem only linear projectors are required. Second, there is another more practical argument. Much of the recent progress in understanding decoherence and

---

[10] The minus sign in (7.25) and the decrease of the relative entropy in (7.26) are a consequence of the sign convention used in (7.22). That definition has the advantage that relative entropy is always positive. But in the literature the other convention, which leads to a negative relative entropy, is also used.

classical properties has been achieved with linear equations for the statistical operator, see Chap. 3. And in the following sections we want to present more about the mathematical foundations of exactly these results. The linearity of the equations for the statistical operator ρ is nevertheless consistent with the nonlinear stochastic Schrödinger equations for the wave function, see Sect. 7.5.

7.3 Generalized Master Equations

In this section we derive an integro-differential equation for the reduced dynamics using the general projection method of Sect. 7.2. These equations are rarely used in their exact form, but they are an important starting point for further approximations. We assume that the time evolution of $S + E$ is given by the unitary dynamics (7.3) with $U(t) = \exp(-itH_{S+E})$ where

$$H_{S+E} = H_S \otimes I_E + I_S \otimes H_E + H_{SE} \tag{7.28}$$

is the Hamiltonian of the total system, leading to the usual Liouville–von Neumann equation for ρ:

$$\frac{d}{dt}\rho(t) = -i[H_{S+E}, \rho(t)] = \widehat{L}\rho(t) \tag{7.29}$$

with the Liouville operator of the total system $\widehat{L}\rho := -i[H_{S+E}, \rho]$. The state of the subsystem S is represented by the reduced statistical operator (7.1) ρ_S. To calculate the dynamics of ρ_S it is convenient to use a projection operator $\hat{P}$ as introduced in Sect. 7.2. Following Zwanzig (1960a, b), equation (7.29) is split thereby into the system of equations:

$$\frac{d}{dt}\hat{P}\rho \quad = \hat{P}\widehat{L}\hat{P}\rho + \hat{P}\widehat{L}(\widehat{I} - \hat{P})\rho,$$
$$\frac{d}{dt}(\widehat{I} - \hat{P})\rho = (\widehat{I} - \hat{P})\widehat{L}\hat{P}\rho + (\widehat{I} - \hat{P})\widehat{L}(\widehat{I} - \hat{P})\rho. \tag{7.30}$$

These equations couple the relevant part of the statistical operator $\rho_{\text{rel}} = \hat{P}\rho$ with the irrelevant part $\rho_{\text{irrel}} = (\widehat{I} - \hat{P})\rho$. Equations (7.30) and their (formal) solutions can be investigated for arbitrary linear projection operators $\hat{P}$. They have been studied in more detail for the diagonalization operators (7.17) and for the subsystem projections (7.19) (Zwanzig 1960b, Haake1973, Fick and Sauermann 1990). We shall maintain the formulation with a general projection operator $\hat{P}$, but finally we are interested in the choice (7.19) with a suitable state $\rho_E \in \mathcal{D}(\mathcal{H}_E)$. Without discussion of the mathematical subtleties of the system (7.30) – for that purpose we refer to (Davies 1974) – we integrate the second equation in (7.30) to:

$$\rho_{\text{irrel}}(t) = e^{(\widehat{I}-\hat{P})\widehat{L}(\widehat{I}-\hat{P})t}\rho_{\text{irrel}}(0) + \int_0^t e^{(\widehat{I}-\hat{P})\widehat{L}(\widehat{I}-\hat{P})(t-\tau)}(\widehat{I} - \hat{P})\widehat{L}\hat{P}\rho_{\text{rel}}(\tau)\, d\tau.$$

Then this result is inserted into the first equation in (7.30). With the choice
(7.19) of the projection, the statistical operator of the system S is $\rho_S =$
$\mathrm{tr}_E\rho = \mathrm{tr}_S\rho_{\mathrm{rel}}$, and we finally obtain the *pre-master equation* for ρ_S

$$\frac{d}{dt}\rho_S(t) = -i[H, \rho_S(t)] + \int_0^t \widehat{K}(t - \tau)\rho_S(\tau)\,d\tau + \widehat{R}(t)\rho(0), \qquad (7.31)$$

where the operators $\widehat{K}$ and $\widehat{R}$ are given below. This equation is an exact
identity at any time. The Hamiltonian part of this evolution equation is
derived from (7.19) and (7.28) as $\mathrm{tr}_E\widehat{P}\widehat{L}\widehat{P}\rho = -i[H, \rho_S]$ with an effective
Hamiltonian of the subsystem S:

$$H = H_S + \mathrm{tr}_E H_{SE}\rho_E. \qquad (7.32)$$

Since a term $V \otimes I_E$ can be added to $H_S \otimes I_E$ and subtracted from H_{SE}
without changing the total Hamiltonian H_{S+E}, the separate operators on
the right hand side of (7.28) are not uniquely determined. Nevertheless, the
effective Hamiltonian H and the other terms in (7.31) do not depend on
this ambiguity. But all terms depend on the reference state ρ_E. The whole
equation (7.31), however, is an exact identity with any choice of ρ_E. The
only criterion, so far, in choosing a specific ρ_E is to simplify the analytic
expressions, but see the discussion following (7.34) below.

The operator $\widehat{K}(\tau)$ is calculated by integration from the second line of
(7.30) as

$$\widehat{K}(\tau)\rho_S = \mathrm{tr}_E\widehat{L}(\widehat{I} - \widehat{P})\,e^{(\widehat{I}-\widehat{P})\widehat{L}(\widehat{I}-\widehat{P})\tau}(\widehat{I} - \widehat{P})\,\widehat{L}\rho_S \otimes \rho_E \qquad (7.33)$$

and accounts for the residual interaction of S with the environment. The
operator

$$\widehat{R}(\tau)\rho = \mathrm{tr}_E\widehat{L}(\widehat{I} - \widehat{P})\,e^{(\widehat{I}-\widehat{P})\widehat{L}(\widehat{I}-\widehat{P})t}\rho = \mathrm{tr}_E\widehat{L}(\widehat{I} - \widehat{P})\,e^{(\widehat{I}-\widehat{P})\widehat{L}(\widehat{I}-\widehat{P})t}\rho_{\mathrm{irrel}}(0),$$

which maps $\mathcal{T}(\mathcal{H}_S \otimes \mathcal{H}_E)$ into $\mathcal{T}(\mathcal{H}_S)$, propagates correlations which exist in
the irrelevant part of the initial state.

In most applications one assumes that the initial state satisfies $\rho_{\mathrm{irrel}}(0) =$
$(\widehat{I} - \widehat{P})\rho(0) = 0$, i. e. for the projection (7.19)

$$\rho(0) = \rho_S(0) \otimes \rho_E. \qquad (7.34)$$

Then the choice of ρ_E acquires physical significance as part of the initial
state, and (7.31) reduces to the homogeneous equation:

$$\frac{d}{dt}\rho_S(t) = -i[H, \rho_S(t)] + \int_0^t \widehat{K}(t - \tau)\rho_S(\tau)\,d\tau. \qquad (7.35)$$

The equation (7.35) is an exact evolution equation for both positive and
negative values of t only if the initial state factorizes according to (7.34). In

that case the solution $\rho_S(t)$ agrees with (7.6) $\rho_S(t) = \mathrm{tr}_E U(t)\rho_S(0)\otimes\rho_E U^\dagger(t)$. The interaction term $(\hat{I} - \hat{P})\hat{L}\hat{P}$ in (7.30) leads to $(\hat{I} - \hat{P})\rho(t) \neq 0$ for $t \neq 0$ (with exception of the recurrences of a closed system $S+E$ in a finite volume). The equations (7.6) and (7.35) therefore break time translation invariance. But since the projection $\hat{P}$ is invariant against the time reflection operator T, if the reference state ρ_E is invariant $T\rho_E T^\dagger = \rho_E$, see Sect. 7.1, the equations are still symmetric under time reflection at $t = 0$. If $\rho_S(t)$ is a solution of (7.6) and (7.35), then also $T\rho_S(-t)T^\dagger$ is a solution. Since (7.6) is completely positive, the relative entropy gets smaller for $t > 0$ and – in agreement with time reflection symmetry – also for $t < 0$ compared to its value at $t = 0$. And the same additional conditions (certainly satisfied for the physical systems we are interested in) which let the von Neumann entropy grow for $t > 0$, also yield an increase of entropy for decreasing $t < 0$. Precisely this remaining symmetry renders the solution completely unrealistic for $t < 0$. Apart from exceptional systems prepared in a laboratory on a small scale, entropy has always been increasing.

As already stated, much of the theory of open systems was developed to understand dissipation and the approach to equilibrium. For many applications the system under consideration is embedded into a heat bath, i. e. ρ_E is chosen as a canonical ensemble (or KMS state) of the environment, see e. g. (Favre and Martin 1968, Davies 1973, Haake 1973). For these models one obtains only one stationary state (or one regular motion under the influence of external fields) which is approached for large t, independently of the initial state ρ_S of S. But it is essential that one has started from the exceptional factorizing initial state of $S + E$ with a canonical ensemble for the environment.[11]

The success of the factorizing ansatz for the initial state within heat-bath models is surely not a sufficient argument on a more fundamental level. An investigation of explicitly soluble models – see Sect. 7.6 – indicates that the emergence of classical behaviour does not depend on correlations between the system and its environment in the initial state. For these models the induced superselection sectors are uniquely determined by the Hamiltonian. The choice of the initial state is essential for the time scale on which these superselection sectors emerge. The time scale depends on some smoothness conditions on that part of the initial state which is related to the environment. In the case of a factorizing initial state this condition is a restriction on the reference state. But under appropriate conditions, which are discussed in Sect. 7.6, the time scale is not modified by correlations $\rho - (\mathrm{tr}_E\rho) \otimes (\mathrm{tr}_S\rho)$ in the initial state. Therefore approximations based on the factorizing initial

---

[11] The dissipative properties of a reduced Hamiltonian dynamics as the consequence of specific initial conditions are also known in classical mechanics, see e. g. the model of radiation damping presented by Lamb (1900). A discussion of this model in a quantum mechanical context has been given by Ford, Lewis and O'Connell (1988).

state, as the Markov approximation – discussed in the next section and widely used in Chap. 3 – yield good results, not only qualitatively but also quantitatively. For a discussion on the cosmological level the ansatz (7.34) can be understood (under an appropriate modification or choice of $\hat{P}$) as a *cosmological* initial condition see (Penrose 1981, Zeh 2001, Halliwell, Pérez-Mercader, and Zurek 1994), and Chap. 4 of this book.

7.4 Markov Approximation and Semigroups

If the correlation time of the environment is much shorter than the typical time scale for the variation of the system, the kernel $\widehat{K}(s)$ in (7.33) is very sharply peaked around $s = 0$ and one can neglect the retardation in (7.35) to obtain in this way a Markovian approximation for the generalized master equation:

$$\frac{d}{dt}\rho(t) = \widehat{M}\rho(t), \quad t \geq 0 \tag{7.36}$$

with a linear operator $\widehat{M}\colon \mathcal{T}(\mathcal{H}_S) \to \mathcal{T}(\mathcal{H}_S)$. This equation is usually called *master equation*. A first guess for $\widehat{M}$ is

$$\widehat{M}\rho = -i[H, \rho] + \int_0^\infty ds\, \widehat{K}(s)\rho, \tag{7.37}$$

but a more detailed analysis is necessary. In the case of standard models the simple ansatz (7.37) violates the positivity of the statistical operator (Dümcke and Spohn 1979). The systematic investigation of such limits goes back to van Hove (1955); a more recent survey can be found in the book of Alicki and Lendi (1987). The resulting time evolution

$$\rho \to \rho(t) = \Lambda_t \rho \tag{7.38}$$

is defined for arbitrary nuclear operators $\rho \in \mathcal{T}(\mathcal{H}_S)$ and maps the state space $\mathcal{D}(\mathcal{H}_S)$ into itself. The family of mappings $\Lambda_t\colon \mathcal{T}(\mathcal{H}_S) \to \mathcal{T}(\mathcal{H}_S)$, $t \geq 0$, satisfies a *semigroup* law

$$\Lambda_0 = \mathrm{id}, \quad \Lambda_{t+s} = \Lambda_t \Lambda_s, \quad t, s \geq 0. \tag{7.39}$$

The operators Λ_t, $t \geq 0$ are not invertible in general (the only exception is the unitary Hamiltonian evolution generated by the Liouville operator $\widehat{L} = -i[H, \,.\,]$). We should like to emphasize that the equation (7.35) already violates time translation invariance and singles out the time $t = 0$ with the special initial state (7.34). The Markovian approximation enhances the irreversibility but does not introduce it. The Markovian approximation can also be applied to the time reversed solution $T\rho(-t)T^\dagger$ of (7.35) with the result $\rho(t) = \Lambda_{-t}\rho(0)$ if $t \leq 0$. As mathematical approximation to (7.35) this relation is of course as reasonable as the approximation (7.38) for $t \geq 0$. But it is again totally unrealistic as evolution of a physical system for the same reasons as discussed in the context of (7.35).

The validity of a Markovian master equation cannot be general and can therefore only be used for specific situations. For example in the case of decoherence arising from scattering, see Sect. 3.1.2, the correlation time of the environment corresponds to the duration of a scattering process, the scattering particles carrying the correlations into open space or into absorbers where they are transformed into inaccessible higher order correlations. For weak coupling or low density more detailed mathematical statements are possible, see (Alicki and Lendi 1987) and the literature given there. The weak coupling approximation (with its additional time scaling) is equivalent to a singular coupling approximation (Palmer 1977, Gorini et al. 1978). Both these approximations indicate that the Markov approximation is a good approximation on an intermediate time scale. In the following it will simply be *assumed* that the system under consideration can be described by a master equation with a dissipative part in the generator $\widehat{M}$, see (7.42) below (in order to let the entropy grow monotonously). This time evolution has been investigated by many authors. For the mathematical foundations see (Kossakowski 1972, Davies 1974, 1976, Gorini, Kossakowski, and Sudarshan 1976, Lindblad 1976). A recent thorough mathematical investigation of the asymptotic behaviour of semigroups and of the superselection sectors induced by semigroups has been given by Olkiewicz (1999, 2000), see also (Blanchard and Olkiewicz 2000). In the context of interpretation of quantum mechanics see e. g. (Ghirardi, Rimini, and Weber 1975, 1986, Barchielli, Lanz and Prosperi 1982, Joos and Zeh 1985, Joos 1987a, Caves and Milburn 1987, Blanchard and Jadczyk 1995, Jadczyk 1995). For the study of phenomenological models and for the use of master equations and semigroups to understand decoherence see Chap. 3 and the literature given there. Moreover, master equations are widely used in quantum optics, see e. g. the book Carmichael (1993). Some of these authors consider the master equations as an approximation to the von Neumann equation of some "total system" (as described above) while others postulate it as a fundamental equation – in particular for solving the measurement problem. These latter proposals will be discussed in Chap. 8. However, for this purpose the formal dynamics of local density matrices does not suffice because of the ambiguous interpretation of density matrices as explained in Sect. 2.4. Any "fundamental" interpretation of the master equation would also require a fundamental specification of the systems of projectors $\hat{P}$ for which it has to hold, going beyond the general nature of a phenomenological discussion. Phenomenological master equations would then again have to be *derived* from this fundamental master equation instead of a von Neumann equation.

The dual Heisenberg dynamics for the observables, $A \rightarrow A(t) = \Lambda_t^+(A)$ is defined by $\mathrm{tr}_S \Lambda_t(\rho)\ A = \mathrm{tr}_S \rho\ \Lambda_t^+(A)$. The family of these operators

$\Lambda_t^+ \colon \; \mathcal{B}(\mathcal{H}_S) \to \mathcal{B}(\mathcal{H}_S)$, $t \geq 0$, forms a semigroup

$$\Lambda_0^+ = \mathrm{id}, \quad \Lambda_{t+s}^+ = \Lambda_t^+ \Lambda_s^+, \quad t, s \geq 0, \tag{7.40}$$

as (7.39) does. Moreover, since $\Lambda_t^+(A)$ is the Markovian limit of the completely positive operators (7.10) $\Phi_t^+(A)$, it is again a completely positive operator, i. e. it has the representation (7.12). The semigroup operators (7.38) for the states are therefore dynamical maps with the representation (7.8). The operators V_t of the Markov approximation differ from those of the exact evolution, but they again satisfy (7.9), and the relations (7.7) and (7.11) apply to the Markov approximations $\Lambda_t(\rho)$ or $\Lambda_t^+(A)$, respectively. The semigroups $\Lambda_t(\rho)$ and $\Lambda_t^+(A)$ are therefore contraction semigroups on Banach spaces with well defined generators $\widehat{M}$ and $\widehat{M}^+$

$$\Lambda_t = \exp \widehat{M} t, \quad \Lambda_t^+ = \exp \widehat{M}^+ t. \tag{7.41}$$

The form of these generators has been derived by Gorini, Kossakowski and Sudarshan (1976) for finite dimensional Hilbert spaces and by Lindblad (1976) for the case of Hilbert spaces of arbitrary dimensions

$$\widehat{M}\rho = -i[H, \rho] - \frac{1}{2} \sum_j (L_j^\dagger L_j \rho + \rho L_j^\dagger L_j - 2 L_j \rho L_j^\dagger), \tag{7.42}$$

and

$$\widehat{M}^+ A = i[H, A] - \frac{1}{2} \sum_j (L_j^\dagger L_j A + A L_j^\dagger L_j - 2 L_j^\dagger A L_j) \tag{7.43}$$

where H is a self-adjoint operator and L_j, $j = 1, 2, \ldots$ is a finite or infinite sequence of operators on $\mathcal{H}_S$.

The mathematical derivation of (7.42) and (7.43) is based on the Heisenberg representation. The original proof of Lindblad has been given for completely positive semigroups which are uniformly continuous in the operator norm. These semigroups have bounded generators, such that the operators H and L_l are bounded and $\sum_j L_j^\dagger L_j$ converges. But in the meantime the extension to semigroups with unbounded operators H and L_j has been given by Holevo (1995, 1998) and finally by Belavkin (1997, 1998). The structure of the generators (7.42) and (7.43) follows from the complete positivity of the reduced dynamics. This structure guarantees that the statistical operator is positive for all times. But another essential property of quantum dynamics, the semiboundedness of the Hamiltonian H_{S+E} of the total system, cannot be encoded in the generators in a simple form. Any derivation of a Markov approximation for a concrete system should recognize this constraint (Accardi, Gough and Lu 1995, Ford, Lewis and O'Connell 1996).

The generators (7.42) and (7.43) do not determine the Hamiltonian H and the Lindblad operators L_j uniquely. Two sets of operators $H, L_1, L_2, \ldots, L_n$ and $H', L_1', L_2', \ldots, L_n'$ are denoted as equivalent if they define the same generator (7.42). The transformations which relate equivalent sets form a group generated by unitary transformations of the Lindblad operators

$$L_j \rightarrow L'_j = \sum_k u_{jk} L_k \tag{7.44}$$

where u_{jk} is a unitary matrix, and by the inhomogeneous transformations

$$L_j \rightarrow L'_j = L_j + a_j$$
$$H \rightarrow H' = H + \frac{1}{2i} \sum_j (a_j^* L_j - a_j L_j^\dagger) + b \tag{7.45}$$

where a_j, $j = 1, \ldots, n$ and b are complex numbers. For self-adjoint Lindblad operators $L_j = L_j^\dagger$ the matrix u_{jk} has to be real. The group of these transformations is a central extension of the Euclidean group, see Parthasarathy (1992).

The semigroup operators Λ_t are dynamical maps in the sense of (7.8). Apart from the exceptional case where there is no interaction with the environment and the generator (7.42) reduces to the von Neumann term $-i[H, \rho]$, the operators are not invertible. The influence of the environment is therefore irreversible and the relative entropy (7.22) decreases, see (7.26). But as already stated in Sect. 7.2 the decrease of relative entropy (7.26) does not necessarily yield an increase $S(\Lambda_t \rho) \geq S(\rho)$ of the von Neumann entropy. The behaviour of the von Neumann entropy depends on more specific details of the semigroup and, possibly, on the initial state of the subsystem S. In the finite dimensional case, a sufficient condition is that all Lindblad operators are normal operators, i. e. $L_j^\dagger L_j = L_j L_j^\dagger$. Then the generator (7.42) annihilates the unit operator, and the uniform distribution is a stationary state of the semigroup. As a consequence of (7.25) the von Neumann entropy is non-decreasing for any state $\rho \in \mathcal{D}(\mathcal{H}_S)$. More generally, the following result has been derived by Olkiewicz (2000): If the Lindblad operators satisfy the inequality $\sum_j L_j L_j^\dagger \leq \sum_j L_j^\dagger L_j$ then the dynamical semigroup Λ_t is contractive in the operator norm $\|\Lambda_t \rho\| \leq \|\rho\|$ and the inequality $S(\Lambda_t \rho) \geq S(\rho)$ follows for all $t \geq 0$.

7.5 Quantum Stochastic Processes

The generators (7.42) or (7.43) deviate by the contribution of the operators L_j from the usual Hamiltonian evolution and, as already emphasized, the resulting semigroup dynamics cannot be restricted to pure states. But nevertheless it is possible to construct a dynamical evolution of randomly disturbed pure states, which – after averaging – leads to the semigroup dynamics (7.38) for the statistical operator. The usual mathematical technique to formulate this evolution are stochastic Schrödinger equations. These equations have been motivated by quite different arguments, which in a rather schematic way, can be summarized as:

1. The stochastic terms in the differential equations result from the coupling to an environment, which in the Markovian approximation can be represented by a white noise field. The stochastic Schrödinger equations are then an effective tool to calculate the statistical operator.

2. Since the stochastic wave functions have rather strong localization properties, these wave functions have been associated to individual systems (Gisin and Percival 1992, 1993a, b). But such an interpretation is highly questionable, not only because of some ambiguities in the derivation of the stochastic equations, but also on a more fundamental level as we shall see below.

3. A repeated measurement of an individual system can be modelled by a stochastic Schrödinger equation. In this approach one can also derive a classical stochastic equation for the measured data. (Barchielli and Belavkin 1991, Belavkin 1989a, b, c, 1995, Belavkin and Staszewski 1992, Diósi 1988a, b, c, Gisin 1984, 1989, Goetsch and Graham 1994, Adler et al. 2001)

4. The Schrödinger equation is modified by stochastic terms which cause a spontaneous collapse mechanism of the wave function (Diósi 1989, Ghirardi, Rimini and Weber 1986, Ghirardi, Pearle and Rimini 1990, Pearle 1993). These models are discussed in detail in Chap. 8.

A quantum stochastic processes is a stochastic processes with $\mathcal{H}_S$ as state space. A trajectory (path) of the process is a function $\Psi(t)\colon \mathbb{R}_+ \to \mathcal{H}_S$ as in the usual Schrödinger picture for pure states; but now we need a whole sample $\Psi(\omega; t)$, $\omega \in \Omega$ of such trajectories to describe the dynamics. Here the set Ω is a probability space; in practically all calculations it is taken to be the space of an n-dimensional classical Markov process, in most cases the real or complex Brownian motion. The dimension n is fixed by the number of independent operators L_j, $j = 1, \ldots, n$ in (7.42). The number of Lindblad operators and therefore the number of components of Brownian motion can even be infinite. But we shall not discuss the mathematical subtleties related to that case.

Let us first assume that the initial state of the system is $\varphi_0 \in \mathcal{H}_S$, $\|\varphi_0\| = 1$, and that the dynamics is given by the semigroup Λ_t with generator (7.42) in the Schrödinger picture or by Λ_t^+ in the Heisenberg picture. Then we construct a (not necessarily normalized) quantum stochastic process $\varphi(\omega; t)$ with $\varphi(\omega; 0) = \varphi_0$ such that

$$\langle \varphi_0 \mid \Lambda_t^+(A)\ \varphi_0 \rangle = \overline{\|\varphi(\omega; t)\|^{-2}\ \langle \varphi(\omega; t) \mid A\ \varphi(\omega; t) \rangle} \qquad (7.46)$$

holds for all observables A on $\mathcal{H}_S$. Here the bar denotes the expectation $\overline{f(\omega)} = \int f(\omega)\, d\mu_t(\omega)$ with the probability measure $d\mu_t$ of the process up to time t. In the Schrödinger picture the identity (7.46) means that the statistical operator (7.38) $\rho(t) = \Lambda_t \rho(0)$ of the system S with initial state

$\rho(0) = |\varphi_0\rangle\langle\varphi_0|$ is decomposed at any time $t \geq 0$ into pure states

$$\rho(t) = \int |\psi(\omega;t)\rangle \langle\psi(\omega;t)| \, d\mu_t(\omega) \text{ with } \psi(\omega;t) = \|\varphi(\omega;t)\|^{-1} \varphi(\omega;t).$$
(7.47)

Such a decomposition is not unique; and one needs additional assumptions – which either follow from the invariance with respect to the transformations (7.44) and (7.45) or which can be motivated within a theory of continuous measurement – to derive a sample of such wave functions $\varphi(\omega;t)$, $\omega \in \Omega$, as solutions of a stochastic differential equation. As we shall see below it is possible to obtain these wave functions as solutions of *linear* stochastic differential equations, which are in some sense minimal stochastic extensions of the Schrödinger equation. The normalization can then only be achieved in mean, $\overline{\|\varphi(\omega;t)\|} = 1$, but not for the individual trajectories. The normalized process $\psi(\omega;t) = \|\varphi(\omega;t)\|^{-1} \varphi(\omega;t)$ always satisfies a *nonlinear* stochastic differential equation.

So far we have made the rather unreasonable assumption that the initial state of the subsystem is a pure state. For a mixed initial state $\rho(0)$ the mathematical problem is, in principle, easy to solve; it is only necessary to decompose $\rho(0)$ into pure states – leading to an integral representation (7.47) also for $t = 0$ – and then to apply the techniques to be described below to each of these pure states. But this first step is highly non-unique and all arguments which motivate a specific type of stochastic differential equation do not remove this arbitrariness. There is no conclusive argument to fix the initial measure apart from mathematical convenience. Therefore the interpretation of the whole sample $\psi(\omega;t)$ as an ensemble of individual events or states – an interpretation which is intended in a great part of the cited literature – is hardly acceptable.

For completeness we give a comparatively simple derivation of a stochastic Schrödinger equation essentially following Ghirardi, Pearle, and Rimini (1990) and Belavkin and Staszewski (1992). As stochastic driving force we take a real Brownian motion $B(t)$ with the expectation values

$$\overline{B(t)} = B(0) = 0, \ \overline{B(s)B(t)} = \min(s,t) \text{ if } s, t \geq 0. \tag{7.48}$$

Let H be the effective Hamiltonian of the system S and L be some operator on $\mathcal{H}_S$ then the linear stochastic differential equation

$$d\varphi(t) = -iH\varphi(t)dt + L\varphi(t)dB(t) \tag{7.49}$$

with initial condition $\varphi(0) = \varphi \in \mathcal{H}_S$, $\|\varphi\| = 1$, defines a stochastic process with values in $\mathcal{H}_S$. The probability measure is the Gaussian measure $dv_t(B)$ of Brownian motion. This process does not conserve the norm, since the stochastic differential[12]

$$d\,\|\varphi\|^2 = \langle \varphi \mid d\varphi \rangle + \langle d\varphi \mid \varphi \rangle + \langle d\varphi \mid d\varphi \rangle$$
$$= \langle \varphi \mid (L + L^\dagger)\varphi \rangle \, dB(t) + \langle \varphi \mid (L^\dagger L)\varphi \rangle \, dt \qquad (7.50)$$

does not vanish. The term $\langle d\varphi(t) \mid d\varphi(t) \rangle = \langle \varphi(t) \mid L^\dagger L \, \varphi(t) \rangle \, dt$ is a consequence of the Itô calculus. If we modify (7.49) to

$$d\varphi(t) = -(iH + \frac{1}{2}L^\dagger L)\varphi(t)dt + L\varphi(t)dB(t), \qquad (7.51)$$

the differential (7.50) reduces to the stochastic term
$d\,\|\varphi(t)\|^2 = \langle \varphi(t) \mid (L + L^\dagger) \, \varphi(t) \rangle \, dB(t)$, and the norm is conserved in mean $\overline{\|\varphi(t)\|^2} = \|\varphi(0)\|^2 = 1$. The equation (7.51) is a minimal linear stochastic extension of the Schrödinger equation $d\varphi(t) = -iH\varphi(t)dt$ within the Itô calculus. The role of this equation for the derivation of stochastic Schrödinger equations was recognized by Belavkin (1989a) within an involved theory of nondemolition measurement.

It is an surprising fact that we can use the stochastic wave functions $\varphi(t)$ to construct the statistical operator without further normalization. The rank one operator $F(t) = |\varphi(t)\rangle \langle \varphi(t)|$ has the stochastic differential

$$dF(t) = (-i\,[H,F] - \frac{1}{2}(L^\dagger L F + F L^\dagger L - 2LFL^\dagger)dt + (LF + FL^\dagger)dB(t)$$

If we define an operator by $\rho(t) = \overline{F(t)}$, where the bar refers to integration with respect to the measure of Brownian motion, this operator is obviously positive, the trace $\mathrm{tr}_S\rho(t) = \overline{\langle \varphi(t) \mid \varphi(t) \rangle} = 1$ is normalized, and it satisfies a master equation with the Lindblad operator L. A generalization to an arbitrary number of Lindblad operators is straightforward.

To obtain a decomposition of $\rho(t)$ into pure states the process $\varphi(t)$ has to be normalized. The stochastic differential of $\psi(t) = \varphi(t)(\|\varphi(t)\|^2)^{-\frac{1}{2}}$ is

$$d\psi(t) = -(iH + \frac{1}{2}L^\dagger L)\psi(t)dt + L\psi(t)dB(t) + R(\frac{3}{2}R - L)\psi(t)dt \quad (7.52)$$

with the quadratic term $R(t) = \frac{1}{2} \langle \psi(t) \mid (L + L^\dagger) \, \psi(t) \rangle$. If we define the operators

$$\widetilde{H} = H - \frac{1}{2i}R(t)(L - L^\dagger), \quad \widetilde{L} = L - R, \qquad (7.53)$$

which differ from H and L by a transformation (7.45), and introduce the differential

$$d\widetilde{B}(t) = dB(t) - 2Rdt, \qquad (7.54)$$

---

[12] We use the Itô calculus, which is explained in App. 7.

then (7.52) can be written as

$$d\psi(t) = -(i\widetilde{H} + \frac{1}{2}\widetilde{L}^\dagger\widetilde{L})\psi(t)dt + \widetilde{L}\psi(t)d\widetilde{B}(t). \tag{7.55}$$

The differential of the projection operator $\sigma(t) = |\psi(t)\rangle\langle\psi(t)|$ follows as

$$d\sigma(t) = (-i[H, \sigma(t)] + \frac{1}{2}L^\dagger L\sigma(t) + \frac{1}{2}\sigma(t)L^\dagger L - L\sigma(t)L^\dagger)dt$$
$$+ (\widetilde{L}\sigma(t) + \sigma(t)\widetilde{L}^\dagger)d\widetilde{B}(t). \tag{7.56}$$

The first term is exactly the Lindblad term (7.42). But the statistical operator $\rho(t) := \overline{\sigma(t)}$, calculated with the measure of Brownian motion, does not satisfy the corresponding master equation, since there is a non vanishing contribution from. $d\widetilde{B}(t)$. For that purpose a numerical process $X(t)$ is defined by

$$dX(t) = 2R(t)dt + dB(t) = \langle\psi(t) | (L + L^\dagger)\,\psi(t)\rangle\,dt + dB(t), \tag{7.57}$$

see e. g. (Belavkin 1989b, c) and (Ghirardi, Pearle, and Rimini 1990). If we replace dB by dX, the differential $d\widetilde{B}$ is replaced by standard Brownian motion dB. As a consequence the last term in (7.56) becomes $(\widetilde{L}\sigma(t) + \sigma(t)\widetilde{L}^\dagger)dB(t)$ and its expectation – calculated with the measure of Brownian motion – vanishes. The formula (7.47) for the statistical operator with the solutions φ of the simple stochastic differential equation (7.51) is therefore only correct after a change of measure; the expectation has to be calculated with the measure $d\mu_t(X)$ of the process $X(t)$. An alternative and most frequently used method is to modify the stochastic differential equation for the normalized vector ψ by the replacement $dB \to dX$ or, equivalently $d\widetilde{B}(t) \to dB(t)$. Then the equation (7.55) is substituted by the non-linear stochastic Schrödinger equation

$$d\psi(t) = (-iH + \frac{1}{2}L^\dagger L - RL + \frac{1}{2}R^2)\psi(t)dt + (L - R)\psi(t)dB(t), \tag{7.58}$$

Here the identities (7.53) have already been inserted. The statistical operator can then be calculated with formula (7.47) using the measure of standard Brownian motion.

The extension to a finite or infinite number of Lindblad operators L_j, $j \in \mathbb{J} \subset \mathbb{N}$, is rather obvious

$$d\psi(t) = -(iH + \frac{1}{2}\sum_j(L_j^\dagger L_j - 2R_j L_j + R_j^2))\psi(t)\,dt$$
$$+ \sum_j(L_j - R_j)\psi(t)\,dB_j(t) \tag{7.59}$$

with the nonlinear terms $R_j = \frac{1}{2}\langle\psi(t) | (L_j + L_j^\dagger)\,\psi(t)\rangle$ and independent real Brownian motion processes B_j, $j \in \mathbb{J}$. This equation is discussed in the context of stochastic collapse models in Chap. 8.

For a projection operator $L = L^\dagger = P$ the equation (7.58) was derived already in 1984 by Gisin (1984). Within the approaches of repeated measurement of an observable $L = L^\dagger$ the classical stochastic process (7.57) $X(t)$ represents the measured values of this observable (Barchielli and Belavkin 1991, Belavkin 1989c, 1995, Belavkin and Staszewski 1992, Diósi 1988b). Here "measurement" means an objective process in the environment, e. g. a track in a bubble chamber.

The generator of the semigroup for the statistical operator is invariant against the transformations (7.44) and (7.45). The stochastic differential equations presented so far have do not have this symmetry.[13] The full invariance (except for a phase factor generated by (7.45)) can be achieved also on the level of the stochastic trajectories if complex Brownian motion is used as driving term. Such stochastic Schrödinger equations have been proposed by Gisin and Ciblis (1992) and by Gisin and Percival (1992, 1993a, b).

Remarks

1) It is also possible to formulate stochastic differential equations for the wave function using counting processes, a case particularly interesting for quantum optics (Belavkin 1989b, Barchielli and Belavkin 1991). Moreover, one can combine a partly deterministic dynamics with jump processes, see e. g. the original spontaneous localization model of Ghirardi, Rimini and Weber (1986), the approach of Blanchard and Jadczyk (Blanchard and Jadczyk 1995, Jadczyk 1995), which combines classical and quantum behaviour, or the models of quantum optics in (Carmichael 1993).

2) The solutions of a stochastic Schrödinger equation can be obtained as solution of Feynman path integrals. (Mensky 1979a, Albeverio, Kolokoltsov and Smolyanov 1997, Smolyanov and Truman 1999)

3) The first implementation of the Heisenberg semigroup (7.40) by stochastic processes was given in (Hudson and Parthasarathy 1984). The calculus of Hudson and Parthasarathy uses *linear* stochastic differential equations with noncommutative creation and annihilation processes on a Fock space. A self-contained presentation of this calculus is given in the book of Parthasarathy (1992). This calculus is the mathematical version of the *quantum white noise* in quantum optics as used e. g. in (Gardiner 1991). The linear quantum stochastic differential equations of Hudson and Parthasarathy are easy to formulate for identical particles and for quantum field theory. An approach to derive this stochastic theory as limit of a quantum field theory can be found in (Accardi, Gough, and Lu 1995).

4) Instead of a quantum white noise it is possible to use an operator valued stochastic process which incorporates already some information, as e. g. the temperature, of the environment. This is done in the quantum Langevin

---

[13] More precisely, if all Lindblad operators are self-adjoint, the equation (7.59) has the restricted symmetry with real matrices u_{jk}. But if at least one Lindblad operator is not self-adjoint, also this restricted symmetry is lost.

equation of Ford, Lewis and O'Connell (1988, 1996, 2001). For further application see (Li, Ford and O'Connell 1990, Dattagupta and Singh 1996, 1997).

5) The formalism of quantum stochastic processes has an interesting counterpart in classical physics. For classical systems the dynamics of open systems is represented by the semigroup of Markov operators, see (Mackey 1992, Lasota and Mackey 1994), and the literature given there. Under rather general conditions this dynamics can be represented by a stochastic flow – a counterpart of the quantum stochastic process.

7.6 Induced Superselection Sectors

7.6.1 General Considerations

Decoherence due to the environment and dynamically induced superselection sectors have already been investigated as the main theme of this book in Chaps. 3–5. In this section we shall reformulate decoherence in the mathematical language of Sects. 7.1-7.4.

To exclude recurrences and to derive statements which are stable for large times $t \to \infty$, we assume that the Hamiltonian of the environment has a continuous spectrum. The initial state of the environment is either a normal state, which can be represented by a statistical operator, or a KMS state of positive temperature.[14] A system, which is weakly coupled to such an environment, usually decays into its ground state, if the environment is in a normal state, or the system approaches a canonical ensemble or KMS state, if the environment is initially in a state with positive temperature. More interesting decoherence effects may occur on an intermediate time scale, and, moreover, if the thermalization is prevented by conservation laws. To select effects on an intermediate time scale one can use a strong coupling between system and environment. This method has some similarity to the singular coupling method of the Markov approximation, which also scales the dynamics at an intermediate time period to large times. The models, which we discuss, have therefore at least one of the following properties: existence of (many) conservation laws and/or strong coupling. Thereby strong coupling means that the spectral properties of the Hamiltonian are modified by the interaction term. But – an essential property of all models – the dynamics is still generated by a semibounded Hamiltonian.

The dynamics of a system embedded in an environment is determined by (7.6) or, equivalently, by the pre-master equation (7.35), if the corresponding initial condition is satisfied. We assume that the system has no kinematical

---

[14] The KMS states of an environment which has a Hamiltonian with a continuous spectrum cannot be represented by a statistical operator in $\mathcal{D}(\mathcal{H}_E)$. In such a case the algebra of observables has to be restricted to the Weyl algebra, which is strictly smaller than $\mathcal{B}(\mathcal{H}_E)$, and the KMS states are continuous functionals on that algebra.

superselection rules of the kind discussed in Chap. 6. Otherwise we just restrict our considerations to a coherent sector. As already explained in the preceding chapters of this book, the dynamics $\rho_S(0) \to \rho_S(t)$, $t \geq 0$ can induce a superselection structure $\rho_S(t) \simeq \sum_n P_n \rho_S(t) P_n$ if $t \to \infty$, i.e. the off-diagonal elements vanish

$$P_m \rho_S(t) P_n \to 0 \text{ if } t \to \infty \text{ and } m \neq n. \tag{7.60}$$

Here $\{P_n\}$, $n \in \mathbb{M} \subset \mathbb{N}$ is a mutually exclusive and exhaustive family of projection operators as used for the definition of kinematical superselection sectors in Sect. 6.1. In a weak sense the statement (7.60) means

$$\text{tr}_S A P_m \rho_S(t) P_n \to 0 \text{ if } t \to \infty \text{ and } m \neq n \tag{7.61}$$

for all observables $A \in \mathcal{B}(\mathcal{H}_S)$ and all initial states $\rho_S(t) \in \mathcal{D}(\mathcal{H}_S)$. Such a behaviour is already possible in the usual Hamiltonian dynamics as can be seen in the following simple example.

Let H be a Hamiltonian on the Hilbert space $\mathcal{H} = \mathcal{H}_b \oplus \mathcal{H}_c$ with a bound state at energy E_0 in the one dimensional subspace $\mathcal{H}_b$ and with an absolutely continuous spectrum in the orthogonal subspace $\mathcal{H}_c$. The dimension of $\mathcal{H}_c$ is necessarily infinite. Take $f_0 \in \mathcal{H}_0$ and $f_c \in \mathcal{H}_c$ with $\|f_0\| = \|f_c\| = 1$, then we calculate as a consequence of the Riemann-Lebesgue Lemma $(U(t)f_b \mid A \, U(t)f_c) = e^{itE_0} \left(A^\dagger f_b \mid U(t)f_c \right) \to 0$ for all $A \in \mathcal{B}(\mathcal{H})$ if $t \to \infty$ Hence (7.61) is true for the projection operators P_b and P_c onto the orthogonal subspaces $\mathcal{H}_b$ and $\mathcal{H}_c$. This statement is still true if H has an arbitrary number of bound states, but it does not apply to subspaces of the absolutely continuous spectrum.

For this example the emergence of superselection sectors is only true in the weak sense (7.61), and a strong operator convergence or even norm convergence in (7.60) is not possible. More refined examples for systems with an infinite number of degrees of freedom have been given in an important paper by Hepp (1972). These examples have been criticized by Bell (1975) because the actual decrease of the off-diagonal matrix elements can be arbitrarily slow due to the weak convergence.

As the main theme of this section we investigate superselection rules induced by the dynamics (7.6) $\rho_S \to \rho_S(t) = \Phi_t(\rho_S)$ of the subsystem S. The subspaces $\mathcal{H}_m = P_m \mathcal{H}_S$, $m \in \mathbb{M}$, are denoted as induced superselection sectors of this dynamics, if for all initial states $\rho_S(0) \in \mathcal{D}(\mathcal{H}_S)$ the off-diagonal matrix elements $P_m \rho_S(t) P_n$ vanish for large times, The type of convergence depends on the on the interaction. Contrary to the Hamiltonian dynamics the dynamics of subsystems allows a norm convergence. We speak about induced continuous superselection sectors, if there exists a continuous family of projection operators (7.100) $\{P(\Delta) \mid \Delta \subset \mathbb{R}\}$, such that

$$P(\Delta_1)\rho_S(t)P(\Delta_2) \to 0 \text{ if } t \to \infty \text{ and } \Delta_1 \cap \Delta_2 = \emptyset \tag{7.62}$$

for all states $\rho_S(t) = \rho_S \in \mathcal{D}(\mathcal{H}_S)$.

As in the other chapters we mainly use the Schrödinger picture. This representation allows to discuss the evolution (7.3) of the system S also for general initial states $W \in \mathcal{D}(\mathcal{H}_S \otimes \mathcal{H}_E)$ with correlations between system and environment. The investigation of the models elucidates the following points:

1. Superselection sectors can be induced in models with a semibounded Hamiltonian. The superselection sectors are determined by the Hamiltonian. They emerge – possibly on a large time scale – for arbitrary initial states $W \in \mathcal{D}(\mathcal{H}_S \otimes \mathcal{H}_E)$ of the total system.
2. The choice of the initial state is essential in order to obtain superselection sectors within a sufficiently short time. For factorizing initial states $W = \rho_S \otimes \rho_E$ the reference state $\rho_E \in \mathcal{D}(\mathcal{H}_E)$ determines this time scale. If the initial state has correlations between system and environment, $W - (\mathrm{tr}_E W) \otimes (\mathrm{tr}_S W) \neq 0$, finally the same superselection sectors emerge, but the estimates of the time scale are more involved.
3. For all models, which are presented in the following subsections, the decoherence of the off-diagonal contributions in (7.60) can be estimated by the trace norm $\|P_m \rho_S(t) P_n\|_1$. Hence superselection sectors induced by the environment differ qualitatively from the reduction of the algebra of observables as considered by Hepp (1972) or Bóna (1973).
4. The case of continuous superselection rules is included.

If the initial state factorizes we can also use the Heisenberg picture with the evolution (7.10) $\Phi_t^+(A)$ of the observables. We shall discuss the Heisenberg picture in more detail at the end of this section.

As actual models we first investigate the prototype for induced superselection rules introduced by Araki (1980) and Zurek (1982). The interaction of these models is rather trivial and commutes with the free Hamiltonian. Then we study systems which interact with a massless Boson field. Here superselection sectors are induced if and only if the Boson fields is infrared divergent. In both these cases – the Araki-Zurek models and the Boson models – the superselection operators are conserved quantities of the full dynamics. But only the reduced dynamics destroys the phase relations between the different sectors of the conservation law. Finally we admit a dynamics with additional scattering processes. Then superselection sectors are still induced, but due to the scattering processes the projection operators onto the superselection sectors do no longer commute with the Hamiltonian.

7.6.2 Hamiltonian Models of Decoherence

The Hamiltonians of the models, which we study first, have the structure

$$
\begin{aligned}
H_{S+E} &= H_S \otimes I_E + I_S \otimes H_E + F \otimes G \\
&= \left(H_S - \tfrac{1}{2}F^2\right) \otimes I_E + \tfrac{1}{2}\left(F \otimes I_E + I_S \otimes G\right)^2 + I_S \otimes \left(H_E - \tfrac{1}{2}G^2\right).
\end{aligned} \tag{7.63}
$$

Thereby H_S is the Hamiltonian of $\mathcal{S}$, H_E is the Hamiltonian of $\mathcal{E}$, and $F \otimes G$ is the interaction term between $\mathcal{S}$ and $\mathcal{E}$ with self-adjoint operators F on

$\mathcal{H}_S$ and G on $\mathcal{H}_E$. The Hamiltonian (7.63) is semibounded from below, if the operators $H_S - \frac{1}{2}F^2$ and $H_E - \frac{1}{2}G^2$ are semibounded operators on $\mathcal{H}_S$ or $\mathcal{H}_E$ respectively. In a later subsection we investigate more complicated Hamiltonians with an additional scattering potential V.

We start with the following assumptions

1) The operators H_S and F commute, $[H_S, F] = 0$, hence
   $[H_S \otimes I_E, F \otimes G] = 0$.
2) The operator G has an absolutely continuous spectrum.

As a consequence of the very restrictive assumption 1) the operator $F \otimes I_E$ is conserved. The aim of the following calculations is to prove that as a consequence of the interaction with the environment the operator F is actually a superselection operator for the system $\mathcal{S}$. The more technical assumption 2) implies that the off-diagonal elements (7.60) vanish in the limit $t \to \infty$. In the case of a point spectrum one would obtain almost periodic functions. But also then the suppression of the off-diagonal elements can be effective for a long time with recurrences beyond any reasonable observation period.

The calculations are straightforward, if F has only a discrete spectrum $\{\lambda_m \in \mathbb{R} \mid m \in \mathbb{M}\}$

$$F = \sum_m \lambda_m P_m, \tag{7.64}$$

where P_m, $m \in \mathbb{M}$, is the spectral family of the operator F. As a consequence of assumption 1) we have $[H_S, P_m] = 0$ for all $m \in \mathbb{M}$. The Hamiltonian (7.63) has therefore the form

$$H_{S+E} = H_S \otimes I_E + \sum_m P_m \otimes (H_E + \lambda_m G). \tag{7.65}$$

The unitary evolution $U(t) := \exp(-iH_{S+E}t)$ of the total system can be written as $U(t) = \left(e^{-itH_S} \otimes I_E\right) \sum_m P_m \otimes e^{-it(H_E + \lambda_m G)}$. For an initial state $\rho(0) = W \in \mathcal{D}(\mathcal{H}_S \otimes \mathcal{H}_E)$ the reduced dynamics (7.3) is

$$\rho_S(t) = e^{-itH_S} \sum_{m,n} P_m \mathrm{tr}_E \left(e^{it(H_E + \lambda_m G)} e^{-it(H_E + \lambda_n G)} W\right) P_n e^{itH_S} \tag{7.66}$$

The emergence of dynamically induced superselection rules depends on an estimate of the traces.

$$\rho_{m,n}(t) = P_m \mathrm{tr}_E \left(e^{it(H_E + \lambda_m G)} e^{-it(H_E + \lambda_n G)} W\right) P_n \in \mathcal{D}(\mathcal{H}_S). \tag{7.67}$$

It is a common feature of the subsequent models that the off-diagonal parts $m \neq n$ vanish in trace norm, $\|\rho_{m,n}(t)\|_1 \to 0$ if $t \to \infty$. But it depends on the details of the models, whether this decrease can be estimated uniformly in the initial state $\rho_S(0) = \mathrm{tr}_E W$ of the system S, and it depends on properties of the initial state W with respect to the environment, whether this decrease is effective on a short time scale.

For factorizing initial states $W = \rho_S \otimes \rho_E$ with $\rho_S \in \mathcal{D}(\mathcal{H}_S)$ and $\rho_E \in \mathcal{D}(\mathcal{H}_E)$ the partial trace (7.67) simplifies to $\rho_{m,n}(t) = \rho_S \chi_{m,n}(t)$ with the trace

$$\chi_{m,n}(t) = \mathrm{tr}_E \left(e^{it(H_E + \lambda_m G)} e^{-it(H_E + \lambda_n G)} \rho_E \right). \tag{7.68}$$

The trace norm of (7.60) is then estimated by

$$\|P_m \rho_S(t) P_n\|_1 \leq |\chi_{m,n}(t)|. \tag{7.69}$$

In the first step of the subsequent calculations we prove that $\chi_{m,n}(t) \to 0$ if $m \neq n$ and $t \to \infty$. Hence the reduced dynamics (7.3) finally yields the superselection sectors $P_m \mathcal{H}_S$, where the projection operators are given by the spectral decomposition (7.64) of the interaction potential F. But it may last a long time until we realize these superselection sectors. An additional task of the following sections is to specify a class of reference states $\rho_E \in \mathcal{D}(\mathcal{H}_E)$ such that (7.68) satisfies bounds of the type

$$|\chi_{m,n}(t)| \leq c \left(1 + (\lambda_m - \lambda_n)^2 \vartheta(t) \right)^{-\gamma} \text{ if } m \neq n \tag{7.70}$$

with a large exponent $\gamma > 0$ and a function $\vartheta(t) \geq 0$ which diverges sufficiently fast for $t \to \infty$. If the eigenvalues λ_m have no accumulation point, such an estimate leads to a fast uniform decrease of (7.69) for any initial state $\rho_S(0) \in \mathcal{D}(\mathcal{H}_S)$ of the system. The extension to initial states with correlations between system and environment is discussed in the next section.

If the operator F has a continuous spectrum $F = \int_{\mathbb{R}} \lambda P_S(d\lambda)$, with weakly continuous family of projection operators $P_S(\Delta)$, $\Delta \subset \mathbb{R}$, see (7.100), the emergence of superselection rules depends on an estimate of

$$\chi(\alpha, \beta; t) = \mathrm{tr}_E \left(e^{it(H_E + \alpha G)} e^{-it(H_E + \beta G)} \rho_E \right). \tag{7.71}$$

In this case we have to specify reference states such that (7.71) and its derivative $\frac{\partial}{\partial \alpha} \chi(\alpha, \beta; t)$ are bounded by $const \left(1 + (\alpha - \beta)^2 \vartheta(t) \right)^{-\gamma}$. If Δ_1 and Δ_2 are intervals with a distance $\delta > 0$, the operator norm of $P_S(\Delta_1) \rho_S(t) P_S(\Delta_2)$ is then estimated as

$$\|P_S(\Delta_1) \rho_S(t) P_S(\Delta_2)\|_1 \leq c \left(1 + \delta^2 \vartheta(t) \right)^{-\gamma}, \tag{7.72}$$

see (Kupsch 2000b), where the calculations have been given in the Heisenberg picture.

7.6.2.1 The Araki-Zurek Models

The first soluble models for the investigation of the reduced dynamics have been given by Emch (1972), Araki (1980) and Zurek (1982). The following construction is essentially based on these papers. In addition to the specifications made above, we demand that

3) the Hamiltonian H_E and the potential G commute, $[H_E, G] = 0$.

Then the interaction $F \otimes G$ commutes with the free Hamiltonian, and the operator $F \otimes G$ is conserved. The surprising point of this model is that despite of the almost trivial dynamics strong decoherence processes occur, such that F does not only yield a conservation law but indeed a superselection rule.

We first evaluate the dynamics for a factorizing initial state $W = \rho \otimes \rho_E$. Under the assumption 3) the trace (7.72) simplifies to $\chi_{m,n}(t) = \text{tr}_E \left(e^{it(\lambda_m - \lambda_n)G} \rho_E \right)$. Let $G = \int_{\mathbb{R}} \lambda P_E(d\lambda)$ be the spectral representation of the operator G. Since G has an absolutely continuous spectrum, the trace $\text{tr}_E \left(P_E(d\lambda) \rho_E \right) = d\mu(\lambda)$ defines an absolutely continuous measure on the real line for all $\rho_E \in \mathcal{D}(\mathcal{H}_E)$. Then the Riemann-Lebesgue Lemma implies that the Fourier integral $\int_{\mathbb{R}} e^{it\lambda} d\mu(\lambda)$ vanishes for $t \to \infty$, i.e.

$$\chi(t) := \text{tr}_E \left(e^{itG} \rho_E \right) = \int_{\mathbb{R}} e^{it\lambda} d\mu(\lambda) \to 0 \text{ if } t \to \infty. \tag{7.73}$$

Hence the superselection sectors $P_m \mathcal{H}_S$ finally emerge for all reference states $\rho_E \in \mathcal{D}(\mathcal{H}_E)$ with the uniform estimate (7.69). But to derive a decrease which is effective in sufficiently short time, we need an additional smoothness condition on ρ_E, which does not impose restrictions on the statistical operator $\rho \in \mathcal{D}(\mathcal{H}_S)$ of the system $\mathcal{S}$. To simplify the arguments we take $\mathcal{H}_E = \mathcal{L}^2(\mathbb{R})$ and the operator G given by the multiplication operator $Gf(x) = \gamma x f(x)$ with a coupling constant $\gamma \in \mathbb{R}$. If the statistical operator ρ_E has a continuous kernel function $\rho_E(x,y)$ the trace (7.73) is the Fourier integral $\text{tr}_E \left(\exp(itG)\rho_E \right) = \int_{\mathbb{R}} e^{it\gamma x} \rho_E(x,x) dx$, which decreases fast for smooth decreasing functions $\rho_E(x,x)$. More explicitly, $|\chi(t)| \leq C \left(1 + |t| \right)^{-n}$ if $\rho_E(x,x) \in \mathcal{L}^1(\mathbb{R})$ and $\left(\frac{d}{dx} \right)^n \rho_E(x,x) \in \mathcal{L}^1(\mathbb{R})$. Moreover, if also $x\rho_E(x,x) \in \mathcal{L}^1(\mathbb{R})$ and $\left(\frac{d}{dx} \right)^n x\rho_E(x,x) \in \mathcal{L}^1(\mathbb{R})$, the bound (7.72) for continuous superselection sectors follows.

So far we have assumed that the initial state factorizes. But exactly the same superselection sectors emerge for non-factorizing initial states $W \in \mathcal{D}(\mathcal{H}_{S+E})$. We can take the more general initial states

$$W = \sum_\mu c_\mu \, \rho_{S\mu} \otimes \rho_{E\mu} \tag{7.74}$$

with $\rho_{S\mu} \in \mathcal{D}(\mathcal{H}_S)$, $\rho_{E\mu} \in \mathcal{D}(\mathcal{H}_E)$ and real numbers c_μ, which satisfy $\sum_\mu |c_\mu| < \infty$ and $\sum_\mu c_\mu = \text{tr}\, W = 1$. Since all the traces $\chi_\mu(t) = \text{tr} \left(e^{iGt} \rho_{E\mu} \right)$ are uniformly bounded and approach zero at sufficiently large times, the partial trace $\text{tr}_E \left(e^{iGt} W \right) = \sum_\mu c_\mu \rho_{S\mu} \chi_\mu(t) \in \mathcal{T}(\mathcal{H}_S)$ decreases to zero in trace norm for any initial state (7.74). As in the discussion given for factorizing initial states this decrease is effective in short times, if W is a smooth operator on $\mathcal{H}_E$, i.e. the operators $\rho_{E\mu} \in \mathcal{D}(\mathcal{H}_E)$ in the decomposition (7.74) have to be represented by smooth functions in the basis (spectral representation) of G. This condition does not impose restrictions on $\rho_S(0) = \text{tr}_E W$.

The set of states (7.74) in dense in $\mathcal{D}(\mathcal{H}_{S+E})$. Since the mapping $W \to \rho_S(t) = \text{tr}_E U(t) W U^t(t)$ is continuous in the respective trace norms, the superselection sectors $P_m \mathcal{H}_S$, $m \in \mathbb{M}$, finally emerge for all initial states $W \in \mathcal{D}(\mathcal{H}_S \otimes \mathcal{H}_E)$ of the total system.

7.6.2.2 Particle Coupled to a Massless Boson Field.

In this section we present a model without the restriction 3) on the Hamiltonian. In addition to the more realistic dynamics this model exhibits an interesting relation between induced superselection sectors and infrared divergence of the environment.

We consider a particle moving on the real line $\mathbb{R}$ coupled to a Boson field which has excitations at arbitrarily low energies. The Hilbert space $\mathcal{H}_S$ can then be chosen as $\mathcal{H}_S = \mathcal{L}^2(\mathbb{R})$. The Hamiltonian and the interaction potential of the particle are

$$H_S = \frac{1}{2} P^2 \text{ and } F = P \tag{7.75}$$

where $P = -i\partial$ is the momentum operator of a particle on the real line $\mathbb{R}$. The property $H_S - \frac{1}{2} F = 0$ guarantees the positivity of the first term in (7.63). The momentum $P \otimes I_E$ is conserved, since it commutes with the Hamiltonian (7.63). As environment we choose a Boson field, which has excitations of arbitrarily small energies. The interesting result of the calculations presented below – and in more detail in (Kupsch 2000b) – is that the momentum of the particle yields a superselection rule exactly when the Boson field is infrared divergent.

As Hilbert space $\mathcal{H}_E$ we take the Fock space of symmetric tensors $\mathcal{H}_E = \mathcal{F}(\mathcal{H}_1)$ based on the one particle Hilbert space $\mathcal{H}_1 = \mathcal{L}^2(\mathbb{R}_+)$. The one-particle Hamiltonian M of the Boson field is $M f(\omega) := \omega f(\omega)$. Here we assume for simplicity that there are no degeneracies at fixed energy. This one-particle Hamilton operator is unbounded and has an unbounded inverse. Let $a_\omega^\dagger$ and a_ω be the standard creation and annihilation operators with commutation relations $[a_\varpi, a_\omega^\dagger] = \delta(\varpi - \omega)$ for $\varpi, \omega \in \mathbb{R}_+$, then the Hamiltonian of the free field is $H_E = \int_0^\infty \omega a_\omega^\dagger a_\omega d\omega$. Let $a^\dagger(f)$ denote the creation operator of the one-particle state $f \in \mathcal{H}_1$ and $a(f) = (a^\dagger(f))^\dagger$ the corresponding annihilation operator, $[a(f), a^\dagger(g)] = (f \mid g)$, where $(f \mid g) = \int \overline{f(\omega)} g(\omega) d\omega$ is the inner product of $\mathcal{H}_1 = \mathcal{L}^2(\mathbb{R}_+)$. The interaction potential G is chosen as the self-adjoint field operator $\Phi(h) = a^\dagger(h) + a(h)$, where the vector $h \in \mathcal{H}_1$ satisfies the additional constraint $\left\| M^{-\frac{1}{2}} h \right\|^2 = \int \omega^{-1} |h(\omega)|^2 d\omega \leq 4^{-1}$. This constraint guarantees that $H_E - \frac{1}{2} \Phi^2(h)$ is bounded from below. Then the Hamiltonian (7.63)

$$
\begin{aligned}
H_{S+E} &= \frac{1}{2} P^2 \otimes I_E + I_S \otimes H_E + P \otimes \Phi(h) \\
&= \frac{1}{2} \left(P \otimes I_E + I_S \otimes \Phi(h) \right)^2 + I_S \otimes \left(H_E - \frac{1}{2} \left(\Phi(h) \right)^2 \right) \tag{7.76}
\end{aligned}
$$

is a well defined semibounded operator on $\mathcal{H}_S \otimes \mathcal{H}_E = \mathcal{L}^2(\mathbb{R}) \otimes \mathcal{F}(\mathcal{H}_1)$.

The operators $H_E + \lambda G$ in (7.71) are then given by $H_\lambda := H_E + \lambda \Phi(h)$, $\lambda \in \mathbb{R}$, which are Hamiltonians of the van Hove model (1952). To derive induced superselection sectors we have to estimate the traces $\chi(\alpha, \beta; t) := \mathrm{tr}_E U_{\alpha\beta}(t) \rho_E$, $\alpha \neq \beta$, where ρ_E is the reference state of the field and the unitary operators $U_{\alpha\beta}(t)$ are given by $U_{\alpha\beta}(t) := \exp(itH_\alpha) \exp(-itH_\beta)$, see (7.71). The results are:

1. If the vector h has only weak contributions at low energies, such that $M^{-1}h = \omega^{-1}h(\omega) \in \mathcal{L}^2(\mathbb{R}_+)$, the Hamiltonians H_λ are (up to an energy shift) unitarily equivalent to the free field Hamiltonian H_E. One can use the standard methods of the van Hove model to evaluate the traces $\chi(\alpha, \beta; t) = \mathrm{tr}_E U_{\alpha\beta}(t) \rho_E$. In general the trace $\chi(\alpha, \beta; t)$ has no limit for $t \to \infty$. If the reference state is a coherent state, the modulus $|\chi(\alpha, \beta; t)|$ decreases to $\exp\left(-(\alpha - \beta)^2 \left\|M^{-1}h\right\|^2\right) > 0$. Hence one can achieve a strong decrease, but there are no induced superselection rules in the strict sense.

2. If the vector h has contributions at low energies such that $M^{-1}h \notin \mathcal{H}_1$ and $\left\|M^{-\frac{1}{2}}h\right\| \leq 2^{-1}$ is still valid, the low energy contributions of the interaction potential dominate, and $\lim_{t \to \infty} \chi(\alpha, \beta; t) = 0$ follows for all $\alpha \neq \beta$. For a large class of reference states – e.g. coherent states – this limit can be approached within a short time, if the vector h satisfies some additional regularity condition at small energies. These result remains valid, if the reference state is a KMS state of positive temperature (Kupsch 2002).

The assumption $M^{-1}h \notin \mathcal{H}_1$ is therefore necessary and sufficient for the emergence of superselection rules, which persist for $t \to \infty$. Exactly this condition $M^{-1}h \notin \mathcal{H}_1$ is related to a remarkable qualitative change of the van Hove model: the Boson field turns into an infrared divergent representation. It is still defined on the Fock space, but the bare photon number diverges and the ground state disappears in the continuum, see (Schroer 1963, Arai and Hirokawa 2000). The importance of infrared divergence for decoherence is known in electromagnetic interactions, see Sect. 4.1.1. It corresponds to an "ohmic" or "subohmic" behaviour of the environment in the models of quantum Brownian of Sect. 3.2.2, see (Leggett et al. 1987). In the infrared divergent case the Hamiltonian (7.76) is no longer unitarily equivalent to the (renormalized) free Hamiltonian $H_0 = \frac{1}{2}P^2 \otimes I_E + I_S \otimes H_E + const$. Therefore the identification of the "physical" or "dressed" particle is problematic. But as in the case of the models discussed in (Arai, Hirokawa and Hiroshima 1999) – and in contrast to models closer to quantum electrodynamics as investigated by Fröhlich (1973) – all state vectors of our model are well defined vectors in the Hilbert space $\mathcal{L}^2(\mathbb{R}) \otimes \mathcal{F}(\mathcal{H}_1)$, and the decay of superpositions of states with different momenta is a dynamical process which only appears on the level of the reduced dynamics.

The system Hamiltonian and the coupling potential in (7.75) can be replaced by other commuting operators. A simple example are spin operators of a spin-$\frac{1}{2}$ system. Moreover one can substitute the Boson field by an infinite set of harmonic oscillators, see e.g. the spin model given in Sect. 8 of (Primas 2000).

7.6.2.3 Models with Scattering So far the projection operators onto the induced superselection sectors $P_m \otimes I_S$ (or $P(\Delta) \otimes I_S$) commute with the total Hamiltonian $[P_m \otimes I_S, H_{S+E}] = 0$ (or $[P(\Delta) \otimes I_S, H_{S+E}] = 0$). We now generalize the Hamiltonian to

$$H = H_{S+E} + V, \tag{7.77}$$

where H_{S+E} is the Hamiltonian (7.63) and V is a scattering potential on $\mathcal{H}_S \otimes \mathcal{H}_E$. There is no constraint on the commutator $[H_{S+E}, V]$. The restriction to scattering potentials means that the wave operator $\Omega = \lim_{t \to \infty} \exp(itH) \exp(-itH_{S+E})$ exists as strong limit. Such a model exhibits the same superselection sectors as the model without the potential (Kupsch 1998, 2000a). But now the projection operators onto the induced superselection sectors do no longer commute with the Hamiltonian.

To simplify the arguments we assume that there are no bound states such that the convergence is valid on the whole space $\mathcal{H}_{S+E}$ with $\Omega^{\dagger} = \Omega^{-1}$. Then the time evolution $U(t) = \exp(-itH)$ behaves asymptotically like $U_0(t)\Omega^{\dagger}$ with $U_0(t) = \exp(-itH_{S+E})$. More precisely, we have for all $W \in \mathcal{D}(\mathcal{H}_{S+E})$

$$\lim_{t \to \infty} \left\| U(t)WU^{\dagger}(t) - U_0(t)\Omega^{\dagger}W\Omega U_0^{\dagger}(t) \right\|_1 = 0 \tag{7.78}$$

in trace norm. Following the calculations for the Araki-Zurek models or for the velocity coupling model the reduced dynamics of $U_0(t)...U_0^{\dagger}(t)$ yields the diagonalization $\mathrm{tr}_E U_0(t)\Omega^{\dagger}W\Omega U_0^{\dagger}(t) \simeq \sum_n P_n U_S(t)\hat{\rho}U_S^{\dagger}(t)P_n$. Thereby the off-diagonal elements vanish in trace norm

$$\lim_{t \to \infty} \left\| P_m \mathrm{tr}_E \left(U_0(t)\hat{\rho}U_0^{\dagger}(t) \right) P_n \right\|_1 = 0 \text{ if } t \to \infty \text{ and } m \neq n, \tag{7.79}$$

where $\hat{\rho} = \mathrm{tr}_E \Omega^{\dagger}W\Omega$ is the statistical operator after the scattering process. The projection operators P_m are again given by the spectral resolution of F. The limit (7.79) together with the scattering asymptotics (7.78) yield the induced superselection sectors $P_n \mathcal{H}_S$ for $\rho_S(t) = \Phi_t(\rho_S) = \mathrm{tr}_E U(t)WU^{\dagger}(t)$

$$\Phi_t(\rho_S) \simeq \sum_m P_m U_S(t) \mathrm{tr}_E \left(\Omega^{\dagger}\rho_S \otimes \rho_E \Omega \right) U_S^{\dagger}(t)P_m \text{ if } t \to \infty. \tag{7.80}$$

Since all estimates are trace norm estimates we still have

$$\| P_m \rho_S(t) P_n \|_1 \to 0 \text{ if } t \to \infty \text{ and } m \neq n. \tag{7.81}$$

But in contrast to the models discussed before we do not obtain a uniform bound with respect to the initial state $\rho_S(0)$, since the limit (7.78) is not uniform in $W \in \mathcal{D}(\mathcal{H}_{S+E})$.

We can derive fast decoherence by additional assumptions on the initial state and on the potential. For that purpose we start with a factorizing initial state $W = \rho_S \otimes \rho_E$ with a smooth ρ_E. The trace $\mathrm{tr}_E U_0(t)\Omega^\dagger W \Omega U_0^\dagger(t)$ can then be estimated as in the case of the Araki-Zurek models, if the statistical operator $\Omega^\dagger (\rho_S \otimes \rho_E) \Omega$ is a sufficiently smooth operator on the tensor factor $\mathcal{H}_E$ for all $\rho_S \in \mathcal{D}(\mathcal{H}_S)$. That is guaranteed if we choose as scattering potential a smooth potential in the sense of Kato (1966). Then both the limits, (7.78) and (7.79) are reached in a sufficiently short time.

Although the estimates are not uniform in $\rho_S(0)$, the bounds on decoherence are nevertheless uniform with respect to the observables of the system. For any given initial state $\rho(0) \in \mathcal{D}(\mathcal{H}_S)$ the norms $\|P_m \rho_S(t) P_n\|_1$ with $m \neq n$ vanish if $t \to \infty$. Consequently, for this initial state the expectation values $\mathrm{tr}_S \rho_S(t) P_n A P_m = \mathrm{tr}_S P_m \rho_S(t) P_n A$, which intertwine between two different superselection sectors, have the upper bound $\|A\| \|P_m \rho_S(t) P_n\|_1$, which decreases uniformly for all observables $A \in \mathcal{B}(\mathcal{H}_S)$.

7.6.2.4 Heisenberg Picture

In quantum field theory superselection rules are usually investigated in the Heisenberg picture, see e. g. (Wightman 1995) or Chap. 6. of this book. We therefore complete this section with a discussion of induced superselection rules in this picture. As already indicated in Sect. 7.1 the Heisenberg picture can be used for the reduced dynamics, if the initial state factorizes $W = \rho_S \otimes \rho_E$. From the calculations in the Schrödinger picture we know that the induced superselection sectors do not depend on the reference state $\rho_E \in \mathcal{D}(\mathcal{H}_E)$. But the choice of this state determines the time scale of decoherence. Here we only discuss models with discrete superselection rules, i. e. we assume a Hamiltonian (7.63) or (7.77) with $[H_S, F] = 0$ and a discrete spectrum of the interaction potential $F = \sum_m \lambda_m P_m$. The generalization to continuous superselection rules is possible, if we use direct integral representations of the Hilbert space, see Sect. 6.1.3 and Sect. 7.7.1.

We assume that the reduced dynamics (7.6) Φ_t induces a block diagonalization of the statistical operator (7.80) $\rho_S(t) = \Phi_t(\rho_S) \simeq P_m \Phi_t(\rho_S) P_m$. The duality relation $\mathrm{tr}_S A \Phi_t(\rho_S) = \mathrm{tr}_S \Phi_t^+(A)\rho_S$ then leads to

$$\mathrm{tr}_S A \Phi_t(\rho_S) \simeq \sum_m \mathrm{tr}_S P_m A P_m \Phi_t(\rho_S) = \mathrm{tr}_S \Phi_t^+(\sum_m P_m A P_m)\rho_S. \quad (7.82)$$

Hence the dynamics induces the superselection sectors $\mathcal{H}_m = P_m \mathcal{H}$ if $\Phi_t^+(A) \simeq \Phi_t^+(\sum_m P_m A P_m)$ for $t \to \infty$.[15] In the language of the Heisenberg

---

[15] As explained in the preceding sections "$t \to \infty$" or "t large" means large with resect to a microscopic time scale and (very) small with respect to macroscopic scales.

picture the criterion for induced superselection sectors therefore is[16]

$$\Phi_t^+(P_m A P_n) \to 0 \text{ for } t \to \infty \text{ and } m \neq n. \tag{7.83}$$

The relation (7.82) also means that for large t only the observables in $\mathcal{A}_{eff} = \{A \mid A = \sum_m P_m A P_m\}$ contribute to the expectation values. But this algebra of *effective observables* should not be confused with the set of observables $\{\Phi_t^+(A) \mid A \in \mathcal{B}(\mathcal{H}_S), t \text{ large}\}$ obtained in the Heisenberg picture at large times. In general there is no simple relation between these two sets.

An exceptional simple case are models with the Hamiltonian (7.63). Then the operator F satisfies the conservation rule $U_{S+E}^\dagger(t)(F \otimes I_E)U_{S+E}(t) = F \otimes I_E$, and the identity $\Phi_t^+(P_m A P_n) = P_m \Phi_t^+(A) P_n$ follows. Hence (7.83) yields a block diagonalization of the observables exactly into the superselection sectors $\mathcal{H}_m = P_m \mathcal{H}$. Therefore the induced superselection rules can be discussed in a transparent way similar to the investigation of the kinematical superselection rules in Chap. 6. The asymptotic behaviour $\Phi_t^+(A) \simeq \Phi_t^+(\sum_m P_m A P_m) = \sum_m P_m \Phi_t^+(A) P_m$ implies $\lim_{t\to\infty} U_S(t)\Phi_t^+(A)U_S^\dagger(t) = \sum_m P_m A P_m$. If we define the set of *observables at large times* by $\mathcal{A}_\infty := \lim_{t\to\infty} U_S(t)\Phi_t^+(\mathcal{B}(\mathcal{H}_S))U_S^\dagger(t)$, then obviously $\mathcal{A}_\infty = \{A \mid A = \sum_m P_m A P_m\} = \{A \mid [F, A] = 0\}$. This set is a $\mathbf{C}^*$-subalgebra of $\mathcal{B}(\mathcal{H}_S)$ and it corresponds to the algebra of observables in the discussion of kinematical superselection rules in Chap. 6. The centre $\mathcal{Z}_\infty$ of $\mathcal{A}_\infty$ is the abelian algebra generated by the projection operators P_m of the spectral resolution of F. Hence the algebra of superselection operators is exactly given by the centre $\mathcal{Z}_\infty$.

These arguments arc no longer valid for models with additional scattering or for other more general models. In the case of the Hamiltonian (7.77) the Heisenberg evolution

$$\Phi_t^+(A) \simeq \mathrm{tr}_E \Omega \left(\sum_m P_m U_S^\dagger(t) A U_S(t) P_m \otimes I_E \right) \Omega^\dagger \rho_E \text{ if } t \to \infty$$

obviously satisfies (7.83). But since $A \to \Phi_t^+(A)$ is a rather general completely positive mapping, the set of observables at large times $\{\Phi_t^+(A) \mid A \in \mathcal{B}(\mathcal{H}_S), t \text{ large}\}$ does not split into blocks within the sectors $\mathcal{H}_m = P_m \mathcal{H}$. To recover these superselection sectors as structures which appear for large times the Schrödinger picture is the adequate frame.

---

[16] For all models discussed here the off-diagonal elements of the statistical operator (7.81) are estimated in trace norm. The Heisenberg evolution therefore leads to $\lim_{t\to\infty} \Phi_t^+(P_m A P_n) = 0$ if $m \neq n$, in the topology of strong operator convergence. Convergence of the norm $\|\Phi_t^+(P_m A P_n)\|$ is possible for those models, for which $\|\rho_S(t)\|_1$ is estimated uniformly with respect to $\rho_S(0)$.

7.7 Mathematical Supplement

In this section we give an overview on some mathematical tools, which are used in the main text. First we present some classes of linear operators. In the second part of this section some results about completely positive mappings are discussed with applications to Zwanzig projectors. In the third part a proof for the entropy inequality (7.21) for all hermitean Zwanzig projectors is given.

7.7.1 Spaces of Linear Operators

We denote by $\mathcal{B}(\mathcal{H})$ the linear space of bounded operators on $\mathcal{H}$ with the norm

$$\|A\| = \sup_{f \in \mathcal{H}, \|f\|=1} \|Af\|. \tag{7.84}$$

In quantum mechanics these operators represent the observables of a theory. In the proper sense only self-adjoint (and possibly unbounded) operators are observables. For technical reasons this class will be restricted to bounded operators. This is only an apparent restriction, since the whole information of an unbounded observable can be recovered from its spectral resolution (involving only bounded operators). Moreover, it is convenient to admit arbitrary complex linear combinations of observables. Hence one is led to the class of all bounded linear operators $\mathcal{B}(\mathcal{H})$ on the Hilbert space $\mathcal{H}$. With the operator norm this is a Banach space. If we include the multiplication of operators and the transition to the adjoint operator as *-operation this space becomes the C^*-algebra of all bounded operators on $\mathcal{H}$.

The linear space of all compact operators $\mathcal{K}(\mathcal{H}) \subset \mathcal{B}(\mathcal{H})$ is the closure of all finite rank operators $\mathcal{F}(\mathcal{H})$ with this operator norm. Hilbert–Schmidt operators are compact operators A with a finite trace of $A^\dagger A$, their linear space is denoted by $\mathcal{T}_2(\mathcal{H})$, their norm is defined by

$$\|A\|_2 = \sqrt{\operatorname{tr} A^\dagger A}. \tag{7.85}$$

The space $\mathcal{T}_2(\mathcal{H})$ is a Hilbert space with the inner product

$$\langle A \mid B \rangle = \operatorname{tr} A^\dagger B. \tag{7.86}$$

Trace class or *nuclear* operators A are those Hilbert–Schmidt operators which have a finite trace of $\sqrt{A^\dagger A}$. Their linear space is denoted by $\mathcal{T}(\mathcal{H})$ and their norm is the *trace norm*

$$\|A\|_1 = \operatorname{tr} \sqrt{A^\dagger A}. \tag{7.87}$$

These spaces obviously satisfy the following inclusions

$$\mathcal{F}(\mathcal{H}) \subset \mathcal{T}(\mathcal{H}) \subset \mathcal{T}_2(\mathcal{H}) \subset \mathcal{K}(\mathcal{H}) \subset \mathcal{B}(\mathcal{H}).$$

Only in the case of finite dimensional Hilbert spaces $\mathcal{H}$ these sets coincide and the topologies are equivalent.

A sequence $A_n \in \mathcal{B}(\mathcal{H})$ converges to an operator $A \in \mathcal{B}(\mathcal{H})$ in the strong operator topology, if $\lim_{n\to\infty} \|A_n f - Af\| = 0$ for all vectors $f \in \mathcal{H}$. The strong operator topology is weaker than the topology induced by the norm (7.84). The closure of the set $\mathcal{F}(\mathcal{H})$ of finite rank operators in the strong operator topology is $\mathcal{B}(\mathcal{H})$. Therefore all the spaces $\mathcal{F}(\mathcal{H})$, $\mathcal{T}(\mathcal{H})$, $\mathcal{T}_2(\mathcal{H})$ and $\mathcal{K}(\mathcal{H})$ are dense subsets of $\mathcal{B}(\mathcal{H})$ with respect to the strong operator topology.

An important class of Hilbert spaces are spaces of square integrable functions. If $\mathbb{T} \subset \mathbb{R}$ is a finite or infinite interval of the real line and $d\mu$ a positive measure on $\mathbb{T}$, then the space $H = \mathcal{L}^2(\mathbb{T}, d\mu)$ is the linear set of all functions $t \in \mathbb{T} \to f(t) \in \mathbb{C}$ with a finite norm $\|f\|^2 = \int_{\mathbb{T}} |f(t)|^2 \, d\mu(t)$. In this case there is a simple criterion to conclude, that an integral operator

$$(Af)(t) = \int_{\mathbb{T}} K(t,s)f(s)d\mu(s) \tag{7.88}$$

with an integrable kernel functions $K(t,s)$ defines a Hilbert-Schmidt operator. If $\int |K(t,s)|^2 \, d\mu(t)d\mu(s) < \infty$ then the operator (7.88) is a Hilbert-Schmidt operator with a Hilbert-Schmidt norm bounded by $\|A\|_2^2 \leq \int |K(t,s)|^2 \, d\mu(t)d\mu(s)$. For the kernel functions of a trace class operator A simple criteria are only available, if one already knows that A is a positive operator. Then continuity of the kernel function $K(t,s)$ and the existence of the integral $\int_{\mathbb{T}} K(t,t)d\mu(t)$ are sufficient to conclude that (7.88) defines a trace class operator with trace $\operatorname{tr} A = \int_{\mathbb{T}} K(t,t)d\mu(t)$, see the Lemma on p. 65 of (Reed and Simon 1979).

The bilinear form $T \in \mathcal{T}(\mathcal{H})$, $A \in \mathcal{B}(\mathcal{H}) \to \operatorname{tr} TA \in \mathcal{T}(\mathcal{H})$ is continuous in the respective norms

$$|\operatorname{tr} TA| \leq \|T\|_1 \cdot \|A\|. \tag{7.89}$$

Moreover, the space $\mathcal{B}(\mathcal{H})$ is the dual space of $\mathcal{T}(\mathcal{H})$ and the norm of the linear functional $T \in \mathcal{T}(\mathcal{H}) \to \lambda(T) = \operatorname{tr} TA$ with a fixed operator $A \in \mathcal{B}(\mathcal{H})$ is exactly the operator norm of A

$$\|A\| = \sup_{T \in \mathcal{T}(\mathcal{H}), \|T\|_1 = 1} |\operatorname{tr} TA|. \tag{7.90}$$

On the other hand, $A \in \mathcal{B}(\mathcal{H}) \to \omega(A) = \operatorname{tr} TA$ with a fixed nuclear operator T is a continuous linear functional on $\mathcal{B}(\mathcal{H})$, and the norm of this functional is the trace norm of the operator T

$$\|T\|_1 = \sup_{A \in \mathcal{B}(\mathcal{H}), \|A\| = 1} |\operatorname{tr} TA|. \tag{7.91}$$

But if the dimension of $\mathcal{H}$ is infinite, the full space of linear functionals on $\mathcal{B}(\mathcal{H})$, which are continuous in the operator norm, is larger than $\mathcal{T}(\mathcal{H})$.

The closed convex set $\mathcal{D}(\mathcal{H}) \subset \mathcal{T}(\mathcal{H})$ of all positive trace class operators $\rho \geq 0$ with $\operatorname{tr} \rho = 1$ represents the "set of physical states". The subset of "pure states" is the space $\mathcal{S}(\mathcal{H}) \subset \mathcal{D}(\mathcal{H})$ of all rank one projection operators. Given a statistical operator $\rho \in \mathcal{D}(\mathcal{H})$, then the expectation of an observable $A \in \mathcal{B}(\mathcal{H})$ in the state ρ

$$\omega(A) = \operatorname{tr} \rho A \tag{7.92}$$

is a linear functional on $\mathcal{B}(\mathcal{H})$ which determines the operator $\rho \in \mathcal{D}(\mathcal{H})$ uniquely, and either the operator ρ or the functional ω can be used to characterize a state. The natural topology for $\mathcal{D}(\mathcal{H})$ is obtained from the operational definition of the distance between two states

$$\operatorname{dist}(\omega_1, \omega_2) = \sup_{\|A\|=1} |\omega_1(A) - \omega_2(A)|. \tag{7.93}$$

Then (7.91) implies that this distance is exactly given by the trace norm

$$\operatorname{dist}(\omega_1, \omega_2) = \|\rho_1 - \rho_2\|_1. \tag{7.94}$$

For pure states represented by rank one projection operators P_1 and P_2 this distance is calculated as

$$\|P_1 - P_2\|_1 = 2\sqrt{1 - \operatorname{tr} P_1 P_2} \tag{7.95}$$

which is determined by the usual transition probability

$$\operatorname{tr} P_1 P_2 = |\langle \psi_1 \mid \psi_2 \rangle|^2 \tag{7.96}$$

of the unit vectors $\psi_i = P_i \psi_i$, $i = 1, 2$.

If ρ_1 and ρ_2 are statistical operators, then

$$\lambda_1 \rho_1 + \lambda_2 \rho_2, \quad \lambda_{1,2} \geq 0, \quad \lambda_1 + \lambda_2 = 1 \tag{7.97}$$

is also a statistical operator. This operation is usually called "mixing" of states. Any mixed state ρ can be decomposed according to

$$\rho = \lambda \rho_1 + (1 - \lambda)\rho_2, \quad 0 \leq \lambda \leq 1 \tag{7.98}$$

into other states, ρ_1 and ρ_2. A pure state, i.e. a rank 1 projection, is characterized by the fact that it cannot be decomposed into other states, and (7.98) can only hold as a trivial identity with $\lambda = 0$ or $\lambda = 1$. The notion of a pure state has therefore a geometrical characterization as extremal point of the convex set of states without any reference to the spectral theorem of self-adjoint operators.

In finite and in infinite dimensional Hilbert spaces $\mathcal{H}$ the finite linear combinations of operators $\rho \in \mathcal{D}(\mathcal{H})$ form the Banach space $\mathcal{T}(\mathcal{H})$. The extension to the larger space $\mathcal{T}_2(\mathcal{H})$ of Hilbert-Schmidt operators is sometimes used for technical convenience. But as seen in (7.93), the metric of the state space is given by the trace norm. Consider as simple example the state space $\mathcal{D}(\mathbb{C}^2)$. Any operator $\rho \in \mathcal{D}(\mathbb{C}^2)$ can be characterized by a real polarization vector $\boldsymbol{n}$ in the unit ball $B^3 = \{\boldsymbol{n} \in \mathbb{R}^3, \; ||\boldsymbol{n}|| \leq 1\}$ as

$$\widehat{\rho}(\boldsymbol{n}) := \frac{1}{2} \left(\mathbf{1} + \boldsymbol{\sigma n}\right) \tag{7.99}$$

where $\mathbf{1}$ is the unit 2x2 matrix and $\boldsymbol{\sigma} = (\sigma_1, \sigma_2, \sigma_3)$ are the Pauli matrices. Thereby the pure states correspond to the points $\boldsymbol{n}$ on the unit sphere $S^2 = \{\boldsymbol{n} \in \mathbb{R}^3, \; ||\boldsymbol{n}|| = 1\}$, which coincides with the set of the extremal points. A simple calculation leads to the identity $||\rho(\boldsymbol{p}_1) - \rho(\boldsymbol{p}_2)||_1 = |\boldsymbol{p}_1 - \boldsymbol{p}_2|$. Hence the trace norm topology of $\mathcal{D}(\mathbb{C}^2)$ agrees with the Euclidean metric of the polarization vectors. The set $\mathcal{D}(\mathbb{C}^2)$ is therefore not only linearly but also metrically isomorphic to the unit ball B^3, if the trace norm is used.

In Sect. 6.1 we have introduced discrete superselection sectors. That scheme can be extended to continuous superselection sectors on infinite dimensional Hilbert spaces $\mathcal{H}$ (Piron 1969, Araki 1980). For most of the continuous kinematical superselection rules which might arise in quantum mechanics or quantum field theory, as e.g. the total mass of non-relativistic quantum mechanics, see Chap. 6, one can restrict the theory to just one superselection sector and a detailed theory is apparently not necessary. On the other hand, approximate dynamical continuous superselection rules emerge by decoherence, see the discussion of localization in Chap. 3; therefore we give here a more general mathematical formulation of continuous superselection rules, which we also need in Sect. 7.6.

A continuous superselection rule is related to a (weakly continuous) family of projection operators $P(\Delta)$ indexed by measurable subsets $\Delta \subset \mathbb{R}$ (or just intervals). These projection operators take over the role of the discrete set $\{P_m\}$. They have to satisfy:

$$\begin{aligned} P(\Delta_1 \cup \Delta_2) &= P(\Delta_1) + P(\Delta_2), \\ P(\Delta_1)P(\Delta_2) &= 0 \quad \text{if} \quad \Delta_1 \cap \Delta_2 = \emptyset, \\ P(\emptyset) &= 0, \quad P(\mathbb{R}) = 1. \end{aligned} \tag{7.100}$$

Any operator $A \in \mathcal{B}(\mathcal{H})$ satisfying $P(\Delta)A = AP(\Delta)$ for all $\Delta \subset \mathbb{R}$ can be most easily represented in a direct integral representation of $\mathcal{H}$ which diagonalizes the operators $P(\Delta)$. The Hilbert space $\mathcal{H}$ is then decomposed into the family of Hilbert spaces $\mathcal{H}_\lambda, \lambda \in \mathbb{R}$ with inner products $(.,.)_\lambda$. A vector $f \in \mathcal{H}$ is represented by the mapping $\lambda \in \mathbb{R} \to f(\lambda) \in \mathcal{H}_\lambda$, and the inner product of $\mathcal{H}$ is given by $(f, g) = \int (f(\lambda), g(\lambda))_\lambda \, d\mu(\lambda)$ with an absolutely continuous measure $d\mu$. The projection operators $P(\Delta)$ are then simply $(P(\Delta)f)(\lambda) = f(\lambda)$ if $\lambda \in \Delta$, and 0 otherwise. Any operator A satisfying $P(\Delta)A = AP(\Delta)$ has the representation

$$(Af)(\lambda) = A(\lambda)f(\lambda) \tag{7.101}$$

where $A(\lambda)$ is a linear operator on $\mathcal{H}_\lambda$. The algebra of observables of this theory is still defined by

$$\mathcal{A} = \{A \in \mathcal{B}(\mathcal{H}) \mid P(\Delta)A = AP(\Delta),\ \Delta \subset \mathbb{R}\}. \qquad (7.102)$$

Hence any $A \in \mathcal{A}$ has the representation (7.101) and consequently a diverging trace. Therefore we cannot impose the condition $P(\Delta)\rho = \rho P(\Delta)$ to restrict the set of statistical operators $\rho \in \mathcal{D}(\mathcal{H}) \subset \mathcal{T}(\mathcal{H})$. One can of course take statistical operators $\sum_n P_n \rho P_n$ where $P_n \equiv P(\Delta_n)$ is defined with the help of a finite or countable set of non-overlapping intervals $\Delta_n \subset \mathbb{R}$, $\Delta_m \cap \Delta_n = \emptyset$ if $m \neq n$, $\cup_m \Delta_m = \mathbb{R}$. But the operators $\sum_n P_n \rho P_n$ have no limit for finer and finer partitions within the set of nuclear operators. In this respect the Heisenberg picture is more adequate to describe continuous superselection sectors.

7.7.2 Complete Positivity

Complete positivity is an essential property of dynamical mappings in the Heisenberg picture. The usual unitary dynamics and the subdynamics (7.10) are completely positive. But in the case of approximations, as e. g. the Markov approximation, complete positivity is an additional constraint.

The general notion of complete positivity refers to mappings between $\mathbf{C}^*$-algebras (Stinespring 1955, Choi 1972, Alicki and Fannes 2001).

We only consider the concrete $\mathbf{C}^*$-algebra $\mathcal{B}(\mathcal{H})$ of all bounded operators on a complex Hilbert space $\mathcal{H}$. A mapping $\Psi : \mathcal{B}(\mathcal{H}) \to \mathcal{B}(\mathcal{H})$ is called *positive* if $\Psi(A) \geq 0$ for all $A \geq 0$. A mapping $\Psi : \mathcal{B}(\mathcal{H}) \to \mathcal{B}(\mathcal{H})$ is called *n-positive* if $\Psi \otimes id_n$ is a positive linear operator on $\mathcal{B}(\mathcal{H} \otimes \mathbb{C}^n)$ for $n \in \{2, 3, ...\}$; a mapping which is n-positive for all $n = 2, 3, ...$ is called *completely positive*. As example consider the case of 2-positivity. The mapping Ψ is 2-positive, if

$$\begin{pmatrix} A_{11} & A_{12} \\ A_{21} & A_{22} \end{pmatrix} \to \begin{pmatrix} \Psi(A_{11}) & \Psi(A_{12}) \\ \Psi(A_{21}) & \Psi(A_{22}) \end{pmatrix} \qquad (7.103)$$

with $A_{jk} \in \mathcal{B}(\mathcal{H})$ is positive. Extending a result of Stinespring the following theorem has been derived by Kraus (1971).

Theorem 1. *The mapping Ψ is completely positive if and only if there exist bounded operators V_j in $\mathcal{H}$ such that $\sum_j V_j^+ V_j$ converges strongly and $\Psi(A)$ has the representation*

$$\Psi(A) = \sum_j V_j^+ A V_j. \qquad (7.104)$$

Take as example the Hilbert space $\mathcal{H} = \mathbb{C}^2$. All operators (matrices) A in $\mathcal{B}(\mathbb{C}^2)$ have the representation

$$T(n) = \frac{1}{2} \sum_\mu n_\mu \sigma_\mu, \qquad (7.105)$$

where $n = (n_0, n_1, ..., n_3) \in \mathbb{C}^4$ is a complex vector, σ_0 is the 2×2 unit matrix and σ_j, $j = 1, 2, 3$, are the Pauli matrices. The positive matrices are characterized by vectors $n = (n_0, \boldsymbol{n}) \in \mathbb{R}^4$ with $|\boldsymbol{n}| \leq n_0$.

1) The mapping $\Psi_1 : T(n) \rightarrow T(n_0, n_1, -n_2, n_3) = T^t(n)$ is positive, but not 2-positive. The proof is similar to the proof given for the following example 2. In general, the transposition of a matrix is not 2-positive (Arveson 1969).

2) The mapping $\Psi_2 : T(n) \rightarrow T(n_0, n_1, 0, n_3) = \frac{1}{2}\left(T(n) + T^t(n)\right)$ is positive,

but not 2-positive. For the proof take the positive 4×4 matrix $\begin{pmatrix} 1 & 0 & 0 & 1 \\ 0 & 0 & 0 & 0 \\ 0 & 0 & 0 & 0 \\ 1 & 0 & 0 & 1 \end{pmatrix}$.

It is easily seen that the image (7.103) of this matrix is not positive. Hence Ψ_2 is not 2-positive.

3) The mapping $\Psi_3 : T(n) \rightarrow T(n_0, n_1, -n_2, -n_3)$ is completely positive. The proof follows from $\sigma_1 \sigma_\mu \sigma_1 = \begin{cases} \sigma_\mu & \text{if } \mu = 0, 1 \\ -\sigma_\mu & \text{if } \mu = 2, 3 \end{cases}$. The mapping Ψ_3 has therefore the representation (7.104) $T(n_0, n_1, -n_2, -n_3) = \sigma_1 T(n) \sigma_1$.

4) The mapping $\Psi_4 : T(n) \rightarrow T(n_0, n_1, 0, 0) = \frac{1}{2}(T(n) + T(n_0, n_1, -n_2, -n_3))$ is completely positive. The proof follows from the case 3.

If we choose $n_0 = 1$ and $\boldsymbol{n} = (n_1, n_2, n_3) \in \mathbb{R}^3$, $|\boldsymbol{n}| \leq 1$, the operators $T(1, \boldsymbol{n}) = \rho(\boldsymbol{n}) \in \mathcal{D}(\mathbb{C}^2)$ are statistical operators. Let K be an orthogonal projection on $\mathbb{R}^3$. Then $\rho(\boldsymbol{n}) \rightarrow \rho(K\boldsymbol{n})$ is a Zwanzig projection, which satisfies the conditions (i) - (iv) of Sect. 7.2 and the hermiticity relation (7.16). Examples are the mappings Ψ_2 and Ψ_4 with $K\boldsymbol{n} = (n_1, 0, n_3)$ or $K\boldsymbol{n} = (n_1, 0, 0)$, respectively. The results derived above imply that the Zwanzig projection Ψ_2 is not completely positive, whereas Ψ_4 is a completely positive mapping. Hence Ψ_2 does neither satisfy (7.104) nor (7.15). The condition (v) is therefore an additional independent demand for Zwanzig projectors.

7.7.3 Entropy Inequalities

Given a Zwanzig projector $\widehat{P}$, which satisfies all the conditions (i) -(v) of Sect. 7.2, the von Neumann entropy $S(\widehat{P}\rho)$ can be larger or smaller than $S(\rho)$, $\rho \in \mathcal{D}(\mathcal{H})$. In this section we prove that $S(\widehat{P}\rho) \geq S(\rho)$ is true for a hermitean Zwanzig projector, which satisfies the conditions (i)-(iv). Following Uhlmann we first define a partial order between statistical operators, which classifies the degree of mixing (Wehrl 1978, Thirring 1983). Let $\rho(k)$, $k = 1, 2$, be two statistical operators with eigenvalues $p_1(k) \geq p_2(k) \geq ... \geq 0$, $k = 1, 2$, ordered according to their magnitude.

Definition 1. *The statistical operator $\rho(2)$ is more mixed (or more chaotic) than the statistical operator $\rho(1)$, if $\sum_{\mu=1}^n p_\mu(2) \leq \sum_{\mu=1}^n p_\mu(1)$ for all $n = 1, 2, ...$*

An important consequence of this mixing-enhancing is the following theorem.

Theorem 2. *If $\rho(2)$ is more mixed than $\rho(1)$, then their von Neumann entropies satisfy*

$$S\left(\rho(2)\right) \geq S\left(\rho(1)\right).\tag{7.106}$$

For the proof see (Wehrl 1978) or (Thirring 1983).

We now derive that hermitean Zwanzig projectors increase mixing. Therefore the inequality (7.106) is true for all hermitean Zwanzig projectors.

Lemma 1. *If $\widehat{P}$ is hermitean with respect to the inner product of $\mathcal{T}_2(\mathcal{H})$, i.e. $\widehat{P} = \widehat{Q}$, then the statistical operator $\widehat{P}\rho$ is more mixed than ρ.*

Proof. Let $P^{(n)} \in \mathcal{T}(\mathcal{H}) \cap \mathcal{B}(\mathcal{H})$ be a rank n projection operator with $n \geq 1$, then

$$\begin{aligned}\left\|\widehat{P}P^{(n)}\right\| &= \left\|\widehat{Q}P^{(n)}\right\| \leq \left\|P^{(n)}\right\| = 1\\\left\|\widehat{P}P^{(n)}\right\|_1 &= \left\|P^{(n)}\right\|_1 = n\end{aligned}\tag{7.107}$$

Hence the eigenvalues λ_μ of $\widehat{Q}P^{(n)}$ satisfy $0 \leq \lambda_\mu \leq 1$ and $\sum_\mu \lambda_\mu = n$. The operator $\widehat{P}\left(\frac{1}{n}P^{(n)}\right)$ is therefore more mixed than the statistical operator $\frac{1}{n}P^{(n)}$.

For any two positive operators A and B with eigenvalues $\alpha_1 \geq \alpha_2 \geq \ldots \geq 0$ and $\beta_1 \geq \beta_2 \geq \ldots \geq 0$, respectively, the inequality

$$\text{tr}AB \leq \sum_\mu \alpha_\mu \beta_\mu \tag{7.108}$$

is true (Gohberg, Goldberg, and Krupnik 2000). Now take a statistical operator ρ with eigenvalues $p_1 \geq p_2 \geq \ldots \geq 0$, $\sum_\mu p_\mu = 1$. The eigenvalues of $\widehat{P}\rho$ are denoted by $q_1 \geq q_2 \geq \ldots \geq 0$, $\sum_\mu q_\mu = 1$. Let $\mathcal{H}_n \subset \mathcal{H}$ be a subspace of dimension n and $P^{(n)}$ the projection operator onto $\mathcal{H}_n$. The trace $\text{tr}_{\mathcal{H}_n}\widehat{P}\rho$ can then be estimated by

$$\text{tr}_{\mathcal{H}_n}\widehat{P}\rho = \left\langle\widehat{P}\rho \mid P^{(n)}\right\rangle = \left\langle\rho \mid \widehat{Q}P^{(n)}\right\rangle \overset{(7.108)}{\leq} \sum_{\mu=1}^\infty p_\mu \lambda_\mu \leq \sum_{\mu=1}^n p_\mu.\tag{7.109}$$

The last inequality follows from the ordering of the p_μ and the restrictions on the eigenvalues λ_μ : $0 \leq \lambda_\mu \leq 1$ and $\sum_\mu \lambda_\mu = n$ of $\widehat{Q}P^{(n)}$. The space $\mathcal{H}_n$ is an arbitrary subspace of dimension n. The supremum with respect to all these subspaces is $\sup_{\mathcal{H}_n} \text{tr}_{\mathcal{H}_n}\widehat{P}\rho = \sum_{\mu=1}^n q_\mu$. Hence (7.109) implies $\sum_{\mu=1}^n q_\mu \leq \sum_{\mu=1}^n p_\mu$ for all $n \geq 1$ and the statistical operator $\widehat{P}\rho$ is more mixed than ρ.

8 Stochastic Collapse Models

I.-O. Stamatescu

8.1 The Question of State Vector Reduction

8.1.1 Two Points of View

In the preceding chapters we have encountered non-Hamiltonian evolution equations describing the reduced dynamics of systems which are physically connected with uncontrollable environments (the global dynamics being unitary). In contrast to this approach we shall be concerned in this chapter with models which postulate non-Hamiltonian dynamics as a fundamental modification of the Schrödinger equation. Without aiming at a comprehensive review we shall try to discuss the conceptual setting and some results of these models such as to permit a comparison with environmental decoherence as it is discussed in this book.

For the sake of the discussion in this chapter we stress that *environmental decoherence* is to be understood here not as a hypothesis which we can make as we wish, but as a well defined quantum mechanical problem of finding out the effects of the interaction between a system of concern and its environment as defined by the given physical situation. Environmental decoherence involves quantum mechanical measurement but does not depend on the particular setting of quantum mechanics – "standard" (unspecified collapse), Everett or even dynamical collapse –, and acts whether we want it or not: it is not a matter of choice (but predicting its effects is a matter of our ability to calculate, therefore this may involve approximations). Decoherence does not provide by itself a solution to interpretational problems, in particular to that of measurement. However, understanding the reach of decoherence phenomena sheds light on various aspects of quantum theory and may allow us to develop a general view exempted from conceptual ambiguities. Such a view, based on unbroken validity of the superposition principle, is proposed in Chap. 2 of this book. The "program of decoherence" is essentially the trial to prove this conjecture – while decoherence itself is just a fact. Collapse models on the other hand are dedicated to the same scope (of achieving a general, objective view), but have a quite different status, since they are based on a dynamical *hypothesis*, which is *added* to quantum theory, can within some limits be tuned to adapt to the phenomena and may be proven wrong if this tuning turns out unsuccessful. On the other hand they directly represent a unifying interpretation, based on real state vector reduction.

A second remark is that decoherence always refers to some (typically local) system, defined by the physical problem, and pertains to the observations which can be made upon this system. The usual working objects are therefore the reduced density matrix and the expectation values of various operators which describe the behavior of the system with respect to local interactions and observations. The question of decoherence is essentially whether or not this behavior reveals the inherent quantum superpositions or can be simulated by classical statistics. This has been developed in the other chapters and we refer to these for detail and more explicit results, while we shall here mostly use general arguments for the purpose of comparison with collapse models. The latter also use sometimes evolution equations for the statistical operator, however this regards now the full and not the reduced dynamics and describes a "proper" mixture.

In the following we refer to the system under observation – the system of interest – as "the system" or "our system", etc. to be distinguished from "systems in the environment", the "total system" (system+environment), etc. – the context should help to disentangle these meanings.

8.1.2 Decoherence, Collapse, Measurement

Since a system in contact with an environment is generally in an entangled state with the latter it does not possess a (pure) state for itself. For the purpose of observations concerning *only* the degrees of freedom of the system one can form a *reduced density matrix* by averaging over the states of the environment. Although this reduced density matrix contains all the information concerning such observations, it cannot be attributed, however, to the system in the same way one attributes a pure state to an isolated system. Indeed, on the one hand, the reduced density matrix depends on the environment, and on the other hand, by performing experiments including also the environment one could in principle observe quantum correlations between the system and the environment which cannot be described by the reduced density matrix (see also App. A4).[1]

The Hamiltonian dynamics of the *total system* (system and environment) usually induces a non-unitary dynamics for the reduced density matrix of the system itself. As discussed in the previous chapters, the question of the emergence of classical properties pertains to those physical situations in which a system is strongly entangled with an environment that cannot be controlled and where the reduced density matrix of the former evolves toward a decohered (diagonal) form with respect to certain properties, i.e., in a certain

---

[1] Since no part of the universe can be absolutely separated from the rest the notion of *isolated system* is an approximation whose justification is a quantitative question. Since this approximation may be very good in specific – typically microscopic – situations it makes sense in such cases to speak of closed systems without always meaning the whole universe – otherwise it would not be understandable why the observed hydrogen spectrum, for instance, can be so well predicted.

basis[2]. Then, *while the system remains entangled with the environment* (and therefore is not in a pure state), all observations at our disposal are compatible with (classical) statistics concerning these properties, and we can say that the system is described by an "apparent" mixture. Since decoherence is a dynamical effect there will be some characteristic time scale τ (which may be as small as 10^{-13}s for a dust particle under thermal radiation at room temperature and as large as 10^{10} years for an atom in interstellar vacuum) for a stable decohered situation to be installed and thereafter preserved – Sects. 3.3, 3.4. We can represent this in the Schrödinger picture by writing:

$$|\Psi_{\text{tot.}}(0)\rangle \xrightarrow[t \gg \tau]{\text{unit. evol.}} |\Psi_{\text{tot.}}(t)\rangle \simeq \sum_n \sqrt{p_n} |\Phi_n^{\text{env.}}\rangle |\varphi_n^{\text{syst.}}\rangle. \tag{8.1}$$

Thereby the states of the system, $|\varphi_n\rangle$, become correlated with environment states $|\Phi_n\rangle$ which came to differ significantly in their *non–accessible* (uncontrollable) degrees of freedom (e.g., systems in the environment which after interacting with our system rapidly leave its neighborhood and are therefore not under our control – such as electromagnetic radiation background). Thus for all observables – and in general, operators – concerning *accessible* degrees of freedom (those of the system and, possibly, the laboratory, etc.) we have:

$$\langle \Phi_n^{\text{env.}} | O^{\text{access.}} | \Phi_m^{\text{env.}} \rangle \simeq 0 \ \text{ for } m \neq n. \tag{8.2}$$

Then no local observation upon our system can reveal differences from a mixture – e.g., interferences –:

$$\langle \Psi | O^{\text{access.}} \otimes O^{\text{syst.}} | \Psi \rangle = \sum_{n,m} \sqrt{p_n p_m} \langle \Phi_n | O^{\text{access.}} | \Phi_m \rangle \langle \varphi_n | O^{\text{syst.}} | \varphi_m \rangle$$

$$= \sum_n \tilde{p}_n \langle \varphi_n | O^{\text{syst.}} | \varphi_n \rangle, \tag{8.3}$$

where, for simplicity, we only considered the case of a direct product between the observables of the accessible environment and those of the system. Note that (8.1–8.3) do not describe a momentary, but a stationary situation which sets in for $t > \tau$ and is maintained thereafter by the continued interaction between system and environment.[3]

---

[2] In many cases this is the position basis, as a consequence of the locality of the interaction. See Sect. 3.2. Therefore macroscopic objects typically appear decohered in positions – cf. 'Einstein's moon'.

[3] Therefore (8.1) implies more than the usual Schmidt decomposition for any two-component system: in the general case the basis will typically change with the time, while the Schmidt decomposition for the decohered situation describes a stable "apparent" mixture according to a basis stabilized by the peculiar and continued interaction between system and environment. See Chaps. 2 and 3, e.g. Sects. 3.3, especially 3.3.3, and 3.4.

As has been stressed in Chaps. 2 and 3, coupling to inaccessible degrees of freedom is nothing exceptional but it is a fact of the rule, the question being a quantitative one. Correspondingly, we shall observe quantum coherence if this coupling is weak (τ is large compared to the observation time), or decohered situations if the coupling is efficient. The consideration of these facts motivated the *program of decoherence* as it has been presented in Chap. 2 and discussed in detail thereafter. It is also referred to as *environmental decoherence* and in the following we shall sometimes use this designation for the sake of definiteness. As already remarked above, since it is a process based on the fundamental quantum dynamics and involves well known interactions it does not depend on the interpretation of the theory, in particular on that of the measurement process. However, since decoherence is relevant for the measurement process, as we shall see below, the interpretation of the *decoherence program* depends on the interpretation of the theory.

We stress that decoherence is derived from unitary evolution. The nonunitarity of the reduced density matrix dynamics results from the "dynamical flow" across the "border" between system and environment, while the total (system+environment) density matrix follows the unitary von Neumann dynamics – see Sects. 3.3, 3.4 and Chap. 7. The non-unitarity is "apparent", in the sense that it results from the restriction to the reduced density matrix. In contrast to this we shall consider in this chapter non-unitary evolution equations which have been proposed as *fundamental modifications* of the Schrödinger and von Neumann equations. In these models the state vector collapse is understood as a fundamental dynamical process. They are therefore concerned with the explanation of the measurement itself as a physical process leading to the emergence of a definite outcome.[4] To make this point clear we shall here briefly discuss the relation between collapse, decoherence and quantum mechanical measurement.[5]

In the "standard" quantum mechanical formalism the emergence of a result in an (ideal) measurement is described as a corresponding collapse of the wave function *of the measured system*.[6] The measurement involves at some stage an *apparatus* whose states become entangled with those of the system

---

[4] We shall not be concerned here with ensemble interpretations of quantum mechanics.

[5] One cannot seriously try to quote all the literature to this point – we shall only mention Bush, Lahti and Mittelstaedt (1996) and Omnès (1999) as more or less recent books, Jammer (1974) for historical presentation and Auletta (2001) for general discussion and recent references.

[6] This, of course, is not necessary if one chooses the point of view of the Everett theory, which instead of collapse postulates separate perception of coexisting but dynamically decoupled branches of the wave function. This is sometimes described as "branching of the consciousness of the observer", who is assumed to be local. Since we are here concerned with collapse models we shall not pursue the Everett approach – see Chap. 2 for more detail. Nevertheless we want to mention that also in the Everett interpretation environmental decoherence is a

during the measurement. This process is governed by the Schrödinger equation and implies a unitary evolution, leading to the state given in first part of the relation below. It is followed by another process, which has to explain the fact that through it we observe with a certain probability a certain property of the system (depending on the measurement design) and for all later proceedings the system can be considered as having been left by the measurement in an eigenstate corresponding to the measured property (second part of the relation):

$$|\Psi_{\text{tot.}}\rangle = \sum_n \sqrt{p_n}\, |\chi_n^{\text{apparat.}}\rangle |\varphi_n^{\text{syst.}}\rangle \overset{\text{non-unit. evol.}}{\longrightarrow} |\chi_{n_0}^{\text{apparat.}}\rangle |\varphi_{n_0}^{\text{syst.}}\rangle. \quad (8.4)$$

For this, the apparatus (which thus represents a *controllable* or *accessible* environment and is usually a macroscopic system) is supposed to experience a collapse (affecting *its* degrees of freedom) and switch into a definite state $|\chi_{n_0}\rangle$. As a consequence of the entanglement, this automatically selects a definite wave function of the measured system (this apparatus is sometimes shifted to the brain of the observer! - see point (vi) in next subsection). The result is that *the system disentangles from the apparatus* and acquires a wave function for itself, namely, $|\varphi_{n_0}\rangle$.

The aim of the collapse models is to provide a dynamics leading to collapse of the wave function of the apparatus (while the system by itself should not easily experience collapse if unmeasured – otherwise we would never see quantum effects). The collapse then ceases to be an abstract, non-physical requirement, and the apparatus is no longer an irreducible concept, but is only a particular kind of quantum system with very many degrees of freedom.

Now, we have noticed that decoherence is a natural process and we expect a macroscopic system such as an apparatus *by excellence* to be subject to decoherence effects. We must account therefore also for the uncontrollable environment in $|\Psi_{\text{tot.}}\rangle$ in (8.4). By this coupling the apparatus states which become correlated to the different system states should in their turn also become correlated with states of the environment which differ significantly in their uncontrollable degrees of freedom and are therefore practically orthogonal. Moreover these correlations should establish rapidly (i.e., τ must be much shorter than the time scale of the measurement) and become irreversible and stabilized by the continuous interaction with the environment. These features can be considered as defining an appropriate apparatus, and due to the normal measurement conditions they appear rather natural. See Zurek (2000).[7] In this case, and in the sense of the previous discussion, we

---

relevant aspect. A further context in which decoherence is relevant is that of *coherent histories* – see Chap. 5.

[7] Not any coupling of an environment to that system which must be designed for playing the role of an apparatus does already imply decoherence according to the envisaged pointer state basis, see, e.g., Bassi and Ghirardi (2000), Bermann

may see the *apparatus* as an *accessible environment*, see (8.2), then

$$\langle \Phi_n^{\text{env.}} | O^{\text{apparat.}} | \Phi_m^{\text{env.}} \rangle \simeq 0 \ \text{ for } m \neq n, \tag{8.5}$$

where "env." indicates apparatus *and* non-accessible degrees of freedom. Therefore, compare (8.3):

$$\langle \Psi | O^{\text{apparat.}} \otimes O^{\text{syst.}} | \Psi \rangle = \sum_{n,m} \sqrt{p_n p_m} \, \langle \Phi_n | O^{\text{apparat.}} | \Phi_m \rangle \langle \varphi_n | O^{\text{syst.}} | \varphi_m \rangle$$

$$= \sum_n \tilde{p}_n \langle \varphi_n | O^{\text{syst.}} | \varphi_n \rangle, \tag{8.6}$$

and thus no apparatus and system observable could put into evidence the difference to a true mixture. One sometimes relates this behavior with the concept of "induced superselection rules" (see Sects. 3.1.3, 3.4 and Chaps. 6 and 7) meaning the applicability of statistical interpretation to the decohered reduced density matrix.

Writing in an illustrative manner[8]:

$$|\Psi_{\text{tot.}}\rangle = \sum_n \sqrt{p_n} \, |\phi_n^{\text{non-access.}}\rangle |\chi_n^{\text{apparat.}}\rangle |\varphi_n^{\text{syst.}}\rangle, \tag{8.7}$$

then due to the orthogonality of the non-accessible environment states:

$$\langle \phi_n^{\text{non-access.}} | \phi_m^{\text{non-access.}} \rangle \simeq 0 \ \text{ for } m \neq n, \tag{8.8}$$

the reduced density matrix of *apparatus+system* becomes diagonal,

$$\text{tr}_{\text{non-access.}} |\Psi_{\text{tot.}}\rangle\langle\Psi_{\text{tot.}}| = \sum_n \langle \phi_n | \Psi \rangle \langle \Psi | \phi_n \rangle$$

$$= \sum_n \tilde{p}_n |\chi_n^{\text{apparat.}}\rangle |\varphi_n^{\text{syst.}}\rangle \langle \varphi_n^{\text{syst.}}| \langle \chi_n^{\text{apparat.}}|, (8.9)$$

and describes an (apparent) mixture. Of course the result (8.6) or (8.9) is not yet quantum mechanical measurement: we still need to explain the selection

---

et al (2001), Grübl (2002), Adler (2002). Since decoherence, however, is not an abstract assumption but a physical fact, the question is to estimate its effect in a measurement set up, and any assumptions have to be physically justified. See also Adler (2001), Omnès (2002). One can in fact design interesting experimental set ups tuning the amount of decoherence in a measurement situation, see e.g. Marshall *et al* (2002).

[8] The essence of the argument is already contained in (8.6). Here we want to make explicit the splitting of the environment according to accessible (apparatus) and non-accessible degrees of freedom. Schmidt-type decompositions with three factors are not always possible – at variance with the usual, two-factor ones. In connection with decoherence see Sect. 3.4. For general Schmidt-type decompositions see also Peres (1995).

of one definite result (out of the improper mixture) and the corresponding fate of the system, corresponding to the selection of only one term in (8.7) – cf (8.4). Hence we must appeal to some collapse or branching hypothesis to define measurement. The above discussion shows, however, that decoherence is very relevant in the measurement and that not only it cannot be neglected but it produces effects which are part of what we think a quantum mechanical measurement actually is. (These aspects are discussed in detail in Chaps. 2, 3, especially Sect. 3.4. See also next subsection.)

The comparison between decoherence and dynamical collapse can be further pursued. Most collapse models use in fact evolution equations which – when written for the density matrix – are of the same form as those obtained in the decoherence approach for the *reduced* density matrix (but, as mentioned, now understood as fundamental, and applying also to isolated systems). Therefore, quite generally, both quantitative and qualitative features of these models show a number of similarities with environmental decoherence. Since decoherence, however, does not lead to genuine collapse and is an elementary quantum mechanical effect compatible with various measurement hypotheses it does not lead by itself to physical effects beyond standard quantum mechanics. As discussed below, this is not the case for dynamical collapse models, which lead to small but in principle observable deviations. Refined experiments should thus be able to support or exclude such models – while with decoherence we have to live on in any case.

Further recent discussion concerning measurement includes the relation to decoherence and experimental proposals (Rozmej, Berej and Arvieu 1997, Gurwitz 1998, Stodolsky 1999, Mayburov 1999, 2000, Castagnino, Laura und Id Betan 2001, Chiao and Kwiat 2002, Marshall *et al* 2002, Klein and Nystrand 2002, Breuer and Petruccione 2002b, Strunz and Haake 2002, Strunz, Haake and Braun 2002), alternatives to collapse (Marchewka and Schuss 1999), relativity aspects and collapse models (Suarez 1998, Berg 1998, Zbinden *et al* 2000, Adler and Brun 2001, Svetlichny 2002, Kent 2002). See also contributions to Breuer and Petruccione (1999), Blanchard et al (2000).

8.1.3 Various Approaches to Collapse

There are many models in the literature which are concerned with these problems. Our discussion will concentrate on a certain class of models which postulate *spontaneous collapse* occurring stochastically and superimposed upon the Schrödinger dynamics. They have a clear conceptual setting and lead to testable predictions (beyond standard quantum mechanics) – at least in principle. Before doing this, however, we shall attempt a sketchy classification of the general discussions and of the various models which have been proposed to describe collapse, with the hope that this may help the reader to build up a general view of this approach. Further discussion is provided in Chap. 9. The following designations are only for the purpose of systematics.

(i) *Unspecified ("kinematical") collapse:* The collapse is postulated to take place at some stage of the measurement process between the interaction system–apparatus and the observation of the result (Heisenberg cut), but no explicit dynamics is provided for it (one usual excuse for this peculiar behavior is the assumed *classicality* of apparatus and observer understood as a fundamentally non-quantum property). This is the *standard* (sometimes only implicit) formulation in most textbooks. It leads to the well known interpretational difficulties (Wigner's friend, etc). The quantitative predictions of quantum mechanics with this collapse postulate are indistinguishable from those of the many worlds interpretation. In the first case the measurement explicitly involves the observer (by the introduction of the Heisenberg cut) and in the second case the branching of the awareness of the local observer correlates her to a definite outcome: in both cases there is no place for an additional observable effect (see also Chap. 2; see Jammer 1974 for the historical discussion; for more recent discussions see, e.g., Balian 1989, Malin 1993, Omnès 1995, 1999).

(ii) *Spontaneous collapse:* A non-unitary evolution, independent of any intentional measurement, is postulated as a *stochastic process* for the wave function. Parameters are chosen such that the effect becomes observable only for systems with very many degrees of freedom. Therefore the collapse is effective only when a "microscopic" system is entangled with a "macroscopic" one, as is typically the case in measurements. Then the collapse occurs in the (macroscopic) apparatus and chooses thereby also a definite state of the system - see (8.4). Since the collapse is now an objective process, it is itself accessible to observation. Therefore these models lead to (in principle) measurable deviations from the predictions of standard quantum mechanics – (i) above. For reviews see, e.g., Pearle (1990, 1993), Ghirardi (1994a,b), Ghirardi, Grassi and Benatti (1995), Ghirardi (1999), Pearle (1999b) and references therein. Some of these models will be discussed in the forthcoming sections.

(iii) *Collapse driven by gravitational effects:* A "universal environment" vaguely related to quantum gravity is postulated either

    (a) to remove the other components of the wave function (see for instance Károlyházy, Frenkel and Lucácz 1986, Penrose 1986, Diósi 1989, Ghirardi, Grassi and Rimini 1990, Unturbe and Sánchez-Gómez 1992), or

    (b) to let them disappear behind some horizon, for instance in connection with "baby universes" (see e.g., Ellis, Mohanty and Nanopoulos 1989, 1990, Chaves, Figueiredo and Nemes 1994, Ellis, Mavromatos and Nanopoulos 1995).

Since the phenomena assumed to lead to these effects (to be described by a quantum theory of space and time) are not well understood at present, the occurrence of the collapse still remains a conjecture. Once this is assumed, however, one can at least give some estimates for the characteristic parameters controlling this process (see, e.g, Ghirardi, Grassi and Rimini 1990). The models of type (a) correspond in this sense to a certain specialization of the approach in (ii) while the models of type (b) could be understood as a special case of *environmental decoherence* with a natural limit for observations and therefore leading to an outstanding type of *apparent collapse*. Because of the special role, however, assumed to be played by quantum gravity we introduced these models as a separate class. See also Penrose (1996), Anandan (1998), compare also class (iv)(b) below.

(iv) *Apparent collapse:* One speaks of apparent collapse if interference effects between the various pointer states are not accessible to observation, although the *apparatus+system* state does not necessarily factorize. This situation is then practically indistinguishable from a mixture. Here again we can distinguish several cases:

(a) In some models this apparent state reduction is achieved non-uniformly with time in the set of observables. Therefore at any large but finite time one can devise measurements involving observables sensitive to the other components of the wave function. We have therefore only an apparent collapse depending on the criteria by which we restrict the observable algebra and define the *local observables* (Hepp 1972, see also Bell 1987, Landsman 1995).

(b) As we have seen, *environmental decoherence* itself acting upon the (macroscopic) apparatus induces in fact an apparent collapse *for the measured system* since it leads to a (usually exponentially) fast decay of the non-diagonal elements of the *apparatus+system* reduced density matrix (8.9). This has been described in the previous subsection, see Chap. 3 for quantitative estimates, see Chaps. 2–7 for extensive discussion, also concerning related concepts (induced superselection rules, coherent histories, etc). The quantum correlations are irreversibly delocalized and hence irretrievable from the point of view of local observations. But even if this decay with the time may be uniform over the set of system observables we would still need a genuine collapse (or branching) to occur if we want only one term in the decomposition of the total wave function to survive (or be available) and the measured system to reach a pure state: otherwise again all components of the wave function are still available and one could *in principle* design experiments to observe their existence (even if such experiments were hardly feasible!). Therefore we speak here also of an *apparent collapse*. See also Anglin *et al* (1995), Breuer (1993), Hakim and Ambegaokar (1985), Landsman (1995), Mayburow (1999, 2000), Castagnino, Laura and Id Betan (2001).

(c) A special case for environmentally induced decoherence has been made in the frame of the algebraic approach to quantum mechanics (see e.g. Pfeifer 1980, Amann 1988, 1991a,b, Primas 1990a,b, compare also Sect. 4.1.1). Here it is discussed in a (very simplified) model of a "small" system coupled to a radiation field how, for infinite time, the total system would be driven into (superpositions of) *disjoint states*, which can be written as products of dressed states of the small system and of the field. The *disjointness* means here by definition that all observables of the algebra have vanishing matrix elements between these states. If the disjointness condition is fulfilled it implies that the center of the algebra contains *classical observables* whose *eigenvalues* designate the dressed states. These states would belong, therefore, to different superselection sectors and superpositions among them would have no physical meaning (could not be distinguished from mixtures). This model is intended to explain the emergence of classical observables, e.g., the existence of chiral states of certain molecules (the "small systems"). It is not intended as a true collapse model: a generic initial state developing in time will have components approximating increasingly well disjoint states (Amann 1988) but no real reduction is achieved. Since the exact disjointness depends on the specification of the observable algebra and arises only in the infinite time limit of the dynamics of the total system, the same remark holds for the apparent collapse obtained here as concerning point (a) above (see also Sect. 9.4).

Although they do not lead to reduction of the wave function – in contrast to the models of classes (ii)–(iii) –, all the effects discussed under this point do, however, induce an *apparent* collapse *from the point of view of some given class of observations*. If this apparent collapse can be shown to hold for all local observations, as environmental decoherence (b) does achieve in certain situations, the difference to a true collapse may be only of cosmological significance. Notice that in contrast with the cases (a) and (c), the environmental decoherence case (b) makes no *ad hoc* assumption concerning the observable algebra but is an effect completely determined by the specific physical situation (which fixes explicitly for which observations the apparent collapse holds). For a general discussion concerning the emergence of *dynamical superselection rules* see Chaps. 3 and 6.

(v) *Non–collapse stochastic models:* Here we want to mention "phenomenological" stochastic models which are not directly intended as collapse models but are often quoted in this context and in connection with the emergence of classical properties. Moreover they use in part stochastic evolution equations for the wave function of the same kind as the true collapse models of class (ii).

(a) A class of *continuous measurement* models also postulate a non-unitary evolution by submitting the wave function to a stochastic process, but unlike class (ii) or (iii) they are claimed to reproduce in this way the effect of a continuous monitoring of the system by the environment, defining thereby an open system evolution which explains measurement and classicality (Srinivas 1977, 1984, Diósi 1992, Belavkin and Staszewski 1992, Chruściński and Staszewski 1992, Mensky 1993, 1995, Belavkin 1994, Bhattacharia, Habib and Jakob 1999). This interpretation requires, however, that the environment itself may be assumed subject to collapse in the sense of (i), (ii) or (iii) such as to produce a time series of factorized states for *system+environment*. For the connection with the decoherent histories approach see Diósi *et al* (1994),

Brun (1997), compare also Chap. 5. See also Berman *et al* (2001, 2002) for numerical simulations of experimental conditions.

(b) The so called "hybrid dynamics" (Peres 1993) describes quantum systems coupled to classical ones (Diósi and Halliwell 1998, Prvanović and Marić 1999, Diósi, Gisin and Strunz 2000) and is connected to the continuum measurement models (which are claimed to represent a subclass of hybrid models – see Diósi 1999). Conceptually, this approach is related to class (i) above and runs in some sense against the philosophy of the spontaneous collapse models and also against that of the decoherence program, since it develops the idea of irreducible classical physics coexisting with quantum physics.

(c) A further collection of *stochastic models* postulates stochastic evolution of the wave function of the system as phenomenological equations or as background (unitary – e.g., orthogonal "jumps" of the wave function –, or non–unitary) stochastic processes behind the master equations for the density matrices – see, e.g., Ford, Lewis and O'Connell (1988), Diósi (1985, 1988b), Baker and Singleton (1990), Milburn (1991, 1993), Finkelstein (1993), Lanz (1994), Garraway and Knight (1994), Habib (1994), Blanchard and Jadczyk (1995). We shall not discuss these approaches but refer the reader to the literature. Here we only note that sometimes it is claimed to reproduce in this way the effects of environmental decoherence (Diósi 1986, 1993b, 1994a, Diósi *et al* 1994, Gisin and Cibils 1992, Gisin and Percival 1993a-c, Percival 1994, 1995). This interpretation is, however, misleading, since as long as they are not understood as mechanisms for the measurement proper (in the sense of (ii) or (iii) above) these models correspond to an open system approach and hence can only refer to the evolution of a (reduced) density matrix. It is not clear, therefore, what the status is of the "system wave function" which these models refer to in their stochastic equations, since the system remains entangled with the environment and does not possess in general a wave function for itself. Therefore these models cannot be considered as describing environmental decoherence, but at most as providing a *formal representation* for its effects.

(d) Measurement models without collapse have been developed to "explain" the statistical predictions of the quantum theory. As an example, we mention the model of Machida and Namiki (1980), who *postulate* the state space of the (macroscopic) apparatus to have a multiple (direct sum) Hilbert space structure over which one has to average. In the limit of infinitely many Hilbert spaces this model was shown by Araki (1980) to lead to ensemble formation. Later papers (Namiki and Pascazio 1990, 1993, Nakazato, Namiki and Pascazio 1996) want to identify this averaging with the real accumulation (in an experiment) of events affected individually by the fluctuations in the apparatus. At variance with the claims of the authors this cannot be achieved, however, by following the usual rules of quantum mechanics (see also Sect. 9.1).

(e) Finally we want to mention the approaches of the *stochastic mechanics* (see e.g., Nelson 1985, Blanchard, Combe and Zheng 1987; see Yasue 1978 for an application to open systems) and of the *stochastic quantization* (see, e.g., Damgaard and Hüffel 1988). In the former the wave function only represents a mathematical means used to define a stochastic process for the degrees of freedom of the system. The second approach starts with the *Euclidean* path integral using an ensemble interpretation and considers the associated stochastic differential equation for the degrees of freedom. These approaches do not lead to different predictions from those of standard quantum mechanics; in fact the second one is only a reformulation of the Euclidean path integral method. Although both use stochastic processes (but not for the state vector) they do not deal explicitly with measurement or classical properties and therefore we shall not discuss them further.

(vi) *Collapse and mind:* We mention this point at the very last, since here the discussion pertains not so much the physics of the collapse, but the question of the intervening of the human mind/brain in the measurement process. Were the mind, however, to be directly involved in some kind of collapse, this immediately raises the question of the quantum nature of the mind processes. The discussion therefore extends over general considerations on the role of mind and the quantum (Lockwood 1989, Snyder 1996, Stapp 2000, Stapp 2001, Samal 2001), to the question of consciousness and the physical, biological and psychological aspects concerning quantum behavior in the brain (Penrose 1994, Jibu and Yasue 1993, Mavromatos and Nanopoulos 1995, Hameroff 1998, Mershin, Nanopoulos and Skoulakis 2000, Georgiev 2002, Khrenikov 2002), including the relevance of decoherence effects for this behavior (Tegmark 2000). Again, we shall not pursue this discussion here, since it is not directly related to dynamical collapse models.

As already stated above, the aim of this chapter is not to provide a detailed review of the dynamical approach to collapse or of the stochastic evolution equations for the wave function, but to discuss certain features of this approach in their relation to environmental decoherence. Therefore we shall concentrate on the *spontaneous localization models* of class (ii) which represent a clearly distinct point of view, while having – both in their formulation and in their predictions – certain similarities with the environmental decoherence (compare the apparent collapse class (iv)(b)). They have also been the most studied and are therefore best suited for an illustration of this approach. In Sect. 8.3 we shall also comment on some related models of class (iii) and (v). For a further discussion of concepts related to decoherence, see also Chap. 9.

8.2 Spontaneous Collapse Models

8.2.1 The Dynamical Collapse Hypothesis

The models we shall review below have been devised as models for the wave function reduction associated with measurements and introduce non-Hamiltonian evolution equations to describe a physical collapse process. Their aim is to provide an objective frame for an individual interpretation, i.e., a quantum mechanical description of the individual events and not only of their statistics, without fundamental assumptions about the peculiar nature of the observer in connection with a splitting consciousness theory. Therefore they acknowledge the collapse as a physical effect, such that the measurement process no longer would suggest the wave function to be prone to subjective influence, or would depend on concepts not describable in the frame of the theory, such as the classical apparatus. This is (or can be) accomplished by modifying the Schrödinger–von Neumann equations such as to arrive at a unified dynamics – which is then no longer unitary. We should note that the primary motivation of these approaches being that of helping the state vector to a realistic status by postulating a physical process for its collapse, their aim is to realize this dynamics in the Hilbert space (and not only for density matrices). This is achieved in most cases by superimposing a stochastic process directly onto the Schrödinger dynamics of the wave function (the corresponding von Neumann equation for the density matrix may then be written down for calculational purposes). If one wants to retain the usual probability rules one has to introduce nonlinear evolution equations for the normalized state vector. The "noise" term is designed such that for isolated microscopic[9] objects its effect will practically never be observed, but under the adequate conditions (e.g., in the presence of a macroscopic apparatus – with many degrees of freedom) it would lead to an effective measurement process. This implies more than the non-observability of quantum coherence effects, since it causes the collapse of the wave function onto narrow wave packets. In particular, in the specific "decoherence situations" (situations prone to decoherence effects, e.g. a system strongly coupled with an uncontrollable environment) it leads to *disentanglement*, that is the system acquires a wave function without being intentionally measured (the collapse takes place in the environment, hence one can say that the latter indeed "measures" the system). Then, – and in contrast to environmental decoherence –, the system density matrix is either of rank one (to a good approximation) or will describe only lack of information about the actual (pure) state of the system – but not the absence of such a state. For a general discussion of the motivation and interpretation of this approach see, e.g., Ghirardi 1992, 1994a,b, Ghirardi, Grassi and Benatti 1995.

---

[9] As in the preceding sections we understand here isolated system in the sense of an approximation whose justification depends on the circumstances.

From the point of view of our discussion of the emergence of classical properties these models belong to the class of what we may call *fundamental decoherence* (because they *postulate* a fundamental non-unitary dynamics). The parameters of this process must be carefully tuned to agree with the observations of quantum or of classical properties according to the circumstances; this puts strong constraints on the parameters, which do not seem to be easy to fulfil. Since on the other hand the *environmental decoherence* is an independent process which always occurs in the relevant physical situations, has no *ad hoc* parameter and explains just those observations (in the sense of presence or absence of effects due to quantum coherence, respectively) one can expect fundamental and environmental decoherence to compete in most cases. It is therefore a refined quantitative question whether such fundamental decoherence effects could be recognized in practice to provide a test for these models. We shall discuss these aspects in sect. 8.3, see also Sect. 3.3.

8.2.2 Spontaneous Localization by a Jump Process

The original model of Ghirardi, Rimini and Weber (1986) is written as a non-Hamiltonian evolution equation for the statistical operator of a particle, of the form

$$\frac{d}{dt}\rho = -\mathrm{i}[H, \rho] - \lambda(\rho - T_\rho) \tag{8.10}$$

where in the coordinate basis T_ρ is:

$$\langle q|T_\rho q'\rangle = \exp\left(-\frac{(q - q')^2}{a^2}\right)\langle q|\rho\, q'\rangle. \tag{8.11}$$

The model uses the evolution equation introduced by Barchielli, Lanz and Prosperi (1982) to describe the effect of a continuous, approximate position measurement. This, however, is now intended to describe the effect of a *spontaneous* reduction of the coordinate space wave function, while eliminating the need for the "process of first kind" of von Neumann. Thereby the parameters a and λ are tuned such as to permit the same dynamics to be valid both for microscopic and for macroscopic systems but leading to different observable behavior in the two cases. The effect on the wave function itself is emphasized by J. Bell (Bell 1987) who reformulates the model to explicitly describe a spontaneous, random collapse of the wave function $\Psi(x_1, \ldots, x_N)$ of, say, an N−particle system in configuration space about some point z:

$$\Psi \to \Psi_{n,z} = L_{n,z}\Psi. \tag{8.12}$$

Here $L_{n,z}$ is a positive, selfadjoint, linear operator whose action is to concentrate the distribution in the coordinate of particle n around the point z:

$$L_{n,z} = \frac{1}{\sqrt{\pi}a}\exp\left(-\frac{(z - x_n)^2}{2a^2}\right), \tag{8.13}$$

assuming a Gaussian distribution centered at z and with width a, while the frequency of these jumps for each degree of freedom is given by $\lambda = 1/\tau$ – see (8.10,8.11). The point z itself is chosen with the probability

$$p(z) = \parallel \Psi_{n,z} \parallel^2 . \tag{8.14}$$

Since the jump (8.12) does not conserve the norm of the state, it is supplemented by a renormalization of the wave function:

$$\Psi_{n,z} \rightarrow \Psi' = \Psi_{n,z} \frac{\parallel \Psi \parallel}{\parallel \Psi_{n,z} \parallel}. \tag{8.15}$$

The full process $\Psi \rightarrow \Psi'$ cannot be described by a linear operator, because of the nonlinear character of its second step (8.15) which was introduced in order to ensure probability conservation while retaining a direct probability interpretation for the (modulus square of the) wave function itself.

Now let us see how this process works in detail. From the requirement of having a unified dynamics compatible with both quantum and classical phenomena a reasonable choice for the two parameters appears to be

$$\tau \simeq 10^{16}\text{s}, \quad a \simeq 10^{-5}\text{cm} \tag{8.16}$$

which would make the localization of genuine, microscopic quantum system as seldom as once in 10^8 years, while a macroscopic system may undergo it as often as $10^{23} \times 10^{-16} = 10^7$ times per second. Similarly the localization range is chosen such as to allow distinct behavior for macroscopic and microscopic objects.

For illustration we shall consider two cases:

(a) *Localization of the center of mass of a macroscopic body.* To fix the ideas let us consider a rigid ensemble of N points at positions r_i relative to a center of mass C and assume a Gaussian wave function in the position of the latter:

$$\Psi(x_1, x_2, \ldots, x_N) \simeq \exp\left(-\frac{(R - R_0)^2}{2c^2}\right), \quad R = \frac{1}{N}\sum_{i=1}^{N} x_i, \quad x_i = R + r_i,$$
$$\tag{8.17}$$

where $R, \{x_i\}$ are the positions of C and of the points in some coordinate system, respectively. As discussed above, even if the average frequency of the spontaneous collapses, $1/\tau$, is exceedingly small, N itself is huge for a macroscopic body and the average time before one of the internal degrees of freedom will collapse, τ/N, can be very short. Hence very rapidly one of the internal coordinates, say x_1, will become localized within the width a around some point z_1, see (8.12,8.13):

$$\Psi(\{x\}) \rightarrow \text{const} \times \exp\left(-\frac{(z_1 - x_1)^2}{2a^2}\right)\Psi(\{x\})$$

$$\simeq \exp\left[-\frac{a^2 + c^2}{2a^2c^2}\left(R - \frac{a^2 R_0 + c^2(z_1 - r_1)}{a^2 + c^2}\right)^2\right]\ldots \tag{8.18}$$

But then also the center of mass will get pinned down, as can be read from (8.18). Assuming that the center of mass had a wide wave function to start with we get localization of the former within a distance a around $z_1 - r_1$. At the next moment some other point, say x_2, will collapse within a of some z_2. If we remember that the collapse positions z_i are selected with the probability given by the momentary $\Psi(x_i = z_i)$, cf. (8.14), we see that the center of mass soon becomes well localized around R_0.

(b)*Disentanglement and decoherence.* Consider the situation of a system coupled to an environment which has many degrees of freedom such as to behave in some sense macroscopically. Assume that the coupling is such that the entangled wave function can be written as a superposition of, say, two terms:

$$\Psi(y, \{x_i\}_N) = \varphi_1(y)\Phi_1(\{x_i\}_N) + \varphi_2(y)\Phi_2(\{x_i\}_N), \quad \langle\Phi_1|\Phi_2\rangle = 0. \quad (8.19)$$

Here y designates (globally) the coordinates of the system, $\{x_i\}_N$ those of the environment and it is assumed that the environmental wave functions Φ_1 and Φ_2 differ macroscopically[10]. Then, because of the rule (8.14), the probability distribution for the collapse point z has two well separated peaks corresponding to the supports of Φ_1 and Φ_2. Because of the large N, again some of the environment coordinates, e.g. x_n, will soon collapse, with the collapse point z, say, in region 1. But Φ_2 evaluated at this point is practically zero, hence as a result

$$\Psi(y, \{x_i\}_N) \to \varphi_1(y)\Phi_1(\{x_i\}_N, x_n = z) \quad (8.20)$$

and the system becomes disentangled from the environment. Repeating the experiment, another point will collapse first, possibly in region 2. The probabilities for collapses in the two regions are proportional to $\| \Phi_k \|^2$, $k = 1, 2$. The statistics of posterior measurements upon the system will therefore just confirm these probabilities and hence agree with the usual decoherence predictions (after tracing out the environment), although the system prior to such a later measurement is now in a very different situation (e.g., (8.20)) from that of the environmental decoherence (in which case (8.19) is valid all the time before this later measurement). Notice that if the environment *is* an apparatus and if the Φ_k correspond to some pointer positions k, the above considerations will describe themselves a measurement process (including the wave function collapse) which proceeds automatically (hence objectively) once we proceed with the experiment, and without the necessity of an observer. Therefore this is a model for the measurement process.

---

[10] Notice that this already *is* decoherence, and all observations upon the system will only produce classical statistics:

$$\langle\Psi|O(y) \otimes \mathbf{1}|\Psi\rangle = \langle\varphi_1|O|\varphi_1\rangle \| \Phi_1 \|^2 + \langle\varphi_2|O|\varphi_2\rangle \| \Phi_2 \|^2 .$$

Although it does not achieve full unification for the dynamics of the wave function (since it assumes two separate processes) this model does incorporate, however, the major physical aspects of the full approach. Since the physical picture it provides is very intuitive we went into some detail in describing it. The ingredient it lacks is the explicit introduction of a noise term in the Schrödinger equation to reproduce the stochastic steps (8.12–8.15). This will be taken care of by the models described in the next subsection; therefore we shall postpone the phenomenological discussion and other questions till the end of the chapter. One thing should already be noted, however, namely the special role of the coordinate space: this is not a specialty of the above formulation but is peculiar to the whole approach. This is, in fact, no accident: we have seen in the environmental decoherence (e.g., Sect. 3.3) that the interactions with the environment typically select a *pointer basis* in the position space. The special role of the position space seems therefore empirically motivated. For a discussion see, e.g., Venugopalan (1994), Dickson (1995).

8.2.3 Continuous Spontaneous Localization

The model of the preceding subsection describes a discrete Markov process in the Hilbert space, but it can be shown (see Pearle 1989, Ghirardi, Pearle and Rimini 1990, see also Diósi 1985, Nakano and Pearle 1994) that it is possible to replace this by a stochastic differential equation. This is similar with going from random walk to Langevin equation – see App. A7. The principal aim of this procedure is to introduce an explicit, uniform dynamics describing an "orderly flow of the Ψ-function during a measurement from a superposition of possible outcomes to a single actual outcome" (Pearle 1990).

In this approach, the full process as described by Eqs. (8.12) and (8.15) is associated with the nonlinear stochastic differential equation:

$$d\Psi = \{[-\mathrm{i}H - \frac{\gamma}{2}(\mathbf{A}^\dagger - \mathbf{R}) \cdot \mathbf{A} + \frac{\gamma}{2}(\mathbf{A} - \mathbf{R}) \cdot \mathbf{R}]dt$$
$$+ \sqrt{\gamma}(\mathbf{A} - \mathbf{R}) \cdot d\mathbf{w}\}\Psi, \tag{8.21}$$
$$\mathbf{R} = \frac{1}{2}\langle\Psi|(\mathbf{A} + \mathbf{A}^\dagger)\Psi\rangle. \tag{8.22}$$

The non-linearity in (8.21,8.22) appears since the stochastic term (the noise) must depend on the wave function if one wants to preserve the normalization of the latter, such as to obtain the usual quantum mechanical probability interpretation starting from the Wiener process distribution. Here $\mathbf{A}$ is a set of operators $\{A_i\}$ and $d\mathbf{w}$ is a real, multidimensional Wiener process[11]:

$$\overline{dw_i} = 0, \quad \overline{dw_i\,dw_j} = \delta_{ij}dt \tag{8.23}$$

---

[11] Similar stochastic equations based on a complex Wiener process have been discussed by Gisin and Cibils (1992) and Gisin and Percival (1992, 1993a,b).

and one must properly deal with the usual inhomogeneity of the stochastic differential equations (notice that $d\mathbf{w}$ is proportional to $\sqrt{dt}$, while the other terms on the RHS of (8.21) are proportional to dt). This equation is written in the Itô calculus – see App. A7 – and has been already discussed in Chap. 7 in the context of open systems.

The evolution equation (8.21,8.22) replaces the Schrödinger dynamics and describes a stochastic process in the Hilbert space (we shall indicate this explicitly sometimes by writing Ψ_w for the series of states on one trajectory). The analysis of Ghirardi, Pearle and Rimini (1990) indicates that if $\{A\}$ is a set of commuting selfadjoint operators, then for large time (8.21,8.22) induces a reduction of the state vector Ψ onto the subspace corresponding to one (set of) *eigenvalue(s)* of the operator(s) A, in the sense that the trajectory Ψ_w gets trapped in this subspace. This is discussed in general in Diósi (1988b), Ghirardi, Pearle and Rimini (1990), Gisin and Percival (1992, 1993a,b), Percival (1994). It is instructive to see how this comes about and therefore we shall present here a simple argument following Ghirardi, Pearle and Rimini (1990) and without aiming at mathematical precision. For this we expand Ψ in the corresponding base (for simplicity we consider here only one operator A with discrete spectrum and no Hamiltonian evolution term; see also Chaps. 3, 7):

$$|\Psi\rangle = \sum_{\alpha} c_{\alpha}|\alpha\rangle, \quad A|\alpha\rangle = a_{\alpha}|\alpha\rangle, \tag{8.24}$$

$$R = \sum_{\alpha} a_{\alpha}|c_{\alpha}|^2 \equiv a, \quad \sum_{\alpha} |c_{\alpha}|^2 = 1 \tag{8.25}$$

– see (8.22) – then (8.21) implies:

$$dc_{\alpha} = \left\{ -\frac{\gamma}{2}(a_{\alpha} - a)^2 dt + \sqrt{\gamma}\,(a_{\alpha} - a)\,dw \right\} c_{\alpha}. \tag{8.26}$$

Since there is no Hamiltonian term, the real and imaginary parts do not interfere. We make the variable transformation

$$u_{\alpha} = |c_{\alpha}|^2, \quad \sum_{\alpha} u_{\alpha} = 1 \tag{8.27}$$

and using (A7.61-A7.63) we obtain:

$$du_{\alpha} = 2\sqrt{\gamma}\,(a_{\alpha} - a)\,u_{\alpha}\,dw. \tag{8.28}$$

First we notice that by summing over α in (8.28) we find:

$$d\left(\sum_{\alpha} u_{\alpha}\right) = 0, \tag{8.29}$$

since $a = \sum_{\alpha} a_{\alpha}u_{\alpha}$. Hence the normalization of the wave function is preserved on each trajectory. Moreover we have from (8.23)

$$\overline{du_{\alpha}} = 0 \tag{8.30}$$

implying
$$\overline{u_\alpha(t)} = u_\alpha(0) = |c_\alpha(0)|^2. \tag{8.31}$$

Using again (A7.61-A7.63) to write the stochastic process for u_α^2 and taking the average over the Wiener process we get:

$$\overline{du_\alpha^2} = \overline{du_\alpha^2} = \gamma \left[2u_\alpha(a_\alpha - a)\right]^2 dt \geq 0. \tag{8.32}$$

Since due to normalization the u_α^2 are always less than 1, they are bound to evolve toward a stable, dispersion–free solution. Notice that the factors multiplying dt above vanish when either $u_\alpha = 0$ or $u_\alpha = 1$ and if one u is 1 all the others are zero. It can be shown (Ghirardi, Pearle and Rimini 1990 and references therein) that each limiting solution corresponds to one of the coefficients $\{u_\alpha\}$ being 1, all the other zero. Considering also (8.31), the probability of a particular choice $u_{\alpha_0} = 1$ is given by $|c_{\alpha_0}(0)|^2$.

Since the evolution of the wave function is no longer deterministic, another way of putting the question is to ask about the probability of finding one or other wave function at a later time, that is to look for the probability distribution of the values of the stochastic variables u_α (this represents therefore a probability distribution over the Hilbert space). The corresponding Fokker–Planck equation for the distribution $p(\{u_\alpha\}, t)$ is (compare (A7.60), see Pearle 1982, 1989):

$$\frac{d}{dt}p(\{u\}, t) = \sum_{\alpha,\beta} \frac{\partial^2}{\partial u_\alpha \, \partial u_\beta} \left\{2\gamma \left(a_\alpha - a\right) u_\alpha \left(a_\beta - a\right) u_\beta \, p(\{u\}, t)\right\}. \tag{8.33}$$

In agreement with the above discussion, and considering also (8.31), $p(\{u\}, t)$ from (8.33) shows increasingly pronounced peaks around the eigenvectors of A with weights $|c_\alpha(0)|^2$.

Hence following a trajectory we obtain a real collapse, the system being driven into just one of the pure states (in general, subspaces) of the decomposition in the pointer basis defined by A. Since in this approach each event in an experiment corresponds to one trajectory, an ideal measurement leads to a definite outcome and leaves the system in the corresponding eigenstate (subspace). As in sect. 8.2.1 we have again a physical realization of von Neumann's process of first kind, proceeding automatically and independent of any observer. The parameter γ does not influence the limiting distribution (but of course the convergence towards it). The time scale of this effect is further controlled by the operators $\{A\}$, and γ must be chosen such that if the $\{A\}$ describe the observables of a macroscopic object (implying therefore very many microscopic degrees of freedom) the collapse is *practically* instantaneous, while for truly microscopic observables the collapsed situation would be reached so slowly as to be *practically* unobservable.[12]

---

[12] Notice that this continuous transition implies that for any finite time the collapse is not complete, i.e. there is a nonvanishing "tail" of the wave function. We have

On each trajectory we can define a statistical operator

$$\sigma_w \equiv \Psi_w \Psi_w^\dagger. \tag{8.34}$$

The "physical" statistical operator is defined as an average over the Wiener process

$$\rho = \overline{\sigma_w}. \tag{8.35}$$

Note the conceptual difference between σ_w and ρ – see also Sect. 7.5. Just as Ψ_w itself, σ_w is a stochastic variable: it refers to a single trajectory and fulfills the nonlinear stochastic differential equation associated to (8.21,8.22) according to the transformation rules for stochastic variables (A7.61-A7.63). It describes the pure state of an isolated system and its dynamics accounts for both the deterministic Hamiltonian dynamics and the nondeterministic collapse. On the other hand, ρ is an ensemble average (over trajectories) and fulfills a linear deterministic evolution equation of the Lindblad form:

$$\frac{d}{dt}\rho = -\mathrm{i}[H, \rho] + \gamma\, \mathbf{A} \cdot \rho \mathbf{A}^\dagger - \frac{1}{2}\gamma\, \{\mathbf{A}^\dagger \cdot \mathbf{A}, \rho\} \tag{8.36}$$

($\{\ \}$ denotes the anticommutator). This is the *diffusion equation* for the expectation values of functions of a stochastic variable (here the state of a system), which is induced by the stochastic process for this variable itself – see (8.21,8.22). Its solution typically evolves towards a fixed point $\rho_{\lim}$. Equation (8.36) in particular has the structure described in Sect. 7.5 and leads to the decay of the non-diagonal elements in the base of the A's, in agreement with the trapping of the trajectories in these subspaces as illustrated above.

The equivalence between the above continuous process and the jump process described in the previous subsection in the limit of infinite frequency is indicated by observing that both lead to the same evolution equation for the statistical operators (but it can also be proven directly at the level of the stochastic differential equations – compare Ghirardi, Pearle and Rimini 1990). Thereby the parameters are related as follows:

$$\lambda \to \infty, \quad a \to \infty, \quad \frac{\lambda}{2a^2} = \gamma \tag{8.37}$$

(see also Sect. 3.2.1 for a comparison with the realistic parameters appearing in the environmental decoherence).

Of course, one can also write down the associated Fokker–Planck equation for probability distributions $p_t[\Psi]$ describing the "flow" in the Hilbert space, as we did in a special case above, (8.33). This allows one to study the

---

therefore a true (not apparent) but only approximate collapse. For a discussion concerning this point see Pearle (1989), Ghirardi *et al* (1993), Karakostas (1994), Ghirardi, Grassi and Benatti (1995).

convergence to equilibrium, the spread, etc. Then the usual density matrix (8.35) is given by

$$\rho(t) = \overline{\Psi_w \Psi_w^\dagger} = \int [D\Psi D\Psi^\dagger] \, p_t[\Psi] \, \Psi \, \Psi^\dagger \tag{8.38}$$

with the integration defined, e.g., by using the decomposition of Ψ in some basis as illustrated above.

Although the process (8.21,8.22) has the appeal of needing no further recipes to compute probabilities, one should observe that one can obtain the same effect by using the *linear* stochastic evolution equation:

$$d\psi = \left\{ \left[-\mathrm{i}H - \frac{\gamma}{2} \mathbf{A}^\dagger \cdot \mathbf{A} \right] dt + \sqrt{\gamma}\, \mathbf{A} \cdot d\mathbf{w} \right\} \psi \tag{8.39}$$

supplemented by the recipe that the state of the system at each moment, Ψ, is given by the normalized vectors but *taken with a probability distribution given by that of the Wiener process times their squared norms* each time we build averages over the Wiener process:

$$\Psi = \psi/||\psi|| \quad \text{with probabilities } ||\psi||^2. \tag{8.40}$$

The construction (8.39,8.40) has been called "raw and cook" (see Ghirardi, Pearle and Rimini 1990) and it directly parallels the one for the jump process (8.12)–(8.15). Equations (8.21,8.22) and (8.39,8.40) are equivalent in the sense that for all times t, to each trajectory in one of the processes leading to a given state at t there corresponds a trajectory in the other process leading to the same state. However in (8.39,8.40) the "noise" term is independent of the actual state while in (8.21,8.22) the probability distribution of the stochastic noise is given by the actual state at each moment. Based on the difference in the stochastic character between the two cases, Ghirardi, Grassi and Pearle (1990), (see also Ghirardi *et al* 1993), argue that it is the "raw and cook" construction (8.39,8.40), and not the non-linear process (8.21,8.22), which is adequate for relativistic generalizations.

One further development in connection with the explicit dynamical models for collapse is related to their formulation for systems of identical particles and to relativistic formulations. Concerning the non-locality, it is shown that the spontaneous collapse models show non-local "output dependence" just like standard quantum mechanics, but no non-local "parameter (or setting) dependence" like hidden variable theories (see Ghirardi *et al* 1993, Ghirardi 1994a). Since collapse models do not, however, form the main subject of this book we shall not discuss these developments here (see, e.g. Ghirardi, Pearle and Rimini 1990, Ghirardi, Grassi and Pearle 1990, Pearle 1990, Ghirardi *et al* 1993, Dove and Squires 1996, Ghirardi 1999, Pearle 1999a, Ghirardi 2000, Zbinden *et al* 2000, Adler and Brun 2001, Dowker and Henson 2002, Kent 2002).

8.3 Spontaneous Localization, Quantum State Diffusion and Decoherence

The attempt to represent non-unitary evolutions by stochastic processes in Hilbert space has a rather long history, see, e.g., Pearle (1982), Gisin (1984), Diósi (1988a), Ghirardi, Pearle and Rimini (1990), Belavkin and Staszewski (1992). The general aspects of constructing a stochastic process in the Hilbert space to reproduce a given generalized master equation have been discussed, e.g., by Gisin and Percival (1992). It amounts essentially to using Itô calculus to relate a Wiener process for a stochastic variable, here the Hilbert space vector Ψ_w, to a diffusion equation for expectation values (over the Wiener process) of functions of the stochastic variable, here the density operator ρ (see App. A7; sometimes Stratonovich calculus is used instead of Itô calculus). We have seen an explicit example of this procedure in the previous section. Of course there may be more stochastic processes associated with a given diffusion equation, usually some simplifying assumptions are made about the former which help define it up to trivial reparameterization, see, e.g., Diósi (1994a). One also attempts to make physical sense of the evolution on the individual trajectories, either at the level of a trajectory density matrix σ_w (Diósi 1988a,b) or of the trajectory state vector Ψ_w (Diósi et al 1994).

For some more general discussion, comparison between various approaches and further phenomenological analyses we refer the reader to the literature (e.g., Joos 1987a, Ghirardi, Rimini and Weber 1987, Damnjanović 1990, Stapp 1991, Ghirardi 1992, Jayannavar 1992, Diósi 1993a,b, Ghirardi 1994a,b, Benatti, Ghirardi and Grassi 1995, Dove and Squires 1995, Jadczyk 1995, Ghirardi 1998, Lewis 1997, Ghirardi and Bassi 1998, Clifton and Monton 1999, Pearle 1999b, Bassi and Ghirardi 2002, Collet and Pearle 2002).

It may be of interest to also quote the motivations for introducing quantum state diffusion processes as they are given by different authors. For instance, Belavkin (1989b,c, 1994, 2002) considers the problem of:

(a) *continuous, non-demolition measurement* of the properties of a quantum system in contact with an environment which continuously *traces* the system by establishing a record of these properties; in this way the state vector of the latter is continuously collapsed.

On the other hand, Diósi (1995b) mentions two points of view, which he calls:

(b) *environmental reduction*, the attempt to represent the non-unitary evolution of a *reduced statistical operator* for a system in contact with an uncontrollable environment by a stochastic process in the Hilbert space *of the system* (this seems to be also the point of view of Gisin and Percival 1992), and

(c) *universal reduction*, which follows from the assumption of a "universal environment" supposed to show up in a *fundamental* semi-group equation for the ensemble evolution and which can then be represented by a stochastic process for the wave function.

Finally, as already discussed above, the approach described in the previous sections, namely

(d) *spontaneous localization* proposes a *fundamental* stochastic process superimposed on the Schrödinger dynamics. This leads in this way to a unified dynamics describing both the unitary, Schrödinger evolution *and* the non-unitary collapse, hence ensuring an objective understanding of the measurement process.

In all these cases classical behavior in certain, specific situations is obtained as a by-product, and this is the reason why we considered these approaches in connection with our discussion of decoherence.

From the point of view of the classification attempted in sect. 8.1.3 motivation (a) belongs to class (v)(a) (*continuous measurement models*) and motivation (b) to class (v)(c) (*phenomenological stochastic models*). As already pointed out in sect. 8.1.3, if the environment does not select a "trace" of the result of the interaction with the system (as assumed under (a)), the wave function of the total system does not factorize, the system always remains entangled with the environment and never acquires a state vector. But such a selection can only happen if a collapse occurs, or if we accept the hypothesis of branching into dynamically decoupled "histories" (or "worlds") to which the potential observer becomes associated. A "reduction" of the kind mentioned in (b) above can only be induced by additionally assuming a measurement process to take place, since without this supplementary supposition the system simply does not possess a state vector. Building the *reduced density matrix* and considering its (non-unitary) evolution makes sense from the point of view of predicting the (statistical) behavior of the system – including the emergence of classical properties. The rewriting of this evolution in terms of a system state vector is, however, either a formal, technical procedure or *assumes* collapse (or branching) – but cannot explain the latter by an interaction with some conventional environment, since this would only lead to an apparent collapse as described under point (iv)(b) (apparent collapse by environmental decoherence). Of course, one could (under motivation (c)) *postulate* a "universal environment" to be able to induce a true reduction, a proposal which has been repeatedly made in connection with quantum gravity (see point (iii)(a) – "genuine collapse driven by gravitational effects" – in sect. 8.1.3). This would obtain if the environment managed by itself to be always in a pure state and to disentangle continuously from the system with which it interacts. This seems, however, to make such an environment of an unconventional character and shift the discussion to other levels.

The motivation (d) corresponds to the class (ii) (spontaneous collapse), is well defined and leads to falsifiable models. We stress again that these models not only describe the loss of coherence between certain states of the system, i.e., the apparent development of a mixture, as would already be the case within environmental decoherence, but also imply that the system is actually driven into just one of the pure states in the mixture (more precisely, the total system is driven in one of the factorized states in the decomposition of its wave function) – see (8.4). They describe therefore true collapse, even if this sets in only after infinitely long time (see sect. 8.2.2): even for finite time the system's wave function becomes more and more concentrated around just *one* of the pure states φ_n. This is to be contrasted with the apparent collapse of class (iv)(b) (environmental decoherence). In the latter case the states of the decomposition become more and more "disjoint" with increasing time (the decomposition simulates more and more a mixture), since for all observables O at our disposal (8.2) holds, but all components are still present and the system is never, even approximately, in a pure state – see (8.1). Because the point of view of the spontaneous localization models represents a clear pendant to the point of view of decoherence, we concentrated the discussion of this chapter on these models. They seem to have an intuitive interpretation, which also includes the choice of basis (the localization takes place in the coordinate space). However, they achieve all that through the *ad hoc* introduction of a stochastic dynamics, which – as long as the implied deviations from the standard quantum mechanical predictions are not experimentally supported – has no further justification.[13] The attempt to find a special kind of interaction responsible for this (motivation (c) above: the models of class (iii)(a), "gravitational effects") remains a conjecture.

A more definite question can be raised, however, concerning quantitative estimation of the effects predicted by these models. This should also allow us to make a direct comparison to environmental decoherence. Firstly, since they replace the Schrödinger equation by a stochastic equation these models represent an alternative to standard quantum mechanics and lead to (possibly measurable) deviations from the predictions of the latter. They predict, among other things, an inherent and continuous energy production (see, e.g., Ballentine 1991a). This already puts strong constraints on the parameterization of the model, see e.g. Ghirardi, Grassi and Rimini (1990), Pearle and Squires (1995), Collet *et al* (1995). Secondly, Gallis and Fleming (1990) already noticed that the evolution of ρ of the system in these models is essentially similar to that of ρ_{red} in environmental decoherence models, and used this observation also to obtain better control on the parameters of the former. In fact, as already remarked in sect. 8.1.3, the situations in which spontaneous reduction is quantitatively effective are also the situations in which environmental decoherence leads to classical behavior in the sense of non-observability of quantum interferences. Tegmark (1993) carries further the calculations of Joos and Zeh (1985) and shows that typically in the situations proposed for tests of the spontaneous collapse models (either with *ad hoc* collapse dynamics or with collapse "due" to quantum gravity) the environmental decoherence leads to effects of the same form, but stronger! We refer to Sects. 3.2, 3.3 for the discussion of these quantitative aspects, here we only note the conclusion, namely that there seems to be as yet no

---

[13] This concerns both the jump and the continuous formulation – sects. 8.2.1, 8.2.2.

experimental indication for the necessity of non–standard effects (i.e., spontaneous collapse) in connection with the explanation of classical properties (one may hope that future experiments could be able to disentangle these effects and provide evidence pro or again collapse models – cf., e.g., Marshall *et al* 2002). Finally we also remark that problems of principle, like energy non–conservation, which are characteristic to the spontaneous collapse models do not appear in the environmental decoherence approach, or are only introduced by further approximations, such as the neglect of some terms in the master equation, which are not essential for the existence of the decohering effects (see, e.g., Sects. 3.2.2, 7.3).

All this seems to indicate that stochastic collapse may be an interesting concept for the purpose of illustrating and discussing the measurement problem in quantum theory, but as for the present date there seems to be no *phenomenological* necessity for its introduction, neither in connection with the observation of classical properties (for which environmental decoherence already appears to provide a satisfactory explanation), nor prompted by new, observed phenomena. It is also not clear that the presupposition of an *ad hoc* random process should be *conceptually* more satisfactory than, say, the branching hypothesis of Everett's theory – as long as this random process can neither be derived from known physics, nor shown to be experimentally relevant. This also affects the discussion on "objectification" (Ghirardi and Bassi 1998, Bassi and Ghirardi 2000), such that other concepts hereto appear as at least serious alternatives (see Chap. 2, see D'Espagnat 1995, 2001). A good feature of the stochastic collapse models resides in fact in the possibility to falsify them, which may raise very interesting experimental problems with exciting implications both in case of a positive and of a negative outcome (Ghirardi 1998).

9 Related Concepts and Methods

H. D. Zeh

There are various contributions in the literature which are either related to the concept of decoherence, or are using this term in a different sense. In this book, "decoherence" is in general understood as environment-induced decoherence (representing an *apparent* or *effective* collapse, as explained in preceding chapters).

Since a *genuine* collapse (that is, a modification of the Schrödinger equation) can also be formulated as a process of decoherence, several collapse models have been reviewed in Chap. 8. It is remarkable that most of them try to mimic precisely those "measurement-like events" which are readily described by the Schrödinger equation as an apparent collapse in the local system if only the environment is taken into account. If a genuine collapse did take place in the observational chain (2.6) somewhere *before* the occurrence of environmental decoherence, it would most probably have been noticed experimentally. If it occurred later, however, environmental decoherence would remain essential for describing the observed quasiclassical world. As soon as phase relations have "irreversibly" been delocalized in a unitary process of ever-growing entanglement, the corresponding state vector components describe dynamically independent "worlds", no matter whether the complete state vector is Everett's or only *part* of it as the result of a genuine collapse. In order to restore local superpositions after their decoherence, the thermodynamical arrow would have to be reversed.

Some other concepts and methods, either related or equivalent to decoherence, will be discussed in this chapter (cf. also Sect. 3.4).

9.1 Phase Averaging in Ensembles ("Dephasing")

While the main obstacle for a theory of quantum measurement is the occurrence of definite *individual* outcomes (instead of their superpositions), their *probabilities* would then be meaningful only as frequencies in *series* of measurements. The ensemble of different quantum states $|\psi_i\rangle$ with corresponding probabilities p_i, describing such series, may be represented for all practical purposes (that is, for the calculation of all expectation values) by a statistical operator $\rho = \sum_i |\psi_i\rangle p_i \langle\psi_i|$ (a "proper mixture" – see Sect. 2.4). In a certain basis $\{|\phi_n\rangle\}$, with $|\psi_i\rangle = \sum_n c_{in} |\phi_n\rangle$, this leads to its representation by means of a density matrix ρ_{nm},

$$\rho = \sum_{nm} |\phi_n\rangle \, \rho_{nm} \, \langle\phi_m| := \sum_{nm} |\phi_n\rangle \left(\sum_i c_{in} p_i c_{im}^* \right) \langle\phi_m| . \qquad (9.1)$$

For a large ensemble of states ψ_i, with coefficients c_{in} which have random phases with respect to the ensemble index i, the non-diagonal elements ρ_{nm} become small. Interference effects are then averaged out *in the ensemble,* and spatial probabilities, $p_n(x) = |\langle x|\phi_n\rangle|^2$, for example, would add incoherently:

$$p(x) = \langle x \, |\rho| \, x \rangle \approx \sum_n \rho_{nn} p_n(x). \qquad (9.2)$$

The wave functions $\phi_n(x) := \langle x|\phi_n\rangle$ may here represent partial waves leaving the n-th slit of an interference device, or resulting from the n-th "path".

An ensemble of superpositions with random phases may be *prepared* by means of a random scattering process,

$$\sum_n c_n \phi_n \to \sum_n c_n S_{in} \phi_n \quad , \qquad (9.3)$$

applied to identically prepared quantum objects (with an initial state $\sum_n c_n \phi_n$). The phases of the scattering amplitudes S_{in} (or of more general unitary evolutions $U_i(t)$) are assumed to vary at random. This stochastic procedure may be realized by means of a macroscopic scatterer that depends uncontrollably on time, or by an ensemble of projectiles which pass the scatterer at slightly different locations (as it may happen for neutrons – see Rauch *et al.* 1990). Such *unusual* experiments can thus easily be described in terms of unitary quantum theory. In the *normal* situation of microscopic scattering, however, the scattering matrix (including its phase) is fixed for the whole series of experiments. Otherwise it could never have been determined experimentally. Nonetheless, there are definite individual outcomes.

Smallness of the non-diagonal elements of the density matrix does *not* necessarily mean that they may be neglected as a whole. They may remain important for calculating the correct ensemble entropy $\mathrm{tr}\{\rho\ln\rho\}$, or for determining the eigenstates of ρ. It is even more important in this context to realize that none of the *individual* superpositions can be expected to collapse into an eigenstate of ρ just because it is *considered* within an ensemble that may either represent incomplete knowledge or a series of similar experiments. This conclusion remains valid even when the nondiagonal elements of a density matrix happen to vanish *exactly.* For example, the individual members of an ensemble of spin-up and spin-down with probabilities $1/2$ are different from those forming an ensemble with spin-right and spin-left – in spite of their identical density matrix.

Dephasing in an ensemble has been claimed to explain the phenomenon of decoherence in general, and even the indeterminism characterizing individual events (see Machida and Namiki 1980, Namiki and Pascazio 1993).

This proposal fails not only because the method does not apply to individual measurements. The process of measurement is fundamentally different from the mere scattering (or "phase shifting") described by a unitary transformation. Its essential aspect is the change of a "pointer" in dependence of the quantity to be measured, as described by von Neumann's interaction. A passive scatterer would usually not even show up as a dynamical element of the description – cf. (9.3). If the pointer does move appropriately (into approximately orthogonal states in order to represent different outcomes), this must lead to strong entanglement in the *individual* quantum states for initial superpositions. No additional ensemble dephasing is required or would be observable at the measured system. This entanglement becomes irreversible as soon as it includes the uncontrollable environment (as it happens extremely fast for a macroscopic pointer).

Phase averaging in *thermal* ensembles has recently been discussed under the name "dissipation-free decoherence" (Ford and O'Connell 2001a, 2002). However, it merely describes the decay of fringe visibility in the statistics of certain measurements, caused by deterministic wave packet dispersion, and/or is based on stochastic classical particle dynamics (interpreting the Wigner function as a "distribution"). A "wave packet at finite temparature", used by these authors, is quantum mechanically an ill-defined concept. This approach does not allow one to transform superpositions into ensembles (not even locally), and it entirely neglects the crucial concept of quantum entanglement – similar to Machida and Namiki's original work.

Random phase approximations have succussfully been used in many applications of quantum theory. Pauli (1928) and van Kampen (1954) used them to derive irreversible master equations and Fermi's golden rule (cf. Sect. 4.1.2 of Zeh 2001) Their ensembles, called "cells of states" (small energy bands), are usually justified by incomplete observation, but can be explained in terms of objective entanglement (improper mixtures) as affecting the individual global states – see Joos (1984) and Sect. 3.3.2.1. Even Omnès (1999) referred occasionally to van Kampen's phase averaging rather than to entanglement when discussing decoherence.

9.2 Ergodicity and Irreversible Amplification

Ensemble dephasing may also be based on averaging over time, in particular for rapidly varying relative phases. This suggests an analogy to classical concepts, where, for example, pressure is defined as a *temporal mean value*. However, this is *not* appropriate for quantum phases: a rapidly varying superposition remains a superposition and does not give rise to an ensemble – again not even locally.

When *classically* describing the measurement of a *microscopic* variable, one would expect the latter to trigger a transition from an initial meta-stable "pointer" state into a thermodynamically stable one, represented by a very large "ergodic" subvolume of phase space. The system would thereby have to approach *one* of various pointer states, depending either on controllable initial conditions (the microscopic variable in genuine measurements), or on uncontrollable ones (in "measurement-like" processes, such as phase transitions). For each *individual* pointer position – not only for their ensemble – would entropy (the size of the ergodic phase space subvolume) have to increase, but in quantum theory this argument remains valid for all their superpositions, too. In order to exclude these unwanted superpositions, some authors used ergodic theorems to replace temporal means by a density matrix. When furthermore using a random phase approximation, non-diagonal elements in the density matrix disappear as described in Sect. 9.1. The authors then interpreted their "mixed states" as representing ensembles of different measurement outcomes (cf. Danieri, Loinger and Prosperi 1962).

These attempts to solve the measurement problem attracted considerable attention at their time because of the authors' ambitious claim that their theory had to be regarded as "an indispensable completion and a natural crowning of the basic structure of present-day quantum mechanics", and since this claim obtained strong support from Leon Rosenfeld (1965), who was respected as representing Bohr's authorized opinion (cf. Jammer 1974, p. 492).

Not surprisingly, this paper was strongly criticized (see Jauch, Wigner and Yanase 1967, Bub 1968, Fine 1970, and Zeh 1970). Evidently, a pointer does not fluctuate in time between different possible outcomes. Since the outcome of each individual measurement is time-independent and fixed, time averaging cannot produce the required ensemble of outcomes. In quantum mechanical terms, the impossibility of dynamically *controlling* phase relations between macroscopic "cells of states" (subspaces) does not imply their non-existence. Regardless of all (thermo)dynamic complexity, each individual measurement would still lead to a superposition of states from *different* ergodic subspaces. Without environmental decoherence, the corresponding phase relations would continue to exist *locally*.

This attempt to explain the occurrence of definite results as an approach to thermal equilibrium leads into particular difficulties for "negative-result measurements", first discussed by Renninger (1953) – cf. Sect. 3.4.1. They occur, for example, when a (non-absorbing) detector is placed at only *one* slit of a two-slit interference device, while merely those events on a subsequent scintillation screen are taken into account for which this which-path counter does *not* click. The outcome of such a "null measurement" (the definite passage through the slit without a detector) would then have to be triggered by amplification processes occurring during the *potential* passage through the other slit (Jauch, Wigner and Yanase 1967). Decoherence, in contrast, applies to null measurements as well, since it dislocalizes *relative* phases between components (paths) that are *all* assumed to exist simultaneously.

9.3 Dressing of States

Many microscopic systems affect their environment reversibly. For example, charged particles polarize surrounding matter and even the vacuum. In this way, they may acquire effective ("renormalized") charge and mass. This is known as a "dressing" of the "bare" particle, since it determines its effective and *observable* properties. Bare particles (that is, fundamental *fields*) remain unobservable, while any *secondary* dressing, due to changed circumstances (such as caused by surrounding matter), can be investigated experimentally. Fundamental dressing of charged particles leads to an "infrared catastrophe", as it requires an infinite number of photons. However, this formal consequence is an artifact of the Fock space (occupation or "particle" number) representation of field functionals (see also Sect. 4.1.1).

Subspaces of an infinite tensor product space are called "unitarily inequivalent" if they differ by an infinite number of factors. Even though this definition depends again on the Fock representation, superpositions of states from inequivalent subspaces have been claimed to be unobservable in principle (and hence not to "exist" in an operational sense), since no appropriate observables can be constructed in Fock space. However, this is a purely formal argument. If physical states have to be described by a mathematically non-trivial dressing, so have the physical observables. The suggestion that non-trivial dressing transforms superpositions into ensembles, and thus explains measurements and classical properties (Hepp 1972, Pfeifer 1980, Primas 1981, Amann 1991a – cf. Sect. 3.2.4), is thus unjustified. For example, different energy eigenstates of the hydrogen atom possess different Lamb shifts, since they polarize the vacuum differently. Nonetheless, the possibility of their superposition can hardly be denied, since the latter remain local. On the other hand, components forming a superposition of many particle states may *dynamically* decouple in practice (except for their collective motion – see Sect. 9.4) if they differ collectively by a large but finite number of single-particle factor states.

Dressing may effectively lead to decoherence if long-range fields are involved. In particular, (local) quantum states representing different charge values decohere because of their entanglement with Coulomb fields (Sect. 4.1.1), which contain "information" about the charge at any distance. Their superposition cannot be observed by any *local* device (cf. Sect. 6.4). This demonstrates once more that the concept of decoherence is based on the locality of the observer. It does not require mathematical idealizations, such as "unitarily inequivalent Hilbert spaces", but depends instead on quantitative aspects characterizing local systems.

Therefore, dressing as a method of determining the correct stable eigenmodes of microscopic objects has to be clearly distinguished from decoherence. It is conceptually remarkable that observable decoherence may distinguish between field theories and action-at-a-distance. The latter would lead to decoherence only after the retarded force has affected the quantum state of an absorber. Decoherence would then *not* be able to explain a *general* charge superselection rule.

9.4 Symmetry Breaking and Collective Motion

In many-particle systems, states of *collective motion* are common. The individual ("fast") particles usually follow these "slow" modes adiabatically. Their states may thus be said to *monitor* the collective variables (or "generator coordinates"). This situation may (but need not) arise according to a Born-Oppenheimer approximation as a consequence of a large mass ratio between different kinds of particles (cf. Sects. 3.2.4 and 4.2.1).

If the many-particle system under consideration is effectively isolated from its environment, it will usually be found in an energy eigenstate. Examples for such collective energy eigenstates are rotational or vibrational states of nuclei or small molecules, phonon eigenstates in solid bodies, plane waves for collective field modes (Goldstone particles), etc. If the collective mode is instead continuously monitored by the environment, its amplitude decoheres and appears to be in a time-dependent *classical* state of motion – as it happens to be the case for the atomic configurations in large molecules or for macroscopic oscillations. In *both* cases, the "fast" variables act as a *dressing* (not as part of decoherence) of the slow ones – cf. Sect. 9.3. A time-dependent (classical) description has also been applied to truly microscopic modes (as in the *cranking model* for rotational states of nuclei), even though this does *not* lead to the (empirically correct) energy eigenstates. In molecular physics one finds both situations: eigenstates for very small molecules, but time-dependent wave packets for the positions of the nuclei in larger ones – as discussed in preceding chapters.

Collective modes may arise as a consequence of an exact or approximate degeneracy. For example, the energy eigenstate of an N-particle system, $H\psi = E\psi$, may be approximated by a Slater determinant $\Phi = \det[\phi_i(q_i)]$, where $\phi_i(q_i)$ are single-particle wave functions, by means of the variational procedure $\langle \delta\Phi | H - E | \Phi \rangle = 0$. This leads to a single-particle Schrödinger equation with a self-consistent (collective) potential. The essential aspect of this "mean field approximation" for the present discussion is its nonlinearity: the sum of two determinants is in general not a determinant any more. Solutions of the variational procedure need then not be approximate eigenstates of symmetry operators which commute with the Hamiltonian (as it would be the case for exact solutions). One may even obtain *strongly* symmetry-violating but still physically meaningful solutions of the variational procedure, such as deformed nuclei or chiral molecules. They cannot be unique, however, since the dynamical symmetry requires the existence of other solutions, for example different orientations of a deformed nucleus, $\Phi(\Omega) = U(\Omega)\Phi$. Here, $U(\Omega)$ is a unitary representation of the rotation group, parametrized by the Euler angles Ω.

In spite of being very different from the expected energy eigenstates, the symmetry-breaking results of the variational procedure are known to describe important properties of them. In order to see how this is possible, one may utilize the degeneracy in order to construct superpositions of the form

$$\Psi(q) = \int d\Omega \, f(\Omega) U(\Omega) \det[\phi_i(q_i)]. \tag{9.4}$$

The coefficients $f(\Omega)$ may here be chosen (or determined from a generalized variational procedure) as the irreducible representations, $D_I(\Omega)$, of the considered group. Note that Ω does here not occur as a dynamical variable, as it would in the quantization of a rigid rotator. Equation (9.4) defines various projections onto subspaces corresponding to different eigenvalues I of the Casimir operators (Peierls and Yoccoz 1957, Zeh 1965). Therefore, a strongly asymmetric "intrinsic" state, such as $\Phi = \det[\phi_i]$, does not only describe the ground state $\Psi_0(q)$ in the form (9.4), but a whole family of collective states $\Psi_I(q)$. For example, this procedure leads to a collective energy band of the form $\mathbf{p}^2/2M$ for translations (where $\Omega \equiv \mathbf{r}$ and $I \equiv \mathbf{p}$), or (approximately) $I(I+1)/2\Theta$ for rotations of axially symmetric objects.

If the symmetry breaking system is macroscopic, and is thus decohered by its environment, it is often appropriately described by the asymmetric state $\Phi(\Omega)$ itself (in general with time-dependent $\Omega(t)$, following a classical equation of motion). Since the states $\Phi(\Omega)$ are not mutually orthogonal, they represent "wave packets" for the collective mode, such as a superposition of different angular momentum eigenstates. In the macroscopic *and* in the microscopic case, the deformed determinant (or another symmetry-violating model state) represents a useful concept, since it describes those robust *correlations* between constituents of the system which form a dressing of the collective mode. The popular claim that (for mathematical reasons) spontaneous symmetry violations occur only in *infinite* systems is absolutely unjustified.

Equation (9.4) is similar to a Born–Oppenheimer approximation, described by a wave function $\Psi(Q,q) = f(Q)\phi(q;Q)$ for massive ("slow") variables Q and "fast" ones, q, obeying a partial wave functions $\phi(q;Q)$ that depends adiabatically on Q. The ϕ's are defined as eigenfunctions of the "zero order" Hamiltonian which neglects the kinetic energy of the Q-variables. The first factor, $f(Q)$, is then assumed to solve a Schrödinger equation with an effective Hamiltonian for Q, defined by the partial expectation value of the exact Hamiltonian, with respect to q only, in the state $\phi(q;Q)$. In contrast to the generator coordinate Ω in (9.4), one does here not integrate over the (now fundamental) dynamical variables Q. This difference between Ω and Q disappears in the abstract ket representation,

$$|\Psi\rangle = \int dQ \, dq \, \Psi(Q,q) \, |Q,q\rangle \quad , \tag{9.5}$$

of the Born–Oppenheimer state, to be compared with $|\Psi\rangle = \int dq\,\Psi(q)\,|q\rangle$ after inserting (9.4). This demonstrates that the collective variables Ω mimic new degrees of freedom if the states $U(\Omega)\Phi$ are approximately orthogonal for different Ω's (see (9.7) below).

There are many other symmetries which lead to similar situations. Consider the following list of examples, where all (usually dynamical) "classical properties" are characterized by a spontaneous intrinsic symmetry breaking:

Dynamical symmetry	Conserved eigenvalue	"Classical" property
translations	momentum	position
rotations	angular momentum	orientation
space reflection	parity	chirality
gauge transformations	charge	Josephson phase
complex conjugation	real or imaginary	sign of phase
time translation	energy	simultaneity

Symmetry under complex conjugation applies to a real Hamiltonian. Its violation is particularly important for recovering classical time from the Wheeler–DeWitt equation of quantized general relativity, as discussed in Sect. 4.2 (see also Barbour 1993).

In a determinant of single-particle wave functions describing a deformed nucleus, for example, each nucleon dynamically "feels" the asymmetric self-consistent potential (plus Coriolis type forces if $I \neq 0$), even though it exists in a superposition of all its orientations. This remains true even for aggregates ("clusters") of nucleons that may exist within the nucleus. One may thus conceive of a "super-nucleus" that contains a super-cluster which is complex enough to describe an *internal observer*. What would such an observer be able to "feel", that is, observe?

Note that this simple picture of a closed quantum world with its participating observer uses only strictly quantum mechanical concepts (wave functions). The physical state of the observer is described by a quasi-stable *relative state wave function* (a factor of the intrinsic state Φ – not of the global state Ψ). There are thus

no particles (hence no "new statistics"),
no classical variables (hence no "complementarity" or "new logic"),
no "observables" (hence no "superselection sectors"),
no "quantum jumps" (no collapse of the wave function), and
no "events" (hence no discontinuous "histories"),

except as derived and secondary concepts. As far as quantum gravity is neglected, physical histories are described by global wave function(al)s, defined on an arbitrarily chosen space-like spacetime slicing (forming a trajectory of quantum states).

In order to discuss the dynamics that affects a quantum observer who is *part* of a thus described quantum world, let us exploit the picture of a deformed super-nucleus further. Since *all* nucleon wave functions ϕ_i of a Hartree product $\Phi = \prod_i \phi_i$, for example, must be deformed because of their deformed self-consistent potential, the inner product between different orientations of the nucleus,

$$\langle \Phi | U(\Omega) | \Phi \rangle = \prod_i \langle \phi_i | U(\Omega) | \phi_i \rangle \quad , \tag{9.6}$$

is a product of many factors that are all smaller than one if $\Omega \neq 0$. Therefore, the expression (9.6) forms a narrow Gaussian peak at $\Omega = 0$, similar to the inner product of two Gaussian wave packets centered at different eigenvalues of an observable Ω_{op}. (The consequence is somewhat reduced for indistinguishable particles, but disappears only if the set of single-particle states spans a symmetric subspace.) This situation of "strong coupling" can also be regarded as an "intrinsic decoherence" of the orientation Ω (meaningful for an internal observer). Note that the narrow peak is here a consequence of the large number of particles contributing to the asymmetry, while it does *not* require an extreme shape of the object ("needle limit"). It applies even to macroscopically spherical objects, where a residual microscopic asymmetry is sufficient to produce a narrow peak, and thus rotational degrees of freedom. An *exactly* symmetric quantum object would not be able to rotate at all, as its moment of inertia vanishes (Peierls and Yoccoz 1957, Zeh 1967, Kübler 1972). This remains true for translations and rotations of a completely symmetric quantum Friedmann universe (Conradi and Zeh 1991).

The formal probability amplitude for the orientation $\Phi(\Omega)$ in a superposition Ψ of (9.4) is thus approximately

$$\langle \Phi(\Omega) | \Psi \rangle \propto f(\Omega), \tag{9.7}$$

as though $f(\Omega)$ were a wave function for a (rigid or non-rigid) rotator. The intrinsic decoherence of orientation, (9.6), will similarly suppress non-diagonal matrix elements of all few-particle operators, such as the Hamiltonian. This leads to an effective *dynamical* decoupling between sufficiently different values of Ω (similar to different Everett branches), except for their relation through collective motion.

A symmetry-breaking vacuum is formally analogous to a deformed nucleus. Heisenberg was led to his concept of a spontaneous symmetry breaking in field theory by the analogy with an oriented spin lattice (Heisenberg 1957). Even if the "ground state of the universe" were a symmetric superposition of all its orientations with respect to all relevant symmetry groups (similar to the nuclear ground state with $I = 0$), an intrinsic observer formed by particles from the corresponding Fock space could only feel (observe) "his" (possibly deformed) relative vacuum, which does *not* represent a global energy eigenstate.

In the case of an exact global symmetry, different (absolute) "orientations" would be *redundant* in the sense of Sect. 6.3 in a Machian theory, while different orientations of a nucleus are just *indistinguishable* to an intrinsic observer. In this non-Machian case, the observer would still be able to feel dynamical consequences of his external inertial frame, such as Coriolis type forces resulting from a specific superposition (that is, the projection onto a specific angular momentum eigenspace). In contrast, redundancies reflect the presence of physically meaningless variables, which may have to be eliminated by means of *constraints*. For example, in general relativity the physical dynamical variable (to be quantized) is coordinate-free spatial geometry on space-like slices (Sect. 4.2), while any coordinate dependence would be unphysical. This *classical* procedure to destill the correct physical variables is often conveniently replaced by "quantum constraints", that is, by restricting physical states to *symmetric superpositions* of all unphysical "symmetry" transforms of the form (9.4) with $f(\Omega)$ = constant.

In another well known example of redundancies, particles (which could be meaningfully permutated) are replaced with spatial fields that are then to be quantized. This approach, which leads to an occupation number (that is, oscillator quantum number) representation, eliminates particle indices *and* the need for a "second quantization". Indeed, the latter would have to quantize wave functions on configuration space rather than the exceptional single particle wave function that is defined on space. Particle configurations *emerge* as an *apparent and approximate* concept by means of decoherence (leading to localization in this case), thus demonstrating that classical concepts need not represent the correct pre-quantum variables (provided the latter do exist). In particular, *there are no particles* even at the correct *pre-quantum* level (cf. Sect. 2.1.1).

As mentioned at the beginning of this section, collective motion is not restricted to generators of a (genuine) dynamical symmetry. This method was initially applied to vibrations of nuclear shapes by means of general generator coordinates (Hill and Wheeler 1953, Griffin and Wheeler 1957). Different shapes are, of course, absolutely distinguished by their dynamics, and could therefore be recognized by *intrinsic* observer subsystems as certain parameters which characterize "their world".

It was this model of collective motion that historically led to the discovery of decoherence as a universal and dynamically relevant concept (Zeh 1970). In order to describe the apparent collapse of the wave function (the branching into dynamically separate world components) along these lines, one may consider an appropriate time-dependence of the wave function. The energy eigenstate of an atomic nucleus with zero angular momentum (possibly the ground state) may be intrinsically deformed. An arbitrary intrinsically *spherical* initial state Ψ^0 would then unitarily develop a component that is a symmetric superposition of intrinsically deformed nuclei with different orientations. If there were internal observers Φ_{obs} in a "super-nucleus" (and if they could survive such a transition) the global state would change as

$$\Phi^0_{\text{obs}}\Psi^0_{\text{rest}} \rightarrow a(t)\Phi^0_{\text{obs}}\Psi^0_{\text{rest}} + b(t)\int d\Omega\, \Phi_{\text{obs}}(\Omega)\Psi_{\text{rest}}(\Omega). \qquad (9.8)$$

According to this quantum description, the intrinsic observer will *split*, and end up in "worlds" (nuclei) with different orientations: each observer, Φ^0_{obs} or $\Phi_{\text{obs}}(\Omega)$, for all values of Ω, feels essentially only his own *relative world*. So one has to abondon the classical prejudice of a unique history of an observer in the real world. This "many worlds" interpretation is the unavoidable consequence of a local (subsystem) observer in an entangled quantum universe. Note that the RHS of (9.8) is only approximately identical with the Schmidt canonical form (2.16), since the states which differ by their "orientation" Ω are not exactly orthogonal.

An evolution similar to (9.8) may also be caused by a time-varying (or expansion-dependent) Higgs potential in the early universe that leads to a phase transition of the vacuum, although observers would here evolve much later in the history of the universe. Conventional measurements are described by *local* evolutions of type (9.8). This description does *not* require any collapse of the wave function as a dynamical process that affects reality. The observed quantum indeterminism originates in the observer – it does not characterize the *real* world.

The concept of decoherence thus emerged as a by-product of the hypothesis that quantum theory is universally valid and complete (so that entanglement must be quite common) – not as a phenomenological (*ad hoc*) supplement to open systems quantum dynamics. Its recent experimental confirmation strongly supports this natural assumption of a universal quantum theory. Non-explanatory pseudo-concepts, such as complementarity, dualism, quantum logic, quantum information and all that, become obsolete, while any modifications of the formalism or its interpretation as describing physical reality would require clear evidence that is *not* simply based on classical prejudice.

A1 Equation of Motion of a Mass Point

E. Joos

In this appendix we outline the derivation of the non-unitary part of the equation of motion (3.71) of a mass point under the influence of scattering processes. As explained in the main text, in most situations the approximation of negligible recoil is entirely sufficient. We therefore limit our presentation to this important case. Treatments which analyze scattering events have been given by several authors, see for example Joos and Zeh (1985), Gallis and Fleming (1990), Tegmark (1993), and Alicki (2001). For the more general case of quantum Brownian motion see Sects. 3.2.2 and 5.1 and references cited there. The following presentation follows largely Gallis and Fleming.

A single scattering process is defined by giving the resulting state when an incoming particle is scattered off the scattering center that is located at a certain position $\boldsymbol{x}$,

$$|\boldsymbol{x}\rangle\,|\chi\rangle \to S\,|\boldsymbol{x}\rangle\,|\chi\rangle = |\boldsymbol{x}\rangle\,|\chi_{\boldsymbol{x}}\rangle\,. \tag{A1.1}$$

The corresponding change in the density matrix of the scattering center is then given by

$$\rho(\boldsymbol{x},\boldsymbol{x}') \to \rho(\boldsymbol{x},\boldsymbol{x}')\,\langle\chi_{\boldsymbol{x}'}|\chi_{\boldsymbol{x}}\rangle = \rho(\boldsymbol{x},\boldsymbol{x}')\,\Big\langle\chi\,\Big|S_{\boldsymbol{x}'}^{\dagger}S_{\boldsymbol{x}}\Big|\chi\Big\rangle\,. \tag{A1.2}$$

The main task is to evaluate the matrix element appearing in (A1.2) which is the overlap of the states scattered off positions $\boldsymbol{x}$ and $\boldsymbol{x}'$, respectively. We will employ the following assumptions:

–   The interaction is invariant under translations, that is, the scattering matrix commutes with total momentum.
–   Recoil is negligible, that is, only the state of the scattered particle is changed during interaction.
–   Many particles are scattered during a short time interval and the incoming particles are distributed isotropically.
–   The incoming particles can be described by an ensemble of momentum eigenstates.

If we use translation invariance we can express the scattering off the state $|\boldsymbol{x}\rangle$ by means of the scattering off a fixed position $|\boldsymbol{x}=0\rangle$, since

$$
\begin{aligned}
|\boldsymbol{x}\rangle\,|\chi\rangle &= \exp(-\mathrm{i}\hat{\boldsymbol{p}}\boldsymbol{x})\,|\boldsymbol{x}=0\rangle\,|\chi\rangle \\
&= \exp\!\Big(-\mathrm{i}\big(\hat{\boldsymbol{p}}+\hat{\boldsymbol{k}}\big)\boldsymbol{x}\Big)\,|\boldsymbol{x}=0\rangle\,\exp\!\big(\mathrm{i}\hat{\boldsymbol{k}}\boldsymbol{x}\big)\,|\chi\rangle
\end{aligned}
\tag{A1.3}
$$

where $\hat{p}$ and $\hat{k}$ are momentum operators for the scattering center and the scattered particle, respectively. Then

$$S(|x\rangle |\chi\rangle) = \exp\left[-i(\hat{p} + \hat{k})x\right] S(|x = 0\rangle \exp(i\hat{k}x) |\chi\rangle \qquad (A1.4)$$

since S commutes with total momentum. If recoil can be neglected, we have furthermore

$$S\left(|x = 0\rangle \exp(i\hat{k}x) |\chi\rangle\right) = |x = 0\rangle S_0 \exp(i\hat{k}x) |\chi\rangle \qquad (A1.5)$$

hence

$$S |x\rangle |\chi\rangle =: |x\rangle |\chi_x\rangle = |x\rangle \exp(-i\hat{k}x) S_0 \exp(i\hat{k}x) |\chi\rangle \qquad (A1.6)$$

and the local density matrix changes according to

$$\rho(x, x') \to \rho(x, x') \left\langle \chi \left| \exp(-i\hat{k}x') S_0^\dagger \exp(-i\hat{k}(x - x')) S_0 \exp(i\hat{k}x) \right| \chi \right\rangle. \qquad (A1.7)$$

It is useful to express the scattering matrix S in terms of the T-matrix in the usual way as

$$S_0 = \mathbb{1} + iT \qquad (A1.8)$$

and to employ the unitarity of the S-matrix (optical theorem) in the form

$$i\left(T - T^\dagger\right) = -TT^\dagger. \qquad (A1.9)$$

With the initial state $|\chi\rangle$ approximated by a momentum eigenstate $|q\rangle$, the matrix element in (A1.7) reads

$$\exp(iq(x - x')) \left\langle q \left| (\mathbb{1} - iT^\dagger) \exp(-i\hat{k}(x - x')) (\mathbb{1} + iT) \right| q \right\rangle. \qquad (A1.10)$$

Inserting a complete set of momentum eigenstates leads to

$$1 + \sum_{q'} (\exp[i(q - q')(x - x')] - 1) |\langle q |T| q'\rangle|^2. \qquad (A1.11)$$

Introducing the scattering amplitude,

$$\langle q |T| q'\rangle = \frac{i}{2\pi q} f(q, q')\delta(q - q') \qquad (A1.12)$$

and the usual replacements

$$\sum_q \to \left(\frac{2\pi}{L}\right)^3 \int d^3q \qquad (A1.13)$$

and

$$\delta^2(q - q') = \delta(q - q')L \qquad (A1.14)$$

for a quantization volume L^3 gives

$$\frac{2\pi}{L^2} \int d\Omega' \left(\exp[\mathrm{i}(\boldsymbol{q} - \boldsymbol{q}')(\boldsymbol{x} - \boldsymbol{x}')] - 1 \right) |f(\boldsymbol{q}, \boldsymbol{q}')|^2 . \tag{A1.15}$$

Because of energy conservation we have $|\boldsymbol{q}| = |\boldsymbol{q}'|$ in all following equations. The flux of particles through the volume L^3 in the time interval Δt is $L^2 v(q) n(q) \Delta t$, where $v(q)$ is the speed and $n(q)$ the density of particles with momentum q. If all contributions are added, the change of ρ in Δt is given by

$$\frac{\Delta\rho(\boldsymbol{x}, \boldsymbol{x}')}{\Delta t} = -\rho(\boldsymbol{x}, \boldsymbol{x}') F(\boldsymbol{x} - \boldsymbol{x}') \tag{A1.16}$$

with

$$F(\boldsymbol{x} - \boldsymbol{x}') = \int dq\, v(q) n(q) \int \frac{d\Omega\, d\Omega'}{2} \left(1 - \exp[\mathrm{i}(\boldsymbol{q} - \boldsymbol{q}')(\boldsymbol{x} - \boldsymbol{x}')] \right) |f(\boldsymbol{q}, \boldsymbol{q}')|^2 . \tag{A1.17}$$

This is the main result, describing the change of ρ for rather arbitrary scattering processes.

For large distances $|\boldsymbol{x} - \boldsymbol{x}'|$, F approaches a constant value,

$$F(\infty) = 2\pi \int dq\, n(q) v(q) \sigma(q), \tag{A1.18}$$

where σ is the total cross section. Obviously, the damping of coherence is governed by the scattering rate in this limit. On the other hand, for small values of $|\boldsymbol{x} - \boldsymbol{x}'|$ we have

$$F(\boldsymbol{x} - \boldsymbol{x}') = \frac{1}{4} \int dq\, v(q) n(q) \int d\Omega\, d\Omega' \left((\boldsymbol{q} - \boldsymbol{q}')(\boldsymbol{x} - \boldsymbol{x}') \right)^2 |f(\boldsymbol{q}, \boldsymbol{q}')|^2 . \tag{A1.19}$$

Performing the angular integrations gives

$$F(\boldsymbol{x} - \boldsymbol{x}') = \int dq\, v(q) n(q) \frac{4\pi^2}{3} q^2 (\boldsymbol{x} - \boldsymbol{x}')^2 \int d\cos\theta\, (1 - \cos\theta) |f(\cos\theta)|^2 , \tag{A1.20}$$

so that for fixed momentum q and particle flux Nv/V interferences are damped according to

$$\rho(\boldsymbol{x}, \boldsymbol{x}', t) = \rho(\boldsymbol{x}, \boldsymbol{x}', 0) \exp\left\{ -\Lambda t (\boldsymbol{x} - \boldsymbol{x}')^2 \right\} \tag{A1.21}$$

with

$$\Lambda = \frac{q^2 N v \sigma_{\mathrm{eff}}}{V} \tag{A1.22}$$

and[1]

$$\sigma_{\mathrm{eff}} = \frac{4\pi^2}{3} \int d\cos\theta\, (1 - \cos\theta) |f(\cos\theta)|^2 . \tag{A1.23}$$

---

[1] The expressions (A1.22) and (A1.23) differ from the result originally given by Joos and Zeh (1985) in the angular dependence of the integrand in (A1.23). For a careful re-evaluation see K. Hornberger and J. E. Sipe, *Phys. Rev.* **A68**, 012105.

A2 Solutions for the Equation of Motion

E. Joos

In this appendix we construct a set of solutions[1] of the equations of motion (3.72) and (3.102) of a mass point under the influence of decoherence. A number of physically interesting quantities are then expressed in terms of these solutions.

A2.1 Gaussian Density Matrices

We seek solutions of the linear master equation

$$i\frac{\partial}{\partial t}\rho(x, x', t) = \left[\frac{1}{2m}\left(\frac{\partial^2}{\partial x'^2} - \frac{\partial^2}{\partial x^2}\right) + \frac{m\omega^2}{2}(x^2 - x'^2) - i\Lambda(x - x')^2\right.$$
$$\left. + i\frac{\gamma}{2}(x - x')\left(\frac{\partial}{\partial x'} - \frac{\partial}{\partial x}\right)\right]\rho(x, x', t) \tag{A2.1}$$

which describes the motion of a test particle under the influence of decoherence (characterized by the constant Λ) and friction (with friction constant γ). The particle is bound harmonically (with bare frequency ω); the case of a "free" particle (discussed in Sect. 3.2) is obtained as the limit $\omega \to 0$. There are several methods for solving an equation of this kind (see, for example, Gardiner 1983). In the following we use a rather elementary, straightforward technique. The structure of the equation of motion suggests a Gaussian ansatz of the form

$$\rho(x, x', t) = \exp - \left[A(t)(x - x')^2 + iB(t)(x - x')(x + x') + C(t)(x + x')^2\right.$$
$$\left. + iD(t)(x - x') + E(t)(x + x') + N(t)\right] \tag{A2.2}$$

where all time-dependent coefficients A, B,... are real if ρ is hermitean.

For the actual construction of solutions it is more convenient to work with another representation of the density matrix, used also by Unruh and Zurek (1989). Following these authors we will call it the "k, Δ"-representation. It is in fact the "characteristic function" associated with the Wigner function (see Hillery *et al.* 1984), and may be defined by

$$\rho(k, \Delta) = \text{tr}(\rho \exp i(k\hat{x} + \Delta\hat{p})) \tag{A2.3}$$

---

[1] For other methods to find solutions see Risken (1984), Roy and Venugopalan (1999), Ford and O'Connell (2001b).

(see also Savage and Walls 1985a). It is related to the position representation by

$$\rho(k, \Delta) = \int dx\, e^{ikx} \rho\left(x + \frac{\Delta}{2}, x - \frac{\Delta}{2}\right) \tag{A2.4}$$

and to the Wigner function $W(x, p)$ via a Fourier transformation,

$$W(x, p) = \left(\frac{1}{2\pi}\right)^2 \int dk \int d\Delta\, e^{-i(kx + \Delta p)} \rho(k, \Delta). \tag{A2.5}$$

In order to derive the equation of motion for $\rho(k, \Delta)$, we write

$$\rho(k, \Delta) = \mathrm{tr}(D\rho) \tag{A2.6}$$

with

$$D = e^{i(k\hat{x} + \Delta\hat{p})} = e^{ik\Delta/2} e^{ik\hat{x}} e^{i\Delta\hat{p}}. \tag{A2.7}$$

It is advantageous to derive some auxiliary relations involving the derivatives of D from the last equation, for example

$$D\hat{p} = -\left(i\frac{\partial}{\partial\Delta} + \frac{k}{2}\right) D, \qquad \hat{x}D = -\left(i\frac{\partial}{\partial k} + \frac{\Delta}{2}\right) D. \tag{A2.8}$$

From these identities one easily finds commutators and anticommutators, which are useful when one finally constructs the equation of motion in the k, Δ-representation by wrapping $\mathrm{tr}(D \ldots)$ over the equation (A2.1) written in operator form,

$$i\dot{\rho} = \frac{1}{2m}\left[p^2, \rho\right] + \frac{m\omega^2}{2}\left[x^2, \rho\right] - i\Lambda[x, [x, \rho]] + \frac{\gamma}{2}[x, \{p, \rho\}]. \tag{A2.9}$$

The resulting equation of motion for $\rho(k, \Delta)$ reads

$$\frac{\partial}{\partial t}\rho(k, \Delta, t) = \frac{k}{m}\frac{\partial}{\partial\Delta}\rho - m\omega^2\Delta\frac{\partial}{\partial k}\rho - \Lambda\Delta^2\rho - \gamma\Delta\frac{\partial}{\partial\Delta}\rho. \tag{A2.10}$$

Two points are noteworthy. First, the equation is real (although in general complex solutions are needed, as can be seen from (A2.4)). Second, the equation is only of first order, so it is much easier to construct solutions.[2] As before, a Gaussian ansatz is appropriate,

$$\rho(k, \Delta) = \exp -[c_1 k^2 + c_2 k\Delta + c_3 \Delta^2 + ic_4 k + ic_5 \Delta + c_6] \tag{A2.11}$$

which leads to simple linear equations for the coefficients (alternatively, one may use the method of characteristics, see Savage and Walls 1985b). The choice of the phases in (A2.11) is again motivated by the hermiticity of ρ. If

---

[2] The equation of motion for the Wigner function can easily be derived from (A2.5) and (A2.10). See (3.153) in the main text.

$\rho = \rho^\dagger$ then $\rho^*(k, \Delta) = \rho(-k, -\Delta)$, hence all coefficients are real. The first three coefficients are related by

$$\dot{c}_1 = \frac{c_2}{m}$$

$$\dot{c}_2 = \frac{2c_3}{m} - 2m\omega^2 c_1 - c_2\gamma$$

$$\dot{c}_3 = \Lambda - m\omega^2 c_2 - 2\gamma c_3 \qquad (A2.12)$$

while

$$\dot{c}_4 = \frac{c_5}{m}$$

$$\dot{c}_5 = -m\omega^2 c_4 - \gamma c_5 \qquad (A2.13)$$

and c_6 is simply a constant. These two sets of equations can be solved by standard methods, but the solutions look awkward for the general case. We present here only the solutions we use in the main text, i. e., for the case of a "free" particle ($\omega = 0$). In terms of initial values the coefficients c then read

$$c_1(t) = c_1(0) + c_2(0)\frac{\Gamma}{m\gamma} + c_3(0)\frac{\Gamma^2}{m^2\gamma^2} - \frac{\Lambda}{m^2\gamma^3}\left(\frac{1}{2}\Gamma^2 + \Gamma - \gamma t\right)$$

$$c_2(t) = c_2(0)e^{-\gamma t} + c_3(0)\frac{2\Gamma e^{-\gamma t}}{m\gamma} + \frac{\Lambda\Gamma^2}{m\gamma^2}$$

$$c_3(t) = \frac{\Lambda}{2\gamma} + \left(c_3(0) - \frac{\Lambda}{2\gamma}\right)e^{-2\gamma t}$$

$$c_4(t) = c_4(0) + c_5(0)\frac{\Gamma}{m\gamma}$$

$$c_5(t) = c_5(0)e^{-\gamma t}, \qquad (A2.14)$$

where we have used

$$\Gamma := 1 - e^{-\gamma t}. \qquad (A2.15)$$

For the important case of decoherence without friction ($\gamma = 0$) these expressions simplify to

$$c_1(t) = c_1(0) + c_2(0)\frac{t}{m} + c_3(0)\left(\frac{t}{m}\right)^2 + \frac{1}{3}\Lambda\frac{t^3}{m^2}$$

$$c_2(t) = c_2(0) + 2c_3(0)\frac{t}{m} + \Lambda\frac{t^2}{m}$$

$$c_3(t) = c_3(0) + \Lambda t$$

$$c_4(t) = c_4(0)$$

$$c_5(t) = c_5(0). \qquad (A2.16)$$

The last step needed is the transformation between the c's and the coefficients A, B,... in the position representation (A2.2). In terms of A, B,...

the c's are given by

$$c_1 = \frac{1}{16C} \qquad c_4 = \frac{E}{4C}$$

$$c_2 = -\frac{B}{4C} \qquad c_5 = D - \frac{BE}{2C} \qquad \text{(A2.17)}$$

$$c_3 = A + \frac{B^2}{4C} \qquad e^{-c_6} = \frac{1}{2}\sqrt{\frac{\pi}{C}} \exp\left(\frac{E^2}{4C} - N\right) = \text{tr}\rho.$$

The inverse relations read

$$A = c_3 - \frac{c_2^2}{4c_1} \qquad D = -\frac{c_2 c_4}{2c_1} + c_5$$

$$B = -\frac{c_2}{4c_1} \qquad E = \frac{c_4}{4c_1}. \qquad \text{(A2.18)}$$

$$C = \frac{1}{16c_1}$$

With the above set of equations it is now straightforward to construct Gaussian solutions of (A2.1) for arbitrary initial conditions. Since we will express all our physical quantities in the following Sect. A2.3 in terms of the coefficients A, B,... from our original ansatz, we give here for completeness also the equations of motion for these functions. They read

$$\dot{A} = \frac{4AB}{m} - 2\gamma A + \Lambda$$

$$\dot{B} = \frac{1}{m}\left(2B^2 - 8AC\right) - \gamma B + \frac{m}{2}\omega^2$$

$$\dot{C} = \frac{4BC}{m}$$

$$\dot{D} = \frac{2}{m}(BD - 2AE) - \gamma D$$

$$\dot{E} = \frac{2}{m}(2CD + BE). \qquad \text{(A2.19)}$$

Of course, these equations are equivalent to the set (A2.12, A2.13).

A2.2 Green Functions

The Gaussian solutions defined by the ansatz (A2.2) or (A2.11) are sufficiently general to allow the construction of a Green function solution of (A2.1). Using the results of the previous subsection it is now a straightforward task to find a solution G satisfying the initial condition

$$G(x, x', 0; x_0, x_0', 0) = \delta(x - x_0)\delta(x' - x_0'). \qquad \text{(A2.20)}$$

G is given by

$$G(x, x', t; x_0, x'_0, 0) = \frac{m\gamma}{2\pi|\Gamma|} \exp\left[\frac{im\gamma}{2\Gamma}\left\{x_0^2 - x_0'^2 + e^{-\gamma t}\left(x^2 - x'^2\right)\right.\right.$$

$$+ \left(1 + e^{-\gamma t}\right)\left(x'x_0' - xx_0\right) + \Gamma(xx_0' - x'x_0)\Big\}$$

$$+ \frac{\Lambda}{2\gamma\Gamma^2}\Big\{(x - x')^2\left(2\Gamma e^{-\gamma t} - \Gamma^2 - 2\gamma t e^{-2\gamma t}\right)$$

$$+ 2(x - x')(x_0 - x'_0)\left(e^{-2\gamma t} + 2\gamma t e^{-\gamma t} - 1\right)$$

$$+ (x_0 - x'_0)^2\left(3\Gamma - \Gamma e^{-\gamma t} - 2\gamma t\right)\Big\}\Bigg] \tag{A2.21}$$

(for a similar expression in terms of rotated momentum variables see Kumar 1984).

For the simpler case of negligible friction (A2.21) reduces to

$$G(x, x', t; x_0, x'_0, 0)\big|_{\gamma=0} = \frac{m}{2\pi|t|} \exp\left\{\frac{im}{2t}\left[(x - x_0)^2 - (x' - x'_0)^2\right]\right.$$

$$\left. - \frac{\Lambda t}{3}\left[(x - x')^2 + (x_0 - x'_0)^2 + (x - x')(x_0 - x'_0)\right]\right\}. \tag{A2.22}$$

A2.3 Some Derived Quantities

We now turn to the calculation of some interesting quantities in terms of the functions defined by the ansatz (A2.2). Among these are position and momentum distributions, the entropy, and the Wigner function.

The position probability distribution is given by

$$P(x) = \langle x\,|\rho|\,x\rangle$$

$$= 2\sqrt{\frac{C}{\pi}} \exp\left[\frac{-(4Cx + E)^2}{4C}\right], \tag{A2.23}$$

the first two moments are

$$\langle x\rangle = \frac{-E}{4C} \tag{A2.24}$$

and

$$\langle x^2\rangle = \frac{2C + E^2}{16C^2}, \tag{A2.25}$$

hence the variance in position is entirely given by the function C as

$$\Delta x^2 = \langle x^2\rangle - \langle x\rangle^2 = \frac{1}{8C}. \tag{A2.26}$$

The coherence length in the position representation is defined by the width in the $x - x'$-direction, hence

$$l_x = \frac{1}{\sqrt{8A}}. \tag{A2.27}$$

The momentum representation of the density matrix can be obtained as usual through a Fourier transformation in the form

$$
\begin{aligned}
\rho(p, p') &= \langle p \,|\rho|\, p' \rangle \\
&= \frac{1}{2\pi} \int dx\, dx'\, e^{-\mathrm{i}(px - p'x')} \rho(x, x') \\
&= \sqrt{\frac{C}{\pi\,(B^2 + 4AC)}} \exp \frac{-1}{4\,(B^2 + 4AC)} \left\{ \frac{(BE - 2CD)^2}{C} + A(p - p')^2 \right. \\
&\quad -\mathrm{i}B(p - p')(p + p') + C(p + p')^2 - 2\mathrm{i}(p - p')(BD + 2AE) \\
&\quad \left. -2(p + p')(BE - 2CD) \right\}.
\end{aligned}
\tag{A2.28}
$$

From this expression the momentum probability distribution can be obtained as

$$
\begin{aligned}
P(p) &= \langle p \,|\rho|\, p \rangle \\
&= \sqrt{\frac{C}{\pi(B^2 + 4AC)}} \exp \left[-\frac{(2Cp - BE + 2CD)^2}{4C(B^2 + 4AC)} \right],
\end{aligned}
\tag{A2.29}
$$

as can the coherence length in momentum (the width in $p - p'$),

$$l_p = \sqrt{\frac{B^2 + 4AC}{2A}}. \tag{A2.30}$$

Expectation values for momentum read

$$\langle p \rangle = \frac{BE}{2C} - D, \tag{A2.31}$$

$$\langle p^2 \rangle = \frac{1}{4C^2} \left[2C(B^2 + 4AC) + (BE - 2CD)^2 \right] \tag{A2.32}$$

and

$$\Delta p^2 = 2A + \frac{B^2}{2C}. \tag{A2.33}$$

The product of position and momentum uncertainties is then

$$\Delta x^2 \Delta p^2 = \frac{B^2 + 4AC}{16C^2}. \tag{A2.34}$$

The coefficient B equals zero for a pure state, as can be seen from the linear entropy which is given by

$$S_{\text{lin}} = \text{tr}(\rho - \rho^2)$$

$$= 1 - \sqrt{\frac{C}{A}}. \tag{A2.35}$$

For a pure state we have $A = C$ and $S_{\text{lin}} = 0$ (in general $0 \leq S_{\text{lin}} < 1$). We use this definition here instead of $S_{\text{lin}} = -\text{tr}\rho^2$, since then S_{lin} has the value zero for pure states, like the logarithmic von Neumann entropy. The time derivative of S_{lin} can be found by using (A2.19) as

$$\frac{d}{dt}S_{\text{lin}} = \frac{1}{2}\sqrt{\frac{C}{A^3}}(\Lambda - 2\gamma A). \tag{A2.36}$$

It is not always positive (see discussion in the main text).

The uncertainty relation is automatically fulfilled, since,

$$\Delta x^2 \Delta p^2 = \frac{B^2 + 4AC}{16C^2} \geq \frac{A}{4C} = \frac{1}{4}\frac{1}{(S_{\text{lin}} - 1)^2} \geq \frac{1}{4}. \tag{A2.37}$$

Not surprisingly, the uncertainty product is minimal for a pure Gaussian state $(A = C)$. The von Neumann entropy

$$S = -\text{tr}\rho \ln \rho \tag{A2.38}$$

can be calculated by diagonalizing the density matrix (A2.2). We quote here only the result given by Joos and Zeh (see also Unruh and Zurek). The entropy can be written in the form

$$S = -\frac{1}{p_0}(p_0 \ln p_0 + q \ln q) \tag{A2.39}$$

where

$$p_0 = \frac{2\sqrt{C}}{\sqrt{A} + \sqrt{C}}, \qquad q = \frac{\sqrt{A} - \sqrt{C}}{\sqrt{A} + \sqrt{C}}. \tag{A2.40}$$

The time derivative is, again with the help of (A2.19), given by

$$\frac{d}{dt}S = \frac{\Lambda - 2\gamma A}{4\sqrt{AC}}\ln\frac{\sqrt{A} + \sqrt{C}}{\sqrt{A} - \sqrt{C}}. \tag{A2.41}$$

In contrast to discrete states, where eigenstates of the density matrix finally coincide with pointer basis eigenstates, the eigenstates of a Gaussian density matrix such as (A2.2) are given by oscillator eigenfunctions. Their width is found to *increase* with time in this case, although the coherence length goes to zero (Joos 1986a).

As our last technical result we give an expression for the Wigner function in terms of our ansatz (A2.2). It reads

$$
\begin{aligned}
W(x,p) &= \frac{1}{\pi} \int dq\, e^{2ipq} \rho(x-q, x+q) \\
&= \frac{1}{\pi} \exp\left\{ -\left[x^2 \left(\frac{B^2}{A} + 4C \right) + \frac{B}{A}px + \frac{1}{4A}p^2 \right.\right. \\
&\qquad \left.\left. +x\left(\frac{BD}{A} + 2E \right) + \frac{D}{2A}p + \frac{D^2}{4A} + \frac{E^2}{4C} \right] \right\}.
\end{aligned} \qquad \text{(A2.42)}
$$

A3 Elementary Properties of Composite Systems in Quantum Mechanics

D. Giulini

This appendix presents self-contained derivations of well known basic properties of quantum systems under composition and separation. We partially follow (Jauch 1968, von Neumann 1932, and d'Espagnat 1976). For the sake of clarity we state the main results as theorems.

The Composite System Is in a Pure State

Let $\mathcal{S}_1$ and $\mathcal{S}_2$ denote two physical systems whose quantum mechanical states are represented by vectors in the Hilbert spaces $\mathcal{H}_1$ and $\mathcal{H}_2$ respectively.[1] The Hilbert space of the composite system, §, is then given by the tensor product $\mathcal{H} = \mathcal{H}_1 \otimes \mathcal{H}_2$ with the naturally inherited inner product. We shall denote all inner products simply by $\langle \cdot | \cdot \rangle$, since it is clear from the entries what Hilbert space is meant. Given two orthonormal bases $\{\varphi_i\}$ and $\{\psi_j\}$ of $\mathcal{H}_1$ and $\mathcal{H}_j$ respectively, a vector $g \in \mathcal{H}$ has a unique expansion

$$g = \sum_{i,j} \gamma_{ij} \varphi_i \otimes \psi_j . \tag{A3.1}$$

Given such a vector g in a tensor product space $\mathcal{H}_1 \otimes \mathcal{H}_2$ it does not define vectors in the individual factor spaces $\mathcal{H}_1$ and $\mathcal{H}_2$ unless it is a pure tensor product $g = \varphi \otimes \psi$. This is sometimes expressed by saying that no (pure) state can be assigned to the subsystems $\mathcal{S}_1$ and $\mathcal{S}_2$, given that the composite system is in a pure state described by g which is not a pure tensor product. However, what g does define are *relative states*: given g and a state of $\mathcal{S}_1$, there is a unique state associated to $\mathcal{S}_2$. In this sense a vector g should be considered as a map from the state space of $\mathcal{S}_1$ to the state space of $\mathcal{S}_2$. In fact, on the corresponding Hilbert spaces this map is easily seen to be given by an antilinear map, which we shall denote by the same symbol g (φ and ψ denote general elements in $\mathcal{H}_1$ and $\mathcal{H}_2$ respectively):

$$g : \mathcal{H}_1 \to \mathcal{H}_2 , \quad \varphi \mapsto g(\varphi) := \sum_j \langle \varphi \otimes \psi_j | g \rangle \, \psi_j = \sum_{i,j} \langle \varphi | \varphi_i \rangle \gamma_{ij} \, \psi_j , \tag{A3.2}$$

---

[1] For convenience, we shall sometimes call (pure) states by the vectors representing them. But when we say two states are different, we always mean that the corresponding rays are different.

with adjoint map

$$g^\dagger : \mathcal{H}_2 \to \mathcal{H}_1 , \quad \psi \mapsto g^\dagger(\psi) := \sum_i \langle \varphi_i \otimes \psi | g \rangle \, \varphi_i = \sum_{i,j} \langle \psi | \psi_j \rangle \gamma_{ij} \, \varphi_i . \quad \text{(A3.3)}$$

It is apparent that these maps are independent of the choice of orthonormal bases $\{\varphi_i\}$ and $\{\psi_j\}$. Note also that the adjoint of an antilinear map is defined via

$$\langle g(\varphi) | \psi \rangle = \langle g^\dagger(\psi) | \varphi \rangle , \quad \text{(A3.4)}$$

which contains an additional complex conjugation compared to the linear case. These maps generally do not preserve the inner products, i.e. they are not isometries.

In standard quantum mechanics, $g(\varphi)$ is the assigned state of $\mathcal{S}_2$, given that $\mathcal{S}$ had been prepared in the state g, and $\mathcal{S}_1$ been measured and found to be in state φ. For the general situation we shall however avoid calling $g(\varphi)$ the relative state of $\mathcal{S}_2$ for the state φ of $\mathcal{S}_1$ (and given state g of $\mathcal{S}$), since nothing prevents $g(\varphi)$ from being the zero-vector, which does not represent any state of $\mathcal{S}_2$. Relative states can only be assigned to states outside the kernels of the maps g and $g^\dagger$. On these it is well defined, since a vector outside the kernel of g is always mapped to a vector outside the kernel of $g^\dagger$ (as can be directly inferred from (A3.16) below). The allowed domains are not linear subspaces, since the sum of two elements outside the kernel of g may be inside it. Moreover, changing φ or g by an overall phase only changes $g(\varphi)$ by an overall phase. Note also that the relation of being relative states is generally non-symmetric, i.e. if $\psi \in \mathcal{H}_2$ represents the relative state to $\varphi \in \mathcal{H}_1$, then φ does not generally represent the relative state to ψ, for this is true if and only if φ is an eigenvector of the composite map $g^\dagger \circ g$. We shall come back to this point below.

Given another vector $h \in \mathcal{H}$ with expansion coefficients η_{ij}, the composite maps $h^\dagger \circ g : \mathcal{H}_1 \to \mathcal{H}_1$ and $h \circ g^\dagger : \mathcal{H}_2 \to \mathcal{H}_2$ can be written

$$h^\dagger \circ g(\varphi) = \sum_{i,j} \langle g | \varphi \otimes \psi_j \rangle \langle \varphi_i \otimes \psi_j | h \rangle \, \varphi_i = \sum_{i,j} (\eta \gamma^\dagger)_{ij} \langle \varphi_j | \varphi \rangle \, \varphi_i , \quad \text{(A3.5)}$$

$$h \circ g^\dagger(\psi) = \sum_{i,j} \langle g | \varphi_i \otimes \psi \rangle \langle \varphi_i \otimes \psi_j | h \rangle \, \psi_j = \sum_{i,j} (\gamma^\dagger \eta)_{ij} \langle \psi_i | \psi \rangle \, \psi_j . \quad \text{(A3.6)}$$

The inner product of h with g reads ($\text{trace}_{1,2}$ denote the traces in $\mathcal{H}_{1,2}$)

$$\langle g | h \rangle = \text{trace}(\gamma^\dagger \eta) = \text{trace}_1(h^\dagger \circ g) = \text{trace}_2(h \circ g^\dagger) . \quad \text{(A3.7)}$$

Operators A_1 and A_2 on $\mathcal{H}_1$ and $\mathcal{H}_2$ correspond to operators $\bar{A}_1 = A_1 \otimes \mathbb{1}$ and $\bar{A}_2 = \mathbb{1} \otimes A_2$ on $\mathcal{H}$; one has

$$\bar{A}_1 g = \sum_{i,j} \gamma_{ij} (A_1 \varphi_i) \otimes \psi_j . \quad \text{(A3.8)}$$

Viewing $\bar{A}_1 g \in \mathcal{H}$ as a map $\mathcal{H}_1 \to \mathcal{H}_2$, this just says that

$$(\bar{A}_1 g)(\varphi) = \sum_{i,j} \gamma_{ij} \langle A_1^\dagger \varphi | \varphi_i \rangle \psi_j \,,$$

which is equivalent to

$$\bar{A}_1 g = g \circ A_1^\dagger \quad \text{and} \quad (\bar{A}_1 g)^\dagger = A_1 \circ g^\dagger \,. \tag{A3.9}$$

In the same way one obtains

$$\bar{A}_2 g = A_2 \circ g \quad \text{and} \quad (\bar{A}_2 g)^\dagger = g^\dagger \circ A_2^\dagger \,. \tag{A3.10}$$

With these notations and formula (A3.7), we obtain expressions for the matrix elements in $\mathcal{H}$ of operators from $\mathcal{H}_1$ and $\mathcal{H}_2$:

$$\langle g | \bar{A}_1 g \rangle = \text{trace}_1(A_1(g^\dagger \circ g)) \,, \tag{A3.11}$$

$$\langle g | \bar{A}_2 g \rangle = \text{trace}_2(A_2(g \circ g^\dagger)) \,. \tag{A3.12}$$

This shows that the individual systems, considered in isolation, are maximally described by the density matrices

$$\rho_1 = g^\dagger \circ g \quad \text{for } \mathcal{S}_1 \,, \tag{A3.13}$$

$$\rho_2 = g \circ g^\dagger \quad \text{for } \mathcal{S}_2 \,. \tag{A3.14}$$

As for linear maps, the anti-linear map g satisfies (Ker: kernel, Im: image, $\perp$: orthogonal complement):

$$\text{Im}(g) = (\text{Ker}(g^\dagger))^\perp \,, \qquad \text{Im}(g^\dagger) = (\text{Ker}(g))^\perp \,, \tag{A3.15}$$

$$\text{Im}(g) \cap \text{Ker}(g^\dagger) = 0 \in \mathcal{H}_1 \,, \quad \text{Im}(g^\dagger) \cap \text{Ker}(g) = 0 \in \mathcal{H}_2 \,, \tag{A3.16}$$

$$\text{Ker}(g) = \text{Ker}(\rho_1) \,, \qquad \text{Ker}(g^\dagger) = \text{Ker}(\rho_2) \,. \tag{A3.17}$$

If g has finite dimensional image, it follows that (dim: dimension)

$$\dim \text{Im}(g) = \dim \text{Im}(g^\dagger) = \dim \text{Im}(\rho_1) = \dim \text{Im}(\rho_2) \,. \tag{A3.18}$$

To each $g \in \mathcal{H}$ is intrinsically associated the number $\dim \text{Im}(g)$ which we call the rank of g. It is equal to the ranks of ρ_1, ρ_2, and the matrix $\{\gamma_{ij}\}$. Let this rank be finite, say n, and $\{\varphi_1, \ldots, \varphi_n\}$ an orthonormal basis of $(\text{Ker}(g))^\perp$ made from the n eigenvectors of ρ_1 with non-vanishing eigenvalues λ_i^2 (positivity follows from (A3.13). The n vectors $\psi_i := \lambda_i^{-1} g(\varphi_i)$ then form an orthonormal basis of $\text{Im}(g)$, where each ψ_i is an eigenvector of ρ_2 with eigenvalue λ_i^2. So ρ_1 and ρ_2 have equal ranks and eigenvalues and we can write (partially using Dirac's "bra-ket" notation):

$$g = \sum_{i=1}^{n} \lambda_i \, \varphi_i \otimes \psi_i \,, \tag{A3.19}$$

$$\rho_1 = \sum_{i=1}^{n} \lambda_i^2 \, |\varphi_i\rangle\langle\varphi_i| \,, \tag{A3.20}$$

$$\rho_2 = \sum_{i=1}^{n} \lambda_i^2 \, |\psi_i\rangle\langle\psi_i| \,. \tag{A3.21}$$

In particular, we have

Theorem 1. *Given that $\mathcal{S}$ is in a pure state $g \in \mathcal{H}$, the following three statements are equivalent: (i) $\mathcal{S}_1$ is in a pure state, (ii) $\mathcal{S}_2$ is in a pure state, (iii) g has rank 1.*

Note the difference to classical systems, where a pure state of the composite system always reduces to pure states for the component systems. In quantum mechanics, the deviation of the rank of g from 1 directly measures the failure of the state of $\mathcal{S}$ to determine pure states for $\mathcal{S}_1$ and $\mathcal{S}_2$. The existence of rank > 1 states of $\mathcal{S}$ is a direct consequence of $\mathcal{H}_1 \otimes \mathcal{H}_2$ being the Hilbert space for $\mathcal{S}$, in which the rank $= 1$ states form a submanifold, but no linear subspace. In a sense (compare definition of tensor product), the tensor product just adds the rank > 1 states to make this a linear space. An example might help to elucidate this point: Let the $\mathcal{H}_i$'s be of finite dimensions n_i, so that the space of pure states is a manifold of $2(n_i - 1)$ real dimensions (the manifold is $\mathbb{C}P^{(n_i-1)}$). The manifold of pure states for $\mathcal{H} = \mathcal{H}_1 \otimes \mathcal{H}_2$ has $2(n_1 \cdot n_2 - 1)$ real dimensions, and not the sum $2(n_1 + n_2 - 2)$, as would be the case for classical systems with the corresponding phase space dimensions[2]. Quantum mechanically, the manifold of pure states for the composite system thus acquires $2(n_1 - 1)(n_2 - 1)$ extra dimensions as compared to a composite classical system.

Let us briefly come back to the concept of relative states. Let $\varphi \in \mathcal{H}_1$, and φ_K its orthogonal projection into $\text{Ker}(\rho_1)$. Then

$$\varphi = \sum_{i=1}^{n} a_i \varphi_i + \varphi_K \,, \tag{A3.22}$$

$$\psi = g(\varphi) = \sum_{i=1}^{n} \lambda_i a_i^* \psi_i \,, \tag{A3.23}$$

$$g^\dagger(\psi) = \rho_1(\varphi) = \sum_{i=1}^{n} \lambda_i^2 a_i \varphi_i \,, \tag{A3.24}$$

---

[2] The points of the phase space of a classical system correspond to pure states. They form a real manifold of dimension $= 2 \times$(number of degrees of freedom). A composite system has as phase space the euclidean product of the individual phase spaces, so that the dimensions simply add.

from which we infer the following

Theorem 2. *Let $\psi \in \mathcal{H}_2$ be a state of $\mathcal{S}_2$ which is relative to the state $\varphi \in \mathcal{H}_1$ of $\mathcal{S}_1$ under $g \in \mathcal{H}$. Then, conversely, φ is also relative to ψ, if and only if either of the following mutually equivalent conditions hold: (i) φ is an eigenvector of ρ_1 of non-zero eigenvalue, (ii) ψ is an eigenvector of ρ_2 of non-zero eigenvalue, (iii) there exists a Schmidt-expansion of g which contains a non-vanishing multiple of $\varphi \otimes \psi$.*

It follows that a continuum of *different* states, each of which is its own "relative-relative" state, exists, if and only if ρ_1 has degenerate non-zero eigenvalues. The states are then represented by all vectors in the corresponding linear eigen-subspace.

In a general orthonormal basis, $\{\varphi_i\}$, which does not necessarily consist of eigenvectors for ρ_1, we obtain, instead of (A3.24), for the "relative-relative" state of, say, φ_1 (w.l.o.g. we choose the first basis vector to be φ)

$$g^\dagger \circ g(\varphi_1) = \rho_1(\varphi_1) = \sum_{i=1}^{n} \langle \varphi_i | \rho_1 | \varphi_1 \rangle \, \varphi_i \,, \tag{A3.25}$$

showing that we only "approximately" come back to φ_1 if the off-diagonal entries in the first column of ρ_1's matrix are "sufficiently" small.

The Composite System is in a Mixed State

Let us now generalize the foregoing to the case where the system $\mathcal{S}$ is in a non-pure state. Calling the density matrix ρ, we have

$$\rho = \sum_M p_M^2 \, |\theta^M\rangle\langle\theta^M| \,. \tag{A3.26}$$

With respect to orthonormal bases $\{\varphi_i\}$ and $\{\psi_j\}$ we expand

$$|\theta^M\rangle = \sum_{M,i,j} \lambda_{ij}^M |\varphi_i\rangle|\psi_j\rangle = \sum_{M,i} |\varphi_i\rangle|\psi_i^M\rangle \,, \tag{A3.27}$$

$$\text{where} \quad |\psi_i^M\rangle := \theta^M(\varphi_i) = \sum_j \lambda_{ij}^M |\psi_j\rangle \,. \tag{A3.28}$$

We now further specialize $\{\varphi_i\}$ by the requirement that it should diagonalize the reduced density matrix ρ_1. Given $\{\varphi_i\}$ and $\{\theta^M\}$, the vectors ψ_i^M are then determined through (A3.28), and we have

$$\rho_1 = \text{trace}_2 \left\{ \sum_{M,i,j} p_M^2 \, |\varphi_i\rangle|\psi_i^M\rangle\langle\psi_j^M|\langle\varphi_j| \right\} = \sum_{M,i,j} p_M^2 \langle\psi_j^M|\psi_i^M\rangle \, |\varphi_i\rangle\langle\varphi_j|$$

$$= \sum_i \lambda_i^2 \, |\varphi_i\rangle\langle\varphi_i| \,,$$

$$\tag{A3.29}$$

so that

$$\sum_M p_M^2 \langle \psi_i^M | \psi_j^M \rangle = \lambda_i^2 \delta_{ij} \,. \qquad (A3.30)$$

As before, let us assume that ρ_1 has finite rank n, so that $\lambda_i \neq 0 \Leftrightarrow i \leq n$. Equation (A3.30) then implies that $\psi_i^M = 0$ if $p_M \neq 0$ and $i > n$, and we have

$$\rho = \sum_M p_M^2 \sum_{i,j=1}^{n} |\varphi_i\rangle |\psi_i^M\rangle \langle \psi_j^M| \langle \varphi_j| \,, \qquad (A3.31)$$

$$\rho_1 = \sum_{i=1}^{n} \left\{ \sum_M p_M^2 \|\psi_i^M\|^2 \right\} |\varphi_i\rangle \langle \varphi_i| \,, \qquad (A3.32)$$

$$\rho_2 = \sum_M \sum_{i=1}^{n} p_M^2 |\psi_i^M\rangle \langle \psi_i^M| \,. \qquad (A3.33)$$

If ρ_1 is pure, one has $n = 1$ and $\rho = \rho_1 \otimes \rho_2$, with $\rho_1 = |\varphi_1\rangle\langle\varphi_1|$ and $\rho_2 = \sum_M p_M^2 |\psi_1^M\rangle\langle\psi_1^M|$. This shows that we can complement Theorem 1 by

Theorem 3. *Let $\mathcal{S}_1$ be in a pure state. Then the density matrix of $\mathcal{S}$ is uniquely determined by the reduced density matrices ρ_1 and ρ_2: $\rho = \rho_1 \otimes \rho_2$. ρ is pure if and only if ρ_2 is pure.*

Conversely, we can show that purity of ρ_1 (or ρ_2) is the only situation where ρ is uniquely determined by ρ_1 and ρ_2. Let us therefore assume that ρ_1 and ρ_2 are genuinely mixed, in which case each of them can be written as a non-trivial convex combination ($a, b, c, d \in \mathbb{R}_+$):

$$\rho_1 = a\rho_{11} + b\rho_{12}\,, \quad \rho_2 = c\rho_{21} + d\rho_{22}\,, \qquad (A3.34)$$
$$\text{with} \quad \rho_{11} \neq \rho_{12} \quad \rho_{21} \neq \rho_{22}\,, \qquad (A3.35)$$
$$\text{and} \quad a + b = c + d = 1\,, \qquad (A3.36)$$

where all the ρ_{ij}'s have unit trace. We then consider the general combination

$$\rho = \alpha\rho_{11} \otimes \rho_{21} + \beta\rho_{11} \otimes \rho_{22} + \gamma\rho_{12} \otimes \rho_{21} + \delta\rho_{12} \otimes \rho_{22}\,, \qquad (A3.37)$$

and require it to produce the reduced density matrices ρ_1 and ρ_2. This is equivalent to

$$\text{trace}_1(\rho) = (\alpha + \gamma)\rho_{21} + (\beta + \delta)\rho_{22} = \rho_2 = c\rho_{21} + d\rho_{22}\,, \qquad (A3.38)$$
$$\text{trace}_2(\rho) = (\alpha + \beta)\rho_{11} + (\gamma + \delta)\rho_{12} = \rho_1 = a\rho_{11} + b\rho_{12}\,, \qquad (A3.39)$$

which, by (A3.35), is equivalent to

$$\alpha + \gamma = c\,, \quad \beta + \delta = d = 1 - c\,, \qquad (A3.40)$$
$$\alpha + \beta = a\,, \quad \gamma + \delta = b = 1 - a\,. \qquad (A3.41)$$

For given a and c, we can thus choose any α and β such that $\alpha + \beta = a$ and then determine γ and δ through $\gamma = c - \alpha$, $\delta = 1 - c - \beta$. In this way we have found a one-parameter family of different ρ's with the same reduced ρ_1 and ρ_2, in case that neither of the latter is pure. Taken together with Theorem 3, we have

Theorem 4. *Let ρ_1 and ρ_2 be the reduced density matrices of ρ. Then ρ is uniquely determined by ρ_1 and ρ_2, if and only if at least one of the reduced density matrices is pure, in which case ρ is of the form $\rho = \rho_1 \otimes \rho_2$.*

Clearly, if the total density matrix ρ decomposes in the form $\rho_1 \otimes \rho_2$, the two systems $\mathcal{S}_1$ and $\mathcal{S}_2$ are uncorrelated. Theorem 4 tells us, in particular, that two systems are always uncorrelated if one of them has a pure reduced density matrix. Let us now be more precise about correlations and ask: given ρ, what is the characterization of the observables A of $\mathcal{S}_1$ which are uncorrelated to *all* observables of $\mathcal{S}_2$? To answer this, let us first recall the definition of a correlation coefficient. If we denote the expectation value by an overbar, the correlation coefficient of two observables A and B, denoted by $\Delta(A, B)$, is defined through

$$\overline{AB} - \overline{A}\,\overline{B} = \overline{(A - \overline{A})(B - \overline{B})} =: \sqrt{\overline{A^2}\,\overline{B^2}}\,\Delta(A, B)\,. \tag{A3.42}$$

Using (A3.31-A3.33) we can evaluate the expressions on the left side, and obtain

$$\overline{AB} = \mathrm{trace}(\rho AB) = \sum_{i,j=1}^{n} \langle \varphi_i | A | \varphi_j \rangle \sum_M p_M^2 \langle \psi_i^M | B | \psi_j^M \rangle\,, \tag{A3.43}$$

$$\overline{A} = \mathrm{trace}_1(\rho_1 A) = \sum_{i=1}^{n} \lambda_i^2 \langle \varphi_i | A | \varphi_i \rangle\,, \quad \lambda_i^2 = \sum_M p_M^2 \|\psi_i^M\|^2\,, \tag{A3.44}$$

$$\overline{B} = \mathrm{trace}_2(\rho_2 B) = \sum_{i=1}^{n} \sum_M p_M^2 \langle \psi_i^M | B | \psi_i^M \rangle\,. \tag{A3.45}$$

A vanishing correlation function $\Delta(A, B)$ is equivalent to (A3.43) being the product of (A3.44) and (A3.45). If, for abbreviation, we introduce the complex hermitean $n \times n$-matrices, $A_{ij} := \langle \varphi_i | A | \varphi_j \rangle$ and $B_{ij} := \sum_M p_M^2 \times \langle \psi_i^M | \psi_j^M \rangle$, the condition for vanishing correlation $\Delta(A, B)$ is equivalent to

$$\sum_{i,j=1}^{n} \left\{ A_{ij} - \left(\sum_{k=1}^{n} \lambda_k^2 A_{kk} \right) \delta_{ij} \right\} B_{ij} = 0\,. \tag{A3.46}$$

A sufficient condition for A to have vanishing correlations with *all* B is that the expression in curly brackets vanishes:

$$A_{ij} = \delta_{ij} \left(\sum_{k=1}^{n} \lambda_k^2 A_{kk} \right) \Leftrightarrow A_{ij} = c\delta_{ij}\,, \quad \text{for some } c \in \mathbb{R}\,. \tag{A3.47}$$

This condition is also necessary in the *generic* case, that is, if the matrices B_{ij} span the whole space of complex hermitean $n \times n$-matrices if B runs through all observables of $\mathcal{S}_2$. In words, (A3.47) states that the operator A, restricted to $Im(\rho_1)$, is a multiple of the identity. If we write $\sum_i |\varphi_i\rangle\langle\varphi_i| = P$ for the projection operator onto the image of ρ_1 in $\mathcal{H}_1$, (A3.47) can thus be expressed in the more geometric form:

$$PAP = cP\,, \tag{A3.48}$$

which leads us to

Theorem 5. *Let $(\mathcal{S}_1, \rho_1)$ and $(\mathcal{S}_2, \rho_2)$ be subsystems of $(\mathcal{S}, \rho)$, where ρ_1 has finite dimensional image with projector P. Then, in the generic case, an observable A of $\mathcal{S}_1$ is uncorrelated with all observables in §2, if and only if*

$$PAP = cP \tag{A3.49}$$

for some real number c. In degenerate cases this condition is always sufficient but might not be necessary.

Let us finally note that the finite-rank condition on ρ_1 is not essential for many arguments, and may be relaxed.

A4 Quantum Correlations

I.-O. Stamatescu

We shall discuss here in some detail the problems involving observations in correlated systems. This appendix is only of didactic character and can be skipped by readers familiar with the matter. We shall again consider the special case of two systems, S_1 and S_2, which after having interacted are left in a total state which by the time of observation may be written as

$$|\Psi\rangle = \sum_{m,n} c_{mn} |\chi_m^{S_2}\rangle |\varphi_n^{S_1}\rangle, \qquad (A4.1)$$

the two systems having by now no interactions with one another (in the following we shall suppress the indexing S_1, S_2). The two bases in (A4.1) are taken to be orthonormal and

$$\sum_{m,n} |c_{mn}|^2 = 1. \qquad (A4.2)$$

We apply now the standard postulates of quantum mechanics and imagine submitting S_1 to a complete measurement which would leave it, say in the state $|\varphi_0\rangle$ (we assume for simplicity that the $\{|\varphi_m\rangle\}$ vectors in (A4.1) were chosen as *eigenvectors* of the prospective measurement operators). The combined system will then undergo a change of state:

$$|\Psi\rangle \rightarrow |\Psi_0\rangle = C \left(\sum_n c_{n0} |\chi_n\rangle \right) |\varphi_0\rangle \qquad (A4.3)$$

with C a normalization factor. As a result the two subsystems now each acquire a wave function, i.e. they will be found (bar later interactions) in pure states: $|\varphi_0\rangle$ for S_1 and

$$|\chi^{(0)}\rangle = C \sum_m c_{m0} |\chi_m\rangle \qquad (A4.4)$$

for S_2.

If we want to ask about the probability of obtaining the result $n = 0$ in the measurement we just performed we must take the norm of the coefficient of $|\varphi_0\rangle$ in (A4.1), i.e.,

$$\sum_{m',m} c_{m'0}^* c_{m0} \langle \chi_{m'} | \chi_m \rangle = \sum_m |c_{m0}|^2. \qquad (A4.5)$$

More generally we can describe the behavior of S_1 to all possible measurements by building its "reduced density matrix"

$$\rho = \text{tr}_{S_2} |\Psi\rangle\langle\Psi| = \sum_{n',n} \left(\sum_m c_{mn'} c_{mn}^* \right) |\varphi_{n'}\rangle\langle\varphi_n| \qquad (A4.6)$$

which in the chosen basis $|\varphi_n\rangle$ is:

$$\rho_{n'n} = \sum_m \langle\chi_m|\Psi\rangle\langle\Psi|\chi_m\rangle = \sum_m c_{mn'} c_{mn}^*. \qquad (A4.7)$$

Then the probability of finding S_1 in some state $|\varphi\rangle$ in a local measurement upon it is given by

$$w_\varphi = \langle\varphi|\rho|\varphi\rangle = \sum_{n',n,m} c_{m,n'} c_{m,n}^* \langle\varphi|\varphi_{n'}\rangle\langle\varphi_n|\varphi\rangle. \qquad (A4.8)$$

Similarly for S_2 we can build

$$\sigma = \text{tr}_{S_1} |\Psi\rangle\langle\Psi| = \sum_{m',m} \left(\sum_n c_{m'n} c_{mn}^* \right) |\chi_{m'}\rangle\langle\chi_m| \qquad (A4.9)$$

and again:

$$\sigma_{m'm} = \sum_n c_{m'n} c_{mn}^*. \qquad (A4.10)$$

For combined measurements the probabilities can be obtained from the full density matrix:

$$\rho^{\text{tot.}} = |\Psi\rangle\langle\Psi|, \quad \rho^{\text{tot.}}_{(m'n'),(mn)} = c_{m'n'} c_{mn}^* \qquad (A4.11)$$

Clearly in general ρ and σ together contain less information than $\rho^{\text{tot.}}$.

To understand better the meaning of these procedures let us consider, in the frame of standard quantum mechanics, the following problems (implying correspondingly designed experiments):

a) Observe $|\varphi\rangle$ upon S_1, no observation concerning S_2:
  – corresponding projector: $|\varphi\rangle\langle\varphi|$;
  – probability for such observations:

$$\langle\Psi|\varphi\rangle\langle\varphi|\Psi\rangle = \sum_{n',n,m} c_{mn'} c_{mn}^* \langle\varphi|\varphi_{n'}\rangle\langle\varphi_n|\varphi\rangle \qquad (A4.12)$$

$$= \langle\varphi|\rho|\varphi\rangle = w_\varphi; \qquad (A4.13)$$

  – posterior state: $S \equiv S_1 + S_2$ is left in the state (C is a normalization factor)

$$|\Psi_\varphi\rangle = C \sum_m \left(\sum_n c_{mn} \langle\varphi|\varphi_n\rangle \right) |\chi_m\rangle|\varphi\rangle. \qquad (A4.14)$$

b) Observe $|\varphi\rangle$ upon $\mathcal{S}_1$ *and* $|\chi_m\rangle$ upon $\mathcal{S}_2$ (fixed m):
  – corresponding projector: $|\chi_m\rangle|\varphi\rangle\langle\varphi|\langle\chi_m|$;
  – probability for such observations:

$$\langle\Psi|\varphi\rangle|\chi_m\rangle\langle\chi_m|\langle\varphi|\Psi\rangle = \sum_{k,n} c_{mk}c_{mn}^* \langle\varphi|\varphi_k\rangle\langle\varphi_n|\varphi\rangle = w_{\chi_m,\varphi}; \qquad (A4.15)$$

  – posterior state: $\mathcal{S}$ is left in the state

$$|\Psi_{\varphi,\chi_m}\rangle = C' \left(\sum_n c_{mn}\langle\varphi|\varphi_n\rangle \right) |\chi_m\rangle|\varphi\rangle. \qquad (A4.16)$$

c) Observe $|\varphi\rangle$ upon $\mathcal{S}_1$ *and* $|\chi_m\rangle$ upon $\mathcal{S}_2$ but accept all m (i.e., do not *look*):
  – corresponding projector: $\sum_m |\chi_m\rangle|\varphi\rangle\langle\varphi|\langle\chi_m| = |\varphi\rangle\langle\varphi|$;
  – probability for such observations:

$$\sum_m \langle\Psi|\varphi\rangle|\chi_m\rangle\langle\chi_m|\langle\varphi|\Psi\rangle = \sum_m w_{\chi_m,\varphi} = w_\varphi; \qquad (A4.17)$$

  – posterior state: $\mathcal{S}$ is left each time in any one of the states $|\Psi_{\chi_m,\varphi}\rangle$ (A4.16) with probability $w_{\chi_m,\varphi}$ (A4.15); we can describe this situation by a *mixture*.

Cases a) and c) are seen to differ concerning the situation in which $\mathcal{S}$ is left, while being indistinguishable from the point of view of observations upon $\mathcal{S}_1$. In this sense the building of the reduced density matrix means renouncing the extraction of further information from observations on $\mathcal{S}_2$, i.e., in repeating the experiments and collecting results of separate measurements upon $\mathcal{S}_1$ to check the statistics (A4.13), (A4.17) we do not try to classify these results according to results of possible, separate measurements upon $\mathcal{S}_2$ – and *this*, of course, cannot depend upon whether such latter measurements were actually performed or not, or on when. On the other hand this also illustrates the irreducibility of quantum correlations to missing information: the difference between case a) and c) can always be put into evidence by subsequent experiments involving the state of the full system, $\mathcal{S}$. Finally, were we interested in a correlated statistics of measurements upon $\mathcal{S}_1$ *and* $\mathcal{S}_2$, we might sample the combined "events" accordingly (see, e.g., case b) above). Further elementary examples can be found in textbooks - see, e.g., Schwabl (1995).

The non-classical character of the correlation in the expectations concerning correlated measurements on two subsystems which do not possess states of their own, i.e., if it is not possible to rewrite (A4.1) as a product of two factors, can be quantitatively exhibited in corresponding experiments (EPR). Assume we measure the properties A, A' on system $\mathcal{S}_1$ *and* B, B' on $\mathcal{S}_2$, that is, we use the observables (Hermitean operators) $\{O\} = \{A\otimes B,\ A'\otimes B,\ \cdots\}$ and construct the quantity:

$$\Delta(A, A'; B, B') \equiv |\mathcal{E}(AB) - \mathcal{E}(AB')| + |\mathcal{E}(A'B) - \mathcal{E}(A'B')| \qquad (A4.18)$$

where $\mathcal{E}$ denote the corresponding expectations in the given state of the total system:

$$\mathcal{E}(O) = \mathrm{tr}(\rho^{\mathrm{tot}} \cdot O) = \langle \Psi | O | \Psi \rangle. \tag{A4.19}$$

Then we have (we choose $\| O \| \leq 1$):

$$\Delta(A, A'; B, B') = |\langle \Psi | A(B - B') | \Psi \rangle| + |\langle \Psi | A'(B + B') | \Psi \rangle| \tag{A4.20}$$
$$= |\langle A\Psi |(B - B')\Psi \rangle| + |\langle A'\Psi |(B + B')\Psi \rangle|$$
$$\leq \| A\Psi \| . \| (B - B')\Psi \| + \| A'\Psi \| . \| (B + B')\Psi \|$$
$$\leq \| (B - B')\Psi \| + \| (B + B')\Psi \|$$
$$\leq \sqrt{2[\| (B - B')\Psi \|^2 + \| (B + B')\Psi \|^2]}$$
$$= \sqrt{4[\| B\Psi \|^2 + \| B'\Psi \|^2]} \leq 2\sqrt{2}. \tag{A4.21}$$

If we had been dealing with a classical problem, that is the expectations were taken with respect to a classical ensemble:

$$\mathcal{E}_c(O) = \int O d\mu \tag{A4.22}$$

with $d\mu$ a (positive semidefinite) probability measure and $\{O\}$ *real* valued functions (assumed to be less than 1 in absolute value) we would have had instead:

$$\Delta_c(A, A'; B, B') = |\mathcal{E}_c(A(B - B'))| + |\mathcal{E}_c(A'(B + B'))| \tag{A4.23}$$
$$\leq \mathcal{E}_c(|A|.|B - B'|) + \mathcal{E}_c(|A'|.|B + B'|) \leq \mathcal{E}_c(|B - B'|) + \mathcal{E}_c(|B + B'|)$$
$$= \mathcal{E}_c(|B - B'| + |B + B'|) \leq 2 \tag{A4.24}$$

since the general inequality:

$$\| a \| + \| b \| \leq \sqrt{2(\| a \|^2 + \| b \|^2)} \tag{A4.25}$$

which was used at the fourth step in (A4.20-A4.21) can be replaced by the equality:

$$|a| + |b| = |a + b.\mathrm{sgn}(ab)| \tag{A4.26}$$

if a, b are *real numbers*, used at the third step in (A4.23-A4.24). Notice that the inequality(A4.21) can be saturated if B, B' (A, A') do not commute and the subsystems are nontrivially correlated, i.e., $|\Psi\rangle$ does not factorize and the subsystems are not in pure states. This can be easily understood from the convexity property of the "space" of density operators which is nontrivial in this case but leads to the "degenerate" relation (A4.24) otherwise. Notice that (A4.24) would also hold if our quantum mechanical problem were reducible to a classical one (hidden variables). These are the well known *Bell's inequalities* and the experimental evidence to date seems to violate the bound (A4.24) and to support (A4.21).

A5 Hamiltonian Formulation
of Quantum Mechanics

D. Giulini

The Schrödinger equation is a deterministic equation for the quantum mechanical state. It can be cast into the form of a Hamiltonian system (familiar from analytical mechanics) on the space of pure states (rays). Quantum states (pure or mixed) then formally become distribution functions This gives, to some extent, an analogy to classical statistical mechanics, where mixed stated are also given by distributions over phase space (the space of pure states). This analogy in itself is quite interesting and has, moreover, been successfully employed to compute the dynamics of open quantum systems when viewed as stochastic process on the space of rays. However, we know that there are essential differences between quantum mechanics and a classical statistical system, which should also show up in this approach and hence render the envisaged analogy incomplete. We think that it is of interest to point out *where* this analogy fails, or, if you like, at what price the analogy can be uphold.

To avoid technical complications, which are not our concern here, we restrict our discussion to finite-dimensional Hilbert spaces. More details can be found in (Cirelli, Manià, and Pizzocchero 1990) and (Gibbons 1992).

Let $\mathcal{H}$ be a finite dimensional Hilbert space of dimension n and orthonormal basis $\{|\psi_i\rangle\}$. Every state has an expansion in the form $|\psi\rangle = z_i|\psi_i\rangle$ with $\bar{z}_i z_i = 1$.[1] We fix this basis and consider the expansion coefficients $\{z_i\}$, taken modulo an overall factor, as projective coordinates on the space of rays in $\mathcal{H}$. We shall frequently denote a ray by its projective coordinates. The space of rays is our state space $\mathcal{S}$ and thus diffeomorphic to the complex projective space $\mathbb{C}P^{n-1} = \mathcal{S}$ which is a complex manifold of $2(n-1)$ real dimensions. The Schrödinger equation is

$$i\hbar\partial_t|\psi\rangle = \hat{H}|\psi\rangle \tag{A5.1}$$

$$\Leftrightarrow \quad \dot{z}_i = -\tfrac{i}{\hbar}H_{ij}z_j \quad \text{with} \quad H_{ij} = \langle\psi_i|\hat{H}|\psi_j\rangle, \tag{A5.2}$$

where the second equation is easily seen to be a Hamiltonian equation with Hamiltonian function H, Hamiltonian vector field X_H and symplectic structure ω, given by (complex-conjugation is denoted by an overbar in this sec-

---

[1] In this appendix we employ the summation convention according to which an automatic summation over repeated indices is understood unless an explicit summation symbol appears. Also, operators carry hats and an overbar denotes complex conjugation.

tion):

$$H \; = \bar{z}_i H_{ij} z_j = \bar{z}_i \langle \psi_i | \hat{H} | \psi_j \rangle z_j \,, \tag{A5.3}$$

$$X_H = -\tfrac{i}{\hbar} H_{ij} z_j \partial_{z_i} + \tfrac{i}{\hbar} \bar{z}_i H_{ij} \partial_{\bar{z}_j} \,, \tag{A5.4}$$

$$\omega \; = -i\hbar \, d\bar{z}_i \wedge dz_i \,, \tag{A5.5}$$

so that $dH = \omega(X_H, \cdot)$. From (A5.5) it is immediately seen that unitary transformations in $\mathcal{H}$ define canonical transformations on $\mathcal{S}$. Generally, every hermitean operator $\hat{F}$ corresponds to a real function F on $\mathcal{S}$ as described by (A5.3), as does any state, mixed or pure, by applying (A5.3) to the corresponding density matrix. In particular, positive hermitean operators correspond to positive functions.

The symplectic structure (A5.5) defines a Poisson bracket for functions on $\mathcal{S}$:

$$\begin{aligned}
\{F, G\} &:= X_G(F) = -\tfrac{i}{\hbar} G_{ij} \left(z_j \partial_{z_i} - \bar{z}_i \partial_{\bar{z}_j} \right) F \\
&= -\tfrac{i}{\hbar} (\partial_{z_i} F \, \partial_{\bar{z}_i} G - \partial_{z_i} G \, \partial_{\bar{z}_i} F).
\end{aligned} \tag{A5.6}$$

Note that for real functions F of the form (A5.3), the vector fields X_F (compare (A5.4)) are tangential to the spheres $\bar{z}_i z_i = $ const. From (A5.6) it follows that the function corresponding to the commutator of two operators $\hat{F}$ and $\hat{G}$ is, up to the usual factor, just the Poisson-bracket of the corresponding functions:

$$\bar{z}_i [\hat{F}, \hat{G}]_{ij} z_j = i\hbar \{F, G\} \,. \tag{A5.7}$$

The space of functions corresponding to observables, i.e. of the form (A5.3), forms a real n^2-dimensional Lie algebra (with Lie multiplication being given by the Poisson bracket), but not in addition an (associative) algebra with respect to just pointwise multiplication, since the product of two functions corresponding to observables is never such a function. The function corresponding to a rank-one projector onto the ray $\{w_i\}$ is given by $W(z) = \sum_i |\bar{w}_i z_i|^2$ with values in $[0, 1]$. It takes the value 0 on all rays perpendicular to $\{w_i\}$ and assumes 1 at the unique point $\{w_i\}$. Any point $\{w_i\}$ can thus be uniquely characterized as the point where W assumes its maximum value one. Since $\mathcal{S}$ has $2(n-1)$ real dimensions, the maximal number of non-constant (Poisson-) commuting functions is $n-1$. A maximal set is e.g. given by the functions $W^{(a)}$, where $\{w_i^{(a)}\}$, $1 \leq a \leq n-1$, denote mutually orthogonal rays. The function for the remaining ray, orthogonal to the previous ones, is not independent since the sum of n mutually orthogonal rank one projectors must equal the identity. Functions of the form (A5.3) suffice to define a coordinate system on $\mathcal{S}$, but clearly they cannot all mutually commute.

Since $\bar{z}_i z_i = 1$ we can split-off the trace as a constant: $\bar{z}_i H_{ij} z_j = \bar{z}_i (H_{ij} - \text{trace}(\hat{H}) \tfrac{1}{n} \delta_{ij}) z_j + \tfrac{1}{n} \text{trace}(\hat{H})$. It is not difficult to show that the trace-free part is an eigenfunction of the Laplace operator on $\mathcal{S}$ corresponding to the first non-zero eigenvalue (which is $4n$). Hence functions which correspond to quantum mechanical observables or states are either constant or eigenfunctions of the Laplace operator with lowest non-zero eigenvalue $4n$. These are strikingly fewer observables than one would expect a classical system with the same state space to have. This implies a major difference to any classical Hamiltonian system, as will be seen below.

Equation (A5.3) says that the expectation value of an operator $\hat{F}$ in a pure state given by the ray $\{z_i\}$ is just the value $F(\{z_i\})$. More generally, for the general mixed state $\hat{\rho} = \sum_a |\psi^{(a)}\rangle p_a \langle \psi^{(a)}|$, where $\psi^{(a)} = w_i^{(a)} \psi_i$, the expectation value $\mathrm{trace}(\hat{\rho}\hat{F})$ is just given by the weighted sum of values: $\sum_a p_a F(\{w^{(a)}\})$. The quantum mechanical expectation values in the state $\hat{\rho}$ can thus be written as integrals over the corresponding functions on $\mathcal{S}$, weighted by the singular distribution function

$$P(\hat{\rho}) = \sum_a p_a \delta_{\{w_i^{(a)}\}} , \qquad (A5.8)$$

where $\delta_{\{w_i^{(a)}\}}$ denotes the Dirac measure at $\{w_i^{(a)}\}$. But note that the assignment $\hat{\rho} \to P(\hat{\rho})$ is highly non-unique. Any other decomposition $\hat{\rho} = \sum_a |\phi^{(a)}\rangle q_a \langle \phi^{(a)}|$, gives rise to other delta-distributions which in general differ in number and location from the previous ones. Note that we did not require the rays $\{w_i^{(a)}\}$ to be mutually orthogonal. For a general classification of mixtures that make up the same density matrix we refer to (Hughston, Jozsa, and Wootters 1993). Singular distribution functions of the kind (A5.8) have recently been used in attempts to formulate the dynamics of open quantum systems as stochastic processes on the space of rays (Breuer and Petruccione 1995; see also Chap. 5 of Breuer and Petruccione 2002a).

It is also possible to find smooth distribution functions. For example, it is not hard to prove the following identity (Gibbons 1992):

$$\mathrm{trace}(\hat{\rho}\hat{F}) = \frac{(n+1)!}{\pi^{n-1}} \int_{\mathcal{S}} \left(\rho - \frac{1}{n+1}\right) F \, d\mu , \qquad (A5.9)$$

where $d\mu$ denotes the (Fubini-Study) measure on $\mathcal{S} = \mathbb{C}P^{n-1}$ and ρ is just the function corresponding to $\hat{\rho}$ according to (A5.3). We thus have a smooth distribution $P(\hat{\rho}) = \frac{(n+1)!}{\pi^{n-1}}(\rho - \frac{1}{n+1})$. However, it is not positive! For the pure state $\{w_i\}$ it equals $\sum_i |\bar{w}_i z_i|^2 - \frac{1}{n+1}$, which is negative for all points $\{z_i\}$ sufficiently perpendicular to $\{w_i\}$. In particular, it only vanishes on a set of measure zero.

A general observation is that for any given density matrix $\hat{\rho}$ there is an infinite dimensional space of real valued functions $P(\hat{\rho})$ on $\mathcal{S}$ such that

$$\mathrm{trace}(\hat{\rho}\hat{F}) = \frac{(n+1)!}{\pi^{n-1}} \int_{\mathcal{S}} P(\hat{\rho}) F \, d\mu , \quad \forall \hat{F} . \qquad (A5.10)$$

The point is that the functions F against which $P(\hat{\rho})$ is integrated are only taken from the n^2-dimensional space of functions of the form (A5.3). There clearly exist infinitely many linearly independent functions on $\mathcal{S}$ which, when integrated against any function of the form (A5.3), yield zero, so that any combination of them may be added to $P(\hat{\rho})$ without disturbing (A5.10). In fact, as already mentioned, the space of functions (A5.3) is spanned by eigenfunctions of the Laplacian on $\mathcal{S}$ for the lowest two eigenvalues (including eigenvalue zero, i.e. the constant functions). We can thus add all combinations of functions taken from the eigenspaces of the higher eigenvalues.

It is natural to ask whether this freedom can be used to redefine $P(\hat{\rho})$ for each $\hat{\rho}$ in such a way that it becomes a non-negative smooth function, at least for mixed $\hat{\rho}$. One may further suspect that by doing this it is indeed possible to smooth out the singular distribution (A5.8) without losing positivity. Let us illustrate the affirmative answer to this question on the simple special case $n = 2$, i.e. a spin-$\frac{1}{2}$ system, where the state space is a two-sphere, $\mathbb{CP}^1 \cong S^2$, which we take to have unit radius. Here, a general density matrix is of the form $\hat{\rho} = \frac{1}{2}(\mathbb{1} + \boldsymbol{a} \cdot \sigma)$, $\|\boldsymbol{a}\| \leq 1$, which, according to (A5.3), defines the function on the two-sphere whose points we label by the unit vector $\boldsymbol{n}$: $\rho(\boldsymbol{n}) = \frac{1}{2}(1 + \boldsymbol{a} \cdot \boldsymbol{n})$. The distribution function in (A5.9) is then given by $P(\hat{\rho})(\boldsymbol{n}) = \frac{1}{4\pi}(1 + 3\boldsymbol{a} \cdot \boldsymbol{n})$. For the moment it is illustrative to think of it as distribution of electric charge. Its total charge is 1 and the dipole moment is just given by $\boldsymbol{a}$. These are fixed by the requirement (A5.10) whereas the remaining freedom resides precisely in the higher multipole moments. Since we can approximate any function by a multipole expansion, our question boils down to whether we can realize the given total charge and dipole moment with a distribution on S^2 of positive charge only. But it is clear that this is the case if and only if the dipole moment does not exceed the value 1, corresponding to all of the charge sitting at one point. This is precisely the condition $\|\boldsymbol{a}\| \leq 1$. So for $n = 2$ it is clear that we can always find a positive $P(\hat{\rho})$ satisfying (A5.10) which can be chosen to be smooth for mixed states. For $\|\boldsymbol{a}\| = 1$, i.e. for pure $\hat{\rho}$, we only have the unique singular δ-distributions. This shows that except for pure $\hat{\rho}$ it remains true that $P(\hat{\rho})$ is highly undetermined by (A5.10), even when it is required to be positive and smooth.

Such attempts to formally express quantum-mechanical expectation values as weighted integrals on complex-projective space suggest a stochastic theory in which a (positive) function $P(\hat{\rho})$ serves as probability distribution and all functions of the form (A5.3) as random variables. Taking this picture seriously would imply that the probability space[2] would be given by

---

[2] We recall that a probability space consists of a triplet $(U, \mathcal{F}, P)$, where U is the sample space (or space of elementary events), $\mathcal{F}$ a set of subsets from U which contains U and the empty set and which is closed under countable unions and taking complements. These properties are summarized by saying that $\mathcal{F}$ forms a σ-algebra or *Borel field* of subsets. The sets contained in $\mathcal{F}$ are called *events*. P is a normalized measure on $(U, \mathcal{F})$, i.e., $P(U) = 1$. It is called a *probability distribution*. Any set of subsets of U generates a σ-algebra just by closing it with respect to countable intersections and taking complements. If U is a topological space, the σ-algebra generated by the open sets is called the Borel field *of U* and

$(\mathcal{S}, \mathcal{B}(\mathcal{S}), P(\hat{\rho}))$. In what sense does this model quantum mechanics? Certainly not in the naive sense, since, for example, a pure state is according to (A5.9) represented by a single δ-distribution which in ordinary probability theory would corresponds to a dispersion free state. But there are no dispersion free states in quantum mechanics! More generally, (A5.10) does not imply that the distribution of the random variable F is the same as the quantum mechanical distribution of $\hat{F}$ in the state $\hat{\rho}$. For this one would have to require that

$$\int_{F^{-1}(\Delta)} P(\hat{\rho}) \, d\mu = \int_{\Delta} \text{trace}\left(\hat{\rho} \, d\hat{P}(\lambda)\right) \tag{A5.11}$$

where $\hat{F} = \int \lambda \, d\hat{P}(\lambda)$, i.e., $d\hat{P}(\lambda)$ denotes the projector-valued measure associated to the self-adjoint operator F. It is a general consequence of theorem 14.1 in (Nelson 1972) that there is always a $\hat{\rho}$ such that (A5.11) cannot hold for all $\hat{F}$, although this is already obvious from the arbitrariness of $P(\hat{\rho})$, since one cannot vary $P(\hat{\rho})$ while keeping all distributions of random variables fixed. For the particular choice (A5.8) the failure of (A5.11) is evident from the remark made above that pure states are not dispersion free, although represented by a δ-distribution. In this respect, note that the random variable corresponding to, say, the square of an operator $\hat{F}$ is *not* given by the squared random variable but rather by the function which is to be determined from $\hat{F}^2$ via (A5.3). As far as quantum mechanics is concerned, no meaning is given to the higher moments $\overline{F^n} = \int P(\hat{\rho}) F^n \, d\mu$ and, in particular, to $\Delta F^2 = \overline{F^2} - (\overline{F})^2$. Quite generally, it is one of the fundamental features of quantum mechanics (of closed systems) that it is *not* equivalent to a classical stochastic theory. A precise statement and proof of this fact is given in theorem 14.1 of the book by Nelson (1972). The basic reason, however, is quite clear: a classical probability theory on the space of rays necessarily regards all n^2 basis random-variables (A5.3) as compatible, in contradiction to quantum mechanics. We refer to (Koopman 1957) for a lucid discussion of the issue of compatibility.

As already seen above, compatible observables correspond to (Poisson-) commuting functions on $\mathcal{S}$ whose maximum number is $n - 1$ (without the constant functions). Compared to the total of n^2 linearly independent functions (A5.3) this makes a big difference in the following sense: Consider a general probability space $(U, \mathcal{F}, P)$. A given set $\{r_1, \ldots, r_m\} = \{r_i\}$ of random variables defines a σ-subalgebra $\mathcal{F}(\{r_i\}) \subset \mathcal{F}$ as the smallest σ-algebra of subsets of U with respect to which all r_i are measurable. It is called the σ-algebra (of events) generated by $\{r_i\}$. In all phenomenological applications of

---

is denoted by $\mathcal{B}(U)$. For $E \in \mathcal{F}$, $P(E)$ is called the probability of the event E. A real valued function X which is measurable, i.e. whose pre-image of each set in $\mathcal{B}(\mathbb{R})$ is in $\mathcal{F}$, is called a random variable. Let $\Delta \in \mathcal{B}(\mathbb{R})$, then the probability for X to take its value in Δ is given by $P(X^{-1}(\Delta))$, where $X^{-1}(\Delta)$ denotes the pre-image of Δ under X. Given X this defines a probability distribution on $(\mathbb{R}, \mathcal{B}(\mathbb{R}))$ called the *distribution of* X.

probability theory one usually takes an operational viewpoint and *defines* the
σ-algebra of events as being given by the σ-algebra generated by all compatible observables. The rationale for this being, that 1) the event space cannot
be smaller than this and that 2) any bigger choice necessarily contains events
that cannot be characterized operationally.

Applied to our case, where $U = S = \mathbb{CP}^{n-1}$, if the set $\{r_i\}$ is taken to
comprise all n^2 linearly independent functions of the form (A5.3), one may
quite easily show that $\mathcal{F}(\{r_i\}) = \mathcal{B}(S)$, i.e., it equals the Borel field of S.
However, if we restrict $\{r_i\}$ to contain only a set of compatible observables,
$\mathcal{F}(\{r_i\})$ is strictly smaller than $\mathcal{B}(S)$. This is also easy to see, since the
maximal number of compatible observables is even less than the dimension
of S. For example, if $n = 2$, the four-dimensional space of observables on the
two-sphere is spanned by some non-zero constant function and the functions
$f_i = \cos\theta_i$, $i = 1, 2, 3$, where θ_i is the polar angle to the i-th axis. The maximal
number of non-constant commuting functions is one; take e.g. $\cos\theta_3$. The σ-
algebra generated by this function thus only consists of those measurable sets
which depend on the polar angle θ_3 but not on the azimuth. The structure
of a probability theory whose sample space is given by the space of rays and
whose event space is its Borel field, $\mathcal{B}(S)$ thus formally assigns probabilities to
– and hence suggests the existence of – events, which, according to quantum
mechanics, cannot exist in any operational sense.

A6 Galilean Symmetry of Non-Relativistic Quantum Mechanics

D. Giulini

Let G denote the (inhomogeneous) Galilei group. A general element $g \in G$ is parameterized by a rotation-matrix R, a boost-velocity $\boldsymbol{v}$ and a space-time translation vector $(\boldsymbol{a}, b)$. The multiplication law is given by

$$
\begin{aligned}
g'g &= (R', \boldsymbol{v}', \boldsymbol{a}', b')(R, \boldsymbol{v}, \boldsymbol{a}, b) \\
&= (R'R, \boldsymbol{v}' + R'\boldsymbol{v}, \boldsymbol{a}' + R'\boldsymbol{a} + \boldsymbol{v}'b, b' + b) \quad (A6.1) \\
g^{-1} &= (R, \boldsymbol{v}, \boldsymbol{a}, b)^{-1}(R^{-1}, -R^{-1}\boldsymbol{v}, -R^{-1}(\boldsymbol{a} - \boldsymbol{v}b), -b). \quad (A6.2)
\end{aligned}
$$

We shall throughout ignore problems that might arise due to the non-simply-connectedness of the rotation group $SO(3)$, since we can always consider its simply-connected double cover $SU(2)$ instead.

We now consider a system of N particles of individual masses m_a, $a = 1, \ldots, N$, interacting through a Galilei-invariant potential V. G acts on $\mathbb{R}^{3N} \times \mathbb{R}$ according to (we write $\{\boldsymbol{x}_1, \ldots, \boldsymbol{x}_N\} = \{\boldsymbol{x}_a\}$)

$$
(\{\boldsymbol{x}_a\}, t) \xrightarrow{g} g(\{\boldsymbol{x}_a\}, t) = (\{R\boldsymbol{x}_a + \boldsymbol{a} + \boldsymbol{v}t\}, t + b). \quad (A6.3)
$$

The Hilbert space is given by $\mathcal{H} = L^2(\mathbb{R}^{3N})$ with respect to the standard measure. We identify $\mathcal{H}$ with the space of initial data at $t = 0$ for the Schrödinger equation:

$$
L\psi := (\mathrm{i}\hbar\partial_t - H)\,\psi = 0
$$

$$
\text{with} \quad H := -\sum_{a=1}^{N} \frac{\hbar^2}{2m_a}\Delta_a + V, \quad (A6.4)
$$

where we set $\Delta_a := \boldsymbol{\nabla}_a \cdot \boldsymbol{\nabla}_a$ (no summation over a). It is integrated by a group of unitary evolution operators $\exp(-\frac{\mathrm{i}}{\hbar}Ht)$, which allows the bijective correspondence between $\mathcal{H}$ and the space of solutions $\psi(\{\boldsymbol{x}_a\}, t)$ to the Schrödinger equation (A6.4):

$$
\psi(\{\boldsymbol{x}_a\}, t) \leftrightarrow \exp\left(\tfrac{\mathrm{i}}{\hbar}Ht\right)\psi(\{\boldsymbol{x}_a\}, t) = \psi(\{\boldsymbol{x}_a\}, t = 0) \in \mathcal{H}. \quad (A6.5)
$$

We now wish to consider the Galilei group as symmetry group acting on $\mathcal{H}$. But it is not obvious how an element $g \in G$ should act on $\mathcal{H}$. The idea is to use the identification (A6.5) and try to determine g's action on the solutions $L\psi = 0$ instead. These are functions of the coordinates $(\{\boldsymbol{x}_a\}, t)$ on which g's action is given by (A6.3). An obvious choice would therefore consist in shifting the function ψ along g by taking the composite function $\psi \circ g^{-1}$. But one may easily check that the shifted function will no longer solve the Schrödinger equation. This can be remedied by also multiplying ψ with a phase-function $\exp(\mathrm{i}f_g)$ which we now determine. We set

$$\psi \overset{g}{\longmapsto} T(g)\psi := \exp(\mathrm{i}f_g)\,(\psi \circ g^{-1}) \tag{A6.6}$$

and impose the requirement that the resulting function again solves the Schrödinger equation. Acting with L on $T(g)\psi$ one finds

$$L(T(g)\psi) = \exp(\mathrm{i}f_g)\,(L\psi \circ g^{-1})$$
$$+ \,\mathrm{i}\hbar \exp(\mathrm{i}f_g) \sum_{a=1}^{N} \left(\frac{\hbar}{m_a} \boldsymbol{\nabla}_a f_g - \boldsymbol{v} \right) \cdot R \cdot \left(\boldsymbol{\nabla}_a \psi \circ g^{-1} \right)$$
$$- \,\hbar \exp(\mathrm{i}f_g) \left(\partial_t f_g + \sum_{a=1}^{N} \frac{\hbar}{2m_a} (\boldsymbol{\nabla}_a f_g)^2 - \mathrm{i} \sum_{a=1}^{N} \frac{\hbar}{2m_a} \Delta_a f_g \right) (\psi \circ g^{-1}) . \tag{A6.7}$$

This shows that $T(g)$ transforms solutions of the Schrödinger equation to solutions, if and only if the extra two terms in (A6.7) vanish. The corresponding equations are easily seen to lead to

$$f_g(\{\boldsymbol{x}_a\}, t) = \tfrac{M}{\hbar} \left(\boldsymbol{v} \cdot \boldsymbol{R} - \tfrac{1}{2}\boldsymbol{v}^2 t + c_g \right) , \tag{A6.8}$$

where $M = \sum_a m_a$ is the total mass and $\boldsymbol{R} = \tfrac{1}{M} \sum_a m_a \boldsymbol{x}_a$ the center-of-mass-vector. c_g is a g-dependent integration constant. A convenient choice is $c_g = (\tfrac{1}{2}\boldsymbol{v}^2 b - \boldsymbol{v} \cdot \boldsymbol{a})$ which leaves us with the transformation law

$$(T(g)\psi)(\{\boldsymbol{x}_a\}, t) = \exp\left\{ \tfrac{\mathrm{i}}{\hbar} M \left[\boldsymbol{v} \cdot (\boldsymbol{R} - \boldsymbol{a}) - \tfrac{1}{2}\boldsymbol{v}^2(t - b) \right] \right\} \psi\left(g^{-1}(\{\boldsymbol{x}_a\}, t) \right). \tag{A6.9}$$

Now, by derivation we know that $T(g)\psi$ satisfies the Schrödinger equation, that is, $(T(g)\psi)(t) = \exp(-\tfrac{\mathrm{i}}{\hbar}Ht)(T(g)\psi)(t = 0)$. Hence we just need to put $t = 0$ in (A6.9) in order to obtain g's action on $\mathcal{H}$, which we call $U(g)$, and which is easily verified to be unitary:

$$U(g)\psi(\{\boldsymbol{x}_a\}) = \exp\left\{ \tfrac{\mathrm{i}}{\hbar} M \left[\boldsymbol{v} \cdot (\boldsymbol{R} - \boldsymbol{a}) + \tfrac{1}{2}\boldsymbol{v}^2 b \right] \right\}$$
$$\cdot \left(\exp(\tfrac{\mathrm{i}}{\hbar}Hb)\psi \right) \left(\{R^{-1}(\boldsymbol{x}_a - \boldsymbol{a} + \boldsymbol{v}b)\} \right) . \tag{A6.10}$$

However, having found the transformation law (A6.10) for each $g \in G$ does not imply a representation of G on $\mathcal{H}$. In fact, one now finds a phase difference between the Galilei transformation $U(g'g)$ and the composite transformation $U(g')U(g)$ (compare (6.38)):

$$U(g')U(g) = \omega(g', g)\,U(g'g) = \exp\left(\tfrac{\mathrm{i}}{\hbar} M \xi(g', g)\right) U(g'g) \tag{A6.11}$$

$$\text{where} \quad \xi(g', g) = \boldsymbol{v}' \cdot R' \cdot \boldsymbol{a} + \tfrac{1}{2}\boldsymbol{v}'^2 b. \tag{A6.12}$$

Due to the correspondence (A6.5) the same relation must hold for the transformations $T(g)$. In fact, the best way to deduce (A6.12) is to first explicitly show (A6.11) for the $T(g)$'s using (A6.9). Furthermore, it is straightforward to check the following condition:

$$\delta\xi(g'', g', g) := \xi(g''g', g) - \xi(g'', g'g) + \xi(g'', g') - \xi(g', g) = 0, \quad (A6.13)$$

which is just (6.39) written for the multiplier exponents, and which is equivalent to associativity of the multiplication law (A6.11). It is clear that the multiplier exponents are only determined up to the arbitrariness induced by the choice of the constants c_g in (A6.8). So if we redefine $c_g \mapsto c'_g = c_g + \delta_g$, then ξ also changes according to

$$\xi(g', g) \mapsto \xi'(g', g) = \xi(g', g) - \delta_{g'g} + \delta_{g'} + \delta_g , \quad (A6.14)$$

which also satisfies (A6.13). This is just (6.41) for the exponents. Now the crucial observation is that $\xi(g', g)$ cannot be made zero by such a redefinition. Indeed, if such a trivializing redefinition existed, it would automatically remove all multipliers on subgroups. But this is not possible as can be easily seen by restricting oneself to the abelian subgroup generated by spatial translations and boosts, which is homomorphic to $\mathbb{R}^3 \times \mathbb{R}^3 \cong \mathbb{R}^6$: $(v', a')(v, a) = (v' + v, a' + a)$. The induced multiplier-phases on this subgroup are

$$\xi\big((v', a'), (v, a)\big) = v' \cdot a. \quad (A6.15)$$

Since this subgroup is abelian, a redefinition (A6.13) only adds terms symmetric in g' and g and therefore never cancels the non-symmetric expression (A6.15). In the terminology of Sect. 6.2.1 the multipliers are not similar to the trivial ones. The various forms of the multipliers that are used in the literature mutually differ by a similarity transformation (A6.14).[1] The same trick immediately shows that multipliers for different overall masses are not similar.

Hence we see that the transformations (A6.6) define a unitary projective representation, with non-trivial multipliers, of the Galilei group on $\mathcal{H}$. Equivalently we may say that there is a proper unitary action of a central $U(1)$-extension $\bar{G}_M$ (indexed by M since it depends on the mass) of G with non-trivial action of the extending $U(1)$ on $\mathcal{H}$. From the general discussion

---

[1] The literature on this subject usually employs two different choices: One is to choose $c_g = 0$ (e.g. Bargmann 1954, Galindo Pascal 1990), the other $c_g = \frac{1}{2}v \cdot (vb - a)$ (e.g. Levy-Leblond 1963, Varadarajan 1985). The multiplier-phases are then given by the more complicated expressions respectively:

$$\xi'(g', g) = -\big(a' \cdot R' \cdot v - v' \cdot R' \cdot v\, b' - \tfrac{1}{2}v'^2\, b'\big),$$
$$\xi'(g', g) = \tfrac{1}{2}\big(v' \cdot R' \cdot a - a' \cdot R' \cdot v - v' \cdot R' \cdot v\, b\big),$$

which are equivalent to the phases (A6.9-A6.10) under the redefinitions (A6.13), where respectively $\delta_g = \big(v \cdot a - \tfrac{1}{2}v^2\, b\big)$ and $\delta_g = \tfrac{1}{2}v \cdot a$.

in Sect. 6.2.1 we infer that the extensions are inequivalent for different mass parameters M. If we parametrize the extending U(1) by $\exp(\frac{i}{\hbar}M\theta)$ and writing the general group element of $\bar{G}_M$ as $\bar{g} = (\theta, g)$ with g as above, we have the multiplication law

$$\bar{g}'\bar{g} = (\theta', g')(\theta, g) = (\theta' + \theta + \xi(g', g), g'g) . \tag{A6.16}$$

Its action on wave functions is given by (compare (6.45))

$$\bar{U}(\bar{g})\psi = \exp(\tfrac{i}{\hbar}M\theta)U(g)\psi , \tag{A6.17}$$

with $U(g)$ as in (A6.10). It is obviously unitary and satisfy the representation property by construction. Note that the periodicity of θ is $2\pi\hbar/M$, i.e. depends on M, so that $\bar{G}_M \ni (\theta, g) \mapsto (\theta, g) \in \bar{G}_{M'}$ is not a homomorphism unless M' is an integer multiple of M.[2] Now instead of having the different U(1)-extensions $\bar{G}_M$ we can consider a single $\mathbb{R}$-extension, called $\hat{G}$, which is defined just as in (A6.16) with θ now $\mathbb{R}$-valued. Then $\hat{G} \ni (\theta, g) \mapsto (\theta, g)) \in \bar{G}_M$ *is a surjective homomorphisms with kernel* $\mathbb{Z}$ generated by $(2\pi\hbar/M, e)$. $\hat{G}$ here plays the rôle of the universal central extension (compare Sect. 6.2.1) and is simply given by the universal cover of $\bar{G}_M$ for some M (if we simultaneously replace the rotation group $SO(3)$ by $SU(2)$). We shall work with $\hat{G}$ rather than the collection of $\bar{G}_M$'s from now on.

From (A6.10) and (A6.17) we can infer the infinitesimal transformation $\hat{U}(\delta\hat{g})$ with infinitesimal parameters $\delta R_{ik} = \varepsilon_{ijk}\delta k_j$, δv, δa, δb, and $\delta\theta$:

$$\hat{U}(\delta\hat{g}) =: \delta\boldsymbol{k} \cdot \mathbf{D} + \delta\boldsymbol{v} \cdot \mathbf{V} + \delta\boldsymbol{a} \cdot \mathbf{A} + \delta b\,\mathbf{B} + \delta\theta\,\mathbf{Z} \tag{A6.18}$$

with

$$\mathbf{D} := -\sum_{a=1}^{N}(\boldsymbol{x}_a \times \boldsymbol{\nabla}_a) , \tag{A6.19}$$

$$\mathbf{V} := \tfrac{i}{\hbar}M\boldsymbol{R} , \tag{A6.20}$$

$$\mathbf{A} := -\sum_{a=1}^{N}\boldsymbol{\nabla}_a , \tag{A6.21}$$

$$\mathbf{B} := \tfrac{i}{\hbar}H , \tag{A6.22}$$

$$\mathbf{Z} := \tfrac{i}{\hbar}M\,\mathbb{1} . \tag{A6.23}$$

---

[2] The map $\bar{G}_M \ni (\theta, g) \mapsto (\frac{M}{M'}\theta, g) \in \bar{G}_M$, which reduces to the identity on the extending U(1), is also not a homomorphism, as is seen form (A6.16). This reflects the inequivalence of the extensions for $M \neq M'$.

The generators (A6.19-A6.23) satisfy the Lie algebra relations for the group $\hat{G}$, whose non-vanishing commutators are given by

$$[\mathbf{D}_i, \mathbf{D}_j] = \varepsilon_{ijk}\mathbf{D}_k \ , \tag{A6.24}$$

$$[\mathbf{D}_i, \mathbf{V}_j] = \varepsilon_{ijk}\mathbf{V}_k \ , \tag{A6.25}$$

$$[\mathbf{D}_i, \mathbf{A}_j] = \varepsilon_{ijk}\mathbf{A}_k \ , \tag{A6.26}$$

$$[\mathbf{V}_i, \mathbf{A}_j] = \delta_{ij}\mathbf{Z} \ , \tag{A6.27}$$

$$[\mathbf{V}_i, \mathbf{B}] \ = \mathbf{A}_i \ . \tag{A6.28}$$

We call this Lie algebra $\hat{L}$. It differs from the corresponding one for the Galilei group, called L, only by (A6.27), whose right hand side is now proportional to the central element $\mathbf{Z}$ instead of being zero. Besides $\mathbf{Z}$ the enveloping algebra has two more central elements (Casimir elements):

$$\boldsymbol{S}^2 = (\mathbf{Z}\mathbf{D} - \mathbf{V} \times \mathbf{A})^2 \ , \tag{A6.29}$$

$$\boldsymbol{K} = \mathbf{A}^2 - 2\mathbf{Z}\mathbf{B} \ . \tag{A6.30}$$

In the representation (A6.19-A6.23) the vector operator $\boldsymbol{S}$ is easily seen to corresponds to $M/\hbar^2$-times internal angular momentum and the operator $\boldsymbol{K}$ to $2M/\hbar^2$-times internal energy.

From the foregoing it may appear as if the motivation to consider the central extension $\hat{G}$ arises only in the quantum theory. But this is not the case for reasons we now wish to explain. In the classical theory we have an action of G by canonical transformations. Expressed in terms of infinitesimal transformations, this implies that there is a Lie homomorphism from L into the Lie algebra of vector fields on phase space, but it does not necessarily imply a Lie homomorphism into the Lie algebra of functions on phase-space. (Their Lie product being given by the Poisson-bracket, henceforth abbreviated PB.) The problem being that the functions are only determined up to a constant by the vector fields, and it may be the case that these constants cannot be arranged in such a way as to result in a true homomorphism. This is precisely what happens for the Galilei group! The phase-space functions that generate rotations, boosts, spatial and time translations are given by total angular momentum, mass times center-of-mass vector, total momentum, and energy respectively. Their PB relations are precisely as in (A6.24-A6.28) and therefore, for non-zero M, represent $\hat{L}$ rather than L (Giulini 1996). Similar to the argument given after equation (A6.15), it can be shown that no redefinition of the generating functions by adding constants can redeem this deficiency. However, the difference is only the constant function M which generates a trivial action. Hence, despite this deficiency, an action of G results. In the terminology of Sudarshan and Mukunda (1974) this is called a PB realization of G *up to neutrals*, where "neutral" refers to the property, shared by M, to have vanishing PBs with *all* phase-space functions. Conversely, a PB realization is called *true*, if – up to redefinitions by adding constants – it can be made homomorphic to the Lie algebra of the acting group. Thus we only have a true PB realization of $\hat{G}$ but not of G.

Following the general arguments of Sect. 6.2.2, one now obtains a superselection rule for the mass from Galilei invariance. This is also easy to see directly. For this we restrict attention to the subgroup of G generated by boosts and translations. Let $g = (\boldsymbol{v}, \boldsymbol{a})$ and $g' = g^{-1} = (-\boldsymbol{v}, -\boldsymbol{a})$, then $U(g^{-1})U(g)$ corresponds to multiplication with the phase factor $\exp(-\frac{i}{\hbar} M \boldsymbol{v} \cdot \boldsymbol{a})$. On the state space it must clearly corresponds to the identity operation. Now consider two copies of $L^2(\mathbb{R}^{3N})$, $\mathcal{H}_M$ and $\mathcal{H}_{M'}$, equipped with the ray representations of the Galilei group corresponding to the mass values M and M' respectively. Consider the superposition $\psi_M + \psi_{M'} \in \mathcal{H}_M \oplus \mathcal{H}_{M'}$. Acting with g and then g^{-1} on this sum yields two different phase factors for ψ_M and $\psi_{M'}$ and therefore a ray different from the one we started from, unless one of the vectors is null. Consequently, if we require the Galilei group to act via some ray representation on the set of vectors representing physical states (the rays), then superpositions of vectors in $\mathcal{H}_M$ and $\mathcal{H}_{M'}$ for $M \neq M'$ cannot be in this set.

Equivalently, using the group $\hat{G}$ and its representations on $\mathcal{H}_M$ and $\mathcal{H}_{M'}$ for $M \neq M'$, the same fact is simply stated by saying that $\mathbf{Z}$ cannot be a multiple of the identity operator on $\mathcal{H}_M \oplus \mathcal{H}_{M'}$. So, from this point of view, what is wrong with non-trivial superpositions in $\mathcal{H}_M \oplus \mathcal{H}_{M'}$ is that they define states (rays) which are not left invariant when acted upon by an element of the extending group $\mathbb{R}$; only fixed points of $\mathbb{R}$ are allowed as states so that the representing vectors can at most be rescaled by $\mathbb{R}$. The set of such vectors is given by $\mathcal{H}_M \cup \mathcal{H}_{M'}$. But whenever the set of "allowed" states is a proper subset of some Hilbert space, we must also restrict the set of observables in order to exclude the possibility to transform "allowed" states into a "disallowed" states. Here this amounts to the requirement that observables should commute with the action of $\mathbb{R}$. Conversely, if we only consider observables commuting with $\mathbb{R}$, then all expectation values of such observables for states which are represented by non-trivial superpositions in $\mathcal{H}_M \oplus \mathcal{H}_{M'}$ can be written as expectation values for mixtures. One says that such superpositions do not define pure states for the given algebra of observables. Note that *purity* is a relative notion. This leads us to the standard statement, made in the algebraic approaches, that the root of the superselection rule is the existence of a non-trivial center in the algebra of observables. This last statement can be directly argued for without mentioning states: Recall that any unitary transformation acts by simple conjugation on the observables. This defines an automorphism on the algebra of observables. We thus have G acting as automorphisms. If we require that there is an action of G on our observables, we must restrict them by hand to those fixed by the action of $\mathbb{R}$, i.e., to the subalgebra commuting with it. Doing this, one forces $\hat{U}(\mathbb{R})$ into the center of the algebra of selected observables.

As discussed in the main text, mass was not a dynamical variable in the present context so that superpositions $\psi_M + \psi_{M'}$ seem to lack a proper interpretation. On the other hand, if mass is treated dynamically, it is the group $\hat{G}$ and not G that appears as the *classical* symmetry group (Giulini 1996). In short, this comes about as follows: The most simple extension of the classical model is to maintain the Hamiltonian, but now regarded as function on an extended, $6N + 2N$ - dimensional phase space with extra "momenta" m_a and conjugate generalized "positions" λ_a. Since the λ_a's do not appear in the Hamiltonian, the m_a's are constants of motion. Hence the equations of motion for the $\boldsymbol{x}_a$'s and their conjugate momenta $\boldsymbol{p}_a$ are unchanged (upon inserting the integration constants m_a) and those of the new positions λ_a are

$$\dot{\lambda}_a(t) = \frac{\partial V}{\partial m_a} - \frac{\boldsymbol{p}_a^2}{2m_a^2} , \tag{A6.31}$$

which, upon inserting the solutions $\{\boldsymbol{x}_a(t), \boldsymbol{p}_a(t)\}$, are solved by quadrature. Now, the new Hamiltonian equations of motion do not allow the Galilei group but only its $\mathbb{R}$-extension $\hat{G}$ as symmetry group. The action of $\hat{G}$ on the extended space of {configurations} $\times$ {times} is now given by

$$\begin{aligned}
\hat{g}\left(\{\boldsymbol{x}_a\}, \{\lambda_a\}, t\right) &= (\theta, R, \boldsymbol{v}, \boldsymbol{a}, b)(\{\boldsymbol{x}_a\}, \{\lambda_a\}, t) \\
&= (\{R\boldsymbol{x}_a + \boldsymbol{v}t + \boldsymbol{a}\}, \{\lambda_a - (\hbar M\theta + \boldsymbol{v} \cdot R \cdot \boldsymbol{x}_a + \tfrac{1}{2}\boldsymbol{v}^2 t)\}, t + b).
\end{aligned} \tag{A6.32}$$

With (A6.16) and (A6.12) it is easy to verify that this defines indeed an action. Hence it also defines an action on curves in the new Hilbert space $\hat{\mathcal{H}} := L^2(\mathbb{R}^{4N}, d^{3N}\boldsymbol{x}\, d^N\lambda)$, simply given by

$$\hat{T}(\hat{g})\psi := \psi \circ \hat{g}^{-1} , \tag{A6.33}$$

which already maps solutions of the new Schrödinger equation to solutions, *without* invoking non-trivial phase factors. The transformation law (A6.33) contains the more complicated one (A6.9) upon writing $\hat{\mathcal{H}}$ via Fourier transformation as direct integral over $\mathbb{R}^N$ (the N mass parameters) of Hilbert spaces each isomorphic to our old $\mathcal{H} \cong L^2(\mathbb{R}^{3N}, d^{3N}\boldsymbol{x})$. Splitting these mass-integrations into one over the total mass M and others allows us to write $\hat{\mathcal{H}}$ as direct integral

$$\hat{\mathcal{H}} = \int dM\, \hat{\mathcal{H}}_M \tag{A6.34}$$

which here corresponds to the decomposition (6.20), i.e. into the (alleged) continuous mass superselection sectors.

A7 Stochastic Processes

I.-O. Stamatescu

A7.1 Random Variables and Distributions

This appendix is not intended as a mathematical background for the discussion in Chaps. 7, 8 but only as a recollection of some elementary definitions and properties of stochastic processes in order to make the text selfcontained. For further reading see e.g. van Kampen (1981), Gnedenko (1982), Feller (1968) or Doob (1953) (in order of increasing mathematical precision) as well as the more specific literature given in Chaps. 7, 8.

We assume in the background an axiomatic definition of probabilities à la Kolmogorov involving the "probability space" $\{U, \mathcal{F}, P\}$ with U: a space of "elementary events", $\mathcal{F}$: a Borel field, i.e. a family of "random events" over U closed under intersection, complement and denumerable unions, and P: a real function on $\mathcal{F}$ taking values in $[0, 1]$, with $P(U) = 1$ and $P(A + B) = P(A) + P(B)$ for $A \cap B = \emptyset$, $A, B \in \mathcal{F}$. Random variables are defined as maps $\xi: e \in U \to f(e) \in \mathbb{R}$ with $f(e)$: measurable with respect to P, i.e.

$$\forall x \in \mathbb{R} \quad A_x \equiv \{e \mid f(e) \leq x\} \in \mathcal{F} \tag{A7.1}$$

$$\text{hence} \quad \exists P(A_x) = P\{\xi \leq x\} = F(x): \text{ the distribution function.}$$

Multidimensional and complex stochastic variables are also easily introduced. It is no loss of generality to take $\mathbb{R}^n$ or $\mathbb{C}^n$ as target space: One can define stochastic variables with values in any linear space over $\mathbb{R}^n$ or $\mathbb{C}^n$ by simply using real or complex stochastic variables as coefficients. Another possibility is to consider $\mathbf{x}$ as indexing some "vectors" $\chi_{\mathbf{x}}$ in some more general state space $\mathcal{X}$. Then the distributions over $\mathbf{x}$ automatically represent distributions over the χ's.

From the general properties of the probability space, the distribution F is a nonnegative, nondecreasing function of x with values in $[0, 1]$ and can have at most a denumerable set of jumps which are approached continuously from the right (above). A continuous distribution can be written with the help of a density $p(x)$ as

$$F(x) = \int_{-\infty}^{x} p(x) \, dx \tag{A7.2}$$

and p is the derivative of F if this exists. It is always useful to remember the distribution itself, since, e.g., the rules for changes of variables are then selfevident:

$$\eta = g(\xi): \quad F_\xi(x) = F_\eta(y = g(x)); \quad \text{hence} \quad p_\xi(x) = \left|\frac{\partial y}{\partial x}\right| p_\eta(y = g(x)).$$
(A7.3)

The introduction of the distributions allows to form "expectation values" for any (F-integrable) function $\phi(\xi)$:

$$\mathcal{E}\{\phi(\xi)\} = \int \phi(x)\,dF(x)$$
(A7.4)

among which the most known are the average and the variance of ξ:

$$\bar\xi \equiv \mathcal{E}\{\xi\}, \quad \Delta\xi \equiv \mathcal{E}\{(\xi - \bar\xi)^2\}.$$
(A7.5)

The "characteristic function" of a distribution is

$$\tilde p(t) = \int e^{itx}\,dF(x)$$
(A7.6)

and coincides with the Fourier transform of $p(x)$ if the latter exists. Of course we can also have purely discrete probability spaces and hence distributions. In this case the integrals are replaced by sums and the discussion above can be done *mutatis mutandis*.

Infinite sums of random variables with distributions satisfying some rather general conditions turn out to show when considered as random variables themselves a limiting distribution which most usually is the Gaussian – or "normal" – distribution (central limit theorem).

A time series of random variables $\{\xi(t)\}_t$ represents a "stochastic process". Its dynamics is fixed by the "transition probabilities" $W(x \to x') \equiv W(x', x)$ accompanying a time step. An actual set of values $\{x(t)\}_t$ realized by the stochastic variables in the course of time is called a trajectory. The distribution of possible values of $\xi(t)$ at given t is described by an "instantaneous" probability distribution $p(x, t)$. There are two ways of looking at a stochastic process: along each trajectory (expectations taken in this way are called "time averages") and at fixed time over the set of all possible values (we then speak of "ensemble averages"). To fix the ideas we shall first consider discrete time processes (also called "chains"). In the following the notation will not always distinguish between a random variable and its values.

A7.2 Markov Chains

We consider a discrete series of instants $t_n = n\Delta t$ and a chain of stochastic variables $\{\xi_n\}$ taking values $x_\alpha \in X$. For simplicity we shall assume the "reservoir" X to be a discrete set. Together with the transition probabilities $W(x_\alpha, x_\beta) = P_{\alpha\beta}$ this defines a Markov chain if the following additional condition is fulfilled: *the conditional probability distributions do not depend on the history*

$$P(x(t_{n+1}), t_{n+1} \mid x(t_n), t_n; x(t_{n-1}), t_{n-1}; \ldots)$$
$$= P(x(t_{n+1}), t_{n+1} \mid x(t_n), t_n) \equiv P_{\alpha\beta} \tag{A7.7}$$
$$\text{where } x(t_{n+1}) = x_\alpha, \ x(t_n) = x_\beta$$

(we use the convention that the first index concerns the later state). This is the so called Markov property and we shall assume it throughout. Notice that the transition probabilities are normalized in the first index (which just means that the chain does not simply disappear)

$$\sum_\alpha P_{\alpha\beta} = 1. \tag{A7.8}$$

The n-step transition probabilities $\left(\mathbf{P}^{(n)}\right)_{\alpha,\beta} \equiv P(x_\alpha, t_n \mid x_\beta, t_1)$ fulfill the evident relations

$$\mathbf{P}^{(n)} = \mathbf{P}^n \tag{A7.9}$$

written here in matrix form ($\mathbf{P}$ is the transition matrix $P_{\alpha\beta}$). This immediately implies the so-called Chapman–Kolmogorov equations:

$$\mathbf{P}^{(n+m)} = \mathbf{P}^{(n)}\mathbf{P}^{(m)}. \tag{A7.10}$$

We can use these $\mathbf{P}^{(n)}$ to define probability distributions for the values of ξ_n at time t_n, $\rho_{n;\alpha}$, which for large n and under very general conditions no longer depend on the initial point and tend toward a limiting distribution π_α. These latter fulfill therefore the "fixed point equation":

$$\pi_\alpha = \sum_\beta P_{\alpha\beta}\pi_\beta \tag{A7.11}$$

and sometimes even the stronger relation of "detailed balance"

$$P_{\beta\alpha}\pi_\alpha = P_{\alpha\beta}\pi_\beta \tag{A7.12}$$

(no summation) which implies (A7.11) but does not result from it. It can be shown quite generally that for $n \to \infty$ time averages along nearly any trajectory approach the ensemble averages taken according to the limiting distribution:

$$\frac{1}{N}\sum_{n=1}^{N} f(x_{\alpha(n)}) = \sum_\alpha \pi_\alpha f(x_\alpha) + \mathcal{O}(\frac{1}{\sqrt{N}}) \tag{A7.13}$$

(this property is sometimes called *ergodicity*). (A7.12) has a number of useful implications for the time analysis, variance estimation, etc.

A very common Markov chain is the "random walk". It is given as the following process:

$$
\begin{aligned}
&\text{if at } t_n: && \xi_n = x \\
&\text{then at } t_{n+1}: && \xi_{n+1} = x \text{ with probability } 1 - q \qquad \text{(A7.14)} \\
& && \xi_{n+1} = x \pm \Delta x \text{ with probability } q P_\pm(x)
\end{aligned}
$$

where

$$
q \in (0, 1], \quad P_+(x) + P_-(x) = 1 \qquad \text{(A7.15)}
$$

The master equation for the probability distribution (density) of $\xi(t)$ is given by counting the ways of achieving a value x:

$$
\rho(x, t + \Delta t) \qquad \text{(A7.16)}
$$
$$
= (1 - q)\rho(x, t) + q P_+(x - \Delta x)\rho(x - \Delta x, t) + q P_-(x + \Delta x)\rho(x + \Delta x, t).
$$

This leads for the time independent limiting distribution $\rho_{\lim}(x)$ to the q-independent, fixed point equation:

$$
\rho_{\lim}(x) = \sum_\pm P_\mp(x \pm \Delta x)\rho_{\lim}(x \pm \Delta x). \qquad \text{(A7.17)}
$$

Let us consider the free, symmetrical random walk for an example

$$
q = 1, \quad P_+ = P_- = \frac{1}{2}. \qquad \text{(A7.18)}
$$

The probability of having arrived $2k - n$ Δx-units further in n Δt-steps is easily seen to be (by just counting the combinations of k forward and $n - k$ backwards steps):

$$
p(k; n) = 2^{-n} C_n^k. \qquad \text{(A7.19)}
$$

For large n this is a Gaussian distribution about $k = \frac{n}{2}$ (De-Moivre–Laplace local limit theorem) and with average square deviation:

$$
\mathcal{E}\{(n - 2k)^2\}_n = n. \qquad \text{(A7.20)}
$$

The above analysis can be performed also for continuous reservoir X, e.g. $x \in \mathbb{R}$. Considering also t as a continuous parameter we arrive at Markov *processes*.

A7.3 Stochastic Processes

Let us consider the continuum limit of the above random walk process. By expanding in (A7.16) to $O(\Delta t)$ and to $O(\Delta x^2)$ we can write:

$$
\dot{\rho}(x, t) = \frac{q}{\Delta t} \left\{ -\Delta x \frac{\partial}{\partial x} \left[(P_-(x) - P_+(x)) \rho(x, t) \right] + \frac{1}{2} \Delta x^2 \frac{\partial^2}{\partial x^2} \rho(x, t) \right\}
$$
$$
\text{(A7.21)}
$$

where we have used the normalization (A7.15). Notice first that assuming Δt and Δx of the same order and $P_+(x) - P_-(x)$ finite leads to an uninteresting limiting distribution $\rho_{\lim}(x) = \frac{\text{const}}{P_-(x) - P_+(x)}$. To describe non–trivial stochastic processes we must take

$$q\Delta x^2 \propto \Delta t \tag{A7.22}$$

and Δx-dependent jump probabilities:

$$P_\pm(x) = P_\pm^{\Delta x}(x) = \frac{1}{2}\left(1 \pm \frac{1}{2}\Delta x K(x)\right). \tag{A7.23}$$

Then in the limit $q\Delta x^2 = 2\Delta t \to 0$ (A7.21) becomes the Fokker–Planck equation:

$$\dot{\rho}(x,t) = -\frac{\partial}{\partial x}\left(K(x) - \frac{\partial}{\partial x}\right)\rho(x,t). \tag{A7.24}$$

Relation (A7.22) characterizes the infinitesimal calculus for stochastic processes. This is not a mathematical subtlety but has simple physical reasons. Consider a particle suffering random, one–dimensional collisions from smaller particles of the environment leading to small jumps in its velocity. With w denoting its velocity and assuming very frequent collisions on a time scale much shorter than the time resolution of the observation Δt, the number of collisions experienced during Δt is proportional to the latter. Assuming collisions that are random and for simplicity of similar strength, the average absolute total shift in the velocity is proportional to the square root of the number of collisions – see (A7.20). From the point of view of our observation Δw is therefore a random variable such that $(\Delta w)^2 \simeq \Delta t$.

Stochastic processes occur in various physical situations and the question is usually that of designing the process which allows to correctly describe the observations or implies the wanted effects (e.g., Brownian motion or the collapse of the wave function!). In general stochastic processes can be defined by giving the equation for the dynamic of the stochastic variables (e.g., the random walk, (A7.14), or the Wiener process below) or by giving the equation for the probability distributions. We shall begin with the latter.

A7.4 The Fokker–Planck Equation

Although the problem can be formulated for the general case we shall assume here the existence of densities. Let us write $p(x, t; y, s)$: the conditional probability density for $\xi = x$ at t if at s we had $\xi = y$, and $\rho(x, t)$: the density of the probability distribution of the values of ξ at fixed time t. Using the Chapman–Kolmogorov equations rewritten for the continuous time case (also called semigroup equations) we can set up two problems:

a) for the conditional transition probabilities:

$$p(x,t;y,s) \qquad = \int dz\, p(x,t;z,r)p(z,r;y,s), \ t>r>s \quad (A7.25)$$

$$\lim_{t\to s} p(x,t;y,s) = \delta(x-y), \qquad\qquad\qquad (A7.26)$$

b) for the distributions at fixed time:

$$\rho(x,t) \ = \int dz\, p(x,t;z,t_0)\rho(z,t_0) \qquad\qquad (A7.27)$$

$$\rho(x,t_0) = \rho_0(x): \text{ given}. \qquad\qquad\qquad (A7.28)$$

Then under the following conditions ($\delta > 0$):

$$\lim_{\Delta t\to 0} \frac{1}{\Delta t} \int_{|x-y|>\delta} p(x,t+\Delta t;y,t) \qquad\quad = 0, \qquad (A7.29)$$

$$\lim_{\Delta t\to 0} \frac{1}{\Delta t} \int_{|x-y|<\delta} (x-y)p(x,t+\Delta t;y,t) \ = a(y,t), \qquad (A7.30)$$

$$\lim_{\Delta t\to 0} \frac{1}{\Delta t} \int_{|x-y|<\delta} (x-y)^2 p(x,t+\Delta t;y,t) \ = b(y,t), \qquad (A7.31)$$

one can derive the so called second (or forward) Kolmogorov equation:

$$\frac{\partial}{\partial t}p(x,t;y,s) = -\frac{\partial}{\partial x}\left(a(x,t)p(x,t;y,s)\right) + \frac{1}{2}\frac{\partial^2}{\partial x^2}\left(b(x,t)p(x,t;y,s)\right) \ (A7.32)$$

(and a similar equation for ρ). With $a = K$ the "drift force" and $b = 2$ the diffusion constant this equation written for $\rho(x,t)$ is the Fokker–Planck equation (A7.24). If the drift is conservative:

$$K(x) = -\frac{\partial}{\partial x}S(x) \qquad\qquad\qquad (A7.33)$$

it can be shown that the Fokker–Planck equation has the solution

$$\rho(x,t) = c_0\, e^{-S(x)} + e^{-\frac{1}{2}S(x)} \sum_{E_n>0} c_n\phi_n e^{-E_n t} \longrightarrow \rho_{\lim}(x) = c_0\, e^{-S(x)},$$

$$(A7.34)$$

for $t\to\infty$. Here E_n (ϕ_n) are the *eigenvalues (eigenvectors)* of the positive semidefinite Fokker–Planck Hamiltonian:

$$H_{\rm FP} = -\partial_x^2 + \frac{1}{4}(\partial_x S)^2 - \frac{1}{2}(\partial_x^2 S)$$

$$= \left(\partial_x + \frac{1}{2}(\partial_x S)\right)^\dagger \left(\partial_x + \frac{1}{2}(\partial_x S)\right) \qquad (A7.35)$$

$$\rho = e^{-\frac{1}{2}S}\phi, \quad \dot\phi = -H_{\rm FP}\,\phi, \qquad E_0 = 0, \quad \phi_0 = e^{-\frac{1}{2}S} \quad (A7.36)$$

For general drift the Fokker–Planck equation still makes sense, it does not lead, however, to time independent asymptotic distributions for nonconservative forces. The general form of the forward equation is

$$\dot{\rho}(x,t) = L\rho(x,t) \tag{A7.37}$$

with various assumptions for the operator L – see Chap. 7.

A7.5 Stochastic Differential Equations

A realization of the Fokker–Planck equation as stochastic process in discretized form is given by the random walk process, as shown in Sect. A7.3. (But notice that (A7.23) implies that for finite Δx the asymptotic distribution of x realized in the random walk is some $\rho_{\lim}^{\Delta x}(x)$ which differs from the fixed point solution $\rho_{\lim}(x)$ of (A7.24) in order Δx^2.) A rather basic stochastic process defined directly in continuous time is the Wiener process $w_t \equiv w(t)$, which can be introduced quite generally as the process with independent, stationary increments:

$$\overline{w_t} = w_0 = 0, \tag{A7.38}$$

$$\overline{(w_{t_2} - w_{t_1})(w_{t_4} - w_{t_3})} = D\left|[t_1, t_2] \cap [t_3, t_4]\right|. \tag{A7.39}$$

By bars we have denoted expectations over the "noise" (i.e., over the actualization of the stochastic variable w_t). It follows for the increments Δw:

$$\overline{\Delta w} - 0, \quad \overline{\Delta w^2} - D\Delta t. \tag{A7.40}$$

Since we can write $w_t = w_0 + \sum_i \Delta w_i$, by central limit theorem w_t is in fact normally distributed in the limit $\Delta t \to 0$. As already noted in Sect. A7.3 we see here again the typical, second relation in (A7.40).[1]

A related process is given by writing formally

$$\xi(t) = \frac{dw}{dt} \simeq \lim_{\Delta t \to 0} \frac{\Delta w}{\Delta t}. \tag{A7.41}$$

---

[1] It is easy to see that (A7.39) is universal. Indeed, stationarity:

$$\overline{(w_{t+s} - w_t)^2} = f(s) \text{ independent of t}$$

and independence of the increments:

$$\overline{(w_{t_2} - w_{t_1})(w_{t_4} - w_{t_3})} = 0 \quad \text{if} \quad [t_1, t_2) \cap [t_3, t_4) = \emptyset$$

imply

$$f(s_1 + s_2) = \overline{(w_{t+s_1+s_2} - w_{t+s_1} + w_{t+s_1} - w_t)^2} = f(s_1) + f(s_2).$$

Assuming continuity of f this implies linearity and hence (A7.39).

We have
$$\overline{\xi(t)} = 0, \quad \overline{\xi(t_1)\xi(t_2)} = D\delta(t_1 - t_2). \tag{A7.42}$$
It is called "white noise" since the Fourier transform of the time correlation function is flat.

Let us start again with the Wiener process, written in differential form:
$$\overline{dw_t} = 0, \quad \overline{dw_t^2} = 2\,dt \tag{A7.43}$$
where we take from now on $D = 2$ (this is no loss of generality since it only means a rescaling). Besides the unproblematic integrals
$$\int_0^t dw_s \quad = w_t - w_0 = w_t \tag{A7.44}$$
$$\int_0^t f(s)\,dw_s = \lim_{N \to \infty} \sum_{i=0}^N f(\tilde{s}_i)[w_{s_{i+1}} - w_{s_i}], \quad \tilde{s}_i \in [s_i, s_{i+1}], \tag{A7.45}$$
we must choose one out of several definitions in the following case
$$\int_0^t w_s\,dw_s = \lim_{N \to \infty} \sum_{i=0}^N w_{\tilde{s}_i}[w_{s_{i+1}} - w_{s_i}]$$
$$= \frac{1}{2}w_t^2 - t, \quad \text{if } w_{\tilde{s}_i} = w_{s_i}, \tag{A7.46}$$
$$= \frac{1}{2}w_t^2 + t, \quad \text{if } w_{\tilde{s}_i} = w_{s_{i+1}}, \tag{A7.47}$$
$$= \frac{1}{2}w_t^2, \qquad \text{if } w_{\tilde{s}_i} = \frac{w_{s_i} + w_{s_{i+1}}}{2} \tag{A7.48}$$

which can be checked by using the noise averages (A7.43). The definition (A7.46) leads to the *Itô calculus*, (A7.48) to the *Stratonovich calculus*. Non–linear processes demand therefore a consistent treatment in the frame of one or other definition.

Considering first the Itô calculus we see that we cannot obtain (A7.46) from (A7.44) by simply replacing w with $\frac{w^2}{2}$ (as in the case of conventional integrals). Therefore stochastic integrals can only be defined as a closed class in the Itô calculus as the stochastic processes:
$$X_t = X_0 + \int_0^t \beta(w_s, s)\,ds + \int_0^t \sigma(w_s, s)\,dw_s. \tag{A7.49}$$

The associated stochastic differential equation is:
$$dX_t = \beta(w_t, t)dt + \sigma(w_t, t)dw_t. \tag{A7.50}$$

For a continuous function $f(x, t)$,
$$Y_t \equiv f(X_t, t) \tag{A7.51}$$

also describes a stochastic process:

$$dY_t = \partial_t f(X_t, t)dt + \partial_x f(X_t, t)dX_t + \frac{1}{2}\partial_x^2 f(X_t, t)dX_t^2. \qquad (A7.52)$$

Assuming for a moment no explicit time dependence the general non-linear stochastic differential equation is:

$$dX_t = A(X_t)dt + B(X_t)dw_t \qquad (A7.53)$$

(remember that in Itô calculus all differentials are taken forward). Let us take:

$$\overline{dX_t} = K(x)dt, \quad \overline{dX_t^2} = 2\,dt \qquad (A7.54)$$

i.e. $A(x) = K(x)$ and $B(x) = 1$, x representing as usual the values of X_t. We realize in this way the process

$$dX_t = K(X_t)dt + dw_t. \qquad (A7.55)$$

By formally dividing by dt it can also be written as:

$$\dot{X}_t = K(X_t) + \xi_t \qquad (A7.56)$$

using thus the basic process (A7.41).

For an arbitrary function $f(x,t)$ we have from (A7.52), (A7.54)

$$\overline{df(X_t, t)} = \partial_t f dt + \partial_x f K dt + \partial_x^2 f dt \qquad (A7.57)$$

which represents the *conditional expectation value* of the increment df, given that $X_t = x$ at t. Assuming that the distribution of the values x can be represented by a density $\rho(x,t)$ we have:

$$\begin{aligned}
\mathcal{E}\{df(x,t))\} &= \int dx\, \rho(x,t)\, \overline{df(x,t)} \\
&= \mathcal{E}\{f(x, t+dt)\} - \mathcal{E}\{f(x,t)\} \qquad (A7.58)\\
&= \int dx\, (\rho(x,t+dt)f(x,t+dt) - \rho(x,t)f(x,t)).
\end{aligned}$$

Using (A7.57) and taking the adjoint equation we obtain the Fokker–Planck equation for ρ:

$$\dot{\rho}(x,t) = -\partial_x(K(x) - \partial_x)\rho(x,t). \qquad (A7.59)$$

For the more general form (A7.53) the associated Fokker–Planck equation for the probability distribution is

$$\dot{\rho}(x,t) = -\partial_x \left(A(x) - \partial_x B(x)^2 \right) \rho(x,t). \qquad (A7.60)$$

The rules for the change of variable:

$$Y_t = f(X_t) \qquad (A7.61)$$

are:

$$A'(Y) = A(X)\partial_X f(X) + B(X)^2 \partial_X^2 f(X), \quad B'(Y) = B(X)\partial_X f(X),$$
$$\tag{A7.62}$$
$$dX_t = A(X_t)dt + B(X_t)dw_t \quad \longrightarrow \quad dY_t = A'(Y_t)dt + B'(Y_t)dw_t. \tag{A7.63}$$

For the Wiener process $K = 0$ in particular (A7.59) has as solution the normal distribution:

$$\rho(w, t) = \frac{1}{\sqrt{4\pi t}} e^{-w^2/4t}. \tag{A7.64}$$

The corresponding conditional probabilities are:

$$p(w + \Delta w, t + \Delta t; w, t) = \frac{1}{\sqrt{4\pi \Delta t}} e^{-\Delta w^2/4\Delta t}. \tag{A7.65}$$

Notice again, that for $dw = \Delta w$, $dt = \Delta t$: finite the discrete process corresponding to (A7.55) leads to a step-dependent asymptotic distribution $\rho_{\lim}^{\Delta t}(x)$ different from the asymptotic solution of (A7.59) in order Δt.

Notice also that expectation values $\overline{f(X)}$ over the Wiener process for not explicitly time dependent functions of the stochastic variables $f(X)$ fulfil the adjoint equation to (A7.59).

The processes (A7.55), (A7.56) stay at the basis of the analysis of the Brownian motion[2]. By generalization a process of the form

$$dX_t = A(X_t, t)dt + B(X_t, t)dw_t \tag{A7.67}$$

is usually called *Langevin equation.*

The differential equation (A7.67) and the rules (A7.61,A7.62) allow one to construct stochastic processes in any space $\mathcal{X}$ over $\mathbb{R}^n$ or $\mathbb{C}^n$ by writing, e.g.:

$$d\chi_t = \tilde{\beta}(\chi_t, t)dt + \tilde{\sigma}(\chi_t, t)d\tilde{w}_t \tag{A7.68}$$

where $d\tilde{w}$ is a multidimensional, real or complex Wiener process defined by correspondingly generalizing (A7.43). In the applications which interest us in this book χ are typically vectors in the Hilbert (state) space or linear operators.

For completeness we shall also briefly recall the stochastic differential equation based on the Stratonovich calculus corresponding to (A7.48). In this

---

[2] Notice that the general process (A7.53) can be formally reduced to the form (A7.55) by the variable transformation

$$y = \int^x \frac{dz}{B(z)} \tag{A7.66}$$

and using (A7.62).

case in the equations of the form (A7.53) the nonlinear term is understood as:

$$B(X) \equiv B\left(\frac{X_t + X_{t+dt}}{2}\right).\tag{A7.69}$$

Then the rules for the variable transformations are just the usual ones (i.e., without the second term in the first equation (A7.62)) but the associated Fokker–Planck equation is now:

$$\dot{\rho}(x,t) = -\partial_x \left(A(x) - B(x)\partial_x B(x)\right)\rho(x,t),\tag{A7.70}$$

instead of (A7.60). This also illustrates the necessity of working consistently within one definition.

References*

Accardi, L., Gough, J., and Lu, Y. G. (1995): "On the stochastic limit for quantum theory." *Rep. Math. Phys.* **36**, 155–187.   [332, 338]

Adler, S.L. (2001): "Environmental influence on the measurement process in stochastic reduction models." Report quant-ph/0109029.   [362]

Adler, S.L. (2002): "Why decoherence has not solved the measurement problem." Report quant-ph/0202095.   [362]

Adler, S.L. and Brun, T.A. (2001): "Generalized stochastic Schrodinger equations for state vector collapse." *J. Phys.* **A34**, 4797–4809.   [363, 377]

Adler, S. L., Brody, D. C., Brun, T. A., and Hughston, L. P. (2001): "Martingale models for quantum state reduction." *J. Phys.* **A34**, 8795–8820.   [334]

Agarwal, G.S. (1971): "Brownian Motion of a Quantum Oscillator." *Phys. Rev.* **A4**, 739–747.   [81]

Agarwal, G.S. and Tewari, S.P. (1994): "An all-optical realization of the quantum Zeno effect." *Phys. Lett.* **A185**, 139–142.   [121]

Agarwal, G.S., Scully, M.O., and Walther, H. (2001): "Inhibition of Decoherence due to Decay in a Continuum." *Phys. Rev. Lett.* **86**, 4271–4274.   [122]

Aharonov, Y. and Susskind, L. (1967a): "Charge Superselection Rule." *Phys. Rev.* **155**, 1428–1431.   [302, 303, 312]

Aharonov, Y. and Susskind, L. (1967b): "Observability of the Sign Change of Spinors under 2π Rotations." *Phys. Rev.* **158**, 1237–1238.   [58, 300]

Aharonov, Y., and Vaidman, L. (1991): "Complete description of a quantum system at a given time." *J. Phys.* **A24**, 2315–2328.   [21]

Aharonov, Y. and Vardi, M. (1980): "Meaning of an individual 'Feynman path'." *Phys. Rev.* **D21**, 2235–2240.   [110, 115, 238]

Aharonov, Y., Bergmann, P.G., and Lebowitz, J.L. (1964): "Time symmetry in the quantum process of measurement." *Phys. Rev.* **134B**, 1410–1416; reprinted in Wheeler and Zurek (1983).   [251, 252]

Albert, D. and Loewer, M. (1988): "Interpreting the Many Worlds Interpretation." *Synthese* **77**, 195–213.   [31]

---

* Page numbers of quotation are added in brackets. Reports such as quant-ph/nnnn can be found on the internet eprint archive *http://arXiv.org/abs/quant-ph/nnnn*.

Albeverio, S., Kolokoltsov, V. N., and Smolyanov, O. G. (1997): "Continuous quantum measurement: local and global approaches." *Rev. Math. Phys.* **9**, 827–840.   [338]

Albrecht, A. (1992): "Investigating decoherence in a simple system." *Phys. Rev.* **D46**, 5504–5520.   [37, 44, 55]

Albrecht, A. (1993): "Following a 'collapsing' wave function." *Phys. Rev.* **D48**, 3768–3778.   [37, 55, 250]

Albrecht, A., Ferreira, P., Joyce, M., and Prokopec, T. (1994): "Inflation and squeezed quantum states." *Phys. Rev.* **D50**, 4807–20. [211, 212, 213, 215]

Alfsen, E.M. (1971): *Compact Convex Sets and Boundary Integrals* (Springer, Berlin).   [265]

Alicki, R. (2001): "A search for a border between classical and quantum worlds." Report quant-ph/0105089.   [67, 77, 395]

Alicki, R. and Fannes, M. (2001): *Quantum Dynamical Semigroups* (Oxford Univ. Press, Oxford).   [354]

Alicki, R. and Lendi, K. (1987): *Quantum Dynamical Semigroups and Applications* (Springer, Berlin).   [321, 326, 330, 331]

Allen, B. (1985): "Vacuum states in de Sitter space." *Phys. Rev.* **D32**, 3136–3149.   [201]

Allen, B. (1997): "The stochastic gravity-wave background: sources and detection." In: *Relativistic gravitation and gravitational radiation*, ed. by J.-A. Marck and J.-P. Lasota (Cambridge University Press, Cambridge).   [210]

Allinger, J. and Weiss, U. (1995): "Non-universality of dephasing in quantum transport." *Z. Phys.* **B98**, 289–296.   [179]

Altenmüller, T.P. and Schenzle, A. (1993): "Dynamics by measurement: Aharonov's inverse quantum Zeno effect." *Phys. Rev.* **A48**, 70–79. [123]

Altenmüller, T.P. and Schenzle, A. (1994): "Quantum Zeno effect in a double-well potential: A model of a physical measurement." *Phys. Rev.* **A49**, 2016–2027.   [126]

Amann, A. (1988): "Chirality as a classical observable in algebraic quantum mechanics." In: *Fractals, Quasicrystals Chaos, Knots and Algebraic Quantum Mechanics.*, ed. by A. Amann et al. (Kluwer), p.305–325. [365]

Amann, A. (1991a): "Chirality: A superselection rule generated by the molecular environment?" *Journal of Mathematical Chemistry* **6**, 1–15.
                                                                       [104, 105, 365, 387]

Amann, A. (1991b): "Molecules coupled to their environment.". In: *Large Scale Molecular Systems*, ed. by W. Gans, A. Blumen and A. Amann (Plenum Press, New York), p. 3–22.   [104, 188, 235, 365]

Amann, A. (1993): "The Gestalt problem in quantum theory: Generation of molecular shape by the environment." *Synthese* **97**, 125–156.   [188]

Amann, A. (1995): "Structure, dynamics and spectroscopy of single molecules: A challenge to quantum mechanics." *J. Math. Chem.* **18**, 247–308.
                                                                                  [101, 104]

Amati, D. and Russo, J. G. (1999): "Fundamental strings as black bodies." *Phys. Lett.* **B454**, 207–212.   [221]

Ambegaokar, V. (1991): "Quantum Brownian Motion and its Classical Limit." *Ber. Bunsenges. Phys. Chem.* **95**, 400–404.   [85]

Anandan, J. (1998): "The quantum measurement problem and the possible role of the gravitational field." Report quant-ph/9808033.   [365]

Anderson, A. (1990): "On predicting correlations from Wigner functions." *Phys. Rev.* **D42**, 585–589.   [97]

Anderson, P.W. (1958): "Absence of diffusion in certain random lattices." *Phys. Rev.* **109**, 1492–1505.   [154]

Anderson, P.W. (1978): " Local moments and localized states." *Rev. Mod. Phys.* **50**, 191–203.   [154]

Anglin, J.R. and Zurek, W.H. (1996a): "Decoherence of Quantum Fields: Pointer States and Predictability." Phys. Rev. **D53**, 7327–7335. [143, 192]

Anglin, J.R. and Zurek, W.H. (1996b): "A precision test of decoherence." In: *Dark matter in cosmology, quantum measurements, experimental gravitation*, ed. by R. Ansari, Y. Giraud-Heraud, and J. Tran Thanh Van (Editions Frontières, Gif-sur-Yvette), p. 263–270.   [186]

Anglin, J.R., Laflamme, R., Zurek, W.H., and Paz, J.P. (1995): "Decoherence and recoherence in an analogue of the black hole information paradox." *Phys. Rev.* **D52**, 2221–2231.   [84, 365]

Arai, A. and Hirokawa, M. (2000): "Ground states of a general class of quantum field hamiltonians." *Rev. Math. Phys.* **12**, 1085–1135.   [346]

Arai, A., Hirokawa, M. and Hiroshima, F. (1999): "On the absence of eigenvectors of Hamiltonians in a class of massless quantum field models without infrared cutoff." *J. Funct. Anal.* **168**, 470–497.   [346]

Araki, H. (1980): "A remark on Machida-Namiki theory of measurement." *Prog. Theor. Phys.* **64**, 719–730.   [341, 343, 353, 367]

Araki, H. and Yanase, M. (1960): "Measurement of quantum mechanical operators." *Phys. Rev.* **120**, 622–626; reprinted in Wheeler and Zurek (1983).   [47, 301, 302]

Arndt, M., Nairz, O., Vos-Andreae, J., Keler, C., van der Zouw, G., and Zeilinger, A. (1999): Wave-particle duality of C_{60} molecules. *Nature* **401**, 680–682.   [10, 67, 77]

Arveson, W. (1969): "Subalgebras of C*-algebras." *Acta Math.* **123**, 141–224.   [355]

Aspect, A. (1976): "Proposed experiment to test the nonseparability of quantum mechanics." *Phys. Rev.* **D14**, 1944–1951; reprinted in Wheeler and Zurek (1983).   [43]

Aspect, A., Dalibard, J., and Roger, G. (1982): "Experimental Test of Bell's Inequalities Using Time-Varying Analyzers." *Phys. Rev. Lett.* **49**, 1804–1807.   [43]

Aspect, A., Grangier, P., and Roger, G. (1981): "Experimental Tests of Realistic Local Theories via Bell's Theorem." *Phys. Rev. Lett.* **47**, 460–463.   [43]

Audretsch, J. and Müller, R. (1994): "Spontaneous excitation of an accelerated atom: The contributions of vacuum fluctuations and radiation reaction." *Phys. Rev.* **A50**, 1755–1763.   [180]

Audretsch, J., Mensky, M., and Müller, R. (1995): "Continuous measurement and localization in the Unruh effect." *Phys. Rev.* **D51**, 1716–1727. [221]

Auletta, G. (2001): *Foundations and Interpretation of Quantum Mechanics* (World Scientific). [1, 360]

Badurek, G., Rauch, H., and Tuppinger, D. (1986): "Polarized Neutron Interferometry." *Ann. New York Acad. Sci.* **480**, 133–146. [76]

Baker, H.C., and Singleton, R.L., Jr. (1990): "Non-Hermitian quantum dynamics." *Phys. Rev.* **A42**, 10–17. [367]

Balachandran, A.P. and Roy, S.M. (2002): "Continuous Time-Dependent Measurements: Quantum Anti-Zeno-Paradox with Applications." Report quant-ph/0102019. [117]

Balian, R. (1989): "On the principles of quantum mechanics and the reduction of the wave packet." *Am. J. Phys.* **57**, 1019–1027. [364]

Ballentine, L.E. (1991a): "Failure of some theories of state reduction." *Phys. Rev.* **A43**, 9–12. [88, 380]

Ballentine, L.E. (1991b): "Comment on 'Quantum Zeno effect'." *Phys. Rev.* **A43**, 5165–5167. [122]

Ballentine, L.E., Yang, Y., and Zibin, J.P. (1994): "Inadequacy of Ehrenfest's theorem to characterize the classical regime." *Phys. Rev.* **A50**, 2854–2859. [150]

Balzer, C., Huesmann, R., Neuhauser, W., and Toschek, P.E. (2000): "The Quantum Zeno Effect – Evolution of an Atom Impeded by Measurement." *Opt. Comm.* **180**, 115–120. [122]

Barbour, J. and Bertotti, B. (1977): "Gravity and Inertia in a Machian Framework." *Il Nuovo Cimento* **B38**, 1–27. [300]

Barbour, J. and Bertotti, B. (1982): "Mach's Principle and the Structure of Dynamical Theories." *Proc. R. Soc. Lond.* **A382**, 295–306. [300]

Barbour, J.B. (1993): "Time and complex numbers in canonical quantum gravity." *Phys. Rev.* **D47**, 5422–5429. [208, 390]

Barbour, J.B. (1994a): "The emergence of time and its arrow from timelessness." In: *Physical Origins of Time Asymmetry*, ed. by J.J. Halliwell, J.P. Pérez-Mercader, and W.H. Zurek (Cambridge University Press), p. 405–414. [197]

Barbour, J.B. (1994b): "The timelessness of quantum gravity: II. The appearance of dynamics in static configurations." *Class. Quantum Grav.* **11**, 2875–2897. [197]

Barbour, J.B. (1999): *The End of Time* (Weidenfeld & Nicolson, London). [31, 38]

Barchielli, A. and Belavkin, V.P. (1991): "Measurements continuous in time and a posteriori states in quantum mechanics." *J. Phys.* **A24**, 1495–1514. [334, 338]

Barchielli, A., Lanz, L., and Prosperi, G.M. (1982): "A model for the macroscopic description and continual observations in quantum mechanics." *Il Nuovo Cimento* **B72**, 79–121. [331, 370]

Bardeen, J., Cooper, L.N., and Schrieffer, J.R. (1957): "Theory of superconductivity." *Phys. Rev.* **108**, 1175–1204. [12]

Bargmann, V. (1954): "On unitary ray representations of continuous groups." *Ann. of Math.* **59** (second series), 1–46.   [285, 297, 427]

Bargmann, V. (1964): "Note on Wigner's theorem on symmetry operations." *J. Math. Phys.* **5**, 862–868.   [280]

Barnett, S.M. and Pegg, D.T. (1990): "Quantum theory of rotation angles." *Phys. Rev.* **41**, 3427–3435.   [104]

Barnett, S.M. and Phoenix, S.J.D. (1989): "Entropy as a measure of quantum optical correlation." *Phys. Rev.* **A40**, 2404–2409.   [55]

Barvinsky, A.O. and Kamenshchik, A.Yu. (1995): "Preferred basis in quantum theory and the problem of classicalization of the quantum Universe." *Phys. Rev.* **D52**, 743–757.   [37]

Barvinsky, A.O., Kamenshchik, A.Yu, and Karmazin, I.P. (1992): "One loop quantum cosmology: Zeta function technique for the Hartle-Hawking wave function of the universe." *Ann. Phys. (N.Y.)* **219**, 201–242.   [203]

Barvinsky, A.O., Kamenshchik, A.Yu, and Mishakov, I.V. (1997): "Quantum origin of the early inflationary universe." *Nucl. Phys.* **B491**, 387–426.   [203]

Barvinsky, A.O., Kamenshchik, A.Yu., Kiefer, C., and Mishakov, I.V. (1999a): "Decoherence in quantum cosmology at the onset of inflation." *Nucl. Phys.* **B551**, 374–396.   [200, 202, 204]

Barvinsky, A.O., Kamenshchik, A.Yu., and Kiefer, C. (1999b): "Origin of the inflationary Universe." *Mod. Phys. Lett.* **A14**, 1083–1088.   [202]

Barvinsky, A.O., Kamenshchik, A.Yu., and Kiefer, C. (1999c): "Effective action and decoherence by fermions in quantum cosmology." *Nucl. Phys.* **B552**, 420–444.   [206]

Bassi, A. and Ghirardi, G. (1999): "Can the decoherent histories description of reality be considered satisfactory?" *Phys. Lett.* **A257**, 247–263.   [258]

Bassi, A. and Ghirardi, G.C. (2000): "A general argument against the universal validity of the superposition principle." Report quant-ph/0009020.   [361, 381]

Bassi, A. and Ghirardi, G.C. (2002): "Dynamical reduction models with general gaussian noise." Report quant-ph/0201122.   [378]

Beige, A. and Hegerfeldt, G.C. (1996): "Projection Postulate and Atomic Quantum Zeno Effect." *Phys. Rev.* **A53**, 53–65.   [123]

Beig, R. and Ó Murchadha, N. (1987): "The Poincaré Group as Symmetry Group of Canonical General Relativity." *Ann. Phys. (N.Y.)* **174**, 463–498.   [311, 315]

Belavkin, V.P. (1989a): "Nondemolition measurements, nonlinear filtering and dynamic programming of quantum stochastic processes." In: *Modeling and Control of Systems. Lect. Notes in Control and Inf. Sciences* **121**, ed. by A. Blaquiére (Springer, Berlin), p. 245-265.   [38, 334, 336]

Belavkin, V.P. (1989b): "A continuous counting observation and posterior quantum dynamics." *J. Phys.* **A22**, L1109–1114.   [334, 337, 338, 378]

Belavkin, V.P. (1989c): "A new wave equation for a continuous nondemolition measurement." *Phys. Lett.* **A140**, 355–358.   [334, 337, 338, 378]

Belavkin, V.P. (1994): "Nondemolition principle of quantum measurement theory." *Found. Phys.* **24**, 685–714.   [366, 378]

Belavkin, V.P. (1995): "The interplay of classical and quantum stochastic diffusion, measurement and filtering." In: *Chaos - The Interplay between Stochastic and Deterministic Behaviour*. Proceedings of the XXXIst Winter School of Theoretical Physics, Karpacz, Poland, ed. by P. Garbaczewski, M. Wolf, and A. Veron (Springer, Berlin), p. 21–46.   [334, 338]

Belavkin, V.P. (1997): "Quantum stochastic positive evolutions: characterization, construction, dilation," *Commun. Math. Phys.* **184**, 533–566.
                                                                        [332]

Belavkin, V.P. (1998): "Quantum stochastic semigroups and their generators." In: *Irreversibility and Causality. Semigroups and Riggid Hilbert Spaces. Lect. Notes in Phys.* **504**, ed. by A. Bohm, H.-D. Doebner, and P. Kielanovski, p. 82–109, (Springer, Berlin).   [332]

Belavkin, V.P. (2002): "Measurement, filtering and control in quantum open dynamical systems." Report quant-ph/0208108.   [378]

Belavkin, V.P. and Staszewski, P. (1992): "Nondemolition observation of a free quantum particle." *Phys. Rev.* **A 45**, 1347–1356.
                                            [334, 335, 338, 366, 378]

Bell, J.S. (1964): "On the Einstein Podolsky Rosen paradox." *Physics* **1**, 195–200; reprinted in Wheeler and Zurek (1983).   [9, 43]

Bell, J.S. (1975): "On wave packet reduction in the Coleman-Hepp model." *Helv. Phys. Acta* **48**, 93–98; reprinted in Bell (1987).   [340]

Bell, J.S. (1981): "Quantum mechanics for cosmologists." In: *Quantum Gravity 2*, ed. by C. Isham, R. Penrose and D. Sciama (Clarendon Press, Oxford), p. 611–637.   [14, 28, 31]

Bell, J.S. (1987): *Speakable and Unspeakable in Quantum Mechanics* (Cambridge University Press).   [17, 365, 370]

Beltrametti, E.G. and Casinelli, G. (1981): *The Logic of Quantum Mechanics* (Addison-Wesley, London).   [263]

Benatti, F. and Floreanini, R. (2002): "Planck's scale dissipative effects in atom interferometry." Report quant-ph/0208164.   [225]

Benatti, F., Ghirardi, G.C., and Grassi, R. (1995): "Testing Macroscopic Quantum Coherence." *Il Nuovo Cimento* **110B**, 593–610.   [103, 378]

Bennett, C.H. (1973): "Logical reversibility of computation." *IBM J. Res. Dev.* **17**, 525–532.   [24, 25]

Bennett, C.H. (1987): "Demons, Engines and the Second Law." *Scientific American* **257**, 88–96.   [145]

Bennett, C.H. (1994): "Night thoughts, dark sight." *Nature* **371**, 479–480.
                                                                        [115]

Bennett, C.H. (1995): "Quantum Information and Computation." *Physics Today* **48**(Oct.), 24–30.   [147]

Bennett, C.H. and Wiesner, S.J. (1992): "Communication via one- and two-particle operators on Einstein-Podolsky-Rosen states." *Phys. Rev. Lett.* **69**, 2881–2884.   [175]

Bennett, C.H., Brassard, G., Crépeau, C., Jozsa, R., Peres, A., and Woot-
ters, W.K. (1993): "Teleporting an Unknown Quantum State via Dual
Classical and Einstein-Podolsky-Rosen Channels." *Phys. Rev. Lett.* **70**,
1895–1899.   [168, 172]
Bennett, C.H., DiVincenco, D.P., Smolin, J.A., and Wootters, W.K. (1996):
"Mixed state entanglement and quantum error correction." *Phys. Rev.*
A54, 3824–3851 – quant-ph/9604024.   [146]
Berg, B.A. (1998): "Quantum Theory and Spacelike Measurements." Report
quant-ph/9801003.   [363]
Berman, G.P., Borgonovi, F., Chapline, G., Gurwitz, S.A., Hammel, P.C.,
Pelekhov, D.V., Suter, A. and Tsifrinovich, V.I. (2001): "Formation and
dynamics of a Schrödinger-cat state in continuous quantum measure-
ment" Report quant-ph/0101035.   [362, 367]
Berman, G.P., Borgonovi, F., Chapline, G., Gurwitz, S.A., Hammel, P.C.,
Pelekhov, D.V., Suter, A. and Tsifrinovich, V.I. (2002): "Single-spin mea-
surement and decoherence in magnetic resonance force microscopy." Re-
port quant-ph/0210043.   [367]
Berman, G.P. and Zaslavsky, G.M. (1978): "Condition of Stochasticity in
Quantum Nonlinear Systems." *Physica* **91A**, 450–460.   [158]
Berry, M.V. (1977): "Semi-classical mechanics in phase space: a study of
Wigner's function." *Phil. Trans. Roy. Soc.* **287**, 237–271.   [97]
Berry, M.V. (1978): "Regular and Irregular Motion." In: *Topics in Nonlinear
Dynamics: A Tribute to Sir Edward Bullard*, ed. by S. Jorna (AIP, New
York).   [150]
Bertet, P., Osnaghi, S., Rauschenbeutel, A., Nogues, G., Auffeves, A., Brune,
M., Raimond, J.M., and Haroche, S. (2001): "A complementarity experi-
ment with an interferometer at the quantum-classical boundary." *Nature*
411, 166–170.   [28]
Bhattacharya, T., Habib, S. and Jacobs, K. (1999): "Continuous quantum
measurement and the emergence of classical chaos." Report quant-ph/
9906092.   [366]
Bixon, M. (1982): "Medium stabilization of localized Born-Oppenheimer
states." *Chem. Phys.* **70**, 199–206.   [105]
Blanchard, Ph. and Jadczyk, A. (1993): "Strongly coupled quantum and
classical systems and Zeno's effect." *Phys. Lett.* **A183**, 272–276.   [124]
Blanchard, Ph. and Jadczyk, A. (1995): "Event-enhanced formalism of quan-
tum theory or Columbus solution to the quantum measurement problem."
In: *Quantum Communications and Measurement*, ed. by V. P. Belavkin,
O. Hirota, and R. L. Hudson (Plenum Press, New York), p. 223–233.
                                                                    [331, 338, 367]
Blanchard, Ph., Combe, Ph., and Zheng, W. (1987): *Mathematical and Phys-
ical Aspects of Stochastic Mechanics* (Springer, Berlin).   [368]
Blanchard, Ph. and Olkiewicz, R. (2000): "Effectively classical quantum
states for open systems." *Phys. Lett.* **A 273**, 223–231.   [331]

Blanchard, Ph., Giulini, D., Joos, E., Kiefer, C., and Stamatescu, I.-O. (2000): *Decoherence: Theoretical, Experimental and Conceptual Problems*, Lectures Notes in Physics **538** (Springer, Berlin).   [3, 363]

Blatt, J.M. (1959): "An Alternative Approach to the Ergodic Problem." *Prog. Theor. Phys.* **22**, 745–756.   [42]

Blatt, R. and Zoller, P. (1988): "Quantum jumps in atomic systems." *Eur. J. Phys.* **9**, 250–256.   [164]

Bloch, F. (1946): "Nuclear Induction." *Phys. Rev.* **70**, 460–474.   [133]

Block, E. and Berman, P.R. (1991): "Quantum Zeno effect and quantum Zeno paradox in atomic physics." *Phys. Rev.* **A44**, 1466–1472.   [123]

Bocko, M.F. and Onofrio, R. (1996): "On the measurement of a weak classical force coupled to a harmonic oscillator: experimental progress." *Rev. Mod. Phys.* **68**, 755–799.   [148]

Bogolubov, N.N., Logunov, A.A., Oksak, A.I., and Todorov, I.T. (1990): *General Principles of Quantum Field Theory* (Kluwer, Dordrecht).   [61, 261, 274, 276, 279, 288, 312]

Bohm, D. (1952): "A Suggested Interpretation of the Quantum Theory in Terms of 'Hidden' Variables." *Phys. Rev.* **85**, 166–193; reprinted in Wheeler and Zurek (1983).   [13]

Bohr, N. (1928): "The Quantum Postulate and the Recent Development of Atomic Theory." *Nature* **121**, 580–590; reprinted in Wheeler and Zurek (1983).   [23]

Bohr, N. (1949): "Discussion with Einstein on epistemological problems in atomic physics.". In: *Albert Einstein: Philosopher-Scientist*, ed. by P.A. Schilpp (Open Court Publishing, Evanston), p. 200–241; reprinted in Wheeler and Zurek (1983).   [76]

Bohr, N. and Rosenfeld, L. (1933): "Zur Frage der Messbarkeit der elektromagnetischen Feldgrössen." *Det Kgl. Danske Videnskabernes Selskab. Mathematisk - fysiske Meddelelser* **XII**, 8, 3–65; English translation in Wheeler and Zurek (1983).   [182, 190]

Bollinger, J.J., Heinzen, D.J., Itano, W.M., Gilbert, S.L., and Wineland, D.J. (1989): "Test of the linearity of quantum mechanics by rf spectroscopy on the $^9Be^+$ ground state." *Phys. Rev. Lett.* **63**, 1031–1034.   [27]

Bóna, P. (1973): "Measurement and irreversibility in infinite systems." *acta phys. slov.* **23**, 149–157.   [341]

Booss, B. and Bleecker, D.D. (1985): *Topology and Analysis* (Springer, New York).   [293]

Bordé, C.J., Courtier, N., du Burck, F., Goncharov, A.N., and Gorlicki, M. (1994): "Molecular interferometry experiments." *Phys. Lett.* **A188**, 187–197.   [76]

Borel, E. (1914a): *Introduction Géométrique à Quelques Théories Physiques* (Gauthier-Villars, Paris).   [41]

Borel, E. (1914b): *Le Hasard* (Alcan, Paris).   [33, 41]

Börner, G. and Gottlöber, S. (eds.) (1997): *The evolution of the Universe* (Wiley, Chichester).   [209]

Born, M. (1926): "Zur Quantenmechanik der Stoßvorgänge." *Z. Phys.* **37**, 863–867; english translation in Wheeler and Zurek (1983).   [23]

Born, M. (1969): *Albert Einstein, Hedwig und Max Born, Briefwechsel 1916–1955* (Nymphenburger Verlagshandlung, München).   [27, 63]

Bouwmeester, D., Ekert A., and Zeilinger, A. (2000): *The Physics of Quantum Information* (Springer, Berlin).   [147]

Bouwmeester, D., Pan, J.-W., Mattle, K., Eibl, M., Weinfurter, H., and Zeilinger, A. (1997): "Experimental quantum teleportation." *Nature* **390**, 575–579.   [172]

Braginski, V.B. and Khalili, F. Ya. (1996): "Quantum nondemolition measurements: the route from toys to tools." *Rev. Mod. Phys.* **68**, 1–11. [148]

Braginski, V.B., Vorontsov, Y.I., and Thorne, K.S. (1980): "Quantum nondemolition measurements." *Science* **209**, 547–557; reprinted in Wheeler and Zurek (1983).   [46, 147]

Brandenberger, R., Laflamme, R., and Mijić, M. (1990): "Classical perturbations from decoherence of quantum fluctuations in the inflationary universe." *Mod. Phys. Lett.* **A5**, 2311–2317.   [216]

Braun, D. (2001): *Dissipative Quantum Chaos and Decoherence* (Springer, Berlin)   [155]

Braun, D., Haake, F., and Strunz, W.T. (2001): "Universality of Decoherence." *Phys. Rev. Lett.* **86**, 2913–2917.   [50, 90]

Braunstein, S.L. (1996): "Quantum teleportation without irreversible detection." *Phys. Rev.* **A53**, 1900–1902.   [174]

Breuer, H.-P. and Petruccione, F. (1995): "Stochastic dynamics of open quantum systems: Derivation of the differential Chapman-Kolmogorov equation." *Phys. Rev.* **E51**, 4041–4054.   [421]

Breuer, H.P. and Petruccione, F. (eds.) (1999): *Open Systems and Measurement in Relativistic Quantum Theory* (Springer, Berlin).   [363]

Breuer, H.-P. and Petruccione, F. (2001): "Destruction of quantum coherence through emission of bremsstrahlung." *Phys. Rev.* **A63**, 032102.   [186]

Breuer, H.-P. and Petruccione, F. (2002a): *The Theory of Open Quantum Systems* (Oxford University Press, Oxford)   [314, 421]

Breuer, H.P. and Petruccione, F. (2002b): "Radiation damping and decoherence in QED." Report quant-ph/0210013.   [363]

Breuer, T. (1993): "Inaccurate Experiments and Environment-Induced Superselection Rules." *Int. J. Theor. Phys.* **32**, 2253–2259.   [365]

Brezger, B., Hackermüller, L., Uttenthaler, S., Petschinka, J., Arndt M., and Zeilinger, A. (2002): "Matter-Wave Interferometer for Large Molecules." *Phys. Rev. Lett.* **88**, 100404.   [67]

Brillouin, L. (1962): *Science and Information Theory* (Academic Press, New York).   [15]

Brillouin, L. (1964): *Scientific Uncertainty, and Information* (Academic Press, New York).   [41]

Brukner, C. and Zeilinger, A. (2000): "Conceptual Inadequacy of the Shannon Information in Quantum Measurements." Report quant-ph/0006087.   [14]

Brun, T.A. (1993): "Quasiclassical equations of motion for nonlinear Brownian systems." *Phys. Rev.* **D47**, 3383–3393.   [232, 246]

Brun, T.A. (1997): "Continuous measurement, quantum trajectories and decoherent histories." Report quant-ph/9710021.   [367]

Brun, T.A. (2002): "A simple model of quantum trajectories." *Am. J. Phys.* **70**, 719–37.   [227]

Brune, M., Hagley, E., Dreyer, J., Maître, X., Maali, A., Wunderlich, C., Raimond, J.M., and Haroche, S. (1996): "Observing the Progressive Decoherence of the 'Meter' in a Quantum Measurement." *Phys. Rev. Lett.* **77**, 4887–4890.   [15, 143]

Brune, M., Haroche, S., Raimond, J.M., Davidovich, L., and Zagury, N. (1992): "Manipulation of photons in a cavity by dispersive atom-field coupling: Quantum-nondemolition measurements and generation of 'Schrödinger cat' states." *Phys. Rev.* **A45**, 5193–5214.   [143]

Bruno, D., Facchi, P., Longo, S., Minelli, P., Pascazio, S., and Scardicchio, A. (2002): "Quantum Zeno effect in a multilevel molecule." Report quant-ph/0206143.   [126]

Bub, J. (1968): "The Danieri-Loinger-Prosperi quantum theory of measurement." *Il Nuovo Cimento* **57B**, 503–520.   [386]

Buchholz, D. (1982): "The Physical State Space of Quantum Electrodynamics." *Comm. Math. Phys.* **85**, 49–71.   [307, 309]

Buchholz, D. (2000): "Algebraic quantum field theory: a status report." Report math-ph/0011044.   [261]

Bunge, M. and Kalnay, A.J. (1983): "Solution of Two Paradoxes in the Quantum Theory of Unstable Systems." *Il Nuovo Cimento* **77B**, 1–8.   [112]

Busch, P, Cassinelli, G., DeVito E., Lahti, P., and Levrero, A. (2001): "Teleportation and Measurements." *Phys. Lett.* **A284**, 141–145 – quant-ph/0102121.   [10, 174]

Busch, P., Lahti, P.J. and Mittelstaedt, P. (1996): *The quantum theory of measurement* (Springer, Berlin).   [360]

Caldeira, A.O. and Leggett, A.J. (1983a): "Path Integral Approach to Quantum Brownian Motion." *Physica* **121A**, 587–616.
                    [14, 76, 80, 81, 223, 229, 235, 236]

Caldeira, A.O. and Leggett, A.J. (1983b): "Quantum Tunnelling in a Dissipative System." *Ann. Phys. (N.Y.)* **149**, 374–456.   [103, 225]

Caldeira, A.O. and Leggett, A.J. (1985): "Influence of damping on quantum interference: An exactly soluble model." *Phys. Rev.* **A31**, 1059–1066.
                    [76, 81, 89]

Callan, C.G., Giddings, S.B., Harvey, J.A., and Strominger, A. (1992): "Evanescent black holes." *Phys. Rev.* **D45**, R1005–1009.   [219]

Calzetta, E. and Hu, B.L. (1994): "Noise and fluctuations in semiclassical gravity." *Phys. Rev.* **D51**, 6636–6655.   [207]

Calzetta, E. and Hu, B.L. (1995): "Quantum fluctuations, decoherence of the mean field, and structure formation in the early Universe." *Phys. Rev.* **D52**, 6770–6788.   [217]

Calzetta, E., Roura, A., and Verdaguer, E. (2001): "Master equation for quantum Brownian motion derived by stochastic methods." *Int. J. Theor. Phys.* **40**, 2317–2332. [237]

Camacho, A. and Camacho-Galván, A. (1999): "Time emergence by self-measurement in a quantum isotropic universe." *Nuovo Cim.* **B114**, 923–938. [208]

Carmichael, H. (1993): *An Open Systems Approach to Quantum Optics* (Springer, Heidelberg). [50, 331, 338]

Carmichael, H.J., Singh, S., Vyas, R., and Rice, P.R. (1989): "Photoelectron waiting times and atomic state reduction in resonance fluorescence." *Phys. Rev.* **A39**, 1200–1218. [163, 165]

Casati, G. and Chirikov, B.V. (1995a): *Quantum Chaos: Between Order and Disorder* (Cambridge University Press). [149]

Casati, G. and Chirikov, B.V. (1995b): "Comment on 'Decoherence, Chaos, and the Second Law'." *Phys. Rev. Lett.* **75**, 350. [151]

Casati, G., Chirikov B.V., Izraelev, F.M., and Ford. J (1979): "Stochastic behaviour of a quantum pendulum under a periodic perturbation." In: *Stochastic Behaviour in Classical and Quantum Hamiltonian Systems*, Lectures Notes in Physics **93** (Springer, Berlin). [153]

Castagnino, M., Laura, R. and Id Betan, R. (2001): "Functional approach to quantum decoherence and the classical limit II." Report quant-ph/0103107. [363, 365]

Caticha, A. (1995): "Construction of exactly soluble double-well potentials." *Phys. Rev.* **A51**, 4264–4267. [102]

Caves, C.M. (1981): "Quantum-mechanical noise in an interferometer." *Phys. Rev.* **D23**, 1693–1708. [148]

Caves, C.M. and Milburn, G.J. (1987): "Quantum-mechanical model for continuous position measurements." *Phys. Rev.* **A36**, 5543–5555. [64, 331]

Caves, C.M., Thorne, K.S., Drever, R.W.P., Sandberg, V.P., and Zimmerman, M. (1980): "On the measurement of a weak classical force coupled to a quantum-mechanical oscillator. I. Issues of principle." *Rev. Mod. Phys.* **52**, 341–392. [148]

Chapman, M.S., Hammond, T.D., Lenef, A., Schmiedmayer, J., Rubenstein, R.A., Smith, E., and Pritchard, D.E. (1995): "Photon Scattering from Atoms in an Atom Interferometer: Coherence Lost and Regained." *Phys. Rev. Lett.* **75**, 3783–3787. [78, 179]

Chaves, A.S., Figueiredo, J.M., and Nemes, M.C. (1994): "Metric fluctuations, thermodynamics, and classical physics: a proposed connection." *Ann. Phys. (N.Y.)* **231**, 174–184. [364]

Chen, X., Kuang, L.-M., and Ge, M.-L. (1995): "Decoherence described by Milburn's theory for the two-photon Jaynes-Cummings model with Stark shift." *J. Phys. A* **28**, L267–L274. [134]

Chiao, R. and Kwiat, P. (2002): "Heisenberg's Introduction of the 'Collapse of the Wavepacket' into Quantum Mechanics." Report quant-ph/0201036. [363]

Chiu, C.B., Sudarshan, E.C.G., and Misra, B. (1977): "Time evolution of unstable quantum states and a resolution of Zeno's paradox." *Phys. Rev.* **D16**, 520–529.   [113]

Choi, M. D. (1972): "Positive linear maps on C*-algebras." *Can. J. Math.* **24**, 520–529.   [354]

Chou, K., Su, S., Hao, B., and Yu, L. (1985): "Equilibrium and nonequilibrium formalism made unified." *Phys. Rep.* **118**, 1–131.   [183]

Chruściński, D. and Staszewski, P. (1992): "On the asymptotic solutions of Belavkin's stochastic wave equation." *Physica Scripta* **45**, 193–199. [366]

Chuang, I.L., Laflamme, R., Shor, P.W., and Zurek, W.H. (1995): "Quantum Computers, Factoring, and Decoherence." *Science* **270**, 1633–1635.
                                                                                    [126, 146]

Chumakov, S.M., Hellwig, K.-E., and Rivera, A.L. (1995): "Quantum Zeno effect in unitary quantum mechanics." *Phys. Lett.* **A197**, 73–82.   [124]

Cina, J.A. and Harris, R.A. (1995): "Superpositions of Handed Wave Functions." *Science* **267**, 832–833.   [102, 106]

Cirelli, R., Manià, A., and Pizzocchero, L. (1990): "Quantum mechanics as an infinite-dimensional Hamiltonian system with uncertainty structure: Part I." *Jour. Math. Phys.* **31**, 2891–2897.   [419]

Clauser, J.F. and Shimony, A. (1978): "Bell's theorem: experimental tests and implications." *Rep. Prog. Phys.* **41**, 1881–1927.   [43]

Claverie, P. and Jona-Lasinio, G. (1986): "Instability of tunneling and the concept of molecular structure in quantum mechanics: The case of pyramidal molecules and the enantiomer problem." *Phys. Rev.* **A33**, 2245–2253.   [104]

Clifton, R. and Monton, B. (1999): "Loosing your marbles in wavefunction collapse theories." Report quant-ph/9905065.   [378]

Cohen-Tannoudji, C. (1992): *Atom-Photon Interactions* (Wiley, New York).
                                                                                      [184]

Cohen-Tannoudji, C. and Dalibard, J. (1986): "Single-Atom Laser Spectroscopy. Looking for Dark Periods in Fluorescence Light." *Europhys. Lett.* **1**, 441–448.   [163]

Coleman, S. (1988a): "Black holes as red herrings: Topological fluctuations and the loss of quantum coherence." *Nucl. Phys.* **B307**, 867–882.
                                                                                    [222, 223]

Coleman, S. (1988b): "Why there is nothing rather than something: A theory of the cosmological constant." *Nucl. Phys.* **B310**, 643–668.   [223]

Collet, B., Pearle, P., Avignone, F., and Nussinov, S. (1995): "Constraint on Collapse Models by Limit on Spontaneous X-Ray Emission in Ge." *Found. Phys.* **25**, 1399–1412.   [380]

Collet, B. and Pearle, P. (2002): " Wavefunction collapse and random walk." Report quant-ph/0208009.   [378]

Conradi, H.D. and Zeh, H.D. (1991): "Quantum cosmology as an initial value problem." *Phys. Lett.* **A154**, 321–326.   [209, 391]

Cook, R.J. (1988): "What are Quantum Jumps?" *Physica Scripta* **T21**, 49–51.   [118, 162]

Cook, R.J. and Kimble, H.J. (1985): "Possibility of Direct Observation of Quantum Jumps." *Phys. Rev. Lett.* **54**, 1023–1026.   [162]

Dalibard, J., Castin, Y., and Mølmer, K. (1992): "Wave-Function Approach to Dissipative Processes in Quantum Optics." *Phys. Rev. Lett.* **68**, 580–583.   [50]

Dalibard, J., Dupont-Roc, J., and Cohen-Tannoudji, C. (1982): "Vacuum fluctuations and radiation reaction: identification of their respective contributions." *J. Physique* **43**, 1617–1638.   [180]

Damgaard, P. and Hüffel, H., eds. (1988): *Stochastic Quantization* (World Scientific, Singapore).   [368]

Damnjanovic, M. (1990): "Quantum evolution disturbed by successive collapses." *Phys. Lett.* **144**, 277–281.   [124, 378]

Danieri, A., Loinger, A., and Prosperi, G.M. (1962): "Quantum theory of measurement and ergodicity conditions." *Nucl. Phys.* **33**, 297–319; reprinted in Wheeler and Zurek (1983).   [386]

Dattagupta, S. (1984): "Brownian motion of a quantum system." *Phys. Rev.* **A30**, 1525–1527.   [81]

Dattagupta, S. and Singh, J. (1996): "Stochastic motion of a charged particle in a magnetic field: II Quantum Brownian treatment." *Pramana* **47**, 211–224.   [339]

Dattagupta, S. and Singh, J. (1997): "Landau diamagnetism in a dissipative and confined system." *Phys. Rev. Lett.* **79**, 961–965.   [339]

Davidovich, L., Brune, M., Raimond, J.M., and Haroche, S. (1996): "Mesoscopic quantum coherences in cavity QED: preparation and decoherence monitoring schemes." *Phys. Rev.* **A53**, 1295–1309.   [10, 145, 191]

Davidovich, L., Maali, A., Brune, M., Raimond, J.M., and Haroche, S. (1993): "Quantum Switches and Nonlocal Microwave Fields." *Phys. Rev. Lett.* **71**, 2360–2363.   [143]

Davies, E.B. (1973): "The harmonic oscillator in a heat bath," *Commun. Math. Phys.* **33**, 171–186.   [329]

Davies, E.B. (1974): "Markovian master equations." *Commun. Math. Phys.* **39**, 91–110.   [327, 331]

Davies, E.B. (1976): *Quantum Theory of Open Systems* (Academic Press, New York).   [319, 323, 331]

Dehmelt, H. (1975): "Proposed $10^{14} \Delta \nu > \nu$ Laser Fluorescence Spectroscopy on Tl$^+$Mono-Ion Oscillator II." *Bull. Am. Phys. Soc.* **20**, 60.   [162]

Dekker, H. (1977): "Quantization of the linearly damped harmonic oscillator." *Phys. Rev.* **A16**, 2126–2134.   [80]

Dekker, H. (1991): "Multilevel tunneling and coherence: Dissipative spin-hopping dynamics at finite temperatures." *Phys. Rev.* **A44**, 2314–2323.   [103]

Demers, J.-G. and Kiefer, C. (1996): "Decoherence of black holes by Hawking radiation." *Phys. Rev.* **D53**, 7050–7061.   [219, 220]

d'Espagnat, B. (1966): "An elementary note about mixtures.". In: *Preludes in theoretical physics*, ed. by De-Shalit, Feshbach and v. Hove (North-Holland Publishing Comp., Amsterdam), p. 185–191.   [36, 43]

d'Espagnat, B. (1976): *Conceptual Foundations of Quantum Mechanics* (W. A. Benjamin, Reading, MA). [22, 43, 407]

d'Espagnat, B. (1979): "The Quantum Theory and Reality." *Scientific American* **241**, 128–140. [43]

d'Espagnat, B. (1989): "Are there realistically interpretable local theories?" *J. Stat. Phys.* **56**, 747–766. [238, 258]

d'Espagnat, B. (1995): *Veiled Reality* (Addison-Wesley, Reading/MA). [1, 22, 43, 238, 241, 242, 258, 381]

d'Espagnat, B. (2001): "A note on measurement." *Phys. Lett.* **A282**, 133–137 – quant-ph/0101141. [27, 381]

Deutsch, D. (1985): "Quantum Theory: The Church-Turing Principle and the Universal Quantum Computer." *Proc. R. Soc. Lond.* **A400**, 97–117. [146]

Deutsch, D. and Jozsa, R. (1992): "Rapid solution of problems by quantum computation." *Proc. R. Soc. Lond.* **A439**, 553–558. [146]

Dewdney, C., Hardy, L., and Squires, E.J. (1993): "How late measurements of quantum trajectories can fool a detector." *Phys. Lett.* **A184**, 6–11. [162]

DeWitt, B.S. (1967): "Quantum theory of gravity I. The canonical theory." *Phys. Rev.* **160**, 1113–1148. [197]

Dicke, R.H. (1986): "On Observing the Absence of an Atom." *Found. Phys.* **16**, 107–113. [162]

Dicke, R.H. (1989): "Quantum Measurements, Sequential and Latent." *Found. Phys.* **19**, 385–395. [126]

Dickson, M. (1995): "What is preferred about the preferred basis?" *Found. Phys.* **25**, 423–440. [373]

Dieks, D. (1982): "Communication by EPR devices." *Phys. Lett.* **A92**, 271–272. [146]

Dieks, D. (1995): "Physical motivation of the modal interpretation of quantum mechanics." *Phys. Lett.* **A197**, 367–371. [37]

Diósi, L. (1985): "Orthogonal Jumps of the Wavefunction in White-Noise Potentials." *Phys. Lett.* **A112**, 288–292. [29, 367, 373]

Diósi, L. (1986): "Stochastic pure state representation for open quantum systems." *Phys. Lett.* **A114**, 451–454. [38, 367]

Diósi, L. (1988a): "Continuous quantum measurement and Itô formalism." *Phys. Lett.* **A129**, 419–423. [334, 378]

Diósi, L. (1988b): "Localized solution of a simple nonlinear quantum Langevin equation." *Phys. Lett.* **A132**, 233–236. [334, 338, 367, 374, 378]

Diósi, L. (1988c): "Quantum stochastic processes as models for state vector reduction." *J. Phys.* **A21**, 2855–2898. [334]

Diósi, L. (1989): "Models for universal reduction of macroscopic quantum fluctuations." *Phys. Rev.* **A40**, 1165–1174. [334, 364]

Diósi, L. (1990a): "Landau's density matrix in quantum electrodynamics." *Found. Phys.* **20**, 63–70. [182]

Diósi, L. (1990b): "Relativistic theory for continuous measurement of quantum fields." *Phys. Rev.* **A42**, 5086–5092. [187]

Diósi, L. (1992): "Coarse graining and decoherence translated into von Neumann language." *Phys. Lett.* **B280**, 71–74.   [366]

Diósi, L. (1993a): "High-Temperature Markovian Equation for Quantum Brownian Motion." *Europhys. Lett.* **22**, 1–3.   [85, 378]

Diósi, L. (1993b): "Continuously diagonalized density operator of open systems." In: *Fundamental Problems in Quantum Physics*, ed. by M. Ferrero and A. van der Merwe (Kluwer).   [367, 378]

Diósi, L. (1994a): " Unique quantum paths by continuous diagonalization of the density operator." *Phys. Lett.* **A185**, 5–8.   [367, 378]

Diósi, L. (1994b): "Quantum Particle As Seen In Light Scattering." In: *Waves and Particles in Light and Matter*, ed. by A. Garuccio and A. v. d. Merwe (Plenum Press, New York).   [180]

Diósi, L. (1995a): "Comments on 'Objectification of classical properties induced by quantum vacuum fluctuations'." *Phys. Lett.* **A197**, 183–184.
                                                                               [188]

Diósi, L. (1995b): "Permanent state reduction: motivations, results, and by-products." In: *Quantum Communications and Measurement*, ed. by V. P. Belavkin, O. Hirota, and R. L. Hudson (Plenum Press), p. 245–250. [378]

Diósi, L. (1999): "Emergence of classicality: from collapse phenomenologies to hybrid dynamics." Report quant-ph/9902087.   [367]

Diósi, L., Gisin, N., Halliwell, J., and Percival, I.C. (1994): "Decoherent Histories and Quantum State Diffusion." *Phys. Rev. Lett.* **74**, 203–207.
                                                                         [366, 367, 378]

Diósi, L., Gisin, N. and Strunz, W.T. (2000): "Royal road to coupling classical and quantum dynamics." *Phys. Rev.* **A61**, 2218 – quant-ph/9902069.
                                                                               [367]

Diósi, L. and Halliwell, J.J. (1998): "Coupling classical and quantum variables using continuous measurement theory" *Phys. Rev.Lett.* **81**, 2846–2849.   [367]

Diósi, L. and Kiefer C. (2000): "Robustness and Diffusion of Pointer States." *Phys. Rev. Lett.* **85**, 3552.   [16, 38]

Diósi, L. and Kiefer, C. (2002): "Exact positivity of the Wigner and P-functions of a Markovian open system." *J. Math. Phys.* **A35**, 2675–2683.
                                                                          [94, 99, 139]

Diósi, L. and Lukácz, B. (1994): *Stochastic Evolution of Quantum States in Open Systems and in Measurement Processes* (World Scientific, Singapore).   [39]

Dirac, P.A.M. (1930, 1947) *The Principles of Quantum Mechanics.* (Clarendon Press, Oxford).   [8, 268]

Dirac, P.A.M. (1964): *Lectures on Quantum Mechanics.* Belfer Graduate School of Science. (Yeshiva University, New York).   [304]

Dittrich, T. and Graham, R. (1990a): "Long Time Behavior in the Quantized Standard Map with Dissipation." *Ann. Phys. (N.Y.)* **200**, 363–421. [155]

Dittrich, T. and Graham, R. (1990b): "Continuous quantum measurements and chaos." *Phys. Rev.* **A42**, 4647–4660.   [155]

Divakaran, P.P. (1990): "Exotic quantum effects in two space dimensions: The role of translation invariance." *Phys. Rev.* **B42**, 6717–6719.   [288]

Divakaran, P.P. (1994): "Symmetries and quantization: structure of the state space." *Rev. Math. Phys.* **6**, 167–205.   [284, 285]

Divakaran, P.P. and Rajagopal, A.K. (1991): "Superconductivity in layered materials." *Physica* **C176**, 457–476.   [288]

Dixmier, J. (1981): *Von Neumann Algebras* (North Holland, Amsterdam)   [275]

Dominguez A.E., Kozameh, C.N., and Ludvigsen, M. (1997): "Superselection sectors in asymptotic quantization of gravity." Report gr-qc/9609071   [315]

Donald, M.J. (1995): "A Mathematical Characterization of the Physical Structure of Observers." *Found. Phys.* **25**, 529–571.   [31]

Doob, J.L. (1953): *Stochastic Processes* (Wiley, New York).   [433]

Dove, C. and Squires, E.J. (1995): "Symmetric Versions of Explicit Wavefunction Collapse Models." *Found. Phys.* **25**, 1267–1282.   [378]

Dove, C. and Squires, E.J. (1996): "A Local Model of Explicit Wavefunction Collapse." Report quant-ph/9605047.   [377]

Dowker, F. and Henson, J. (2002): "A spontaneous collapse model on a lattice." Report quant-ph/0209051.   [377]

Dowker, F. and Kent, A. (1995): "Properties of consistent histories." *Phys. Rev. Lett.* **75**, 3038–3041.   [258]

Dowker, F. and Kent, A. (1996): "On the consistent histories approach to quantum mechanics." *J. Stat. Phys.* **82**, 1575–1646.   [258]

Dowker, F. and Halliwell, J.J. (1992): "The quantum mechanics of history: The decoherence functional in quantum mechanics." *Phys. Rev.* **D46**, 1580–1609.   [240, 243, 248]

Dümcke, R. and Spohn, H. (1979): "The proper form of the generator in the weak coupling limit." *Z. Physik* **B34**, 419–422.   [330]

Dum, R., Zoller, P., and Ritsch, H. (1992): "Monte Carlo simulation of the atomic master equation for spontaneous emission." *Phys. Rev.* **A45**, 4879–4887.   [50]

Dürr, D. and Spohn, H. (2000): "Decoherence through coupling to the radiation field." In: Blanchard *et al.* (2000), p. 77–86.   [186]

Dürr, D., Fusseder, W., Goldstein, S., and Zanghi, N. (1993): "Comment on 'Surrealistic Bohm Trajectories'." *Z. Naturforsch.* **48a**, 1261–1262. [162]

Dürr, S., Nonn, T., and Rempe, G. (1998): "Origin of quantum-mechanical complementarity probed by a 'which-way' experiment in an atom interferometer." *Nature* **395**, 33–37.   [78]

Dyakonov, M.I. (2001): "Quantum Computing: A View from the Enemy Camp." to appear in: *Future Trends in Microelectronics: The Nano Millenium"*, ed. by S. Luryi, J. Xu and A. Zaslavsky (Wiley) – cond-mat/0110326.   [147]

Dyson, F. (1949): "The S Matrix in Quantum Electrodynamics." *Phys. Rev.* **75**, 1736–1755.   [11]

Eberly, J.H., Narozhny, N.B., and Sanchez-Modragon, J.J. (1980): "Periodic Spontaneous Collapse and Revival in a Simple Quantum Model." *Phys. Rev. Lett.* **44**, 1323–1326.   [140]

Einstein, A. (1917): "Zum Quantensatz von Sommerfeld und Epstein." *Verhandlungen der Deutschen Physikalischen Gesellschaft* **19**, 82–92.   [149]

Einstein, A., Podolsky, B., and Rosen, N. (1935): "Can quantum-mechanical description of physical reality be considered complete?" *Phys. Rev.* **47**, 777–780; reprinted in Wheeler and Zurek (1983).   [42]

Ekert, A. and Knight, P.L. (1995): "Entangled quantum systems and the Schmidt decomposition." *Am. J. Phys.* **63**, 415–423.   [44]

Ekstein, H. and Siegert, A.J.F. (1971): "On a Reinterpretation of Decay Experiments." *Ann. Phys. (N.Y.)* **68**, 509–520.   [112]

Elattari, B. and Gurvitz, S.A. (2000): "Influence of measurement on the lifetime and the line-width of unstable systems." *Phys. Rev.* **A62**, 032102.
                                                                        [132]

Elby, A. and Bub, J. (1994): "Triorthogonal uniqueness theorem and its relevance to the interpretation of quantum mechanics." *Phys. Rev.* **A49**, 4213–4216.   [44]

Elitzur, A.C. and Vaidman, L. (1993): "Quantum Mechanical Interaction-Free Measurements." *Found. Phys.* **23**, 987–997.   [115, 162]

Ellis, J., Hagelin, J., Nanopoulos, D.V., and Srednicki, M. (1984): "Search For Violations Of Quantum Mechanics." *Nucl. Phys.* **B241**, 381–405.
                                                                   [224, 225]

Ellis, J., Kanti, P., Mavromatos, N.E., Nanopoulos, D.V., and Winstanley, E. (1998): "Decoherent scattering of light particles in a D-brane background." *Mod. Phys. Lett.* **A13**, 303–320.   [221, 225]

Ellis, J., Mavromatos, N.E., and Nanopoulos, D.V. (1996): "CPT and superstring." Report hep-ph/9607434 (unpublished).   [224, 225]

Ellis, J., Mavromatos, S., and Nanopoulos, D.V. (1995): "A non-critical string approach to black holes, time and quantum dynamics." Lecture at *Int. School of Subnuclear Physics: 31st Course: From Supersymmetry to the Origin of Space-Time*, Erice, Italy, 4-12 July 1993, Report CERN-TH-7195-94.   [364]

Ellis, J., Mohanty, S., and Nanopoulos, D.V. (1989): "Quantum Gravity and the Collapse of the Wave Function." *Phys. Lett.* **B221**, 113–119. [30, 364]

Ellis, J., Mohanty, S., and Nanopoulos, D.V. (1990): "Wormholes violate quantum mechanics in SQUIDs." *Phys. Lett.* **B235**, 305–312.
                                                          [66, 223, 224, 364]

Emch, G. G. (1972): "On quantum measurement processes." *Helv. Phys. Acta* **45**, 1049–1056.   [343]

Englert, B.-G., Schwinger, J., and Scully, M.O. (1988): "Is Spin Coherence Like Humpty-Dumpty? I. Simplified Treatment." *Found. Phys.* **18**, 1045–1056.   [49]

Englert, B.-G., Scully, M.O. and Walther, H. (1995): "Complementarity and uncertainty." *Nature* **375**, 367–368.   [78]

Englert, B.G., Scully, M.O., Süssmann, G., and Walther, H. (1992): "Surrealistic Bohm trajectories." *Z. Naturf.* **47a**, 1175–1186.   [14, 162]

Englert, B.-G., Scully, M.O., Süssmann, G., and Walther, H. (1993): "Reply to Comment on 'Surrealistic Bohm Trajectories'." *Z. Naturforsch.* **48a**, 1263–1264.   [162]

Eppley, K. and Hannah, E. (1977): "The necessity of quantizing the gravitational field." *Found. Phys.* **7**, 51–68.   [193]

Epstein, S.T. (1960): "On the Origin of Super-Selection Rules." *Il Nuovo Cimento* **16**, 362–364.   [60]

Erber, T., Hammerling, P., Hockney, G., Porrati, M., and Putterman, S. (1989): "Resonance Fluorescence and Quantum Jumps in Single Atoms: Testing the Randomness of Quantum Mechanics." *Ann. Phys. (N.Y.)* **190**, 254–309.   [165]

Everett, H., III. (1957): "'Relative State' Formulation of Quantum Mechanics." *Rev. Mod. Phys.* **29**, 454–462; reprinted in Wheeler and Zurek (1983).   [14]

Facchi, P. (2002): "Quantum Zeno effect, adiabaticity and dynamical superselection rules." Report quant-ph/0202174.   [124]

Facchi, P. and Pascazio, S. (2002): "Quantum Zeno subspaces and dynamical superselection rules." Report quant-ph/0207030.   [117, 126]

Facchi, P., Mariano, A., and Pascazio, S. (2002): "Mesoscopic interference." *Recent Res. Devel. Physics* **3**, 1–29.   [77]

Facchi, P., Nakazato, H., and Pascazio S. (2001): "From the Quantum Zeno to the Inverse Quantum Zeno Effect." *Phys. Rev. Lett.* **86**, 2699–2703.   [132]

Facchi, P., Pascazio, S., Scardicchio, A., and Schulman, L.S. (2001a): "Zeno dynamics yields ordinary constraints." *Phys. Rev.* **A65**, 012108-1–5. [118]

Favre, C., Martin, P.A. (1968): "Dynamique quantique des systèmes amortis ≪ non markoviens ≫." *Helv. Phys. Acta* **41**, 333–361. [318, 319, 323, 329]

Fearn, H., Cook, R.J., and Milonni, P.W. (1995): "Sudden replacement of a Mirror by a Detector in Cavity QED: Are Photons Counted Immediately?" *Phys. Rev. Lett.* **74**, 1327–1330.   [39]

Feller, W. (1968): *An Introduction to the Theory of Probability and its Applications* (Wiley, New York).   [433]

Ferrari, R., Picasso, L.E., and Strocchi, F. (1974): "Some Remarks on Local Operators in Quantum Electrodynamics." *Comm. Math. Phys.* **35**, 25–38.   [313]

Ferrari, R., Picasso, L.E., and Strocchi, F. (1977): "Local Operators and Charged States in Quantum Electrodynamics." *Il Nuovo Cimento* **A35**, 1–8.   [313]

Feynman, R.P. and Vernon, F.L., jr. (1963): "The Theory of a General Quantum System Interacting with a Linear Dissipative System." *Ann. Phys. (N.Y.)* **24**, 118–173.   [37, 80, 229, 233]

Feynman, R.P., Leighton, R.B., and Sands, M. (1965): *The Feynman Lectures on Physics* (Addison-Wesley, Reading).   [8, 76]

Fick, E. and Sauermann, G. (1990): *The Quantum Statistics of Dynamic Processes*, Springer Series in Solid-State Sciences **86** (Springer, Berlin). [327]

Fine, A. (1970): "Insolubility of the quantum measurement problem." *Phys. Rev.* **D2**, 2783–2787. [386]

Finkelstein, J. (1993): "Comment on 'Intrinsic decoherence in quantum mechanics'." *Phys. Rev.* **A47**, 2412–2414. [134, 367]

Fischer, B.R. and Mittelstaedt, P. (1990): "Chirality as a quasi-classical property of molecular systems." *Phys. Lett.* **A147**, 411–416. [104]

Fischer, M.B., Gutiérrez-Madina, and Raizen, M.G. (2001): "Observation of the Quantum Zeno and Anti-Zeno Effects in an Unstable System." *Phys. Rev. Lett.* **87**, 040402-1–4. [121]

Flores, J.C. (1995): "Milburn theory of decoherence using a random kicked dynamics." *Phys. Rev.* **A51**, 2774–2776. [134]

Fonda, L., Ghirardi, G.C., and Rimini, A. (1978): "Decay theory of unstable quantum systems." *Reports on Progress in Physics* **41**, 587–631. [113]

Ford, G.W. and O'Connell, R.F. (2001a): "Decoherence without dissipation." *Phys. Lett.* **A286**, 87–90. [385]

Ford, G.W. and O'Connell, R.F. (2001b): "Exact solution of the Hu-Paz-Zhang master equation." *Phys. Rev.* **D64**, 105020-1–13. [399]

Ford, G.W. and O'Connell, R.F. (2002): "Wave packet spreading: Temperature and squeezing effects with applications to quantum measurement and decoherence." *Am. J. Phys.* **70**, 319–324. [385]

Ford, G.W., Kac, M. and Mazur, P. (1965): "Statistical Mechanics of Assemblies of Coupled Oscillators." *J. Math. Phys.* **6**, 504–515. [80]

Ford, G.W., Lewis, J.T., and O'Connell, R.F. (1988): "Quantum Langevin equation." *Phys. Rev.* **A37**, 4419–4428. [80, 329, 339, 367]

Ford, G.W., Lewis, J.T. and O'Connell, R.F. (1996): "Master equation for an oscillator coupled to the electromagnetic field." *Ann. Phys.* **252**, 362–385. [332]

Ford, G.W., Lewis, J.T. and O'Connell, R.F. (2001): "Quantum measurement and decoherence." *Phys. Rev.* **A64**, 032101-1–4. [339]

Ford, J. (1983): "How random is a coin toss?" *Physics Today* **36**(April), 40–47. [149]

Ford, J. and Ilg, M. (1992): "Eigenfunctions, eigenvalues, and time evolution of finite, bounded, undriven, quantum systems are not chaotic." *Phys. Rev.* **A45**, 6165–6173. [150]

Ford, J. and Mantica, G. (1992): "Does quantum mechanics obey the correspondence principle? Is it complete?" *Am. J. Phys.* **60**, 1086–1098. [150]

Ford, L.H. (1993): "Electromagnetic vacuum fluctuations and electron coherence." *Phys. Rev.* **D47**, 5571–5580. [186]

Frerichs, V. and Schenzle, A. (1991): "Quantum Zeno effect without collapse of the wave packet." *Phys. Rev.* **A44**, 1962–1968. [123]

Friedman, J. and Sorkin, R. (1980): "Spin 1/2 from Gravity." *Phys. Rev. Lett* **44**, 1100-1103. [300]

Friedman, J.R., Patel, V., Chen, W., Tolpygo, S.K., and Lukens, J.E. (2000): "Quantum superposition of distinct macroscopic states." *Nature* **406**, 43–45.   [10, 46, 103]

Friedrich, H. (1998): *Theoretical Atomic Physics*, 2nd edition (Springer, Berlin)   [151]

Fröhlich, J. (1973): "On the infrared problem in a model of scalar electrons and massless, scalar bosons," *Ann. Inst. Henri Poincaré* **19**, 1–103. [346]

Fröhlich, J., Morchio, G., and Strocchi, F. (1979): "Charged Sectors and Scattering States in Quantum Electrodynamics." *Ann. of Phys. (N.Y.)* **119**, 241–284.   [307]

Fu, Y. and Ramaswami, A. (1991): "Transient response in quantum transport of noninteracting electrons in nanostructures." *Phys. Rev.* **B44**, 10884–10887.   [113]

Gagen, M.J., Wiseman, H.M., and Milburn, G.J. (1993): "Continuous position measurements and the quantum Zeno effect." *Phys. Rev.* **A48**, 132–142.   [126]

Galindo, A. and Martín-Delgado, M.A. (2002): "Information and Computation: Classical and Quantum Aspects." *Rev. Mod. Phys* **74**, 347–423 – quant-ph/0112105.   [147]

Galindo, A. and Pascual, P. (1990): *Quantum Mechanics I* (Springer, Berlin).   [297, 427]

Galindo, A., Morales, A., and Nuñez-Lagos, R. (1962): "Superselection Principle and Pure States of n Identical Particles" *Jour. Math. Phys.* **3**, 324–328.   [273, 274]

Gallis, M.R. (1992): "Spatial correlations of random potentials and the dynamics of quantum coherence." *Phys. Rev.* **A45**, 47–53.   [86]

Gallis, M.R. (1993): "Model for local Ohmic quantum dissipation." *Phys. Rev.* **A48**, 1028–1034.   [86]

Gallis, M.R. (1996): "The Emergence of Classicality via Decoherence Described by Lindblad Operators." *Phys. Rev.* **A53**, 655–660.   [138]

Gallis, M.R. and Fleming, G.N. (1990): "Environmental and spontaneous localization." *Phys. Rev.* **A42**, 38–48.   [78, 82, 380, 395]

Gallis, M.R. and Fleming, G.N. (1991): "Comparison of quantum open-system models with localization." *Phys. Rev.* **A43**, 5778–5786.   [88]

Gangui, A., Mazzitelli, F.D., and Castagnino, M.A. (1991): "Loss of coherence in multidimensional minisuperspaces." *Phys. Rev.* **D43**, 1853–1858.   [208]

Gardiner, C.W. (1983): *Handbook of Stochastic Methods* (Springer, Berlin).   [80, 399]

Gardiner, C.W. (1991): *Quantum Noise* (Springer, Heidelberg). [50, 133, 338]

Garraway, B.M. and Knight, P.L. (1994): "Evolution of quantum superpositions in open environments: Quantum trajectories, jumps, and localization in phase space." *Phys. Rev.* **A50**, 2548–2563.   [50, 135, 143, 367]

Garraway, B.M., Knight, P.L., and Steinbach, J. (1995): "Dissipation of quantum superpositions." *Appl. Phys* **B60**, S63-S68.   [50, 97]

Gea-Banacloche, J. (1990): "Collapse and Revival of the State Vector in the Jaynes-Cummings Model: An Example of State Preparation by a Quantum Apparatus." *Phys. Rev. Lett.* **65**, 3385–3388.   [140]

Gell-Mann, M. and Hartle, J.B. (1990): "Quantum mechanics in the light of quantum cosmology.". In: *Complexity, Entropy, and the Physics of Information*, ed. by W.H. Zurek (Addison-Wesley, Reading), p. 425–458.
[239]

Gell-Mann, M. and Hartle, J.B. (1993): "Classical equations for quantum systems." *Phys. Rev.* **D47**, 3345–3382.   [229, 241, 243, 246]

Gell-Mann, M. and Hartle, J.B. (1994): "Time symmetry and asymmetry in quantum mechanics and quantum cosmology." In: *Physical Origins of Time Asymmetry*, ed. by J.J. Halliwell, J. Pérez-Mercader, and W.H. Zurek (Cambridge University Press), p. 311–345.   [253]

Gell-Mann, M. and Hartle, J.B. (1995): "Strong decoherence." In: *Proceedings of the 4th Drexel Symposium on Quantum Non-Integrability*, ed. by D.-H. Feng and B.-L. Hu (International Press, Boston/Hongkong). [242]

Gell-Mann, M. and Pais, A. (1955): "Behavior of Neutral Particles under Charge Conjugation." *Phys. Rev.* **97**, 1387–1389.   [10]

Georgiev, D.D. (2002): "Where do the photons collapse - in the retina or in the brain cortex?" Report quant-ph/020805.   [368]

Gervais, J.-L. and Zwanziger, D. (1980): "Derivation from First Principles of the Infrared Structure of Quantum Electrodynamics." *Phys. Lett.* **B94**, 389–393.   [304, 305, 309]

Ghirardi, G.C. (1992): "An attempt at a macrorealistic quantum worldview." Report ICTP Trieste, IC/92/392.   [369, 378]

Ghirardi, G.C. (1994a): "Spontaneous wave packet reduction.", Lecture at *Conf. of Fundamental Problems of Quantum Theory*, Baltimore, MD, June 18-22, 1994, Report ICTP Trieste, IC/94/117. [364, 369, 377, 378]

Ghirardi, G.C. (1994b): "Testing Quantum Mechanics at the Macroscopic Level." Lecture at *First Int. Conf. on Phenomenology of Verification from Present to Future*, Rome, March 23-26, 1994, Report ICTP Trieste IC/94/165.   [364, 369, 378]

Ghirardi, G.C. (1998): "Quantum superpositions and definite perceptions." Report quant-ph/9810028.   [378, 381]

Ghirardi, G.C. (1999): "Some Lessons from Relativistic Reduction Models." In: *Open Systems and Measurement in Relativistic Quantum Theory* ed. by H.P. Breuer and F. Petruccione (Springer, Berlin).   [364, 377]

Ghirardi, G.C. (2000): "Local measurements of nonlocal observables and the relativistic reduction process." Report quant-ph/0003149.   [377]

Ghirardi, G.C. and Bassi, A. (1998): "Do dynamical reduction models imply that arithmetic does not apply to ordinary macroscopic objects?" Report quant-ph/9810041.   [378, 381]

Ghirardi, G.C. and Marinatto (2002): "Entanglement and properties." Report quant-ph/0206021.   [46]

Ghirardi, G.C., Grassi, R., and Benatti, F. (1995): "Describing the macroscopic world: closing the circle within the dynamical reduction program." *Found. Phys.* **25**, 5–38.   [364, 369, 376]

Ghirardi, G.C., Grassi, R., and Pearle, P. (1990): "Relativistic dynamical reduction models and nonlocality." *Joensuu Found. Mod. Phys.*, 1990:109–123.   [377]

Ghirardi, G.C., Grassi, R., and Rimini, A. (1990): "Continuous spontaneous reduction model involving gravity." *Phys. Rev.* **A42**, 1057–1064.   [364, 380]

Ghirardi, G.C., Grassi, R., Butterfield, J., and Fleming, G.N. (1993): "Parameter dependence and outcome dependence in dynamical models for state vector." *Found. Phys.* **23**, 341–364.   [376, 377]

Ghirardi, G.C., Pearle, P., and Rimini, A. (1990): "Markov processes in Hilbert space and continuous spontaneous localization of systems of identical particles." *Phys. Rev.* **A42**, 78–89.
                              [334, 335, 337, 373, 374, 375, 376, 377, 378]

Ghirardi, G.C., Rimini, A., and Weber, T. (1975): "Quantum dynamical semi-groups and the reduction process." *Il Nuovo Cimento* **30B**, 133–144.   [331]

Ghirardi, G.C., Rimini, A., and Weber, T. (1986): "Unified dynamics for microscopic and macroscopic systems." *Phys. Rev.* **D34**, 470–491.
                                    [29, 66, 78, 331, 334, 338, 370]

Ghirardi, G.C., Rimini, A., and Weber, T. (1987): "Disentanglement of quantum wave functions: Answer to 'Comment on "Unified dynamics for microscopic and macroscopic systems"'." *Phys. Rev.* **D36**, 3287–3289.
                                                                  [79, 378]

Gibbons, G. (1992): "Typical states and density matrices." *Jour. Geom. Phys.* **8**, 147–162.   [419, 421]

Giddings, S.B. and Strominger, A. (1988a): "Axion induced topology change in quantum gravity and string theory." *Nucl. Phys.* **B306**, 890–907. [222]

Giddings, S.B. and Strominger, A. (1988b): "Loss of incoherence and determination of coupling constants in quantum gravity." *Nucl. Phys.* **B307**, 854–866.   [223]

Gisin, N. (1984): "Quantum measurement and stochastic processes". *Phys. Rev. Lett.* **52**, 1657–1660.   [334, 338, 378]

Gisin, N. (1989): "Stochastic quantum dynamics and relativity." *Helv. Phys. Acta* **62**, 363–371.   [334]

Gisin, N. and Cibils, M.B. (1992): "Quantum diffusions, quantum dissipation and spin relaxation." *J. Phys.* **A25**, 5165–5176.   [338, 367, 373]

Gisin, N. and Percival, I. (1992): "The quantum-state diffusion model applied to open systems." *J. Phys.* **A25**, 5677–5691.
                                      [38, 50, 334, 338, 373, 374, 378]

Gisin, N. and Percival, I. (1993a): "Quantum state diffusion, localization and quantum dispersion entropy". *J. Phys.* **A26**, 2233–2243.
                                              [334, 338, 367, 374]

Gisin, N. and Percival, I.C. (1993b): "The quantum state diffusion picture of physical processes". *J. Phys.* **A26**, 2245–2260.   [334, 338, 367, 373, 374]

Gisin, N. and Percival, I.C. (1993c): "Stochastic wave equations versus parallel world components." *Phys. Lett.* **A175**, 144–145.   [367]

Giulini, D. (1995a): "Asymptotic Symmetry Groups of Long-Ranged Gauge Configurations." *Mod. Phys. Lett.* **A10**, 2059–2070.   [293, 295, 300]

Giulini, D. (1995b): "Quantum Mechanics on Spaces with Finite Fundamental Group." *Helv. Phys. Acta* **68**, 438–469.   [274, 298]

Giulini, D. (1996): "On Galilei Invariance in Quantum Mechanics and the Bargmann Superselection Rule." *Ann. Phys. (N.Y.)* **249**, 222–235.
    [279, 299, 429, 431]

Giulini, D. and Kiefer, C. (1995): "Consistency of semiclassical gravity." *Class. Quantum Grav.* **12**, 403–411.   [199]

Giulini, D., Kiefer, C., and Zeh, H.D. (1995): "Symmetries, superselection rules, and decoherence." *Phys. Lett.* **A199**, 291–298. [13, 15, 188, 302, 312]

Glauber, R.J. (1963): "Coherent and Incoherent States of the Radiation Field." *Phys. Rev.* **131**, 2766–2788.   [9, 138]

Gleason, A.M. (1957): "Measures on the Closed Subspaces of a Hilbert Space." *Jour. Math. Mech.* **6**, 885–893.   [263]

Gnedenko, B.V. (1982): *The Theory of Probability* (MIR Publishers, Moscow).
    [433]

Goetsch, P. and Graham, R. (1993): "Quantum trajectories for nonlinear optical processes." *Ann. Physik* **2**, 706–719.   [50]

Goetsch, P. and Graham, R. (1994): "Linear stochastic wave equations for continuously measured quantum systems." *Phys. Rev.* **A50**, 5242–5255.
    [50, 334]

Goetsch, P., Graham, R., and Haake, F. (1995): "Schrödinger cat states and single runs for the damped harmonic oscillator." *Phys. Rev.* **A51**, 136–142.   [50, 143]

Gohberg, I., Goldberg, S., and Krupnik, N. (2000): *Traces and Determinants of Linear Operators* (Birkhäuser, Basel).   [356]

Goldstein, S. and Page, D.N. (1995): "Linearly positive histories: Probabilities for a robust family of sequences of quantum events." *Phys. Rev. Lett.* **74**, 3715–3719.   [243]

Gordon, J.P. (1967): "Quantum Theory of a Simple Maser Oscillator." *Phys. Rev.* **161**, 369–386.   [133]

Gorini, V., Frigerio, A., Verri, M., Kossakowski, A., and Sudarshan, E.C.G. (1978): "Properties of quantum markovian master equations." *Rep. Math. Phys.* **13**, 149–173.   [331]

Gorini, V., Kossakowski, A., and Sudarshan, E.C.G. (1976): "Completely positive dynamical semigroups of N-level systems." *J. Math. Phys.* **17**, 821–825.   [133, 331, 332]

Grabert, H. and Weidlich, W. (1974): "Masterequation, H-theorem and stationary solution for coupled quantum systems." *Z. Physik* **268**, 139–143.
    [326]

Grabert, H., Schramm, P., and Ingold, G.-L. (1988): "Quantum Brownian Motion: the functional integral approach." *Phys. Rep.* **168**, 115–207.
[86, 231]

Graham, N. (1970): *The Everett Interpretation of Quantum Mechanics* (Univ. North Carolina, Chapel Hill). [32]

Greenberger, D.M. (1983): "The neutron interferometer as a device for illustrating the strange behavior of quantum systems." *Rev. Mod. Phys.* **55**, 875–905. [58, 287]

Greenberger, D.M., Horne, M.A., and Zeilinger, A. (1993): "Multiparticle interferometry and the superposition principle." *Physics Today* **46** (Aug.), 22–29. [78]

Greenberger, D.M., Horne, M., Shimony, A., and Zeilinger, A. (1990): "Bell's theorem without inequalities." *Am. J. Phys.* **58**, 1131–1143. [43]

Greenland, P.T. and Lane, A.M. (1989): "Effect of disturbances and decay: Zeno paradox without observations." *Phys. Lett.* **A134**, 429–434. [124]

Grempel, D.R., Prange, R.E. and Fishman, S. (1984): "Quantum dynamics of a nonintegrable system." *Phys. Rev.* **A29**, 1639–1647. [154]

Griffin, J.J. and Wheeler, J.A. (1957): "Collective Motions in Nuclei by the Method of Generator Coordinates." *Phys. Rev.* **108**, 311–327. [392]

Griffiths, R.B. (1984): "Consistent Histories and the Interpretation of Quantum Mechanics." *J. Stat. Phys.* **36**, 219–272. [38, 238, 241, 257]

Griffiths, R.B. (1987): "Correlations in separated quantum systems: A consistent histories analysis of the EPR problem." *Am. J. Phys.* **55**, 11–17.
[239]

Griffiths, R.B. (1993): "Consistent interpretation of quantum mechanics using quantum trajectories." *Phys. Rev. Lett.* **70**, 2201–2204. [239, 253]

Griffiths, R.B. (1994): "A consistent history approach to the logic of quantum mechanics." In: *Symposium on the Foundations of Modern Physics 1994*, ed. by K.V. Laurikainen, C. Montonen, and K. Sunnarborg (Edition Frontières, Gif-sur-Yvette), p. 85. [238, 239, 242]

Griffiths, R.B. and Hartle, J.B. (1998): "Comment on 'Consistent sets yield contrary inferences in quantum theory'." *Phys. Rev. Lett.* **81**, 1981. [258]

Grishchuk, L.P. and Sidorov, Yu.V. (1989): "On the quantum state of relic gravitons." *Class. Quantum Grav.* **6**, L161–165. [213]

Gross, D.J. (1995): "Symmetry in Physics: Wigner's Legacy." *Phys. Today* **48**, (Dec.) 46–50. [10]

Grössing, G. and Zeilinger, A. (1991): "Zeno's paradox in quantum cellular automata." *Physica* **D50**, 321–326. [121]

Grübl, G. (2002): "The quantum measurement problem enhanced." Report quant-ph/0202101. [362]

Gurvitz, S.A.. (1998): "Dephasing and collapse in continuous measurement of a single system." Report quant-ph/9808058. [363]

Gutzwiller, M.C. (1990): *Chaos in Classical and Quantum Mechanics* (Springer, New York). [149, 150]

Haag, R. (1963): "Bemerkungen zum Nahwirkungsprinzip in der Quantenphysik." *Ann. Phys.* (Leipzig) **11**, 29–34. [313]

Haag, R. (1992): *Local Quantum Physics* (Springer, Berlin).   [21, 298]

Haake, F. (1973): "Statistical treatment of open systems by generalized master equations." In: *Springer Tracts in Modern Physics* **66**, ed. by G. Höhler (Springer, Berlin), p. 98–168.   [318, 327, 329]

Haake, F. (2001): *Quantum Signatures of Chaos*, 2nd edition (Springer, Berlin).   [149, 151]

Haake, F. and Walls, D.F. (1987): "Overdamped and amplifying meters in the quantum theory of measurement." *Phys. Rev.* **A36**, 730–739.   [50]

Haake, F. and Zukowski, M. (1993): "Classical motion of meter variables in the quantum theory of measurement." *Phys. Rev.* **A47**, 2506–2517. [50]

Habib, S. (1990): "Classical limit in quantum cosmology: Quantum mechanics and the Wigner function." *Phys. Rev.* **D42**, 2566–2576.   [97]

Habib, S. (1994): "Quantum Diffusion." Report hep-th/9410181.   [367]

Habib, S., Kluger, Y., Mottola, E., and Paz, J.P. (1996): "Dissipation and decoherence in mean field theory." *Phys. Rev. Lett.* **76**, 4660-4663. [189, 191]

Habib, S., Shizume, K. and Zurek, W.H. (1998): "Decoherence, Chaos, and the Correspondence Principle." *Phys. Rev. Lett.* **80**, 4361–4365 – quant-ph/9803042.   [157]

Hahn, E.L. (1950): "Spin Echos." *Phys. Rev.* **80**, 580–594.   [49]

Haken, H. (1984): *Laser Theory* (Springer, Heidelberg).   [133]

Haken, H. (2002): *Brain Dynamics* (Springer, Berlin).   [108]

Hakim, V. and Ambegaokar, V. (1985): "Quantum theory of a free particle interacting with a linearly dissipative environment." *Phys. Rev.* **A32**, 423–434.   [365]

Haller, K. (1978): "Gupta-Bleuler condition and Infrared-Coherent States." *Phys. Rev.* **D18**, 3045-3051.   [309]

Halliwell, J.J. (1987): "Correlations in the wave function of the Universe." *Phys. Rev.* **D36**, 3626–3640.   [97]

Halliwell, J.J. (1989): "Decoherence in quantum cosmology." *Phys. Rev.* **D39**, 2912–2923.   [207]

Halliwell, J.J. (1991): "Introductory lectures on quantum cosmology." In: *Quantum cosmology and baby universes*, ed. by S. Coleman, J.B. Hartle, T. Piran, and S. Weinberg (World Scientific, Singapore), p. 159–243.   [197, 201]

Halliwell, J.J. (1998): "Effective theories of coupled classical and quantum variables from decoherent histories: A new approach to the backreaction problem." *Phys. Rev.* **D57**, 2337–2348.   [193]

Halliwell, J.J. (2002): "Decoherent histories for spacetime domains." In: *Time in quantum mechanics*, ed. by J.G. Muga, R. Sala Mayato, and I.L. Egusquiza (Springer, Berlin), p. 153–182.   [228]

Halliwell, J.J. and Hawking, S.W. (1985): "Origin of structure in the Universe." *Phys. Rev.* **D31**, 1777–1791.   [200]

Halliwell, J.J. and Ortiz, M.E. (1993): "Sum-over-histories origin of the composition laws of relativistic quantum mechanics and quantum cosmology." *Phys. Rev.* **D48**, 748–768.   [256]

Halliwell, J.J., Pérez-Mercader, J., and Zurek, W.H., eds. (1994): *Physical Origins of Time Asymmetry* (Cambridge University Press).   [325, 330]

Hamermesh, M. (1989): *Group Theory and its Application to Physical Problems* (Dover Publications, New York).   [288]

Hartle, J.B. and Taylor, J.R. (1969): "Quantum Mechanics of Paraparticles." *Phys. Rev.* **178**, 2043-2051.   [274]

Hameroff, S. (1998): "Quantum computing in brain microtubules? The Penrose-Hameroff Orch Or model of consciousness." *Phil. Tr. Royal Sc., London, A*, 1-28.   [368]

Haroche, S. (1995): "Mesoscopic Coherences in Cavity QED." *Il Nuovo Cimento* **110B**, 545–556.   [143]

Haroche, S. (1998): "Entanglement, Decoherence and the Quantum/Classical Boundary." *Physics Today* **51** (July) 36–42.   [145]

Haroche, S. and Kleppner, D. (1989): "Cavity quantum electrodynamics." *Physics Today* (Jan.), 24–30.   [49, 113, 133]

Haroche, S. and Raimond, J.-M. (1996): "Quantum computing: dream or nightmare?" *Physics Today* **49** (Aug.) 51–52 , letters **49** (Nov.) 107–108.   [147]

Haroche, S., Brune, M., Raimond, J.-M., and Davidovich, L. (1993): "Mesoscopic quantum coherences in cavity QED." In: *Fundamentals of Quantum Optics III*, ed. by F. Ehlotzky (Springer, Berlin), p. 223–236. [95, 145]

Harris, R.A. and Stodolsky, L. (1981): "On the time dependence of optical activity." *J. Chem. Phys.* **74**, 2145–2155.   [57, 105]

Harris, R.A. and Stodolsky, L. (1982): "Two state systems in media and 'Turing's paradox'." *Phys. Lett.* **166B**, 464–468.   [105]

Hartle, J.B. (1991): "The quantum mechanics of cosmology." In: *Quantum Cosmology and Baby Universes*, ed. by S. Coleman *et al.* (World Scientific, Singapore), p. 65–157.   [241, 255]

Hartle, J.B. (1995): "Spacetime quantum mechanics and the quantum mechanics of spacetime." In: *Gravitation et Quantifications*, ed. by B. Julia and J. Zinn-Justin (North-Holland, Amsterdam), p. 285–480.   [255]

Hasselbach, F., Kiesel, H., and Sonnentag, P. (2000): "Exploration of the fundamentals of quantum mechanics by charged particle interferometry." In: Blanchard *et al.* 2000, p. 201–12.   [185]

Hawking, S.W. (1982): "The unpredictability of quantum gravity." *Comm. Math. Phys.* **87**, 395–415.   [66]

Hawking, S.W. (1987): "Quantum Coherence down the Wormhole." *Phys. Lett.* **195B**, 337–343.   [30]

Hawking, S.W. and Ross, S. (1997): "Loss of quantum coherence through scattering off virtual black holes." *Phys. Rev.* **D56**, 6403–6415. [220, 223]

Hegerfeldt, G. (1993): "How to reset an atom after a photon detection: Applications to photon-counting processes." *Phys. Rev.* **A47**, 449–455. [50]

Hegerfeldt, G. (1994): "Causality Problems for Fermi's Two-Atom System." *Phys. Rev. Lett.* **72**, 596–599.   [39]

Hegerfeldt, G. and Kraus, K. (1968): "Critical Remark on the Observability of the Sign Change of Spinors under 2π Rotations." *Phys. Rev.* **170**, 1185–1186.   [287, 300]

Hegerfeldt, G.C., Kraus, K., and Wigner, E.P. (1968): "Proof of the Fermion Superselection Rule without the Assumption of Time-Reversal Invariance." *J. Math. Phys.* **9**, 2029–2031.   [58, 260]

Heisenberg, W. (1927): "Über den anschaulichen Inhalt der quantentheoretischen Kinematik und Mechanik." *Z. Phys.* **43**, 172–198; English translation in Wheeler and Zurek (1983).   [179]

Heisenberg, W. (1957): "Quantum Theory of Fields and Elementary Particles." *Rev. Mod. Phys.* **29**, 269–278.   [391]

Heisenberg, W. (1958): *Die physikalischen Prinzipien der Quantentheorie* (Bibliographisches Institut, Mannheim).   [179]

Heisenberg, W. (1973): *Der Teil und das Ganze. Gespräche im Umkreis der Atomphysik* (Deutscher Taschenbuchverlag).   [107]

Hendry, P.C., Lawson, N.S., Lee, R.A.M., McClintock, P.V.E., and Williams, C.D.H. (1994): "Generation of defects in superfluid ^{4}He as an analogue of the formation of cosmic strings." *Nature* **368**, 315–317.   [106]

Hepp, K. (1972): "Quantum theory of measurement and macroscopic observables." *Helv. Phys. Acta* **45**, 236–248.   [37, 340, 341, 365, 387]

Herzog, Th.J., Kwiat, P.G., Weinfurter, H., and Zeilinger, A. (1995): "Complementarity and The Quantum Eraser." *Phys. Rev. Lett.* **75**, 3034–3037.
                                                                       [29, 78]

Hill, D.L. and Wheeler, J.A. (1953): "Nuclear Constitution and the Interpretation of Fission Phenomena." *Phys. Rev.* **89**, 1102–1145.   [392]

Hillery, M. and Scully, M.O. (1983): "On State Reduction and Observation in Quantum Optics: Wigner's Friends and Their Amnesia.". In: *Quantum Optics, Experimental Gravitation, and Measurement Theory*, ed. by P. Meystre and M.O. Scully (Plenum Press, New York), p. 65.   [78]

Hillery, M., O'Connell, R.F., Scully, M.O., and Wigner, E.P. (1984): "Distribution functions in physics: Fundamentals." *Phys. Rep.* **106**, 121–167.
                                                                      [91, 399]

Hioe, F.T. and Eberly, J.H. (1981): "N-Level Coherence Vector and Higher Conservation Laws in Quantum Optics and Quantum Mechanics." *Phys. Rev. Lett.* **47**, 838–841.   [119]

Holevo, A.S. (1995): "On the structure of covariant dynamical semigroups." *J. Funct. Anal.* **131**, 255–278.   [332]

Holevo, A.S. (1998): "Covariant quantum dynamical semigroups: Unbounded generators," in *Irreversibility and Causality. Semigroups and Riggid Hilbert Space* Lect. Notes Phys. **504**, ed. by A. Bohm, H.-D. Doebner, and P. Kielanovski (Springer, Berlin), p. 67–81.   [332]

Home, D. and Whitaker, M.A.B. (1986): "Reflections on the quantum Zeno paradox." *J. Phys.* **A19**, 1847–1854.   [113]

Home, D. and Whitaker, M.A.B. (1992): "Negative-result experiments, and the requirement of wavefunction collapse." *J. Phys.* **A25**, 2387–2394.
                                                                        [162]

Home, D. and Whitaker, M.A.B. (1993): "A unified framework for quantum Zeno processes." *Phys. Lett.* **A173**, 327–331.   [114]

Horowitz, G.T. (1997): "Quantum states of black holes." In: *Black holes and relativistic stars*, ed. by R.M. Wald (The University of Chicago Press, Chicago), p. 241–66.   [221]

Houtappel, R.M.F., van Dam, H., and Wigner, E.P. (1965): "The Conceptual Basis and Use of the Geometric Invariance Principles." *Rev. Mod. Phys.* **37**, 595–632.   [60]

Hu, B.L. and Matacz, A. (1994): "Quantum Brownian motion in a bath of parametric oscillators: A model for system-field interactions." *Phys. Rev.* **D49**, 6612–6635.   [86]

Hu, B.L. and Matacz, A. (1995): "Back reaction in semiclassical gravity: The Einstein-Langevin equation." *Phys. Rev.* **D51**, 1577–1586.   [207]

Hu, B.L., Paz, J.P., and Zhang, Y. (1992): "Quantum Brownian motion in a general environment: Exact master equation with nonlocal dissipation and colored noise." *Phys. Rev.* **D45**, 2843–2861.   [85, 86, 231, 235, 237]

Hu, B.L., Paz, J.P., and Zhang, Y. (1993): "Quantum Brownian motion in a general environment: II. Nonlinear Coupling and Perturbative Approach." *Phys. Rev.* **D47**, 1576–1594.   [86, 247]

Hudson, R.L. and Parthasarathy, K.R. (1984): "Quantum Ito's formula and stochastic evolutions." *Commun. Math. Phys.* **93**, 301–332.   [338]

Hughston, L.P., Jozsa, R., and Wootters, W.K. (1993): "A Complete Classification of Quantum Ensembles Having a Given Density Matrix." *Physics Letters* **A183**, 14–18.   [267, 421]

Hund, F. (1927): "Zur Deutung der Molekelspektren. III." *Z. Phys.* **43**, 805–826.   [13, 103]

Hunziker, W. (1972): "A note on symmetry operations in quantum mechanics." *Helv. Phys. Acta* **45**, 233–236.   [279]

Inagaki, S., Namiki, M., and Tajiri, T. (1992): "Possible observation of the quantum Zeno effect by means of neutron spin-flipping." *Phys. Lett.* **A166**, 5–12.   [121]

Isham, C.J. (1981): "Topological θ-sectors in canonically quantized gravity." *Phys. Lett.* **B106**, 188–192.   [269, 293]

Isham, C.J. (1992): "Canonical Quantum Gravity and the Problem of Time." In: *Integrable Systems, Quantum Groups, and Quantum Field Theories*, ed. by L.A. Ibart and M.A. Rodrigues (Kluwer, Dordrecht), p. 157–287. [197]

Isham, C.J. (1994a): "Prima facie questions in quantum gravity." In: *Canonical gravity: From classical to quantum*, ed. by J. Ehlers and H. Friedrich (Springer, Berlin), p. 1–21.   [195]

Isham, C.J. (1994b): "Quantum logic and the histories approaches to quantum theory." *J. Math. Phys.* **35**, 2157–2185.   [255]

Isham, C.J., Linden, N., and Savvidon, K. (1998): "Continuous time and consistent histories." *J. Math. Phys.* **39**, 1818–34.   [243]

Itano, W.M., Heinzen, D.J., Bollinger, J.J., and Wineland, D.J. (1990): "Quantum Zeno effect." *Phys. Rev.* **A41**, 2295–2300.   [118, 121]

Jackiw, R. (1988): "Analysis on infinite-dimensional manifolds - Schrödinger representation for quantized fields." In: *Field Theory and Particle Physics*, ed. by O. Eboli, M. Gomes, and A. Santano (World Scientific, Singapore).   [190]

Jackson, J.D. (1975): *Classical Electrodynamics.* (Wiley, New York)   [308]

Jadczyk, A. (1995): "On quantum jumps, events, and spontaneous localization models." *Found. Phys.* **25**, 220–232.   [331, 338, 378]

Jammer, M. (1974): *The Philosophy of Quantum Mechanics* (Wiley, New York).   [1, 360, 364, 386]

Jauch, J.M. (1960): "Systems of Observables in Quantum Mechanics." *Helv. Phys. Acta* **33**, 711–726.   [268, 270, 271]

Jauch, J.M. (1968): *Foundations of Quantum Mechanics* (Addison-Wesley, Reading, MA).   [19, 407]

Jauch, J.M. and Báron, J.G. (1972): "Entropy, information and Szilard's paradox." *Helv. Phys. Acta* **45**, 220–232.   [325]

Jauch, J.M. and Misra, B. (1961): "Supersymmetries and Essential Observables." *Helv. Phys. Acta* **34**, 699–709.   [272]

Jauch, J.M., Wigner, E.P., and Yanase, M.M. (1967): "Some comments concerning measurements in quantum mechanics." *Il Nuovo Cimento* **48B**, 144–151.   [386]

Jayannavar, A.M. (1992): "Comment on some theories of state reduction." *Phys. Lett.* **A167**, 433–434.   [179, 378]

Jibu, M. and Yasue, K. (1993): *Quantum brain dynamics and consciousness* (J. Benjamin Publ. Comp., Amsterdam/Philadelphia).   [368]

Joos, E. (1983): *Das Verhalten quantenmechanischer Systeme unter ständiger Messung: Watchdog-Effekt und Master-Gleichung.* Dissertation, Heidelberg.   [124, 127, 129]

Joos, E. (1984): "Continuous Measurement: Watchdog Effect versus Golden Rule." *Phys. Rev.* **D29**, 1626–1633. [32, 38, 123, 125, 127, 129, 166, 385]

Joos, E. (1986a): "Quantum Theory and the Appearance of a Classical World." *Ann. New York Acad. Sci.* **480**, 6–13.   [405]

Joos, E. (1986b): "Why do we observe a classical spacetime?" *Phys. Lett.* **A116**, 6–8.   [16, 193]

Joos, E. (1987a): "Comment on 'Unified dynamics for microscopic and macroscopic systems'." *Phys. Rev.* **D36**, 3285–3286.   [27, 79, 331, 378]

Joos, E. (1987b): "Time Arrow in Quantum Theory." *J. Non-Equil. Thermodyn.* **12**, 27–44.   [106]

Joos, E. (1989): "Motion of a laser-cooled three-level ion in a harmonic trap." *Z. Phys.* **D14**, 301–306.   [119]

Joos, E. (2000): "Elements of Environmental Decoherence." In: Blanchard *et al.* (2000) – quant-ph/9908008.   [45]

Joos, E. (2003): "Decoherence Unlimited: From Zeno to Classical Motion." Proceedings of *"Time and Matter"*, Venice 2002, ed. by I. Bigi and M. Faessler (to appear).   [128]

Joos, E. and Zeh, H.D. (1985): "The Emergence of Classical Properties Through Interaction with the Environment." *Z. Phys.* **B59**, 223–243.
[3, 14, 15, 16, 53, 65, 78, 88, 185, 237, 331, 380, 395, 397]

Jordan, P. and Klein, O. (1927): "Zum Mehrkörperproblem in der Quantentheorie." *Z. Phys.* **45**, 751–765.   [9]

Kadison, R.V. (1965): "Transformation of states in operator theory and dynamics." *Topology* **3**. Suppl. 2, 177–198.   [279]

Kaempffer, F.A. (1965): *Concepts in Quantum Mechanics* (Academic Press, New York).   [297]

Karakostas, V. (1994): "Limitations on Stochastic Localization Models of State Vector Reduction." *Int. J. Theor. Phys.* **33**, 1645–1659.   [376]

Karkuszewski, Z.P., Jarzynski, Ch., and Zurek, W.H. (2001): "Quantum Chaotic Environments, the Butterfly Effect, and Decoherence." Report quant-ph/0111002.   [160]

Karkuszewski, Z.P., Zakrzeswki, J. and Zurek, W.H. (2001): "Breakdown of correspondence in chaotic systems: Ehrenfest versus localization times." Report nlin.CD/0012048.   [155]

Károlyházy, F., Frenkel, A., and Lukács, B. (1986): "On the possible role of gravity in the reduction of the wave function." In: *Quantum Concepts in Space and Time*, ed. by C.J. Isham and R. Penrose (Clarendon Press), p. 109–128.   [196, 364]

Kato, T. (1966): "Wave operators and similarity for some non-selfadjoint operators." *Math. Annalen* **162**, 258–279.   [348]

Kaulakys, B. and Gontis, V. (1997): "Quantum anti-Zeno effect." *Phys. Rev.* **A56**, 1131–1137.   [156]

Keller, M. and Mahler, G. (1994): "Nanostructures, entanglement and the physics of quantum control." *Journ. Mod. Opt.* **41**, 2537–2555.   [50]

Kent, A. (1997): "Consistent sets yield contrary inferences in quantum theory." *Phys. Rev. Lett.* **78**, 2874–77.   [258]

Kent, A. (1998): "Reply on the comment by Griffiths and Hartle." *Phys. Rev. Lett.* **81**, 1982.   [258]

Kent, A. (2002): "Causal Quantum Theory and the Collapse Locality Loophole." Report quant-ph/0204104.   [363, 377]

Khalfin, L.A. (1958): "Contribution to the Decay Theory of a Quasi-Stationary State." *Sov. Phys. JETP* **6**, 1053–1063.   [112, 113]

Khalfin, L.A. (1968): "Phenomenological Theory of K^0 Mesons and the Non-Exponential Character of the Decay." *JETP Letters* **8**, 65–68.   [111]

Khrennikov, A. (2002): "On the cognitive experiments to test quantum-like behaviour of mind." Report quant-ph/0205092.   [368]

Kiefer, C. (1987): "Continuous measurement of mini-superspace variables by higher multipoles." *Class. Quantum Grav.* **4**, 1369–1382.   [16, 200]

Kiefer, C. (1988): "Wave packets in minisuperspace." *Phys. Rev.* **D38**, 1761–1772.   [197]

Kiefer, C. (1991a): "On the meaning of path integrals in quantum cosmology." *Ann. Phys. (N.Y.)* **207**, 53–70.   [255]

Kiefer, C. (1991b): "Interpretation of the decoherence functional in quantum cosmology." *Class. Quantum Grav.* **8**, 379–391.   [255, 257]

Kiefer, C. (1992): "Decoherence in quantum electrodynamics and quantum cosmology." *Phys. Rev.* **D46**, 1658–1670.   [16, 189, 191, 204, 205, 207]

Kiefer, C. (1993): "Topology, decoherence, and semiclassical gravity." *Phys. Rev.* **D47**, 5414–5421.   [199, 207]

Kiefer, C. (1994): "The semiclassical approximation to quantum gravity." In *Canonical gravity: From classical to quantum*, ed. by J. Ehlers and H. Friedrich (Springer, Berlin), p. 170–212.   [199]

Kiefer, C. (1998a): "Interference of two independent sources." *Am. J. Phys.* **66**, 661–62.   [185]

Kiefer, C. (1998b): "Towards a full quantum theory of black holes." In: *Black holes: Theory and observation*, ed. by F.W. Hehl, C. Kiefer, and R. Metzler (Springer, Berlin), p. 416–450.   [220]

Kiefer, C. (1999): "Thermodynamics of black holes and Hawking radiation." In: *Classical and quantum black holes*, Studies in High Energy Physics, Cosmology and Gravitation, ed. by P. Fré, V. Gorini, G. Magli, and U. Moschella (IOP Publishing, Bristol), p. 17–74.   [218]

Kiefer, C. (2000): "Conceptual issues in quantum cosmology." In: *Towards quantum gravity*, ed. by J. Kowalski-Glikman (Springer, Berlin), p. 158–187.   [194, 197]

Kiefer, C. (2001): "Hawking radiation from decoherence." *Class. Quantum Grav.* **18**, L151–154.   [220]

Kiefer, C. (2003): *Quantum gravity* (Oxford University Press, forthcoming).   [193, 194, 197]

Kiefer, C. and Polarski, D. (1998): "Emergence of classicality for primordial fluctuations: Concepts and analogies." *Ann. Phys. (Leipzig)* **7**, 137 158.   [214, 216]

Kiefer, C. and Zeh, H.D. (1995): "Arrow of time in a recollapsing quantum universe." *Phys. Rev.* **D51**, 4145–4153.   [197, 208, 253]

Kiefer, C., Padmanabhan, T., and Singh, T.P. (1991): "A comparison between semiclassical gravity and semiclassical electrodynamics." *Class. Quantum Grav.* **8**, L185–L192.   [189]

Kiefer, C., Lesgourgues, J., Polarski, D., and Starobinsky, A.A. (1998a): "The coherence of primordial fluctuations produced during inflation." *Class. Quantum Grav.* **15**, L67–72.   [215]

Kiefer, C., Polarski, D., and Starobinsky, A.A. (1998b): "Quantum-to-classical transition for fluctuations in the early universe." *Int. J. Mod. Phys.* **D7**, 455-462.   [215]

Kiefer, C., Polarski, D., and Starobinsky, A.A. (2000): "Entropy of gravitons produced in the early Universe." *Phys. Rev.* **D62**, 043518.   [218]

Kijowski, J., Rudolph, G., Thielmann, A. (1997): "Algebra of Observables and Charge Superselection Sectors for QED on the Lattice." *Commun. Math. Phys.* **188**, 535-564.   [309]

Kimble, H.J., Cook, R.J., and Wells, A.L. (1986): "Intermittent atomic fluorescence." *Phys. Rev.* **A34**, 3190–3195.   [163]

Klavetter, T.S. (1989): "Rotation of Hyperion. II. Dynamics." *The Astro-nomical Journal* **98**, 1855–1874.   [158]

Klein, S.R. and Nystrand, J. (2002): "Does particle decay cause wave function collapse: An experimental test." Report quant-ph/0206060.   [363]

Knight, P. (1992): "Practical Schrödinger cats." *Nature* **357**, 438–439. [143]

Knill, E. and Laflamme, R. (1997): "A theory of quantum error-correcting codes." *Phys. Rev.* **A55**, 900–911 – quant-ph/9604034.   [146]

Knill, E., Laflamme, R., and Zurek W.H. (1998): "Resilient quantum com-putation." *Science* **279**, 342–345 – quant-ph/9702058.   [147]

Kofman, A.G. and Kurizki, G. (2000): "Acceleration of quantum decay pro-cesses by frequent observations." *Nature* **405**, 546–550.   [113]

Kofman, A.G. and Kurizki, G. (2002): "Frequent observations accelerate de-cay: The anti-Zeno effect." Report quant-ph/0102002.   [132]

Koopman, B.O. (1957): "Quantum theory and foundations of probability." In: *Proceedings of Symposia in Applied Mathematics*, Volume **VII**, 97–102, ed. by L.A. MacColl. (American Mathematical Society, McGraw-Hill, New York).   [264, 423]

Kossakowski, A. (1972): "On quantum statistical mechanics of non-hamiltonian systems." *Rep. Math. Phys.* **3**, 247–288.   [331]

Kraus, K. (1971): "General state changes in quantum theory." *Ann. Phys.* **64**, 311–335.   [321, 354]

Kraus, K. (1981): "Measuring Processes in Quantum Mechanics I. Continu-ous Observation and the Watchdog Effect." *Found. Phys.* **11**, 547–576.
                                                               [110, 123, 124]

Kraus, K. (1983) *States, Effects and Operations* Lecture Notes in Physics **190** (Springer, Berlin).   [321]

Krenn, G., Summhammer, J., and Svozil, K. (1996): "Interaction-free prepa-ration." *Phys. Rev.* **A53**, 1228–1231.   [115]

Kreuzer, H.J. (1981): *Nonequilibrium Thermodynamics and its Statistical Foundations* (Clarendon Press, Oxford).   [127]

Krzywicki, A. (1993): "Coherence and decoherence in radiation off colliding heavy ions." *Phys. Rev.* **D48**, 5190–5195.   [86]

Kübler, O. (1972): "Projection method, cranking condition and collective energy." *Nucl. Phys.* **A196**, 113–134.   [391]

Kübler, O. and Zeh, H.D. (1973): "Dynamics of Quantum Correlations." *Ann. Phys. (N.Y.)* **76**, 405–418.   [16, 35, 37, 53, 140, 192, 259, 318]

Kuchař, K.V. (1992): "Time and Interpretations of Quantum Gravity." In: *Proceedings of the 4th Canadian Conference on General Relativity and Relativistic Astrophysics*, ed. by G. Kunstatter, D. Vincent, and J. Williams (World Scientific, Singapore), p. 211–314.   [197]

Kullback, S. and Leibler, R.A. (1951): "On information and sufficiency." *Ann. Math. Stat.* **22**, 79–86.   [325]

Kumar, D. (1984): "Brownian motion of a quantum particle." *Phys. Rev.* **A29**, 1571–1573.   [403]

Kupsch, J. (1998): "The structure of the quantum mechanical state space and induced superselection rules." *Pramana - J. Phys.* **51**, 615–624 – quant-ph/9612033.   [347]

Kupsch, J. (2000a) "Mathematical aspects of decoherence." In: Blanchard *et al.* (2000), p. 125–136.   [347]

Kupsch, J. (2000b): "The role of infrared divergence for decoherence." *J. Math. Phys.* **41**, 5945–5953. Extended version math-ph/9911015v3.
                                                                      [343, 345]

Kupsch, J. (2002): "Decoherence and infrared divergence." *Pramana - J. Phys.* **59**, 195–202 – math-ph/0209010.   [346]

Kupsch, J., Smolyanov, O.G., and Sidorova, N.A. (2001): "States of quantum systems and their liftings." *J. Math. Phys.* **42**, 1026–1037.   [319]

Kurtsiefer, Ch., Pfau, T., Spälter, S., Ekstrom, C.R., and Mlynek, J. (1995): "A Heisenberg microscope for atoms." *Ann. New York Acad. Sci.* **755**, 162–172.   [179]

Kwiat, P.G., Mattle, K., Weinfurter, H., Zeilinger, A., Sergienko, A.V., and Shih, Y.H. (1995): "New High-Intensity Source of Polarization-Entangled Photon Pairs." *Phys. Rev. Lett.* **75**, 4337–4341.   [43]

Kwiat, P.G., Steinberg, A.M., and Chiao, R.Y. (1992): "Observation of a 'quantum eraser': A revival of coherence in a two-photon interference experiment." *Phys. Rev.* **A45**, 7729–7739.   [78]

Kwiat, P.G., Weinfurter, H., and Zeilinger, A. (1996): "Quantum Seeing in the Dark." *Scientific American* **275** (Nov.), 52–58.   [115]

Kwiat, P.G., Weinfurter, H., Herzog, T., Zeilinger, A., and Kasevich, M.A. (1995): "Interaction-Free Measurement." *Phys. Rev. Lett.* **74**, 4763–4766.
                                                                      [115]

Kwiat, P.G., White, A.G., Mitchell, J.R., Nairz, O., Weihs, G., Weinfurter, H., and Zeilinger A. (1999): "High-efficiency quantum interrogation measurements via the quantum Zeno effect." *Phys. Rev. Lett.* **83**, 4725–4728.
                                                                      [116]

Laflamme, R. and Louko, J. (1991): "Reduced density matrices and decoherence in quantum cosmology." *Phys. Rev.* **D43**, 3317–3331.   [204]

Lamb, H. (1900): "On a pecularity of the wave-system due to the free vibrations of a nucleus in an extended medium." *Proc. London Math. Soc.* **XXXII**, 208–211.   [329]

Landauer, R. (1961): "Irreversibility and heat generation in the computing process." *IBM J. Res. Dev.* **5**, 183–191; reprinted in Leff and Rex (1990).
                                                                      [145]

Landauer, R. (1995): "Is quantum mechanics useful?" *Philos. Trans. R. Soc. London* **A353**, 367-376.   [147]

Landauer, R. (1996): "The physical nature of information." *Phys. Lett.* **A217**, 188–193.   [147]

Landau, L. (1927): "Das Dämpfungsproblem in der Wellenmechanik." *Z. Phys.* **45**, 430–441.   [182, 184]

Landsman, N.P. (1990): "Quantization and superselection sectors I, II." *Rev. Math. Phys.* **2**, 45–72, 73–104.   [288]

Landsman, N.P. (1995): "Observation and Superselection in Quantum Mechanics." *Stud. Hist. Phil. Mod. Phys.* **26**, 45–73.   [365]

Lange, L. (1885): "Über das Beharrungsgesetz." *Sitzungsberichte der mathematischen Klasse der königlich Sächsischen Gesellschaft der Wissenschaften*, 333–351.   [303]

Lanz, L. (1994): "Quantum Physics and Objective Description." *Int. J. Theor. Phys.* **33**, 19–29.   [367]

Laskar, J. (1989): "A Numerical Experiment on the Chaotic Behavior of the Solar System.", *Nature* **338**, 237–238.   [157]

Lasota, A. and Mackey, M.S. (1994): *Chaos, Fractals, and Noise* (Springer, New York).   [325, 339]

Lax, M. (1966): "Quantum Noise. IV. Quantum Theory of Noise Sources." *Phys. Rev.* **145**, 110–129.   [133]

Lee, T.D. and Yang, C.N. (1956): "Question of Parity Conservation in Weak Interactions." *Phys. Rev.* **104**, 254–258.   [10]

Leff, H.S. and Rex, A.F. (1990): *Maxwell's Demon: Entropy, Information, Computing* (Princeton University Press).   [145]

Leggett, A.J. (1980): "Macroscopic Quantum Systems and the Quantum Theory of Measurement." *Supp. Prog. Theor. Phys.* **69**, 80–100.   [45]

Leggett, A.J. (1984): "Schrödinger's Cat and her Laboratory Cousins." *Contemp. Physics* **25**, 583–598.   [103]

Leggett, A.J. (1987): "Macroscopic Quantum Tunnelling and Related Matters." *Japanese Journal of Applied Physics* **26, Supplement 26–3**, 1986–1993.   [45, 103]

Leggett, A.J. and Garg, A. (1985): "Quantum Mechanics versus Macroscopic Realism: Is the Flux There when Nobody Looks?" *Phys. Rev. Lett.* **54**, 857–860.   [103]

Leggett, A.J., Chakravarty, S., Dorsey, A.T., Fisher, M.P.A., Garg, A., and Zwerger, W. (1987): "Dynamics of the dissipative two-state system." *Rev. Mod. Phys.* **59**, 1–85.   [103, 346]

Levitan, J. (1988): "Small time behaviour of an unstable quantum system." *Phys. Lett.* **A129**, 267–272.   [113]

Levy-Leblond, J-M. (1963): "Galilei Group and Nonrelativistic Quantum Mechanics." *Jour. Math. Phys.* **4**, 776–788.   [297, 427]

Lewenstein, M., Mossberg, T.W., and Glauber, R.J. (1987): "Dynamical Suppression of Spontaneous Emission." *Phys. Rev. Lett.* **59**, 775–778.   [113]

Lewis, P. (1997): "Quantum Mechanics, Orthogonality and Counting." *British Journal for Philosophy of Science* **48** 313.   [378]

Liddle, A.R. and Lyth, D.H. (2000): *Cosmological inflation and large-scale structure* (Cambridge University Press, Cambridge).   [209]

Lindblad, G. (1973): "Entropy, information and quantum measurements." *Commun. Math. Phys.* **33**, 305–322.   [325]

Lindblad, G. (1975): "Completely positive maps and entropy inequalities." *Commun. Math. Phys.* **40**, 147–151.   [326]

Lindblad, G. (1976): "On the Generators of Quantum Dynamical Semigroups." *Commun. Math. Phys.* **48**, 119–130.   [133, 331, 332]

Li, X.L., Ford, G.W., and O'Connell, R.F. (1990): "Magnetic-field effects on the motion of a charged particle in a heat bath," *Phys. Rev.* **A41**, 5287–5289.   [339]

Lockwood, M. (1989): *Mind, Brain and the Quantum: The Compound 'I'* (Blackwell, Oxford).   [31, 368]

London, F. and Bauer, E. (1939): *La théorie d'observation en Mécanique Quantique* (Hermann, Paris); english translation in Wheeler and Zurek (1983).   [29]

Lorenz, E.N. (1963): "Deterministic nonperiodic flow." *J. Atmos. Sci.* **20**, 130–141.   [149]

Loudon, R. (1983): *The quantum theory of light* (Clarendon Press, Oxford).
                                                                       [133]

Lubkin, E. (1970): "On Violation of the Superselection Rules." *Ann. Phys. (N.Y.)* **56**, 69–80.   [302, 312, 314]

Lüders, G. (1951): "Über die Zustandsänderung durch den Meßprozeß." *Ann. Phys. (Leipzig)* **8**, 322–328.   [20, 263]

Lüders, G. and Zumino, B. (1958): "Connection between spin and statistics." *Phys. Rev.* **110**, 1450–1453.   [288]

Ludwig, G. (1990): *Die Grundstrukturen einer physikalischen Theorie*, 2nd edn. (Springer, Berlin).   [27]

Machida, S. and Namiki, M. (1980): "Theory of Measurement in Quantum Mechanics." *Progr. Theor. Phys.* **63**, 1457–1473.   [20, 367, 384]

Mackey, G.W. (1963): *The Mathematical Foundations of Quantum Mechanics* (Benjamin, New York).   [320]

Mackey, M.C. (1992): *Time's Arrow: The Origins of Thermodynamic Behaviour* (Springer, New York).   [339]

Malin, S. (1993): "The Collapse of Quantum States: A New Interpretation." *Found. Phys.* **23**, 881–893.   [364]

Mandelstam, S. (1962): "Quantum Electrodynamics Without Potentials." *Ann. Phys. (N.Y.)* **19**, 1–24.   [274, 312]

Marchewka, A. and Schuss, Z. (1999): "Measurement as absorption of Feynman trajectories." Report quant-ph/9906078.   [363]

Marshall, W., Simon, C., Penrose, R. and Bouwmeester, D. (2002): "Towards quantum superpositions of a mirror." Report quant-ph/0210001.
                                                                  [362, 381]

Martin, P.Λ. (1979): *Modèles en Mécanique Statistique des Processus Irréversibles* Lecture Notes in Physics **103** (Springer, Berlin).   [319, 323]

Mavromatos, N.E. and Nanopoulos, D.V. (1995): "Critical String (liouville) Approach to Brain Microtubules: State Vector Reduction, Memory Coding and Capacity." Report quant-ph/951202.   [368]

Mayburov, S.N. (1999): "Quantum information, irreversibility and state collaps in some microscopic models of measurement." Report quant-ph/9911105.   [363, 365]

Mayburov, S.N. (2000): "Relational quantum measurements, information and state collapse." Report quant-ph/0006104.   [363, 365]

Meekhof, C., Monroe, C., King, B.E., Itano, W.M., and Wineland, D.J. (1996): "Generation of nonclassical motional states of a trapped atom." *Phys. Rev. Lett.* **76**, 1796–1799. [214]

Mensky, M.B. (1979): "Quantum restrictions for continuous observation of an oscillator." *Phys. Rev.* **D20**, 384–387. [14, 229, 338]

Mensky, M.B. (1993): *Continuous Quantum Measurements and Path Integrals* (IOP Publishing, Bristol). [229, 366]

Mensky, M.B. (1995): "Continuous Quantum Measurements: Restricted Path Integrals and Master Equations." *Phys. Lett.* **A196**, 159–167. [366]

Mensky, M.B. (2000): *Quantum Measurements and Decoherence* (Kluwer Academic). [9]

Mermin, N.D. (1990a): "What's wrong with these elements of reality?" *Physics Today* **43** (June), 9–11. [43]

Mermin, N.D. (1990b): "Quantum mysteries revisited." *Am. J. Phys.* **58**, 731–734. [43]

Mermin, N.D. (1994): "What's wrong with this temptation?" *Physics Today* **47**(June), 9–11. [43]

Mershin, A., Nanopoulos, D.V. and Skoulakis, E.F.C. (2000): "Quantum Brain?" Report quant-ph/0007088. [368]

Meyers, R. (1997): "Pure states don't wear black." *Gen. Rel. Grav.* **29**, 1217–1222. [221]

Meystre, P. and Sargent, M., III. (1991): *Elements of Quantum Optics* (Springer, Berlin). [180]

Meystre, P. and Scully, M.O., eds. (1983): *Quantum Optics, Experimental Gravity and Measurement Theory* (Plenum Press, New York). [148]

Mihokova, E, Pascazio, S., and Schulman, L.S. (1997): "Hindered decay: Quantum Zeno effect through electromagnetic field domination." *Phys. Rev.* **A56**, 25–32. [123]

Mijič, M. (1998): "Particle production and classical condensates in de Sitter space." *Phys. Rev.* **D57**, 2138 2146. [210]

Milburn, G.J. (1988): "The quantum Zeno effect and motional narrowing in a two-level system." *J. Opt. Soc. Am.* **B5**, 1317–1322. [124]

Milburn, G.J. (1991): "Intrinsic decoherence in quantum mechanics." *Phys. Rev.* **A44**, 5401–5406. [134, 367]

Milburn, G.J. (1993): "Reply to 'Comment on "Intrinsic decoherence in quantum mechanics"'." *Phys. Rev.* **A47**, 2415–2416. [134, 367]

Milburn, G.J. and Walls, D.F. (1983a): "Quantum solutions of the damped harmonic oscillator." *Am. J. Phys.* **51**, 1134–1136. [138]

Milburn, G.J. and Walls, D.F. (1983b): "Quantum nondemolition measurements via quadratic coupling." *Phys. Rev.* **A28**, 2065–2070. [148]

Milburn, G.J. and Walls, D.F. (1988): "Effect of dissipation on interference in phase space." *Phys. Rev.* **A38**, 1087–1090. [89, 135]

Milonni, P.W. and Smith, W.A. (1975): "Radiation reaction and vacuum fluctuations in spontaneous emission." *Phys. Rev.* **A11**, 814–824. [180]

Mirman, R. (1969): "Coherent Superposition of Charge States." *Phys. Rev.* **186**, 1380–1383. [302]

Mirman, R. (1970): "Analysis of the Experimental Meaning of Coherent Superposition and the Nonexistence of Superselection Rules." *Phys. Rev.* **D1**, 3349–3363.   [18, 287, 300, 302]

Misra, B. and Sudarshan, E.C.G. (1977): "The Zeno's paradox in quantum theory." *Journal of Mathematical Physics* **18**, 756–763.   [110, 117]

Mølmer, K., Castin, Y., and Dalibard, J. (1993): "Monte Carlo wave-function method in quantum optics." *J. Opt. Soc. Am.* **B10**, 524–537.   [50]

Monroe, C., Meekhof, D.M., King, B.E., and Wineland, D.J. (1996): "A 'Schrödinger Cat' Superposition State of an Atom." *Science* **272**, 1131–1136.   [145]

Monteoliva, D. and Paz, J.P. (2000): "Decoherence and the rate of entropy production in chaotic quantum systems." Report quant-ph/0007052.   [159]

Mooij, J.E., Orlando, T.P., Levitov, L., Tian, L., van der Wal, C.H., and Lloyd, S. (1999): "Josephson Persistent-Current Qubit." *Science* **285**, 1036–1039.   [10]

Morikawa, M. (1990): "Quantum decoherence and classical correlations in quantum mechanics." *Phys. Rev.* **D42**, 2929–2932.   [73]

Mott, N.F. (1929): "The wave mechanics of α-ray tracks." *Proc. R. Soc. Lond.* **A126**, 79–84; reprinted in Wheeler and Zurek (1983).   [89]

Moya-Cessa, H., Buzek, V., Kim, M.S., and Knight, P.L. (1993): "Intrinsic decoherence in the atom-field interaction." *Phys. Rev.* **A48**, 3900–3905.   [134]

Munro, W.J. and Gardiner, C.W. (1996): "Non-rotating-wave master equation." *Phys. Rev.* **A53**, 2633–2640.   [85]

Murray, N. and Holman, M. (1999): "The Origin of Chaos in the Outer Solar System." *Science* **283**, 1877-1881.   [157]

Myatt, C.J., King, B.E., Turchette, Q.A., Sackett, C.A., Kielpinski, D., Itano, W.M., Monroe, C., and Wineland, D.J. (2000): "Decoherence of quantum superpositions through coupling to engineered reservoirs." *Nature* **403**, 269–273.   [145, 179]

Nagourney, W., Sandberg, J., and Dehmelt, H. (1986): "Shelved Optical Electron Amplifier: Observation of Quantum Jumps." *Phys. Rev. Lett.* **56**, 2797–2799.   [164]

Nairz, O., Brezger, B., Arndt, M., and Zeilinger, A. (2001): "Diffraction of Complex Molecules by Structures Made of Light." *Phys. Rev. Lett.* **87**, 160401.   [67]

Nakajima, S. (1958): "On quantum theory of transport phenomena." *Prog. Theor. Phys.* **20**, 948–959.   [321]

Nakano, A. and Pearle, P. (1994): "Statevector reduction in discrete time: a random walk in Hilbert space." *Found. Phys.* **24**, 363–377.   [373]

Nakazato, H., Namiki, M., and Pascazio, S. (1994): "Exponential Behavior of a Quantum System in a Macroscopic Medium." *Phys. Rev. Lett.* **73**, 1063–1066.   [113]

Nakazato, H., Namiki, M., and Pascazio, S. (1996): "Temporal behavior of quantum mechanical systems." *Int. Journal of Modern Physics* **B10**, 247–295.   [124, 367]

Nakazato, H., Namiki, M., Pascazio, S., and Rauch, H. (1995): "On the quantum Zeno effect." *Phys. Lett.* **A199**, 27–32.   [121]

Namiki, M. and Pascazio, S. (1990): "On the possible reduction of the interference term due to statistical fluctuations." *Phys. Lett.* **A147**, 430–434.
                                                                              [367]

Namiki, M. and Pascazio, S. (1993): "Quantum Theory of Measurement Based on the Many-Hilbert-Space Approach." *Phys. Rep.* **232**, 301–411.
                                                                          [367, 384]

Nelson, E. (1967): *Dynamical Theories of Brownian Motion* (Princeton University Press).   [264, 423]

Nelson, E. (1985): *Quantum Fluctuations* (Princeton University Press). [368]

Nemes, M.C. and de Toledo Piza, A.F.R. (1986): "Effective dynamics of quantum subsystems." *Physica* **137A**, 367–388.   [53]

Netterfield, C. B. et al. (2002). "A measurement by BOOMERANG of multiple peaks in the angular power spectrum of the cosmic microwave background." *Astrophys. J.* **571**, 604–614.   [218]

Nicklaus, M. and Hasselbach, F. (1993): "Wien filter: A wave-packet-shifting device for restoring longitudinal coherence in charged-matter-wave interferometers." *Phys. Rev.* **A48**, 152–161.   [185, 186]

Nielsen, M.A. (2002): "Rules for a Complex Quantum World." *Scientific American* **287** (Nov.), 66–75.   [145]

Nielsen, M.A. and Chuang, I.L. (2000): *Quantum Computation and Quantum Information* (Cambridge University Press).   [44, 147]

Olkiewicz, R. (1999): "Environment-induced superselection rules in Markovian regime." *Commun. Math. Phys.* **208**, 245–265.   [331]

Olkiewicz, R. (2000): "Structure of the algebra of effective observables in quantum mechanics." *Ann. Phys.* **286**, 10–22.   [324, 331, 333]

Omnès, R. (1988): "Logical reformulation of quantum mechanics I-III." *J. Stat. Phys.* **53**, 893–975.   [239]

Omnès, R. (1992): "Consistent interpretation of quantum mechanics." *Rev. Mod. Phys.* **64**, 339–382.   [38, 239]

Omnès, R. (1995): "A New Interpretation of Quantum Mechanics and Its Consequences in Epistemology." *Found. Phys.* **25**, 605.   [38, 364]

Omnès, R. (1999): *Understanding Quantum Mechanics* (Princeton University Press, Princeton).   [39, 239, 360, 364, 385]

Omnès, R. (2002): "Decoherence, Irreversibility and Selection by Decoherence of Exclusive Quantum States with Definite Probabilities." *Phys. Rev.* **A65**, 052119.   [362]

Orzalesi, C.A. (1970): "Charges and Generators of Symmetry Transformations in Quantum Field Theory." *Rev. Mod. Phys.* **42**, 381–408.   [313]

Ott, E. (2002): *Chaos in Dynamical Systems* 2nd edition (Cambridge University Press).   [149]

Ott, E., Antonsen, T.M., and Hanson, J. (1984): "Effect of noise on time-dependent quantum chaos." *Phys. Rev. Lett.* **53**, 2187–2190.   [150]

Page, D.N. (1993): "No time asymmetry from quantum mechanics." *Phys. Rev. Lett.* **70**, 4034–4037.   [253]

Page, D.N. (1995): "Sensible quantum mechanics: are only perceptions probabilistic?" Report quant-ph/9506010.   [31]

Palmer, P. F. (1977): "The singular coupling and weak coupling limits." *J. Math. Phys.* **18**, 527–529.   [331]

Parthasarathy, K.R. (1992): *An Introduction to Quantum Stochastic Calculus* (Birkhäuser, Basel).   [263, 333, 338]

Pascazio, S. and Namiki, M. (1994): "Dynamical quantum Zeno effect." *Phys. Rev.* **A50**, 4582–4592.   [123, 170]

Pascazio, S., Namiki, M., Badurek, G., and Rauch, H. (1993): "Quantum Zeno effect with neutron spin." *Phys. Lett.* **A179**, 155–160.   [121]

Pauli, W. (1928): "Über das H-Theorem vom Anwachsen der Entropie vom Standpunkt der neuen Quantenmechanik.". In: *Probleme der Modernen Physik. Arnold Sommerfeld zum 60. Geburtstage, gewidmet von seinen Schülern* (Hirzel, Leipzig), p. 30–45.   [127, 385]

Paz, J.P. and Sinha, S. (1992): "Decoherence and back reaction in quantum cosmology: Multidimensional minisuperspace examples." *Phys. Rev.* **D45**, 2823–2842.   [204, 205]

Paz, J.P. and Zurek, W.H. (1993): "Environment-induced decoherence, classicality, and consistency of quantum histories." *Phys. Rev.* **D48**, 2728–2738.   [172, 247]

Paz, J.P., Habib, S., and Zurek, W.H. (1993): "Reduction of the wave packet: Preferred observables and decoherence time scale." *Phys. Rev.* **D47**, 488–501.   [76, 86, 95, 97]

Pearle, P. (1976): "Reduction of the state vector by a nonlinear Schrödinger equation." *Phys. Rev.* **D13**, 857–868.   [29]

Pearle, P. (1979): "Toward Explaining Why Events Occur." *Int. J. Theor. Phys.* **18**, 489–518.   [37]

Pearle, P. (1982): "Might God toss coins?" *Found. Phys.* **12**, 249–263.   [375, 378]

Pearle, P. (1989): "Combining stochastic dynamical state-vector reduction with spontaneous localization." *Phys. Rev.* **A39**, 2277–2289.   [123, 373, 375, 376]

Pearle, P. (1990): "Toward a relativistic theory of statevector reduction." In: *Sixty-Two Years of Uncertainty*, ed. by A.I. Miller (Plenum Press, New York), p. 193–214.   [364, 373, 377]

Pearle, P. (1993): "Ways to describe dynamical state-vector reduction," *Phys. Rev.* **A48**, 913–923.   [334, 364]

Pearle, P. (1999a): "Relativistic Collapse Model with Tachyonic Features" *Phys. Rev.* **A59**, 80–101.   [377]

Pearle, P. (1999b): " Collapse models." Report quant-ph/9901077. [364, 378]

Pearle, P. and Squires, E. (1994): "Bound State Excitation, Nucleon Decay Experiments, and Models of Wave Function Collapse." *Phys. Rev. Lett.* **73**, 1–5.   [27, 66]

Pearle, P. and Squires, E. (1995): "Gravity, Energy Conservation and Parameter values in Collapse Models." Report quant-ph/9503019.   [380]

Pegg, D.T. (1991): "Wavefunction collapse time." *Phys. Lett.* **A153**, 263–264.   [165]

Pegg, D.T. and Knight, P.L. (1988a): "Interrupted fluorescence, quantum jumps, and wave-function collapse." *Phys. Rev.* **A37**, 4303–4308.   [165]

Pegg, D.T. and Knight, P.L. (1988b): "Interrupted fluorescence, quantum jumps and wavefunction collapse: II." *J. Phys.* **D21**, S128-S130.   [165]

Pegg, D.T., Loudon, R., and Knight, P.L. (1986): "Correlations in light emitted by three-level atoms." *Phys. Rev.* **A33**, 4085–4091.   [163]

Peierls, R.E. and Yoccoz, J. (1957): "The Collective Model of Nuclear Motion." *Proc. Phys. Soc. London* **A70**, 381–387.   [389, 391]

Penrose, R. (1979): "Singularities and time-asymmetry." In: *General Relativity*, ed. by S. W. Hawking and W. Israel (Cambridge University Press, Cambridge), p. 581–638.   [252]

Penrose, R. (1981): "Time asymmetry and quantum gravity." In: *Quantum Gravity Vol. 2*, ed. by Isham, C.J., Penrose, R. and Sciama, D.W. (Clarendon, Oxford).   [208, 330]

Penrose, R. (1986): "Gravity and state vector reduction." In: *Quantum Concepts in Space and Time"*, ed. by C.J. Isham and R. Penrose (Clarendon Press), p. 129-146.   [196, 364]

Penrose, R. (1994): *Shadows of the Mind* (Oxford University Press).
                                                                                [108, 115, 368]

Penrose, R. (1996): "On Gravity's Role in Quantum State Reduction." *Gen. Rel. and Grav.* **28**, 581.   [365]

Percival, I.C. (1994): "Localisation of wide open quantum systems." *J. Phys.* **A27**, 1003–1020.   [135, 367, 374]

Percival, I.C. (1995): "Quantum space-time fluctuations and primary state diffusion." Report quant-ph/9508021.   [367]

Peres, A. (1980): "Zeno paradox in quantum theory." *Am. J. Phys.* **48**, 931–932.   [114]

Peres, A. (1993): *Quantum Theory: Concepts and Methods* (Kluwer, Dordrecht).   [159, 367]

Peres, A. (1995): "Higher order Schmidt decompositions." *Phys. Lett.* **A202**, 16–17.   [44, 362]

Peres, A. (1996): "Separability Criterion for Density Matrices." Phys. Rev. Lett. **77**, 1413–1415.   [36]

Peres, A. (2002): "How the No-Cloning Theorem Got its Name." Report quant-ph/0205076.   [146]

Peres, A. and Ron, A. (1990): "Incomplete 'collapse' and partial quantum Zeno effect." *Phys. Rev.* **A42**, 5720–5722.   [122]

Perkins, D.H. (1982): *Introduction to High Energy Physics* (Addison-Wesley, Reading).   [103]

Petrosky, T., Tasaki, S., and Prigogine, I. (1990): "Quantum Zeno effect."
  Phys. Lett. **A151**, 109–113.   [123]
Petrosky, T., Tasaki, S., and Prigogine, I. (1991): "Quantum Zeno Effect."
  Physica **A170**, 306–325.   [123]
Petzold, J. (1959): "Wie gut gilt das Exponentialgesetz beim α-Zerfall?" *Z.
  Phys.* **155**, 422–432.   [113]
Pfau, T., Spälter, S., Kurtsiefer, C., Ekstrom, C.R., and Mlynek, J. (1994):
  "Loss of Spatial Coherence by a Single Spontaneous Emission." *Phys.
  Rev. Lett.* **73**, 1223–1226.   [76, 179]
Pfeifer, P. (1980): *Chiral Molecules - a Superselection Rule Induced by the Ra-
  diation Field.* Dissertation #6551, ETH Zürich. [101, 104, 188, 365, 387]
Phoenix, S.J.D. (1990): "Wave-packet evolution in the damped oscillator."
  Phys. Rev. **A41**, 5132–5138.   [135, 143]
Piron, C. (1969): "Les règles de supersélection continues." *Helv. Phys. Acta*
  42, 330–338.   [353]
Poincaré, H. (1908): *Science et Méthode* (Ernest Flammarion, Paris).   [149]
Polarski, D. (1999): "Primordial fluctuations from inflation: A consistent
  histories approach." *Phys. Lett.* **B446**, 53–57.   [216]
Polarski, D. and Starobinsky, A.A. (1996): "Semiclassicality and decoherence
  of cosmological perturbations." *Class. Quantum Grav.* **13**, 377–392. [215]
Porrati, M. and Putterman, S. (1987): "Wave-function collapse due to null
  measurements: The origin of intermittent atomic fluorescence." *Phys.
  Rev.* **A36**, 929–932.   [165]
Power, W.L. and Knight, P.L. (1996): "Stochastic simulations of the quantum
  Zeno effect." *Phys. Rev.* **53**, 1052–1059.   [123]
Poyatos, J.F., Cirac, J.I., and Zoller, P. (1996): "Quantum reservoir engi-
  neering with cooled trapped ions." *Phys. Rev. Lett.* **77**, 4728–4731. [135]
Preskill, J. (1998): "Reliable quantum computers." *Proc. R. Soc. London*
  A454, 385–410.   [147]
Prezhdo, O. V. (2000): "Quantum Anti-Zeno Acceleration of a Chemical
  Reaction." *Phys. Rev. Lett.* **85**, 4413–4417.   [121]
Primas, H. (1981): *Chemistry, Quantum Mechanics and Reductionism*
  (Springer, Berlin).   [12, 387]
Primas, H. (1990a): "Mathematical and philosophical questions in the the-
  ory of open and macroscopic quantum systems." In: *Sixty-Two Years of
  Uncertainty.*, ed. by A.I. Miller (Plenum Press, New York), p. 233-257.
                                                                      [365]
Primas, H. (1990b): "Induced nonlinear time evolution of open quantum
  objects." In: *Sixty-Two Years of Uncertainty.*, ed. by A.I. Miller (Plenum
  Press, New York), p. 259-280.   [19, 365]
Primas, H. (2000): "Asymptotically disjoint quantum states." In: Blanchard
  et al. (2000), p. 161–178.   [275, 320, 347]
Prokopec, T. (1993): "Entropy of the squeezed vacuum." *Class. Quantum
  Grav.* **10**, 2295–2306.   [217]
Prvanović, S. and Marić, Z. (1999): "Toward the Collapse of State." Report
  quant-ph/9910020.   [367]

Quack, M. (1986): "On the measurement of the parity violating energy difference between enantiomers." *Chem. Phys. Lett.* **132**, 147–153.   [106]

Raghunathan, M.S. (1994): "Universal Central Extensions." *Rev. Math. Phys.* **6**, 207–225.   [282, 283, 284, 285]

Rauch, H. (1984): "Wave-particle dualism in matter wave interferometry." In: *The Wave-Particle Dualism*, ed. by S. Diner, D. Fargue, G. Lochak and F. Selleri (Reidel Publishing Company, Dordrecht), p. 69–83.   [76]

Rauch, H., Summhammer, J., Zawisky, M., and Jericha, E. (1990): "Low-contrast and low-counting-rate measurements in neutron interferometry." *Phys. Rev.* **A42**, 3726–3732.   [384]

Rauch, H., Wilfing, A., Bauspiess, W., and Bonse, U. (1978): "Precise Determination of the 4π-Periodicity Factor of a Spinor Wave Function." *Z. Phys.* **B29**, 281–284.   [58, 287]

Rauch, H., Zeilinger, A., Badurek, G., Wilfing, A., Bauspiess, W., and Bonse, U. (1975): "Verification of Coherent Spinor Rotations of Fermions." *Phys. Lett.* **54A**, 425–427.   [12, 58, 287]

Reed, M. and Simon, B. (1979): *Methods of Modern Mathematical Physics III, Scattering Theory.* (Academic Press, New York).   [351]

Regge, T. and Teitelboim, C. (1974): "Role of Surface Integrals in the Hamiltonian Formulation of General Relativity." *Ann. Phys. (N.Y.)* **88**, 286–318.   [311]

Reißner, H. (1914): "Über die Relativität der Beschleunigung in der Mechanik." *Physikalische Zeitschrift* **15**, 371–375; english translation in *Mach's Principle*, ed. by J. Barbour and H. Pfister (Birkhäuser, Basel 1995).   [300]

Rempe, G., Walther, H., and Klein, N. (1987): "Observation of Quantum Collapse and Revival in a One-Atom Maser." *Phys. Rev. Lett.* **58**, 353–356.   [32, 140]

Renninger, M. (1953): "Zum Wellen-Korpuskel-Dualismus." *Z. Phys.* **136**, 251–261.   [386]

Renninger, M. (1960): "Messungen ohne Störung des Meßobjekts." *Z. Phys.* **158**, 417–421.   [161]

Rényi, A. (1961): "On measures of entropy and information." *Proc. 4th Berkeley Symposium on Math. Stat. and Probability*, Vol. 1 (Univ. Calif. Press, Berkeley) p. 547–561.   [325]

Rhim, W.-K., Pines, A., and Waugh, J.S. (1971): "Time-Reversal Experiments in Dipolar-Coupled Spin Systems." *Phys. Rev.* **B3**, 684–695.   [32, 49]

Rieffel, E. and Polak, W. (2000): "An Introduction to Quantum Computing for Non-Physicists." to appear in *ACM Computing Surveys* – quant-ph/9809016.   [147]

Risken, H. (1984): *The Fokker-Planck Equation* (Springer, Berlin). [80, 399]

Roberts, J.E. and Roepstorff, G. (1969): "Some Basic Concepts of Algebraic Quantum Theory." *Comm. Math. Phys.* **11**, 321–338.   [296]

Rosenfeld, L. (1965): "The measuring process in quantum mechanics." *Progr. Theor. Phys.* **Suppl.** (extra number), 222–231.   [386]

Roy, S.M. and Venugopalan, A. (1999): "Exact Solution of the Caldeira-Leggett Master Equation: A Factorization Theorem For Decoherence." Report quant-ph/9910004.   [399]

Rozmej, P., Berej, W. and Arvieu, R. (1997): "New mechanism of Collapse and Revival in Wave Packet Dynamics Due to Spin-Orbit Interaction." Report quant-ph/9702044.   [363]

Ruelle, D. (1991): *Chance and Chaos* (Princeton University Press).   [149]

Ruseckas, J. and Kaulakys, B. (2001): "Real measurements and Quantum Zeno effect." *Phys. Rev.* **A63**, 062103 – quant-ph/0105138.   [129]

Sakagami, M. (1988): "Evolution from Pure States into Mixed States in de Sitter Space." *Progr. Theor. Phys.* **79**, 442–453.   [216]

Samal, M.K. (2001): "Speculations on a Unified Theory of Matter and Mind." Report quant-ph/0111035.   [368]

Santos, E. (1994): "Objectification of classical properties induced by quantum vacuum fluctuations." *Phys. Lett.* **A188**, 198–204.   [188]

Santos, E. (1995): "Reply to comments on 'Objectification of classical properties induced by quantum vacuum fluctuations'." *Phys. Lett.* **A197**, 185–186.   [188]

Savage, C.M. and Walls, D.F. (1985a): "Damping of quantum coherence: The master-equation approach." *Phys. Rev.* **A32**, 2316–2323.   [400]

Savage, C.M. and Walls, D.F. (1985b): "Quantum coherence and interference of damped free particles." *Phys. Rev.* **A32**, 3487–3492.   [75, 89, 400]

Savage, C.M., Braunstein, S.L., and Walls, D.F. (1990): "Macroscopic quantum superpositions by means of single-atom dispersion." *Opt. Lett.* **15**, 628–630.   [171]

Schenzle, A. and Brewer, R.G. (1986): "Macroscopic quantum jumps in a single atom." *Phys. Rev.* **A34**, 3127–3142.   [163]

Schenzle, A., DeVoe, R.G., and Brewer, R.G. (1986): "Possibility of quantum jumps." *Phys. Rev.* **A33**, 2127–2130.   [163]

Schleich, W. and Wheeler, J.A. (1987): "Oscillations in photon distribution of squeezed states and interference in phase space." *Nature* **326**, 574–577.   [91]

Schleich, W.P. (2001): *Quantum Optics in Phase Space* (Wiley).   [92]

Schleich, W., Pernigo, M., and Fam Le Kien (1991): "Nonclassical state from two pseudoclassical states." *Phys. Rev.* **A44**, 2172–2187.   [143]

Schleich, W., Walls, D.F., and Wheeler, J.A. (1988): "Area of overlap and interference in phase space versus Wigner pseudoprobabilities." *Phys. Rev.* **A38**, 1177–1186.   [92]

Schlienz, J. and Mahler, G. (1995): "Description of entanglement." *Phys. Rev.* **A52**, 4396–4404.   [55, 119]

Schlögl, F. (1966): "Zur statistischen Theorie der Entropieproduktion in nicht abgeschlossenen Systemen." *Z. Phys.* **191**, 81–90.   [325]

Schmidt, E. (1907a): "Zur Theorie der linearen und nichtlinearen Integralgleichungen. I. Teil: Entwicklung willkürlicher Funktionen nach Systemen vorgeschriebener." *Math. Annalen* **63**, 433–476.   [37, 44]

Schmidt, E. (1907b): "Zur Theorie der linearen und nichtlinearen Integral-gleichungen. Zweite Abhandlung: Auflösung der allgemeinen linearen Integralgleichung." *Math. Annalen* **64**, 161–174.   [44]

Schrödinger, E. (1925): "Die Erfüllbarkeit der Relativitätsforderung in der klassischen Mechanik." *Ann. Phys.* (Leipzig) **77**, 325–336; English translation in *Mach's Principle*, ed. by J. Barbour and H. Pfister (Birkhäuser, Basel 1995).   [300]

Schrödinger, E. (1926): "Der stetige Übergang von der Mikro- zur Makro-mechanik." *Naturwiss.* **14**, 664–666.   [138]

Schrödinger, E. (1935a): "Die gegenwärtige Situation in der Quantenmecha-nik." *Naturwiss.* **23**, 807–812, 823–828, 844–849; English translation in Wheeler and Zurek (1983).   [45]

Schrödinger, E. (1935b): "Discussion of probability relations between sepa-rated systems." *Proc. Cambridge Phil. Soc.* **31**, 555–563.   [37]

Schroer, B. (1963): "Infrateilchen in der Quantenfeldtheorie." *Fortschr. Physik* **11**, 1–32.   [346]

Schulman, L.S. (1996): "Destruction of interference by entanglement." *Phys. Lett.* **A211**, 75–81.   [78]

Schulman, L.S. (1998): "Continuous and pulsed observations in the quantum Zeno effect." *Phys. Rev.* **A57**, 1509–1515.   [124, 126]

Schumacher, B. (1995): "Quantum Coding." *Phys. Rev.* **A51**, 2738–2747.   [145]

Schumaker, B. (1986): "Quantum mechanical pure states with Gaussian wave functionals." *Phys. Rep.* **135**, 317–408.   [213]

Schwabl, F. (1995): *Quantum Mechanics* (Springer, Heidelberg).   [417]

Schwinger, J., Scully, M.O., and Englert, B.-G. (1988): "Is spin coherence like Humpty-Dumpty? II. General Theory." *Z. Phys.* **D10**, 135–144. [49]

Scully, M.O. and Zubairy, M.S. (1997): *Quantum Optics* (Cambridge University Press).   [90, 133]

Scully, M.O., Englert, B.-G., and Walther, H. (1991): "Quantum optical tests of complementarity." *Nature* **351**, 111–116.   [78]

Seiberg, N. and Witten, E. (1999): "String theory and noncommutative geometry". *JHEP* **9909**, 032.   [195]

Senitzky, I.R. (1960): "Dissipation in Quantum Mechanics. The Harmonic Oscillator." *Phys. Rev.* **119**, 670–679.   [80]

Shaisultanov, R.Zh. (1995): "Backreaction in scalar QED, Langevin equation and decoherence functional." Report hep-th/9509154 (unpublished). [191]

Shiokawa, K. and Hu, B.L. (1995): "Decoherence, delocalization, and ir-reversibility in quantum chaotic systems." *Phys. Rev.* **E52**, 2497–2509.   [155]

Shi, Y. (2000): "Early gedanken experiments of quantum mechanics revis-ited." *Ann. Phys.* (Leipzig) **9**, 637–648 – quant-ph/9811050.   [28]

Shor, P.W. (1994): "Algorithms for Quantum Computation: Discrete Log-arithms and Factoring." In: *35th Annual Symposium on Foundations of Computer Science*, ed. by S. Goldwasser (IEEE Computer Society Press).   [145]

Shor, P.W. (1995): "Scheme for reducing decoherence in quantum computer memory." *Phys. Rev.* **A52**, 2493–2496.   [146]

Simonius, M. (1978): "Spontaneous Symmetry Breaking and Blocking of Metastable States." *Phys. Rev. Lett.* **40**, 980–983.   [105, 126]

Slusher, R.E. and Yurke, B. (1988): "Squeezed Light." *Scientific American* (May), 50–56.   [141]

Smolyanov, O.G. and Truman, A. (1999): "Schrödinger-Belavkin equations and associated Kolmogorov and Lindblad equations." *Theor. Math. Phys.* **120**, 973–984.   [338]

Snyder, D.M. (1996): *The Mind and the Physical World: a Psychologist's Exploration of Modern Physical Theory* (Tailor Press, Los Angeles). [368]

Spohn, H. (1978): "Entropy production for quantum dynamical semigroups." *J. Math. Phys.* **19**, 1227–1230.   [325]

Squires, E. (1990): *Conscious Mind in the Physical World* (IOP Publishing, Bristol).   [31, 108]

Srinivas, M.D. (1977): "Quantum counting processes." *J. Math. Phys.* **18**, 2138–2145.   [366]

Srinivas, M.D. (1984): "Quantum theory of continuous measurements." In: *Quantum Probability and Applications*, ed. by L. Accardi *et al.* (Springer, Berlin).   [20, 366]

Stapp, H.P. (1991): "EPR and Bell's Theorem: a Critical Review.", *Found. Phys.* **21**, 1–23.   [378]

Stapp, H.P. (1993): *Mind, Matter, and Quantum Mechanics* (Springer, Berlin).   [31, 108]

Stapp, H.P. (2000): "From quantum non-locality to mind-brain process." Report quant-phys/0009062.   [368]

Stapp, H.P. (2001): "Quantum Theory and the Role of Mind in Nature." Report quant-ph/0103043.   [368]

Starobinsky, A.A. (1979): "Spectrum of relict gravitational radiation and the early state of the universe." *JETP Lett.* **30**, 682–685.   [201]

Steane, A.M. (1996): "Error correction codes in quantum theory." *Phys. Rev. Lett.* **77**, 793–797.   [146]

Stern, A., Aharanov, Y., and Imry, Y. (1990): "Phase uncertainty and loss of interference: A general picture." *Phys. Rev.* **A41**, 3436–3448.   [78]

Stinespring, W.F. (1955): "Positive functions on C*-algebras." *Proc. Amer. Math. Soc.* **6**, 211–216.   [354]

Stöckmann, H.J. (1999): *Quantum Chaos. An Introduction* (Cambridge University Press).   [149, 151]

Stodolsky, L. (1999): "Decoherence - Fluctuation Relation and measurement Noise." Report quant-ph/9903072.   [363]

Storey, P., Tan, S., Collett, M. and Walls, D. (1994): "Path detection and the uncertainty principle." *Nature* **367**, 626–628.   [78]

Streater and Wightman (1964): *TCP, Spin and Statistics, and All That* (Addison-Wesley, Reading).   [11, 62]

Strocchi, F. and Wightman, A.S. (1974): "Proof of the Charge Superselection Rule in Local Relativistic Quantum Field Theory." *Jour. Math. Phys.* **15**, 2198–2224; Erratum *Ibid.* **17** (1976), 1930–1931. [313]

Strunz, W.T. and Haake, F. (2002): "Decoherence scenarios from micro-to macroscopic superpositions." Report quant-ph/0205108. [363]

Strunz, W.T., Haake, F. and Braun, D. (2002): "Universality of Decoherence in the Macroworld." Report quant-ph/0204129. [363]

Suarez, A. (1998): "Does the quantum collapse make sense?" Report quant-ph/9812009. [363]

Sudarshan, E.C.G. and Mukunda, N. (1974): *Classical Dynamics: A Modern Perspective* (Wiley, New York). [429]

Sudbery, A. (1988): "The Rapid-Dispersal Approximation in Radiative Atomic Processes." *Ann. Phys. (N.Y.)* **188**, 1–19. [113]

Summhammer, J., Badurek, G., Rauch, H., Kischko, U., and Zeilinger, A. (1983): "Direct observation of fermion spin superposition by neutron interferometry." *Phys. Rev.* **A27**, 2523–2532. [49, 170]

Sussmann, G.J. and Wisdom, J. (1992): "Chaotic Evolution of the Solar System." *Science* **257**, 56–62. [157]

Svetlichny, G. (2002): "Causality implies formal state collapse." Report quant-ph/0207180. [363]

Szilard, L. (1929): "Über die Entropieverminderung in einem thermodynamischen System bei Eingriffen intelligenter Wesen." *Z. Physik* **53**, 840–856; english translation in Wheeler and Zurek (1983). [23, 25, 145]

Takesaki, M. (1979): *Theory of Operator Algebras I* (Springer, New York). [265]

Tan, S.M. and Walls, D.F. (1993): "Loss of coherence in interferometry." *Phys. Rev.* **47**, 4663–4676. [78]

Tegmark, M. (1993): "Apparent Wave Function Collapse Caused by Scattering." *Found. Phys. Lett.* **6**, 571–590. [66, 380, 395]

Tegmark, M. (2000): "Importance of quantum decoherence in brain processes." *Phys. Rev.* **E61**, 4194–4206 – quant-ph/9907009. [30, 108, 368]

Tegmark, M. and Shapiro, H.S. (1994): "Decoherence produces coherent states: An explicit proof for harmonic chains." *Phys. Rev.* **E50**, 2538–2547. [138]

Tegmark, M., and Wheeler, J.A. (2001): "100 Years of Quantum Mysteries." *Scientific American* **284** (Febr.), 54–61. German translation "100 Jahre Quantentheorie." *Spektrum der Wissenschaft* (April), 68–76 – quant-ph/0101077. [1, 9]

Teich, W.G. and Mahler, G. (1992): "Stochastic dynamics of individual quantum systems: Stationary rate equations." *Phys. Rev.* **A45**, 3300–3318. [50]

Teitelboim, C. (1991): "Hamiltonian formulation of general relativity." In: *Quantum Cosmology and Baby Universes*, ed. by S. Coleman *et al.* (World Scientific, Singapore), p. 1–63. [237]

Tesche, C.D. (1986): "Schrödinger's Cat: A Realization in Superconducting Devices." *Ann. New York Acad. Sci.* **480**, 36–50. [103]

Tessieri, L., Vitali, D., and Grigolini, P. (1995): "Quantum jumps as an objective process of nature." *Phys. Rev.* **A51**, 4404–4414.   [29]

Thirring, W. (1983): *A Course in Mathematical Physics*, Vol. 4: *Quantum Mechanics of Large Systems* (Springer, New York).   [325, 355, 356]

Tittel, W., Brendel, T., Zbinden, H., and Gisin, N. (1998): "Violation of Bell Inequalities by Photons More Than 10 km Apart." *Phys. Rev. Lett.* **81**, 3563–3566.   [10]

Twamley, J. (1993): "Phase-space decoherence: A comparison between consistent histories and environment-induced superselection." *Phys. Rev.* **D48**, 5730–5745.   [247]

Ulfbeck, O., and Bohr, A. (2001): "Genuine fortuitousness: Where did that click come from?" *Found. Phys.* **31**, 757–774.   [27]

Unruh, W.G. (1995): "Maintaining coherence in quantum computers." *Phys. Rev.* **A51**, 992–997.   [86]

Unruh, W.G. and Zurek, W.H. (1989): "Reduction of a wave packet in quantum Brownian motion." *Phys. Rev.* **D40**, 1071–1094.   [81, 237, 399]

Unturbe, J. and Sánchez-Gómez, J.L. (1992): "On the Role of Gravitation in the Possible Breakdown of the Quantum Superposition Principle for Macroscopic Systems." *Il Nuovo Cimento* **107B**, 211–237.   [364]

Vacchini, B. (2001): "Brownian Motion: the Quantum Perspective." *Z. Naturforsch.* **56a**, 230-233.   [86]

Vaidman, L. (1998): "Teleportation: Dream or Reality?" Report quant-ph/9810089.   [39, 174]

van der Wal, C.H., ter Haar, A.C.J, Wilhelm, F.K., Schouten, R.N., Harmans, C.J.P.M., Orlando, T.P., Lloyd, S., Mooij, J.E. (2000): "Quantum Superposition of Macroscopic Persistent-Current States." *Science* **290**, 773–777.   [46, 103]

van Enk, S.J. (2002): "Entanglement of photons." Report quant-ph/0206135.   [46]

van Enk, S.J. and Kimble, H.J. (2001): "On the classical character of control fields in quantum information processing." Report quant-ph/0107088.   [152]

van Hove, L. (1952): "Les difficultés de divergences pour un modèle particulier de champ quantifié." *Physica* **18**, 145–159.   [346]

van Hove, L. (1955): "Quantum–mechanical perturbations giving rise to a statistical transport equation." *Physica* **21**, 517–540.   [330]

van Hove, L. (1957): "The Approach to Equilibrium in Quantum Statistics." *Physica* **23**, 441–480.   [127]

van Kampen, N.G. (1954): "Quantum Statistics of Irreversible Processes." *Physica* **20**, 603.   [385]

van Kampen, N.G. (1981): *Stochastic Processes in Physics and Chemistry* (North-Holland, Amsterdam).   [433]

Varadarajan, V.S. (1985): *Geometry of Quantum Theory*, 2nd edition (Springer, Berlin).   [263, 427]

Vedral, V., Plenio, M.B., Rippin, M.A., Knight, P.L. (1997): "Quantifying Entanglement." *Phys. Rev. Lett.* **78**, 2275–2279.   [36]

Venugopalan, A. (1994): "Preferred basis in a measurement process." *Phys. Rev.* **A50**, 2742–2745.   [83, 373]

Venugopalan, A. (1998): "Energy basis via decoherence." *Pramana – J. Phys.* **51**, 625–631.   [83]

Venugopalan, A. (1999): "Pointer states via Decoherence in a Quantum Measurement." Report quant-ph/9909005.   [138]

Venugopalan, A., Kumar, D., and Ghosh, R. (1995a): "Environment-induced decoherence I. The Stern–Gerlach measurement." *Physica* **A220**, 563–575.   [86]

Venugopalan, A., Kumar, D., and Ghosh, R. (1995b): "Environment-induced decoherence II. Effect of decoherence on Bell's inequality for an EPR pair." *Physica* **A220**, 576–584.   [33, 179]

Vilkovisky, G.A. (1984): "The unique effective action in quantum field theory." *Nucl. Phys.* **B234**, 125–137.   [204]

von Neumann, J. (1931): "Über Funktionen von Funktionaloperatoren." *Ann. Math. (Princeton)* **32**, 191-226.   [272]

von Neumann, J. (1932): *Mathematische Grundlagen der Quantenmechanik* (Springer, Berlin); reprinted 1981. English translation by R.T. Beyer (1955): *Mathematical Foundations of Quantum Mechanics* (Princeton University Press).   [20, 22, 27, 48, 107, 110, 407]

Wadia, S. (1977): "Hamiltonian Formulation of Non-Abelian Gauge Theory with Surface Terms: Application to the Dyon Solution." *Phys. Rev.* **D15**, 3615–3628.   [304]

Wald, R.M. (1984): *General relativity* (University of Chicago Press).   [196]

Walls, D.F. (1983): "Squeezed states of light." *Nature* **306**, 141–146.   [141]

Walls, D.F. and Milburn, G.J. (1994): *Quantum Optics* (Springer, Berlin).   [90, 133, 147, 148, 185]

Walls, D.F., Collet, M.J., and Milburn, G.J. (1985): "Analysis of a quantum measurement." *Phys. Rev.* **D32**, 3208–3215.   [148]

Walmsley, I.A. and Raymer, M.G. (1995): "Detecting quantum superpositions of classically distinguishable states of a molecule." *Phys. Rev.* **A52**, 681–685.   [105]

Wehrl, A. (1978): "General properties of entropy." *Rev. Mod. Phys.* **50**, 221–260.   [325, 355, 356]

Weinberg, S. (1989): "Precision Tests of Quantum Mechanics." *Phys. Rev. Lett.* **62**, 485–488.   [27]

Weinberg, S. (1996): *The Quantum Theory of Fields (Vol I)* (Cambridge University Press, Cambridge)   [260]

Weisskopf, V. and Wigner, E. (1930): "Berechnung der natürlichen Linienbreite auf Grund der Diracschen Lichttheorie." *Z. Phys.* **63**, 54–73. [113]

Weiss, U. (1999): *Quantum Dissipative Systems*, 2nd edition (World Scientific).   [133]

Werner, S.A., Collella, R., Overhauser, A.W., and Eagen, C.F. (1975): "Observation of the phase shift of a neutron due to precession in a magnetic field." *Phys. Rev. Lett.* **35**, 1053–1055.   [58, 287]

Wess, J. and Zumino, B. (1971): "Consequences of anomalous Ward identities." *Phys. Lett.* **37B**, 95–97.   [12]

Wheeler, J.A. (1968): "Superspace and the nature of quantum geometrodynamics." In: *Battelle rencontres*, ed. by C.M. De Witt and J.A. Wheeler (Benjamin, New York), p. 242–307.   [197]

Wheeler, J.A. (1978): "The 'Past' and the 'Delayed-Choice' Double-Slit Experiment." In: *Mathematical Foundations of Quantum Theory*, ed. by A.R. Marlow (Academic Press, New York), p. 9–48.   [77]

Wheeler, J.A. and Zurek, W.H. (1983): *Quantum Theory and Measurement* (Princeton University Press).   [1]

Whitaker, A. (2000): "Many Minds and Single Mind Interpretations of Quantum Theory." In: Blanchard *et al.* (2000)   [108]

White, A.G., Mitchell, J.R., Nairz, O, and Kwiat, P.G. (1998): "'Interaction-Free' Imaging." *Phys. Rev.* **A58**, 605–613.   [116]

Wick, G.C., Wightman, A.S., and Wigner, E.P. (1952): "The Intrinsic Parity of Elementary Particles." *Phys. Rev.* **88**, 101–105.   [57, 259]

Wick, G.C., Wightman, A.S., and Wigner, E.P. (1970): "Superselection Rule for Charge." *Phys. Rev.* **D1**, 3267–3269.   [11, 60]

Wightman, A.S. (1959): "Relativistic Invariance and Quantum Mechanics" (Notes by A. Barut). *Nuovo Cimento, Suppl.* **14**, 81–94.   [272]

Wightman, A.S. (1995): "Superselection Rules; Old and New." *Il Nuovo Cimento* **110B**, 751–769.   [16, 57, 104, 188, 259, 348]

Wightman, A.S. and Glance, N. (1989): "Superselection rules in molecules." *Nucl. Phys. B (Proc. Suppl.)* **6**, 202–206.   [104, 278]

Wigner, E. (1931): *Gruppentheorie und ihre Anwendung auf die Quantenmechanik der Atomspektren* (Vieweg, Braunschweig).   [280]

Wigner, E.P. (1932): "On the quantum correction for thermodynamic equilibrium." *Phys. Rev.* **40**, 749–759.   [91]

Wigner, E.P. (1952): "Die Messung quantenmechanischer Operatoren." *Z. Phys.* **131**, 101–108.   [47]

Wigner, E.P. (1962): "Remarks on the Mind-Body Question." In: *The Scientist Speculates*, ed. by L.G. Good (Heinemann, London), p. 284–302; reprinted in Wheeler and Zurek (1983).   [29, 108]

Wigner, E.P. (1963): "The Problem of Measurement." *Am. J. Phys.* **31**, 6–15; reprinted in Wheeler and Zurek (1983).   [49, 241]

Wigner, E.P. (1964): "Events, Laws of Nature, and Invariance Principles." In: *Les Prix Nobel en 1963* (The Noble Foundation, Stockholm).   [10]

Wigner, E.P. (1983): "Review of the quantum mechanical measurement problem." In: *Quantum Optics, Experimental Gravitation, and Measurement Theory*, ed. by P. Meystre and M.O. Scully (Plenum Press, New York), p. 43.   [65]

Wilkinson, S.R. *et al.* (1997): "Experimental evidence for non-exponential decay in quantum tunneling." *Nature* **387**, 575–577.   [121]

Winter, R.G. (1961): "Evolution of a Quasi-Stationary State." *Phys. Rev.* **123**, 1503–1507.   [113]

Wisdom, J., Peale, S.J., and Mignard, F. (1984): "The Chaotic Rotation of Hyperion." *Icarus* **58**, 137–152.   [158]

Wisdom, J. (1987): "Urey price lecture: chaotic dynamics in the solar system." *Icarus* **72**, 241–375.   [158]

Wiseman, H.M. and Milburn, G.J. (1993a): "Quantum theory of field-quadrature measurements." *Phys. Rev.* **A47**, 642–662.   [50]

Wiseman, H.M. and Milburn, G.J. (1993b): "Interpretation of quantum jump and diffusion processes illustrated on the Bloch sphere." *Phys. Rev.* **A47**, 1652–1666.   [50]

Wolfenstein, L. (1978): "Neutrino oscillations in matter." *Phys. Rev.* **D17**, 2369–2374.   [10]

Wolsky, A.M. (1976): "Why 'A Watched Pot Never Boils.'" *Found. Phys.* **6**, 367–369.   [110]

Woolley, R.G. (1976): "Quantum theory and molecular structure." *Advances in Physics* **25**, 27–52.   [104]

Woolley, R.G. (1982): "Natural Optical Activity and the Molecular Hypothesis." *Structure and Bonding* **52**, 1–35.   [104]

Woolley, R.G. (1986): "Molecular Shapes and Molecular Structures." *Chem. Phys. Lett.* **125**, 200–205.   [12, 104]

Wootters, W.K. (1987): "A Wigner-Function Formulation of Finite-State Quantum Mechanics." *Ann. Phys. (N.Y.)* **176**, 1–21.   [90, 119]

Wootters, W.K. and Zurek, W.H. (1979): "Complementarity in the double-slit experiment: Quantum nonseparability and a quantitative statement of Bohr's principle." *Phys. Rev.* **D19**, 473–484; reprinted in Wheeler and Zurek (1983).   [78]

Wootters, W.K. and Zurek, W.H. (1982): "A single quantum cannot be cloned." *Nature* **299**, 802–803.   [146, 168]

Yanase, M. (1961): "Optimal Measuring Apparatus." *Phys Rev.* **123**, 666–668.   [302]

Yasue, K. (1978): "Quantum Mechanics of Nonconservative Systems." *Ann. Phys. (N.Y.)* **114**, 479–496.   [368]

Yourgrau, W. (1968): "A Budget of Paradoxes in Physics." In: *Problems in the Philosophy of Science*, ed. by A.M.I. Lakatos (North-Holland Publishing Company, Amsterdam).   [110]

Yuen, H.P. (1976): "Two-photon coherent states of the radiation field." *Phys. Rev.* **A13**, 2226–2243.   [16]

Yurke, B. and Stoler, D. (1988): "The dynamic generation of Schrödinger cats and their detection." *Physica* **B151**, 298–301.   [143]

Zbinden, H., Brendel, J., Tittel, W. and Gisin, N. (2000): "Experimental Tests of Relativistic Quantum State Collapse with Moving Reference Frames." Report quant-ph/0002031.   [363, 377]

Zeh, H.D. (1965): "Symmetry Violating Trial Wave Functions." *Z. Phys.* **188**, 361–373.   [389]

Zeh, H.D. (1967): "Symmetrieverletzende Modellzustände und kollektive Bewegungen." *Z. Phys.* **202**, 38–48.   [391]

Zeh, H.D. (1970): "On the Interpretation of Measurement in Quantum Theory." *Found. Phys.* **1**, 69–76; reprinted in Wheeler and Zurek (1983).
                                                    [3, 14, 15, 16, 31, 259, 386, 392]
Zeh, H.D. (1971): "On the Irreversibility of Time and Observation in Quantum Theory." In: *Foundations of Quantum Mechanics*, Varenna 1970, ed. by B. d'Espagnat (Academic Press, NewYork), p. 263–273.   [29, 31, 37]
Zeh, H.D. (1973): "Toward a Quantum Theory of Observation." *Found. Phys.* **3**, 109–116.   [14, 29, 37]
Zeh, H.D. (1975): "Symmetry-Breaking Vacuum and State Vector Reduction." *Found. Phys.* **5**, 371–373.   [106]
Zeh, H.D. (1979): "Quantum Theory and Time Asymmetry." *Found. Phys.* **9**, 803–818.   [30, 37]
Zeh, H.D. (1981): "The Problem of Conscious Observation in Quantum Mechanical Description." *Epist. Letters* **63.0** (Ferd. Gonseth Ass., Biel) – for revised version see Zeh (2000).   [31]
Zeh, H.D. (1986): "Emergence of classical time from a universal wave function." *Phys. Lett.* **A116**, 9–12.   [16, 197, 199]
Zeh, H.D. (1988): "Time in Quantum Gravity." *Phys. Lett.* **A126**, 311–317.
                                                                       [16, 162]
Zeh, H.D. (1993): "There are no quantum jumps, nor are there particles!" *Phys. Lett.* **A172**, 189–192.   [162]
Zeh, H.D. (1999): "Why Bohm's Quantum Theory?" *Found. Phys. Lett.* **12**, 197–200   quant-ph/9812059.   [14]
Zeh, H.D. (2000): "The Problem of Conscious Observation in Quantum Mechanical Description." *Found. Phys. Lett.* **13**, 221–233 – quant-ph/-9908084.   [31]
Zeh, H.D. (2001): *The Physical Basis of the Direction of Time*, 4th edn. (Springer, Berlin) – see also www.time-direction.de.
           [15, 16, 25, 31, 34, 36, 37, 40, 196, 208, 215, 231, 252, 253, 330, 385]
Zeilinger, A., Gähler, R., Shull, C.G., Treimer, W., and Mampe, W. (1988): "Single- and double-slit diffraction of neutrons." *Rev. Mod. Phys.* **60**, 1067–1073.   [76]
Zoller, P., Marte, M., and Walls, D.F. (1987): "Quantum jumps in atomic systems." *Phys. Rev.* **A35**, 198–207.   [163, 164]
Zurek, W.H. (1981): "Pointer basis of quantum apparatus: Into what mixture does the wave packet collapse?" *Phys. Rev.* **D24**, 1516–1525.
                                                         [3, 14, 20, 50, 166]
Zurek, W.H. (1982): "Environment-induced superselection rules." *Phys. Rev.* **D26**, 1862–1880.   [15, 58, 59, 259, 341, 343]
Zurek, W.H. (1984a): "Maxwell's demon, Szilard's engine, and quantum measurements." In: *Frontiers in Nonequilibrium Statistical Mechanics*, ed. by G.R. Moore and M.O. Scully (Plenum, New York) – quant-ph/0301076.   [25]

Zurek, W.H. (1984b): "Destruction of Coherence in Nondemolition Monitoring: Quantum 'Watchdog Effect' in Gravity Wave Detectors." In: *The Wave-Particle Dualism*, ed. by S. Diner, D. Fargue, G. Lochak and F. Selleri (Reidel, Dordrecht), p. 515.   [148]

Zurek, W.H. (1986): "Reduction of the wave packet: How long does it take?" In: *Frontiers in Nonequilibrium Statistical Physics*, ed. by G.T. Moore and M.T. Sculley (Plenum, New York).   [89]

Zurek, W.H. (1991): "Decoherence and the Transition from Quantum to Classical." *Physics Today* **44** (Oct.), 36–44.   [3, 9, 68]

Zurek, W.H. (1993a): "Preferred States, Predictability, Classicality and the Environment-Induced Decoherence." *Progr. Theor. Phys.* **89**, 281–312.   [136]

Zurek, W.H. (1993b): "Negotiating the tricky border between quantum and classical." *Physics Today* (letters) **46** (April), 13.   [3]

Zurek, W.H. (2000), "Decoherence and Einselection" In: Blanchard *et al* (2000). "Decoherence, Einselection and the Origin of the Classical." Report quant-ph/0105127.   [361]

Zurek, W.H., (2001): "Sub-Planck structure in phase space and its relevance for quantum decoherence." *Nature* **412**, 712–717.   [93]

Zurek. W.H. (2003): "Decoherence, Einselection, and the Quantum Origin of the Classical." *Rev. Mod. Phys.* (to appear) – quant-ph/0105127.   [3, 9]

Zurek, W.H. and Paz, J.P. (1994): "Decoherence, Chaos and the Second Law." *Phys. Rev. Lett.* **72**, 2508–2511.   [13, 151, 159]

Zurek, W.H. and Paz, J.P. (1995a): "Decoherence, Chaos, the Quantum and the Classical." *Il Nuovo Cimento* **110B**, 611–624.   [159]

Zurek, W.H. and Paz, J.P. (1995b): "Reply to Comment on 'Decoherence, Chaos, and the Second Law'." *Phys. Rev. Lett.* **75**, 351.   [151, 158]

Zurek, W.H. and Paz, J.P. (1995c): "Quantum chaos: a decoherent definition." *Physica* **D83**, 300–308.   [159]

Zurek, W.H. and Paz, J.P. (1996): "Why we don't need planetary dynamics: Decoherence and the correspondence principle for chaotic systems." Proceedings of the Fourth Drexel Meeting – quant-ph/9612037.   [158]

Zurek, W.H., Habib, S., and Paz, J.P. (1993): "Coherent States via Decoherence." *Phys. Rev. Lett.* **70**, 1187–1190.   [16, 136, 138]

Zwanziger, D. (1976): "Physical States in Quantum Electrodynamics." *Phys. Rev.* **D14**, 2570-2589.   [307, 309]

Zwanziger, D. (1978): "Gupta-Bleuler and Infrared-Coherence Subsidiary Conditions." *Phys. Rev.* **D18**, 3051-3057.   [309]

Zwanzig, R. (1960a): "Ensemble method in the theory of irreversibility." *J. Chem. Phys.* **33**, 1338–1341.   [318, 321, 327]

Zwanzig, R. (1960b): "Statistical Mechanics of Irreversibility." In: *Boulder Lectures in Theoretical Physics*, Vol. III, (Interscience), p. 106–141.   [318, 321, 327]

Printing: Strauss GmbH, Mörlenbach
Binding: Schäffer, Grünstadt

Printed by Books on Demand, Germany